The Classical Stefan Problem

Basic Concepts, Modelling and Analysis

NORTH-HOLLAND SERIES IN

APPLIED MATHEMATICS AND MECHANICS

VOLUME 45

ELSEVIER

AMSTERDAM – BOSTON – HEIDELBERG – LONDON – NEW YORK – OXFORD
PARIS – SAN DIEGO – SAN FRANCISCO – SINGAPORE – SYDNEY – TOKYO

The Classical Stefan Problem

Basic Concepts, Modelling and Analysis

S.C. GUPTA

Professor (Retd), Department of Mathematics,
Indian Institute of Science, Bangalore, India

ELSEVIER
2003

AMSTERDAM – BOSTON – HEIDELBERG – LONDON – NEW YORK – OXFORD
PARIS – SAN DIEGO – SAN FRANCISCO – SINGAPORE – SYDNEY – TOKYO

ELSEVIER SCIENCE B.V.
Sara Burgerhartstraat 25
P.O. Box 211, 1000 AE Amsterdam, The Netherlands

First edition 2003

Library of Congress Cataloging in Publication Data
A catalog record from the Library of Congress has been applied for.

British Library Cataloguing in Publication Data
A catalogue record from the British Library has been applied for.

ISBN: 0-444-51086-9
ISSN: 0167-5931 (Series)

Transferred to Digital Printing 2009

Dedicated to

Free Boundary Fraternity

Contents

List of Symbols

Some symbols and abbreviations which are common throughout the book are given below. They have also been explained in the text, occasionally, but not every time they occur. In addition to these symbols, several other symbols have been used in the text, and they have been explained at the places they occur. The parameters occurring in the equations could be dimensionless quantities or may have dimensions. At those places where the equations are in the dimensionless form, the method of dimensionalization has been mentioned.

For the notations used for the function spaces, the reader is referred to Appendices A–D. The same have been explained in the text, occasionally.

Standard notations have been used for the numbering of equations, figures, definitions and propositions. For example, (7.2.7) refers to the seventh equation in the second section of chapter 7.

$c(x,t)$	Concentration, $[kg/m^3]$
C	Specific heat, $[Jkg^{-1}K^{-1}]$
C_V	Specific heat at constant volume
C_P	Specific heat at constant pressure
$\hat{C}$	Heat capacity, $[JK^{-1}/m^3]$
e	Specific internal energy, $[Jkg^{-1}]$; also used for energy per unit volume (indicated in the text)
H	Enthalpy, $[J]$; also enthalpy per unit volume (indicated in the text)
h	Specific enthalpy, $[Jkg^{-1}]$
h_t	Heat transfer coefficient, $[WK^{-1}/m^2]$
k	Thermal diffusivity, $[m^2s^{-1}]$
K	Thermal conductivity in the isotropic case, $[Jm^{-1}s^{-1}K^{-1}]$
K_{ij}	Thermal conductivity coefficients in an anisotropic case; $i = 1,2,3$ and $j = 1,2,3$
$\bar{K}_c$	Mean curvature of the free boundary, $[m^{-1}]$
l	Latent heat of fusion, $[Jkg^{-1}]$
l_m	Latent heat per unit mole, $[J/k\,mol]$
$\hat{l}$	$l + (C_L - C_S)T_m$
R^n	$n \geq 1$, real n-dimensional space, R or R^1 used for real line
$S(t)$	$x = S(t)$ ($x = S(y,z,t)$), equation of the phase-change boundary in one-dimension (three-dimensions)

$\hat{s}$	Specific entropy, $[Jkg^{-1}K^{-1}]$
$\hat{S}$	Entropy, $[JK^{-1}]$
t	real time, $[s]$
T	Temperature, [°K]
T_m	Ideal equilibrium melting/freezing temperature, also taken as 0 or 1
T_m^c	Equilibrium phase change temperature in supercooling/superheating
V_m	Molar volume, $[m^3/k\,mol]$
$\vec{n}$	unit normal vector

Subscripts

(L, S, M)	Stand for liquid, solid and mushy regions, respectively
$i = 1, 2$	quantities in the two phases

Greek symbols

ρ	Density, $[kg/m^3]$
σ	Surface tension, $[Nm^{-1}]$

Some other symbols

$'$	Differentiation with respect to the argument
$\cdot$	Time derivative
∇f	Gradiant of a scalar function
∇^2	Laplacian operator
erf(x)	Error function
erfc(x)	1-erf(x)

Abbreviations

meas (A)	measure of the set A
CEF	Classical enthalpy formulation
CES	Classical enthalpy solution
CODP	Constrained oxygen-diffusion problem
CSS	Classical Stefan solution

HSP	Hele-Shaw problem
ODP	Oxygen-diffusion problem
QSSP	Quasi steady-state problem
SPF	Standard phase-field model
SSP	Supercooled Stefan problem
UODP	Unconstrained ODP
WS	Weak solution

Preface

This volume emphasises studies related to classical Stefan problems. The term "Stefan problem" is generally used for heat transfer problems with phase-changes such as from the liquid to the solid. Stefan problems have some characteristics that are typical of them, but certain problems arising in fields such as mathematical physics and engineering also exhibit characteristics similar to them. The term "classical" distinguishes the formulation of these problems from their weak formulation, in which the solution need not possess classical derivatives. Under suitable assumptions, a weak solution could be as good as a classical solution. In hyperbolic Stefan problems, the characteristic features of Stefan problems are present but unlike in Stefan problems, discontinuous solutions are allowed because of the hyperbolic nature of the heat equation. The numerical solutions of inverse Stefan problems, and the analysis of direct Stefan problems are so integrated that it is difficult to discuss one without referring to the other. So no strict line of demarcation can be identified between a classical Stefan problem and other similar problems. On the other hand, including every related problem in the domain of classical Stefan problem would require several volumes for their description. A suitable compromise has to be made.

The basic concepts, modelling, and analysis of the classical Stefan problems have been extensively investigated and there seems to be a need to report the results at one place. This book attempts to answer that need. Within the framework of the classical Stefan problem with the emphasis on the basic concepts, modelling and analysis, I have tried to include some weak solutions and analytical and numerical solutions also. The main considerations behind this are the continuity and the clarity of exposition. For example, the description of some phase-field models in Chapter 4 arose out of this need for a smooth transition between topics. In the mathematical formulation of Stefan problems, the curvature effects and the kinetic condition are incorporated with the help of the modified Gibbs-Thomson relation. On the basis of some thermodynamical and metallurgical considerations, the modified Gibbs-Thomson relation can be derived, as has been done in the text, but the rigorous mathematical justification comes from the fact that this relation can be obtained by taking appropriate limits of phase-field models. Because of the unacceptability of some phase-field models due their so-called thermodynamical inconsistency, some consistent models have also been described. This completes the discussion of phase-field models in the present context.

Making this volume self-contained would require reporting and deriving several results from tensor analysis, differential geometry, non-equilibrium thermodynamics, physics and functional analysis. I have chosen to enrich the text with appropriate references so as not to enlarge the scope of the book. The proofs of propositions and theorems are often lengthy and different from one another. Presenting them in a condensed way may not be of much help to the reader. Therefore only the main features of proofs and a few results have been presented to suggest the essential flavour of the theme of investigation.

However at each place, appropriate references have been cited so that inquisitive readers can follow them on their own.

Each chapter begins with basic concepts, objectives and the directions in which the subject matter has grown. This is followed by reviews–in some cases quite detailed–of published works. In a work of this type, the author has to make a suitable compromise between length restrictions and understandability. I have followed my best judgement in this regard. I hope the readers will appreciate my efforts.

S.C. Gupta, Bangalore
Email : sgupta@math.iisc.ernet.in

Acknowledgements

I gratefully acknowledge the financial support from the Department of Science and Technology, Ministry of Science and Technology, Government of India, without which it would not have been possible for me to undertake this book-writing project. The administration of the Indian Institute of Science, Bangalore, and the Department of Mathematics, provided infrastructural support for which I am extremely thankful. During my visit to the University of Oulu, Finland, in 1997, I collected lot of material relevant for this book. I sincerely thank my host, Professor Erkki Laitinen and Department of Mathematics, Oulu, for their kind hospitality. The idea of writing this book nucleated during my visit to University of Rosario, Argentina, in 1991, and fructified eleven years later. I thank Professor D.A. Tarzia, Department of Mathematics, Universidad Austral, Rosario, for sending me the bibliography prepared by him, and some reprints which I could not get from other sources.

I am grateful to Professor Adimurthi, Department of Mathematics, TIFR, at IISc, Bangalore, and professor A.K. Lahiri, Department of Metallurgy, Indian Institute of Science, Bangalore, for reading and responding to some portions of this book. My grateful thanks to Dr. Thomas Chacko, Professor of English and Communication Skills, Department of Management Studies, for style editing this volume with a great deal of attention. There are many old friends and well-wishers who sent reprints and preprints that proved very useful for me. Without naming them individually, I thank all of them for their selfless help.

I thank the Publishing Editor and staff of Elsevier and the Series Editor Professor J.D. Achenbach of the book series 'Applied Mathematics and Mechanics' for their cooperation.

The excellent typing in Latex and drawing of figures in computer graphics were done by Susheel Graphics, Bangalore. Without their cooperation the work would not have taken the final shape.

Chapter 1

The Stefan Problem and its Classical Formulation

1.1 Some Stefan and Stefan-like Problems

The term *Stefan problem* can be best understood with the help of an example for which the reader is referred to § 1.3. Our concern in the present section is to understand the notion of a *free boundary* which is a typical feature of the Stefan and Stefan-like problems. Therefore this chapter begins with some examples of Stefan and Stefan-like problems which demonstrate the existence of an unknown boundary, commonly known in the literature as a 'free boundary' or a *moving boundary.* In the context of solidification/ melting problems, with which Stefan problems are commonly associated, the free boundary is also called a *phase-change boundary* or a *melting front* or a *freezing front.* Some authors use the term free boundary when the unknown boundary is static and moving boundary when it is time dependent. In this volume we do not make any distinction between 'free' and 'moving' boundaries. The term boundary is used for a surface also. In many of the examples considered in this section, the identification of the free boundary and the mathematical formulation of the problem are rather easy but in some of them even the identification of the free boundary is difficult. The problems given here are from various fields of mathematics, physics and engineering and demonstrate the existence of free boundaries. Our interest in examples given in this section is more on demonstrating the existence of a free boundary and its typical characteristics than justifying the formulation.

Problem 1.1.1. Steady-state heat conduction with a free boundary

Find the steady-state temperature $T(x, y)$ satisfying the equation

$$\frac{\partial^2 T}{\partial x^2} + \frac{\partial^2 T}{\partial y^2} = 0, \tag{1.1.1}$$

in an open bounded region $D \subset R^2$. The boundary ∂D of D consists of two disjoint parts R_1 and R_2, i.e., $\partial D = R_1 \cup R_2$, where R_1 is unknown and R_2 is known. On R_2, the temperature is prescribed as

$$T(x, y)|_{R_2} = f(x, y). \tag{1.1.2}$$

If $f(x, y)$ is known throughout the plane, then one boundary conditions on R_1 will be

$$T(x, y)|_{R_1} = f(x, y), \tag{1.1.3}$$

and another boundary condition can be imposed as [1],

$$\nabla (T - f)|_{R_1} = 0. \tag{1.1.4}$$

The problem is to determine the temperature $T(x, y)$ in D, and the unknown boundary R_1. Two boundary conditions are to be prescribed on R_1. One, because it is a boundary and one more boundary condition is required to determine an unknown boundary. Note that R_1 can be determined only by solving the system (1.1.1)–(1.1.4). Such an unknown boundary is called a *free boundary* or a *moving boundary*. In 'boundary value problems' of mathematical physics, the boundary of the region under consideration is completely known. Conditions (1.1.3) and (1.1.4) are called *free boundary conditions* and (1.1.2) is a *fixed boundary condition*. Although the equation (1.1.1) is linear, free boundary problems are nonlinear problems because of the nonlinearity of the boundary conditions at the free boundary.

Problem 1.1.2. Steady-State Free Surface Flow with Surface Tension

Consider a two-dimensional steady, incompressible, irrotational flow in a long channel. x and y axes are taken along the length and depth of the channel, respectively, with the bottom of the channel taken as $y = 0$ and the upper surface of water, as a free surface or a free boundary denoted by $y = \eta(x)$, where $\eta(x)$ is unknown. If $u(x, y)$ and $v(x, y)$ are the velocity components in x and y directions, respectively, then

$$\frac{\partial u}{\partial x} + \frac{\partial v}{\partial y} = 0 \quad \text{(equation of continuity)}, \tag{1.1.5}$$

$$\frac{\partial v}{\partial x} - \frac{\partial u}{\partial y} = 0 \quad \text{(irrotationality condition)}. \tag{1.1.6}$$

The pressure p is given by the following Bernoulli equation [2]

$$H = y + p/(\rho g) + |\vec{q}|^2/(2g), \tag{1.1.7}$$

where, H is the given total water head of water, ρ is the density, g the acceleration due to gravity and $\vec{q}$ is the velocity vector. If the bottom of the channel is a rigid boundary, then

$$v = 0, \quad \text{on} \quad y = 0. \tag{1.1.8}$$

At the free boundary $y = \eta(x)$,

$$H = \eta(x) + p/(\rho g) + |\vec{q}|^2/(2g), \; p = \bar{p} - \sigma/R_c \tag{1.1.9}$$

and

$$\vec{q} \cdot \vec{n} = 0. \tag{1.1.10}$$

Here, $\bar{p}$ is the atmospheric pressure (known), σ is the surface tension (known) and R_c is the radius of curvature of the free surface, taken as positive when the center of curvature is situated above the free surface and $\vec{n}$ is the unit outward normal to the free surface. R_c can be expressed in terms of derivatives of $\eta(x)$. Free boundary has been taken as static in this problem. Equation (1.1.10) implies that the normal component of the fluid velocity at the free boundary is zero.

Problem 1.1.3. Free Surface Flow with Time Dependent Free Surface

Let the free surface in Problem 1.1.2. be time dependent and represented by $y = \eta(x,t)$ with $\eta|_{t=0}$ being given. Surface tension effects will be neglected. If the velocity field $\vec{q}$ is expressed as

$$\vec{q} = \nabla\phi, \tag{1.1.11}$$

then from (1.1.5) and (1.1.6) it is easy to conclude that $\phi(x,t)$ satisfies the equation

$$\frac{\partial^2\phi}{\partial x^2} + \frac{\partial^2\phi}{\partial y^2} = 0, \; b(x) < y < \eta(x,t). \tag{1.1.12}$$

Here, $y = b(x)$ is the equation of the bottom of the channel. The momentum equation can be written as (cf. [3])

$$\frac{\partial\vec{q}}{\partial t} + \frac{1}{2}\text{grad}|\vec{q}|^2 - \vec{q} \wedge \text{curl}\vec{q} = \vec{F} - \text{grad}(p/\rho), \tag{1.1.13}$$

$\vec{F}$ represents body forces. On substituting $\vec{q}$ from (1.1.11) in (1.1.13) and integrating with respect to x, we obtain

$$\frac{\partial\phi}{\partial t} + \frac{1}{2}|\nabla\phi|^2 + gy + p/\rho = \psi(t), \tag{1.1.14}$$

provided the density is taken as constant and the gravitational field is the only force field. The arbitrary function $\psi(t)$ can be absorbed in $\phi(t)$ and (1.1.14) becomes

$$\frac{\partial\phi}{\partial t} + \frac{1}{2}|\nabla\phi|^2 + gy + p/\rho = 0. \tag{1.1.15}$$

If $y = b(x)$ is taken as a rigid boundary, then

$$\frac{\partial\phi}{\partial n} = 0 \;\text{ on }\; y = b(x), \tag{1.1.16}$$

where $\vec{n}$ denotes the unit outward normal to $y = b(x)$. On the unknown free boundary $y = \eta(x,t)$, the two conditions are given by

$$\frac{\partial \phi}{\partial t} + \frac{1}{2}|\nabla\phi|^2 + g\eta(x,t) + p/\rho = 0, \tag{1.1.17}$$

and

$$\frac{\partial \phi}{\partial n} = \vec{V}\cdot\vec{n}, \tag{1.1.18}$$

where $\vec{V}$ is the velocity of the free boundary and $\vec{n}$ is the unit outward drawn normal on it. Equation (1.1.18) can be expressed in terms of quantities already defined. Let,

$$F(x,y,t) = y - \eta(x,t) = 0. \tag{1.1.19}$$

Then

$$DF = \frac{\partial F}{\partial x}dx + \frac{\partial F}{\partial y}dy + \frac{\partial F}{\partial t}dt = 0, \tag{1.1.20}$$

or

$$-\frac{\partial F}{\partial t} = -\frac{\partial \eta}{\partial x}\frac{dx}{dt} + \frac{dy}{dt} = \vec{V}\cdot\nabla F = (\vec{V}\cdot\vec{n})|\nabla F|,\ \vec{n} = \nabla F/|\nabla F|. \tag{1.1.21}$$

From (1.1.18) and (1.1.21), we get

$$\begin{aligned} -\frac{\partial F}{\partial t} &= \frac{\partial \phi}{\partial n}|\nabla F| = (\nabla\phi\cdot\vec{n})|\nabla F| \\ &= \left(\nabla\phi\cdot\frac{\nabla F}{|\nabla F|}\right)|\nabla F| = (\nabla\phi\cdot\nabla F) \end{aligned} \tag{1.1.22}$$

or

$$\frac{\partial \eta}{\partial t} = -\frac{\partial \phi}{\partial x}\frac{\partial \eta}{\partial x} + \frac{\partial \phi}{\partial y}. \tag{1.1.23}$$

Equation (1.1.12) is to be solved using the boundary conditions (1.1.17) and (1.1.23), the fixed boundary condition (1.1.16) and the prescribed $\eta(x,t)$ at $t = 0$. In this problem the velocity of the fluid is not time dependent but the free boundary is time dependent. Such problems are called *quasi-steady state free boundary problems or degenerate free boundary problems.* We shall see later that the term degenerate free boundary problem is used for other types of problems also.

Linearization of the above problem

Let $y = y_0$ be the flat upper surface of water. When the deviation of the free boundary from the flat surface is small, i.e., if $|\partial\eta/\partial x| << 1$, then the above problem can be linearized as follows. Let

$$y = \eta(x,t) = y_0 + \varepsilon f(x,t), \quad \phi = \varepsilon u,\ \varepsilon << 1. \tag{1.1.24}$$

$\varepsilon f(x,t)$ can be considered as the disturbance at the flat surface $y = y_0$. Substituting (1.1.24) in (1.1.12), (1.1.16), (1.1.17) and (1.1.23), and in the changed equations retaining only linear terms in ε and dropping higher order terms of ε, we obtain

$$\nabla^2 u = 0 \quad , \quad b(x) < y < y_0, \tag{1.1.25}$$

$$\frac{\partial u}{\partial n} = 0 \;\text{ on }\; y = b(x), \tag{1.1.26}$$

$$\frac{\partial u}{\partial t} + gf = 0 \;\text{ on }\; y = y_0, \tag{1.1.27}$$

$$\frac{\partial u}{\partial y} = \frac{\partial f}{\partial t} \quad \text{on} \quad y = y_0. \tag{1.1.28}$$

In (1.1.27), we have taken $p = -\rho g y_0$ which comes from the contribution of the zeroth order terms in (1.1.17).

On eliminating f from (1.1.27) and (1.1.28), we obtain

$$\frac{\partial^2 u}{\partial t^2} + g\frac{\partial u}{\partial y} = 0 \quad \text{on} \quad y = y_0. \tag{1.1.29}$$

Since $y = y_0$ is not a free boundary, only one boundary condition is to be prescribed on it. For further information about the Problems 1.1.2. and 1.1.3., see ([3, 4, 5]).

Problem 1.1.4. A Problem of Reproductive Toxic Mass Diffusion

Let $u(x,t)$ be the concentration of a toxic mass which is diffusing in a region Ω, where

$$\Omega = \{x \in R : 0 \le x \le 1\}. \tag{1.1.30}$$

If the concentration exceeds a certain value u_v in a portion of Ω, then it is called a toxic region. Let the reproduction rate of toxic mass in the toxic region be P and in the non-toxic region αP, $0 < \alpha < 1$. The toxic and non-toxic regions are separated by a surface S, where

$$S = \{(x,t) \in \Omega_{t_*} \mid x = \phi(t)\,,\; u = u_v\} \tag{1.1.31}$$

and

$$\Omega_{t_*} = \{(x,t) | \, x \in \Omega\,,\, 0 < t \le t_*\}. \tag{1.1.32}$$

$u(x,t)$ and the free boundary $x = \phi(t)$ are to be obtained by solving the following system of equations. For $u > u_v$

$$u_t - u_{xx} + d_0 + d_1 u = P, \;\; (x,t) \in \Omega_{t_*}\,,\; 0 < x < \phi(t),\; 0 < t \le t_*. \tag{1.1.33}$$

For $u < u_v$

$$u_t - u_{xx} + d_0 + d_1 u = \alpha P, \;\; (x,t) \in \Omega_{t_*},\;\; \phi(t) < x < 1,\;\; 0 < t \le t_*. \tag{1.1.34}$$

The term d_0+d_1u accounts for the mass loss due to the bottom leakage, and other similar factors. d_0, d_1, α, and P are positive constants. Mass diffusion coefficient has been taken to be unity which is possible by suitably defining the time and/or length scales.

At the free boundary

$$u_x(\phi(t)-,t) = u_x(\phi(t)+,t), \quad \text{(continuity of mass flux)} \tag{1.1.35}$$

and

$$u(\phi(t),t) = u_v. \tag{1.1.36}$$

It may be noted that the velocity of the free boundary is not explicitly occurring in (1.1.35) which was so in Problems 1.1.2 and 1.1.3. The boundary conditions at the free boundary in which the velocity of the free boundary is not occurring explicitly are known as *implicit free boundary conditions.* Equations (1.1.33)–(1.1.36) are to be supplemented with a suitable initial condition at $t = 0$ and with boundary conditions at the fixed boundaries $x = 0$ and $x = 1$. Some results on the existence of solution of the above problem and the regularity of the free boundary can be found in [6].

Problem 1.1.5. Gas Flow Through Porous Media

The equation of state for an isentropic (constant entropy) flow of an ideal gas in a homogeneous porous media is given by [7]

$$\rho(x,y) = \rho_0 p^\alpha \geq 0, \tag{1.1.37}$$

where $\rho(x,y)$ is the density and $p(x,y)$ is the pressure. $\rho_0 \in R^+$ and $\alpha \in (0,1]$ are constant. The conservation of mass gives

$$\text{div}\,(\rho\vec{V}) = -\gamma\frac{\partial\rho}{\partial t}, \tag{1.1.38}$$

where γ is the porosity of the medium. According to Darcy's law [8], the velocity $\vec{V}$ of the gas flow in a porous medium is given by

$$\vec{V} = -(\beta/\eta)\,\text{grad}\,p, \tag{1.1.39}$$

$\beta \in R^+$ is the permeability of the medium and $\eta \in R^+$ is the viscosity of the gas. $\vec{V}$ and p can be eliminated from (1.1.38) and (1.1.39) and we obtain

$$\frac{\partial\rho}{\partial t} = \frac{\beta}{\eta\gamma\rho_0^{1/\alpha}(1+\alpha)}\nabla^2(\rho^m)\ , \quad \rho \geq 0\,, \tag{1.1.40}$$

where $m = 1 + 1/\alpha$. The diffusion is called 'fast' if $m > 1$, and 'slow' if $m < 1$.

By suitably choosing the time and/or length scales, the following equation can be obtained from (1.1.40).

$$\frac{\partial\rho}{\partial t} = \nabla^2(\rho^m), \quad \rho \geq 0. \tag{1.1.41}$$

If $\alpha \in (0,1]$, then $m \in [2,\infty)$. Equation (1.1.41) is the porous media equation and it arises also in other contexts such as population dynamics and plasma physics [7]. In order to calculate the mass flux of the gas, the r.h.s. of (1.1.41) can be written as

$$\nabla^2(\rho^m) = \mathrm{div}(m\rho^{m-1}\,\mathrm{grad}\rho), \tag{1.1.42}$$

$m\rho^{m-1}\,\mathrm{grad}\,\rho$ is the mass flux and $m\rho^{m-1}$ can be taken as diffusivity. Diffusivity vanishes as the density ρ tends to zero. Therefore (1.1.41) is a *nonlinear degenerate equation* in the neighbourhood of any point where $\rho = 0$ but is non-degenerate and *uniformly parabolic* (see [9] and (7.3.26)) in the neighbourhood of any point at which ρ is away from zero. Such problems are called *degenerate parabolic-elliptic* problems. An important consequence of nonlinear degeneracy is that there is a finite speed of propagation of a disturbance from rest which is in contrast to the parabolic heat equation in which the speed of heat propagation is infinite. The finite speed of propagation may give rise to *waiting time solutions.* Equation (1.1.41) is to be supplemented with an initial condition if the region is infinite and with both initial and boundary conditions if the region considered is finite. The existence of a free boundary in such problems can be illustrated with the help of the following example. Let

$$\frac{\partial \rho}{\partial t} = \nabla^2(\rho^m), \quad -\infty < x < \infty, \; t > 0, \tag{1.1.43}$$

$$\rho(x,0) = \begin{cases} > 0 \text{ for } x \in R_I = (a_1, a_2), -\infty < a_1 < a_2 < \infty \\ = 0 \text{ for } x \in R\backslash R_I. \end{cases} \tag{1.1.44}$$

This problem, generally, does not have a *classical solution.* The classical solution of a problem can be roughly stated to be a solution in which the dependent variable possesses continuous derivatives of the order required in the problem formulation. The mathematical definition of a classical solution will be discussed later but at present it would suffice to state that the solution $\rho(x,t)$ of (1.1.43) and (1.1.44) may not possess the required continuous derivatives. For $t > 0$, gas will be diffusing to the right of $x = a_2$ and to the left of $x = a_1$ and thus giving rise to two moving boundaries $x = S_i(t), i = 1,2$. Let S_1 be moving towards $+\infty$ and S_2 moving towards $-\infty$. Using a *weak formulation* of the above problem in (1.1.43) and (1.1.44), several interesting results on the behavior of $S_i(t)$, $i = 1,2$ have been obtained in [10] and [11]. The following proposition indicates that in some cases, the interface $S_i(t)$ starts moving only after an elapse of time $t^* > 0$.

Proposition 1.1.1. There exist numbers $t_i^* \in [0,+\infty)$ for $i = 1,2$ such that $S_i(t)$ is strictly monotone for $t \in (t_i^*, +\infty)$ and

$$S_i(t) = a_i, \quad i = 1,2 \text{ for } t \in [0, t_i^*]. \tag{1.1.45}$$

If $t_i^* > 0$, then $S_i(t)$, $i = 1,2$ remain stationary for t_i^* units of time [11].

In this case t_i^* is called a *waiting time.* It has been proved in [11] that the interface is Hölder continuous under certain conditions and if the interface is in motion, then one expects it to move with the velocity $\vec{V}$ of the gas, and

$$\frac{dS_i}{dt} = -\lim_{x \to S_i(t)\pm} W_x(x,t), \; i = 1,2; \; W = \frac{m}{m-1}\rho^{m-1}. \tag{1.1.46}$$

The limits in (1.1.46) are taken as x approaches the boundary of the region from within the region in which $\rho(x,t) > 0$. Equation (1.1.46) can also be obtained from the mass balance condition at the interface which states that the jump in the density at the interface multiplied by the velocity of the interface is equal to the jump in the mass flux across the interface.

In a problem symmetrical with respect to x, it is sufficient to consider the region $0 \leq x < \infty$ with a single interface $x = S_1(t)$ and $a_1 = a_2 > 0$. $S_1(t)$ should satisfy (1.1.46) for $i = 1$, and another condition to determine the unknown $S_1(t)$ may be prescribed as

$$\int_0^{S_1(t)} \rho(x,t)dx = \int_0^{a_1} \rho(x,0)dx. \tag{1.1.47}$$

Conditions of the type (1.1.47) are called *non-local boundary conditions* at the free boundary. $x = 0$ is now a fixed boundary and the boundary condition on it is given by

$$\left.\frac{\partial \rho}{\partial x}\right|_{x=0} = 0. \tag{1.1.48}$$

Note that we have two conditions prescribed at the free boundary, viz., (1.1.46) and (1.1.47).

Problem 1.1.6. Shock Propagation

The solution of Burger's equation (1.1.49) with boundary condition (1.1.50) has been discussed in [12].

$$u_y + uu_x = u_y + \frac{1}{2}\frac{\partial}{\partial x}(u^2) = 0, \; (x,y) \in R^2, \; y \geq 0. \tag{1.1.49}$$

On $y = 0$,

$$u(x,y)|_{y=0} = f(x), \quad x \in R, \tag{1.1.50}$$

where

$$\left.\begin{aligned} f(x) &= 1, \quad x \leq 0, \\ &= 1 - x, \quad 0 \leq x \leq 1, \\ &= 0, \quad x \geq 1. \end{aligned}\right\} \tag{1.1.51}$$

The characteristic equations of (1.1.49) in parametric form in terms of a parameter t are given by

$$\frac{dx}{dt} = 0, \; \frac{dy}{dt} = 1, \text{ and } \frac{du}{dt} = 0. \tag{1.1.52}$$

Let $x(s,0) = s,\ y(s,0) = 0$ and $u(s,0) = f(s)$.

The solution of (1.1.52) is given by

$$x(s,t) = \begin{cases} s+t, & s \le 0, \\ (1-s)t+s, & 0 \le s \le 1, \\ s, & s \ge 1. \end{cases} ; \quad y = t \tag{1.1.53}$$

$$u(s,t) = \begin{cases} 1, & s \le 0, \\ (1-s), & 0 \le s \le 1, \\ 0, & s \ge 1. \end{cases} \tag{1.1.54}$$

The characteristic curves and $u(x,y)$ are shown in Fig. 1.1.1. It is clear from the figure that the characteristic curves intersect in the region $x \ge 1,\ y \ge 1$ and therefore $u(x,t)$ is not defined (is not single valued) in this region. $u(x,y)$ defined by (1.1.54) ceases to be the classical solution of (1.1.49) in the region $x \ge 1,\ y \ge 1$ and (1.1.49) is satisfied in this region only in the weak sense. It may be noted that the solution $u(x,y)$ could be discontinuous in some portion of the region even if $f(x)$ is a continuous and differentiable function throughout the region. If (1.1.49) is written as

$$\frac{\partial P}{\partial y} + \frac{\partial Q}{\partial x} = 0, \quad P = u, \quad Q = \frac{1}{2}u^2, \tag{1.1.55}$$

then it can be shown (cf. [12]) that the weak solution is discontinuous across a curve which is called a *shock* and whose equation is given by

$$[P]dx - [Q]dy = 0. \tag{1.1.56}$$

The notation $[f]$ denotes the difference between the limits of a function f as the shock or a surface of discontinuity of f is approached from both sides. Conditions of the type (1.1.56) are called *Rankine-Hugoniot conditions.* In the present problem $[P] = 1$ and $[Q] = 1/2$, and therefore the shock is given by $y = 2x + d$. The constant d can be determined if we know the point from which the shock is emanating, for example, in the present problem this point is (1.1) (see Fig. 1.1.2.). Shock is the oldest form of the free boundary and its origin lies in the study of gas dynamics [13].

If (1.1.49) is multiplied by $u^n, n \ge 1$, then P and Q in (1.1.55) will change. This change will result in the non-uniqueness of the shock and infinitely many shocks can be obtained. For nonlinear hyperbolic equations of the type (1.1.49), further information in the form of physical arguments involving stability, entropy, dissipation or continuous dependence on the initial data is needed to ensure uniqueness (cf. [14, 15]).

There are some important differences between the present problem and the free boundary Problems 1.1.1. to 1.1.5. In the previous five problems, the free boundaries can be

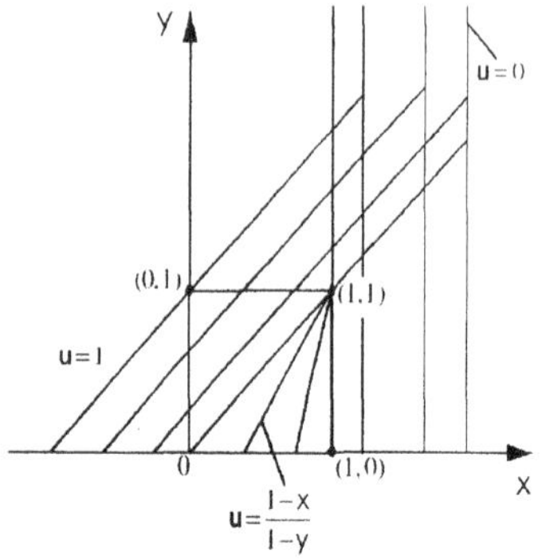

Fig. 1.1.1. Characteristic curves and $u(x, y)$

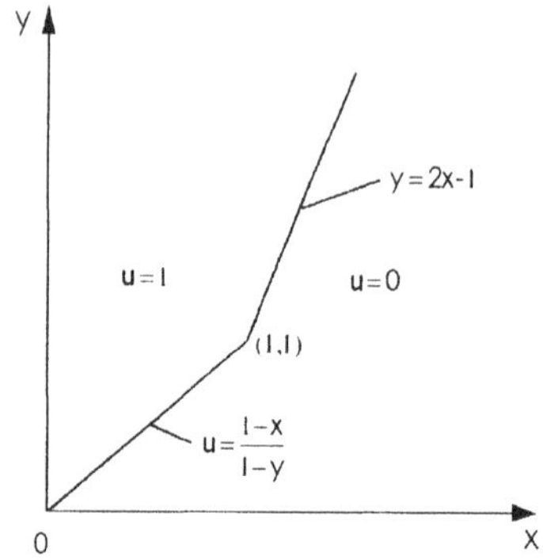

Fig. 1.1.2. Geometry of the shock

identified without much difficulty and the boundary conditions on them can be imposed after some thought. It is a different matter that the classical solution may or may not exist. In the solution of equations (1.1.49)–(1.1.50), there is no indication of a free boundary. It is only through the construction of the solution that we come to know about the magnitudes of jumps in P and Q. The equation (1.1.56) can be obtained only through the weak solution and not through the classical approach, which was followed in other problems.

Problem 1.1.7. Free Boundary Associated with a Frictional Oscillator Problem

This interesting free boundary problem has been reported in [15]. As shown in Fig. 1.1.3., a block of mass m rests on a conveyer belt moving with a constant velocity V. The forces acting on the mass are: (1) spring force with a spring modulus S_m, (2) prescribed force $F(t)$ assumed to be sufficiently smooth, and (3) Coulomb frictional force with coefficient of friction μ. The motion of the block with mass m is governed by the equation

$$\frac{d^2x}{dt^2} + \mu mg \left\{ sgn\left(\frac{dx}{dt} - V\right)\right\} + S_m x = F(t), \tag{1.1.57}$$

where

$$\begin{aligned} sgn(z) &= \quad 1 \text{ if } z > 0, \\ &= -1 \text{ if } z < 0, \end{aligned} \tag{1.1.58}$$

and g is the acceleration due to gravity.

At time $t = 0$, x and dx/dt are prescribed. Both analytical and numerical solutions of (1.1.57) are extremely difficult as at each instant of time information is required whether $dx/dt > V$, $= V$, or $< V$. The unknown instants of time $t = t_i$, $i = 1, 2, \ldots$ or the points $x_i = x(t_i)$, $i = 1, 2, \ldots\ldots$ at which $dx/dt = V$ can be regarded as free boundaries.

When $dx/dt = V$, $d^2x/dt^2 = 0$. The second derivative becomes discontinuous at $t = t_i$. Once dx/dt becomes V at $t = t_i$, it will remain so in some interval $t_i \le t \le t_i^*$ until at $t = t_i^*$, $|F(t) - S_m x| > \mu mg$. Again at $t = t_{i+1}$, dx/dt may be equal to V. If so, $t = t_{i+1}$ is a free boundary. If it is assumed that mass does not jump and there is no abrupt change in the velocity of the mass, then it can be assumed that $x(t)$ and dx/dt are continuous functions of time and $dx/dt = V$ at the free boundary. These assumptions provide boundary conditions at each free boundary.

There is a fundamental difference between this problem and the free boundary Problems 1.1.1. to 1.1.6. In the earlier problems, free boundaries arose due to the physical situations or the mathematical nature of the solution such as a weak solution, but in the present problem free boundaries arise due to discontinuities in the physical properties of the solution.

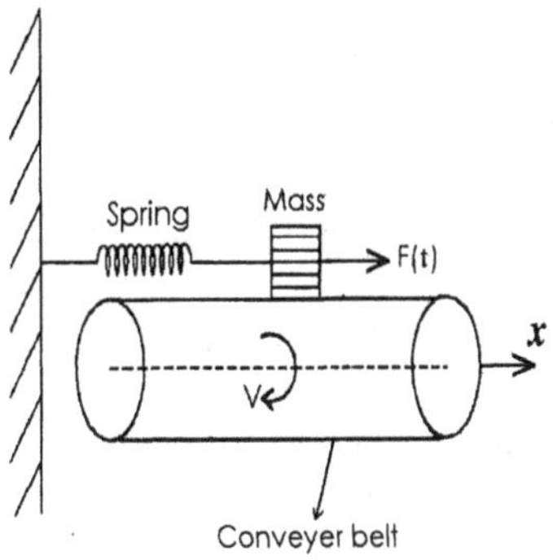

Fig. 1.1.3. Frictional oscillator

Problem 1.1.8. Impact of a Visco-Plastic Bar on a Rigid Wall

A bar of length b made up of visco-plastic incompressible material moving with constant velocity $-V_0$ hits a rigid wall at time $t = 0$. We consider a one-dimensional problem in which x-axis is taken opposite to the initial motion of the bar and the rigid wall is taken at $x = 0$. After the bar hits the wall, compressional stresses develop in the bar giving rise to visco-plastic flow of the material in the region. If $V(x,t)$ is the velocity of the bar for $t > 0$, then the gradient of $V(x,t)$ or the velocity of deformation can be expressed as follows (cf. [16]).

$$\frac{\partial V}{\partial x} = \begin{cases} \frac{\tau + \tau_0}{\mu}, & |\tau| \ge \tau_0, \ \tau_0 > 0, \ \tau \le 0, \\ 0\ . & |\tau| \le \tau_0. \end{cases} \tag{1.1.59}$$

Here, τ represents compressional stress, which is negative as the x-axis is oriented opposite to the direction of motion of the bar, τ_0 is the stress at the limit point and μ is the coefficient of viscosity of the material.

It will be assumed that any disturbance is propagated over the whole bar instantaneously. The equation of motion in the visco-plastic region can be easily obtained by using Newton's second law of motion. We have

$$\rho\frac{\partial V}{\partial t}=\frac{\partial \tau}{\partial x}=\mu\frac{\partial^2 V}{\partial x^2},\ 0\le x\le S_0(t),\ 0<t<t_*,\ S_0(0)=0, \tag{1.1.60}$$

where ρ is the density of the material and $x=S_0(t)$ is the interface between elastic and visco-plastic regions. In the elastic region

$$\frac{\partial V}{\partial x}=0,\quad |\tau|\le\tau_0,\ S_0<x<b. \tag{1.1.61}$$

Integrating (1.1.61), we obtain

$$V=-X(t),\ \ X(0)=V_0. \tag{1.1.62}$$

In obtaining (1.1.62) it has been assumed that the elastic portion of the rod moves like a rigid body.

Using again Newton's second law of motion at $x=S_0$, we obtain

$$-M\frac{dX}{dt}=\tau|_{S_0+}F_0, \tag{1.1.63}$$

where M is the mass of the elastic part and F_0 is the area of cross section of the rod assumed to be uniform. Since $M=F_0\rho(b-S_0(t))$, equation (1.1.63) becomes

$$\frac{dX}{dt}=-\frac{\tau_0}{\rho(b-S_0(t))}. \tag{1.1.64}$$

In obtaining (1.1.64) it has been assumed that stress is continuous at the interface $x=S_0(t)$. Using the condition that $V(x,t)$ is also continuous at $x=S_0(t)$, we get

$$V(S_0,t)=-X(t). \tag{1.1.65}$$

Also

$$\frac{\partial}{\partial x}V(S_0,t)=0. \tag{1.1.66}$$

The initial and boundary conditions for $V(x,t)$ are

$$V(x,0)=-V_0,\ 0\le x\le b;\ \ V(0,t)=0,\ \ t>0. \tag{1.1.67}$$

The functions $V(x,t)$, $S_0(t)$ and $X(t)$ are to be determined. An approximate solution of the above problem has been discussed in [16].

The free boundary in this problem arises due to the change in the physical properties of the system.

Problem 1.1.9. A Problem with Discontinuous Moving Boundary

Discontinuity of the free boundary in multi-dimensional problems can be easily demonstrated geometrically. For example, a piece of ice floating in water breaks into two pieces after melting for some time. A discontinuous moving boundary in a one-dimensional problem does not commonly occur. In [17] an interesting problem of discontinuous moving boundary which is associated with the diffusion of moisture in a porous capillary tube of length unity is discussed. At time $t = 0$, the portion of the tube $0 \leq x \leq x_0,\ x_0 < 1$ is filled with moisture and the remaining portion is dry. The temperature $T(x,t)$ of the moisture is less than the boiling temperature $T = 0$, except at $x = x_0$ where $T = 0$. The temperature of the dry air in some neighbourhood of $x = x_0$ on the right is greater than zero and in the remaining portion, the temperature of the air is less than zero. There is a continuous flow of moisture into the tube at $x = 0$. At $x = 1$, the dry air is getting heated, causing evaporation to take place and making the moisture advance into the dry air. Let $T_1(x,t)$ and $T_2(x,t)$ be the temperatures of the moisture and the dry air respectively, $W_1(x,t)$ the concentration of the moisture, and $x = S(t)$, the equation of the free boundary which is the interface between moisture and the dry air.

Under certain initial conditions it may happen that at some instant of time, say $t = m,\ m > 0$, the temperature in some neighbourhood δ_m of the moving boundary becomes less than or equal to zero. In this case the moisture will advance with a jump into the dry part, i.e.,

$$S(m+) - S(m-) = \delta_m > 0. \tag{1.1.68}$$

The problem is concerned with finding T_1, T_2, W_1 and $S(t)$. The formulation of this problem is as follows:

Differential Equations

$$\frac{\partial T_i}{\partial t} = a_i^2 \frac{\partial^2 T_i}{\partial x^2},\quad (x,t) \in D_i,\ \ i = 1,2, \tag{1.1.69}$$

$$\frac{\partial W_1}{\partial t} = d\frac{\partial^2 W_1}{\partial x^2},\quad (x,t) \in D_1, \tag{1.1.70}$$

$$D_1 = \{(x,t) : 0 < x < S(t),\ 0 < t < m\} \cup \{(x,t) : 0 < x < S(t),\ m < t < t_*\},$$

$$D_2 = \{(x,t) : S(t) < x < 1,\ 0 < t < m\} \cup \{(x,t) : S(t) < x < 1,\ m < t < t_*\}.$$

Here a^2 is the thermal diffusivity and d is the mass diffusivity. Equation (1.1.69) is the heat conduction equation and (1.1.70), the mass diffusion equation (Fick's law).

Initial conditions

$$T_1(x,0) = \phi_1(x) \leq 0,\ \ W_1(x,0) = \psi_1(x) \geq 0;\ 0 \leq x \leq x_0, \tag{1.1.71}$$

$$T_2(x,0) = \phi_2(x),\ x_0 \leq x \leq 1\,;\ \phi_2(x_0) > 0 \text{ and } \phi_2(1) > 0. \tag{1.1.72}$$

Boundary conditions at $x = 0$ and $x = 1$

$$T_1(0,t) = f_1(t) \le 0, \;\; W_1(0,t) = g_1(t) \ge 0; \;\; 0 < t \le t_*, \tag{1.1.73}$$

$$T_2(1,t) = f_2(t) \ge 0. \tag{1.1.74}$$

Boundary conditions at the moving boundary $x = S(t)$, $t \ne m$

$$\{T_1(x,t),\; W_1(x,t)\}_{x=S(t)} = 0,\; t > 0,\; S(0) = x_0, \tag{1.1.75}$$

$$K_2 \left.\frac{\partial T_2}{\partial x}\right|_{x=S(t)} = \alpha(T_2(S(t)+0,t) - T_1(S(t)-0,t)),\; t > 0. \tag{1.1.76}$$

Here, K is the thermal conductivity and α is the heat transfer coefficient. The convective boundary condition in (1.1.76) arises because of the discontinuity of temperatures at $x = S(t)$, $t \ne m$. Heat balance at $x = S(t)$ implies

$$\begin{aligned} C_2\rho_2 T_2(S(t)+0,t)\frac{dS}{dt} + K_2\frac{\partial T_2}{\partial x}(S(t)+0,t) - K_1\frac{\partial T_1}{\partial x}(S(t)-0,t) \\ = -d\,q_1\frac{\partial W}{\partial x}(S(t)-0,t). \end{aligned} \tag{1.1.77}$$

Here, C is the specific heat, ρ the density and q_1 the latent heat of evaporation. Definitions of different parameters have been given later in § 2.1.3. The derivation of (1.1.77) is based on the law of conservation of energy (see § 1.4.7)

A sufficiently small real number $\varepsilon > 0$ exists such that

$$T_2(x,t) > 0 \text{ for } S(t) < x < S(t) + \varepsilon, \;\; t \ne m. \tag{1.1.78}$$

For $t = m$,

$$\left.\begin{aligned} T_1(x,m+) = T_2(x,m-), \text{ and } W_1(x,m+) = 0; S(m-) \le x \le S(m+), \\ m\pm = m \pm 0, \end{aligned}\right\} \tag{1.1.79}$$

$$T_2(x,m) \le 0, \;\; S(m-) \le x \le S(m+). \tag{1.1.80}$$

The existence and uniqueness of this problem under suitable assumptions have been discussed in [17] and sufficient conditions for the existence of a discontinuous moving boundary are given in [18]. Discontinuity in the temperature and in the free boundary is a typical feature of this problem.

Problem 1.1.10. Penetration of Solvents in Polymers

Consider a slab of a glassy polymer, such as methyl methacrylate in contact with a solvent, n-alkyl alcohol [19]. If the solvent concentration exceeds a threshold value, say,

$q \geq 0$, then the solvent moves into the polymer, creating a swollen layer in which the solvent diffuses according to Fick's law for mass diffusion. If $W(x,t)$ is the concentration of the solvent in the polymer and $x = S(t)$ is the free boundary representing the penetration depth of the solvent in the polymer, then W and S satisfy the following equations.

$$W_t - W_{xx} = 0, \ \ (x,t) \in D_{t_*} = \{(x,t) : 0 < x < S(t),\ 0 < t < t_*\}, \tag{1.1.81}$$

$$W(0,t) = 1, \quad 0 < t < t_*, \tag{1.1.82}$$

$$S(t) = f\{W(S(t))\}, \ \ S(0) = 0, \ \ 0 < t < t_*, \tag{1.1.83}$$

$$W_x(S(t),t) = -\dot{S}(t)[W(S(t),t) + q], \ \ 0 < t < t_*. \tag{1.1.84}$$

In these equations, normalized solvent concentration is represented by $W + q$. W represents the excess concentration, normalized to 1 at $x = 0$. Equation (1.1.83) describes the prescribed penetration law and (1.1.84) arises from the mass conservation at the free boundary. The well-posedness, existence, uniqueness, etc., of the solution of the above problem have been shown in [19]. A numerical method for its solution has also been suggested in [19]. Some more mathematical models describing the crystallization of polymers and their mathematical analysis are presented in [20]. In (1.1.83), $S(t)$ is a function of the concentration.

Problem 1.1.11. Filtration of Water Through Oil in a Porous Medium

Consider a one-dimensional problem in a semi-infinite porous medium $x \geq 0$ of porosity m. At time $t = 0$, the region $0 \leq x \leq b$ is filled with oil and the region $b \leq x < \infty$ is filled with water. Water percolates into oil, and so for $t > 0$ there will be three regions. Water-filled region, $b < x < \infty$, will have hundred percent saturation of water; oil-bearing region, $0 \leq x < S(t)$, will have hundred percent saturation of oil; in the region $S(t) < x < b$, both water and oil mixture will be present. This region can be called an intermediate zone. $S(t)$ is the free boundary. The oil content of this intermediate zone changes with time, and tends to reach a limiting state called the residual oil saturation state. This stage is characterized by the fact that if time is counted from the moment of passage of the oil-water contact boundary past a fixed element of volume, the degree of oil saturation of this volume will not depend on time and the flow in this region will be of one phase. It will be assumed that water filters through the intermediate zone while the oil in this zone remains stationary [21]. However the rate of penetration of water in the intermediate zone is lower than in the water-filled zone as the oil concentration in the intermediate zone is greater than zero. Under appropriate assumptions, the following one-dimensional model is obtained.

$$u_0 = -\frac{\lambda_0}{\mu_0}\frac{\partial p_0}{\partial x}; \ \ \alpha_0^2\frac{\partial^2 p_0}{\partial x^2} = \frac{\partial p_0}{\partial t}, \ \ b < x < \infty, \ t > 0, \tag{1.1.85}$$

$$u_1 = -\frac{\lambda_1}{\mu_1}\frac{\partial p_1}{\partial x}; \ \ \alpha_1^2\frac{\partial^2 p_1}{\partial x^2} = \frac{\partial p_1}{\partial t}, \ \ S(t) < x < b, \ t > 0, \tag{1.1.86}$$

$$u_2 = -\frac{\lambda_2}{\mu_2}\frac{\partial p_2}{\partial x}; \quad \alpha_2^2 \frac{\partial^2 p_2}{\partial x^2} = \frac{\partial p_2}{\partial t}, \quad 0 < x < S(t), \ t > 0. \tag{1.1.87}$$

The subscripts $0, 1, 2$ stand for the water-filled, intermediate and oil-rich zones, respectively. u_i and p_i, $i = 0, 1, 2$, denote the filtration velocities and pressures in the three different zones. The constants λ_i, α_i and μ_i, $i = 0, 1, 2$, denote the coefficients of permeability, piezoconductivity and viscosity of the ith zone. The equation of velocity in terms of pressure gradient is the result of Darcy's law [8] and the pressure obeys the equation of piezoconductivity.

Boundary conditions at $x = b$ and $x = S(t)$ follow from the continuity of velocities and pressures and are given by

$$u_0 = u_1, \ p_0 = p_1; \ \text{at } x = b, \tag{1.1.88}$$

$$u_1 = u_2, \ p_1 = p_2; \ \text{at } x = S(t). \tag{1.1.89}$$

At the free boundary, an additional condition is required which is obtained from the 'mass balance' consideration

$$m(1 - \delta_1 - \delta_2)\frac{dS}{dt} = u_1. \tag{1.1.90}$$

Here, δ_1 and δ_2 are the contents of water and the residual oil saturation, respectively in the transition zone, and m is the porosity of the stratum. To complete the formulation of the problem, suitable initial and boundary conditions for pressure should be prescribed. For example in [21], the initial pressure in the entire stratum is assumed to be constant. Similarly it is assumed that pressure has a constant value at the boundary $x = 0$. The zero reference point for calculation of pressure is so chosen that $p_2(0, t) = 0$. With the above assumptions the initial and boundary conditions can be written as

$$p_i|_{t=0} = P = \text{ constant }, \ i = 0, 1, 2; \ p_2|_{x=0} = 0. \tag{1.1.91}$$

An approximate solution under the quasi-steady approximation, valid at the initial stages has been obtained in [21] with constant parameter values . The boundary conditions (1.1.89) do not contain the velocity of the free boundary.

Problem 1.1.12. Obstacle Problem for a String

This problem belongs to a class of problems which have a variational inequality formulation. Three different types of formulations of this obstacle problem are possible and their equivalence is discussed in Chapter 7. A new notion of codimensionality-two of the free boundary is associated with this problem and this will be discussed briefly in § 1.2. Consider a weightless elastic string which is held tight between two fixed points $P_1 = (0, 0)$ and $P_2 = (b, 0)$, $b > 0$ in R^2-plane. This string is displaced upwards by a rigid body called an obstacle (cf. [22, 15]). A view of the cross section of this system of string and obstacle from above in R^2-plane is shown in Fig. 1.1.4. Let $y = \psi(x)$ be the equation of the cross section of the obstacle in R^2, i.e., the equation of the curve Q_1ABQ_2. $\psi(x)$

is assumed to be sufficiently smooth. We shall relax this smoothness condition later (see § 7.2.5). Let $y = u(x)$ be the equation of the string in the equilibrium position or the equation of the curve P_1ABP_2. The problem is to find the function $u(x)$ and the arc AB of the string which is in contact with the obstacle and on which $u(x) = \psi(x)$. The points A and B are free boundaries and in the present case we can call them 'free points'. Once A and B are determined, the arc AB is also determined because $\psi(x)$ is known.

The formulation of the problem is as follows:

$$u(0) = u(b) = 0, \quad \text{(fixed end conditions)}, \tag{1.1.92}$$

$$u(x) = \psi(x) \text{ on } AB; \;\; u(x) \geq \psi(x) \;\text{ on } P_1ABP_2, \tag{1.1.93}$$

$$u''(x) \leq 0, \tag{1.1.94}$$

$$u(x) > \psi(x) \Rightarrow u''(x) = 0. \tag{1.1.95}$$

The second equation in (1.1.93) implies that the string does not penetrate the obstacle. The concavity of the string as viewed from the x axis implies (1.1.94) and the tightness of the elastic string implies that the portions P_1A and BP_2 are straight lines and so $u''(x) = 0$. The two boundary conditions at the free boundary can be obtained from the continuity of u and du/dx, i.e.,

$$[u]_A = [u]_B = \left[\frac{du}{dx}\right]_A = \left[\frac{du}{dx}\right]_B = 0, \tag{1.1.96}$$

where $[f]$ stands for the jump in the quantity under consideration at the given point. Equation (1.1.95) is equivalent to the following condition

$$[u(x) - \psi(x)]u''(x) = 0. \tag{1.1.97}$$

In view of (1.1.93)–(1.1.95), equation (1.1.97) is valid. When $u(x) = \psi(x)$, (1.1.97) is satisfied and when $u(x) \neq \psi(x)$ then the second equation in (1.1.93) implies $u(x) > \psi(x)$ and from (1.1.95), $u''(x) = 0$ and therefore (1.1.97) is satisfied. If (1.1.97) holds and if $u(x) > \psi(x)$, then $u''(x) = 0$, which is (1.1.95). The formulation (1.1.92)–(1.1.95) is equivalent to the formulation given by (1.1.92)–(1.1.94) and (1.1.97). Let us call the formulation (1.1.92)–(1.1.95) as Problem (R_1) and the formulation equivalent to Problem (R_1) as Problem (R_2). Consider the following minimization problem which is concerned with the minimization of the energy of the above string.

$$\min_{u \geq \psi} \int_{P_1}^{P_2} \left(\frac{du}{dx}\right)^2 dx \;, \quad \forall\, u(x) \in \mathcal{D}, \tag{1.1.98}$$

where $\mathcal{D} = \{u : u(0) = u(b) = 0,\;\; u \text{ and } \partial u/\partial x$ are continuous, and u satisfies relations (1.1.94) and (1.1.95)$\}$.

We shall call this formulation in (1.1.98) as Problem (R_3). This is a fixed domain formulation as it does not consider separate formulations on different portions of P_1ABP_2. In view of the equivalence of Problems (R_2) and (R_3) which will be established in Chapter 7, the formulation (R_2) is also called 'variational formulation' of the obstacle problem. A variational inequality formulation of this problem has been given in Chapter 7.

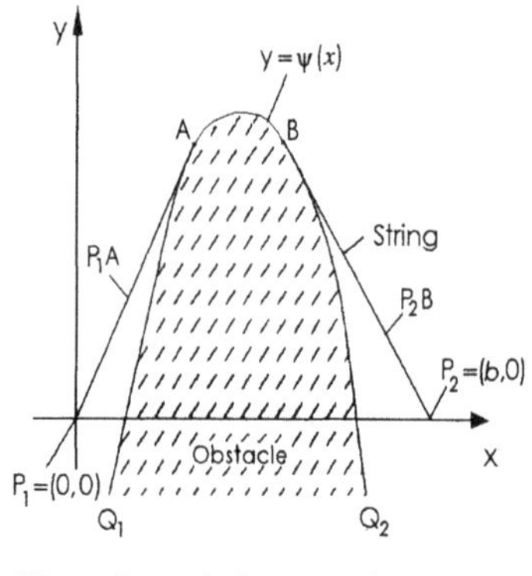

Fig. 1.1.4. Obstacle and the stretching of the string.

1.2 Free Boundary Problems with Free Boundaries of Codimension-two

In Problem 1.1.12., the free boundary consists of just two points A and B. Therefore the dimension of the free boundary is zero, whereas the spacial dimension of the problem is two. Another example of a free boundary of codimensional-two (or free points) is the tip of a propagating line crack in an elastic medium. We shall see later in § 3.2.2 that in the problems concerning initiation of solidification/melting along a known surface, codimension-two free boundaries could be time dependent. If the spacial dimension of the problem is n (the problem could be transient) and the dimension of the free boundary is $n-2$, then we say that the free boundary is of codimension-two. Several problems with free boundaries of codimension-two have been discussed in [23]. In Problems (1.1.1)-(1.1.11), the dimension of the free boundary in only one less than the spacial dimension of the problem. The distinction between the free boundaries of codimension-two and the free boundaries of codimension-one is not superficial but it raises some serious questions [23], some of which are mentioned below.

1. Can these problems with free boundaries of codimension-two, in some sense such as linearization, be modelled as limits of free boundary problems of codimension-one?

2. What can be said about the existence, uniqueness and regularity of solutions to models as stated ? In particular, is any of the information redundant ?

3. What methodology is available to solve these problems explicitly? No systematic methodology has yet been proposed for nonlinear problems of this type.

4. If solutions exist, are they stable to perturbations in the direction parallel to the free curve?

5. Is there a possibility of generating the models either to make their analysis or numerical solutions easier?

1.3 The Classical Stefan Problem in One-dimension and the Neumann Solution

Stefan problems are free boundary problems with some special features which can be easily explained with the help of the following one-dimensional solidification problem. At time $t = 0$, a pool of liquid at constant temperature $T_0 > T_m$, occupies a semi-infinite region $x \geq 0$. T_m is the temperature at constant pressure at which the liquid solidifies. It is called *freezing temperature* or *phase-change temperature.* In the case of pure metals such as copper, T_m is called *melting temperature* also because in the case of pure metals freezing and melting temperatures are the same. T_m is also called *equilibrium phase-change temperature* as both solid and liquid phases can stay together in thermodynamic equilibrium at $T = T_m$ (cf. § 2.1.4.). The fixed boundary $x = 0$ is being cooled, for example, temperature $T_b < T_m$ is prescribed at $x = 0$. In this case solidification starts instantaneously at $x = 0$. For any $t > 0$, the region $0 \leq x < \infty$ will consist of solid and liquid phases with the solid phase occupying the region $0 \leq x < S(t)$ and the liquid phase the region $S(t) < x < \infty$. $S(t)$ is the free boundary or the phase-change boundary/interface. This solidification problem can be formulated mathematically as follows:

In the solid region

$$C_S \rho_S \frac{\partial T_S}{\partial t} = K_S \frac{\partial^2 T_S}{\partial x^2}, \quad 0 < x < S(t), \ t > 0, \tag{1.3.1}$$

$$T_S(x,t)|_{x=0} = T_b < T_m \, , \quad t > 0. \tag{1.3.2}$$

In the liquid region

$$C_L \rho_L \frac{\partial T_L}{\partial t} + C_L \rho_L u_x \frac{\partial T_L}{\partial x} = K_L \frac{\partial^2 T_L}{\partial x^2}, \quad < S(t) < x < \infty, \quad t > 0, \tag{1.3.3}$$

$$T_L(x,t)|_{t=0} = T_0 \, , \ T_L(x,t)|_{x \to \infty} = T_0. \tag{1.3.4}$$

At $x = S(t)$

$$T_S = T_L = T_m, \tag{1.3.5}$$

$$K_S \frac{\partial T_S}{\partial x} - K_L \frac{\partial T_L}{\partial x} = \{l\rho_S + (\rho_L C_L - \rho_S C_S)T_m\}\frac{dS}{dt}, \tag{1.3.6}$$

$$S(0) = 0. \tag{1.3.7}$$

Equations (1.3.1)–(1.3.6) are not dimensionless. Temperatures T_b, T_0, T_m and thermophysical parameters ρ, C, K, and l are known quantities. The subscripts S and L stand for solid and liquid regions.

The symbols used in (1.3.1)–(1.3.6) are explained in the 'List of Notations'. The definitions of thermophysical parameters C, K, l, etc., will be discussed later in § 2.1.3. At present, our interest is to highlight some basic features of Stefan problems. Equation (1.3.1) is the Fourier heat conduction equation derived on the basis of Fourier's law given in (1.3.8) for heat conduction in a homogeneous isotropic medium.

$$\vec{q} = -K \text{ grad } T. \tag{1.3.8}$$

Here, $\vec{q}$ is called a heat flux vector and gives the heat flow at a spacial point at any instant of time $t\cdot$ $\vec{q}\cdot\vec{n}$ gives heat flow per unit area, per unit time in the direction of the outward drawn unit normal $\vec{n}$ to the surface of the region under consideration. The derivation of heat conduction equation is given in many books such as [24] and is obtained on the basis of principle of conservation of energy (cf. § 1.4.7). The heat equation (1.3.1) is based on a *macroscopic model* but it can also be derived on the basis of a *microscopic model.* The transport of heat energy takes place due to the random motion of molecules. The parabolic heat equation as well as hyperbolic heat equations can also be derived by considering discrete random walks of molecules [25].

If the densities of solid and liquid phases are not equal, then liquid may acquire motion whose velocity u_x in the one-dimensional case can be obtained from the mass balance condition at the interface given below.

$$\rho_L u_x = (\rho_L - \rho_S)\frac{dS}{dt}. \tag{1.3.9}$$

If $\rho_S > \rho_L$, then there will be a shrinkage of total volume in solidification, and u_x will be negative as dS/dt is positive in the above problem. If $\rho_S < \rho_L$, then there will be an expansion of the volume and u_x will be positive as dS/dt is positive. Equation (1.3.3) is the Fourier's heat conduction equation for the liquid in motion. Boundary conditions at the fixed boundary $x = 0$ could differ from that in (1.3.2) (see § 1.4.4.) but in the 'Neumann solution', a constant temperature is prescribed at $x = 0$. At $t = 0$ it was only the liquid phase therefore the initial condition has been prescribed only for the liquid. In the present problem T_0 is a constant. If T_m is a constant, then the conditions in (1.3.5) are called *isotherm conditions*. Equation (1.3.6) is the *dynamic energy balance condition*. Both (1.3.5) and (1.3.6) are based on the assumption of instantaneous achievement of local thermodynamic equilibrium at the phase-change interface. $-K_L \partial T_L/\partial x$ is the heat flux from the liquid outwards (the outward normal to liquid at $S(t)$ points into the solid) and $-K_S \partial T_S/\partial x$ is the heat flux from the solid outwards (the outward normal to solid at $S(t)$ points into the liquid).

Equation (1.3.6) can be derived by using the energy conservation law. It gives the energy balance at the phase-change boundary. In § 1.4.1, the derivation of the equation (1.3.6) has been obtained in a more general context. When the phase-change takes place from liquid to solid, the latent heat l released by the system is removed by conduction. Equation (1.3.6) is a dynamic compatibility condition and is commonly known as the *Stefan condition*. It is called a condition of Rankine-Hugoniot type. Equation (1.3.6) can also be stated in the form

$$\left[K\frac{\partial T}{\partial x}\right]_{\text{liquid}}^{\text{solid}} = -[H]_{\text{liquid}}^{\text{solid}} V_n. \tag{1.3.10}$$

Here, H is the enthalpy per unit volume, which is the sum of the latent heat plus the sensible heat, and V_n is the normal component of the velocity of the interface with the unit normal $\vec{n}$ pointing into the liquid.

Melting Problem

The problem discussed in (1.3.1)–(1.3.6) is a solidification problem as the phase changes from liquid to solid. The melting problem in which the phase changes from solid to liquid can be easily formulated by making appropriate changes in (1.3.1)–(1.3.6). For example, (1.3.3) is still valid for liquid but liquid region is now $0 < x < S(t)$. Similarly other equations can be modified. During solidification, the system releases latent heat but during melting the system 'absorbs' latent heat. Since we take $l > 0$, a minus sign should be put for heat absorbed. Equation (1.3.10) is still valid for the problem of melting provided the changed directions of normals to the two regions at the free boundary are taken into account. Since the formulation of both solidification and melting problems has the same mathematical structure, they are said to be mathematically analogous. If a solution to one of them is obtained, then the solution to the other problem can be obtained by making some minor changes in it. It is for this reason that it suffices to study either of them.

Neumann Solution

The method of analytical solution of (1.3.1)–(1.3.7) which is discussed below is based on some special type of solutions of heat equations (1.3.1) and (1.3.3), which can be written in terms of error and complementary error functions [26]. An exact analytical solution of the Stefan problem even in the one-dimensional case is possible only in few simple cases. If in (1.3.2), T_0 is time dependent or if a constant flux is prescribed at $x = 0$ then the method discussed below fails to provide a solution. The solution of (1.3.1) can be written as

$$T_S = A\,\mathrm{erf}\,\frac{x}{2(k_S t)^{1/2}} + T_b, \tag{1.3.11}$$

$$T_L = T_0 - B\,\mathrm{erf}\left\{\frac{x}{2(k_L t)^{1/2}} + \frac{\delta(\rho_S - \rho_L)k_S^{1/2}}{\rho_L k_L^{1/2}}\right\}. \tag{1.3.12}$$

A and B are constants to be determined. It is easy to see that T_S satisfies (1.3.1) and T_L satisfies (1.3.3) with u_x given in (1.3.9) and the initial condition given by the first condition in (1.3.4). Let $S(t)$ be given by

$$S(t) = 2\delta(k_S t)^{1/2},\ \delta \text{ an unknown constant.} \tag{1.3.13}$$

On $x = S(t)$ (given in (1.3.13)), both T_S and T_L are constant, and $S(0) = 0$. Now we have three unknowns, namely, A, B and δ, and three conditions, i.e., (1.3.6) and isotherm conditions for T_S and T_L. After some algebraic manipulations, the unknown constants can be obtained thus.

$$A = (T_m - T_b)/\mathrm{erf}(\delta), \tag{1.3.14}$$

$$B = (T_0 - T_m)/\mathrm{erfc}(\delta\rho_S k_S^{1/2}/(\rho_L k_L^{1/2})), \tag{1.3.15}$$

$$\frac{e^{-\delta^2}}{\mathrm{erf}\delta} - \frac{(T_0 - T_m)K_L k_S^{1/2} e^{-(\delta^2\rho_S^2 k_S/\rho_L^2 k_L)}}{T_m K_S k_L^{1/2}\mathrm{erfc}\left((\delta\rho_S k_S^{1/2}/(\rho_L k_L^{1/2})\right)} = \frac{\delta\hat{l}\pi^{1/2}}{C_S T_m}, \tag{1.3.16}$$

$$\hat{l} = l + (C_L\rho_L/\rho_S - C_S)T_m. \tag{1.3.17}$$

This completes the solution of (1.3.1)–(1.3.7). When $\rho_S = \rho_L = \rho$, this solution is called *Neumann solution* which was given by Franz Neumann [27] in 1860. The first published discussion of such problems seems to be that by Stefan [28] in a study of the thickness of polar ice.

The problem discussed above is called a *two-phase Stefan problem* as there are two distinct phases. If the liquid is initially at the equilibrium temperature T_m, then heat conduction will take place only in the solid phase. In principle two phases will be present, but since there is no temperature gradient in the liquid, such problems are called *one-phase Stefan problems*. If there are $n\,(> 2)$ distinct phases separated by $(n-1)$ distinct phase-change boundaries, then we have an n-phase Stefan problem. It is not possible to give a precise definition of a Stefan problem but in the light of the above discussion some of the characteristic features of Stefan problems can be described as follows: (1) the transport phenomenon is that of heat transfer governed by parabolic equations, (2) any two distinct phases are separated by a sharp phase-change interface, (3) at the phase-change boundary, temperature is prescribed which is known, (4) heat flux is discontinuous across the phase-change boundary and the latent heat is released or absorbed at the interface. Phase changes in which the heat flux is discontinuous are known as *first-order phase transitions.* We shall learn more and more about these characteristic features of Stefan problems as we proceed further. In many physical problems although a free boundary exists, there is some deviation of features in them from the characteristic features of the Stefan problem. It has become almost customary to indicate such problems by a suitable nomenclature. For example, if a hyperbolic equation is considered instead of a parabolic equation, then it is called 'hyperbolic Stefan problem'. Similarly, we have a 'Stefan problem with kinetic condition' if a kinetic condition is introduced at the free boundary or 'Stefan problem with supercooling' if the liquid is supercooled.

The term 'classical' is generally used to distinguish the classical formulation from the *weak formulation* (§ 5.2 for the weak formulation of the classical Stefan problem). In a classical formulation it is assumed that there exists a sufficiently regular sharp interface, which separates two distinct phases. Two phases will be considered different if they differ in their composition, structure, or properties. At the phase-change interface, two boundary conditions are prescribed, one in the form of temperature and the other arising from the energy balance at the interface.

1.4 Classical Formulation of Multi-dimensional Stefan Problems

1.4.1 Two-phase Stefan problem in multipledimensions

We shall now extend the mathematical formulation given in (1.3.1)–(1.3.7) to multi-dimensional problems. For simplicity we consider R^3. Consider a heat-conducting open-bounded region $G \subset R^3$ whose boundary ∂G is smooth, for example $\partial G \in C^{2+\eta}$, for some $\eta > 0$. Let $G(t) = G \times \{t\}$ and $\partial G(t) = \partial G \times \{t\}$, $0 \le t \le t_* < \infty$ for some suitable real number t_*. $G^{t_*} = \cup_{0 \le t \le t_*} G(t)$ is a cylinder over G (see Figs. 1.4.1.–1.4.4.). G consists of two phases, for example, solid and liquid which for the sake of convenience will be denoted by subscripts 1 (solid) and 2 (liquid). Let $G_i(t)$, $i = 1, 2$ be the region occupied by the ith phase at time $t \in [0, t_*]$. $G(t)$ admits disjoint decomposition as

$$G(t) = G_1(t) \cup S(t) \cup G_2(t), \; \forall\, t \in [0, t_*], \tag{1.4.1}$$

where

$$S(t) = \{(x,t) \in \bar{G}^{t_*} \,|\, \Phi(x,t) = 0, x \in G \text{ and } t \in [0, t_*]\}, \tag{1.4.2}$$

$\Phi \in C^1(\bar{G}^{t_*})$ and $\nabla_x \Phi(x,t) \neq 0$ on $S(t)$. $\Phi(x,t) < 0$ is the phase 1 and $\Phi(x,t) > 0$ is the phase 2. $\bar{G}$ is the closure of G.

We assume that G^{t_*} also admits disjoint decomposition

$$G^{t_*} = G_1^{t_*} \cup \Gamma(t) \cup G_2^{t_*},$$

where

$$G_1^{t_*} = \cup_{0 \le t \le t_*} G_1(t), \quad G_2^{t_*} = \cup_{0 \le t \le t_*} G_2(t), \quad \Gamma(t) = \cup_{0 \le t \le t_*} S(t).$$

Here, $\Gamma(t)$ and $S(t)$ are both called free boundaries. $\Gamma(t)$ is the free boundary in $G \times [0, t]$ and $S(t)$ is the free boundary in G. It may be noted that we are still dealing with the same solidification process discussed in (1.3.1)–(1.3.7). The only difference is that for the solid and the liquid regions and the phase-change boundary, some set theoretic notations have been introduced for rigorous mathematical analysis. In problem (1.3.1)–(1.3.7), $\Phi(x,t) = x - S(t)$.

Let $\mathcal{P}_i$, $i = 1, 2$ be the elliptic operators [9] defined as

$$\mathcal{P}_i = \sum_{J,k=1}^{3} a^i_{Jk}(x,t)\frac{\partial^2}{\partial x_J \partial x_k} + \sum_{J=1}^{3} b^i_J(x,t)\frac{\partial}{\partial x_J} + r^i(x,t), \quad i = 1, 2. \tag{1.4.3}$$

Here, $a^i_{Jk}, \nabla_x a^i_{Jk}, \nabla^2_x a^i_{Jk}, b^i_J, \nabla_x b^i_J$, and r^i are continuous in $\bar{G}^{t_*}$; $x \in G$ and $x = (x_1, x_2, x_3)$. Consider the following solidification problem in $G(t)$.

Conservation of energy in $G_i(t)$

$$C_i \rho_i \frac{\partial T_i}{\partial t} = \mathcal{P}_i T_i, \; (x,t) \in G_i(t), \; i = 1, 2, \; 0 < t < t_*. \tag{1.4.4}$$

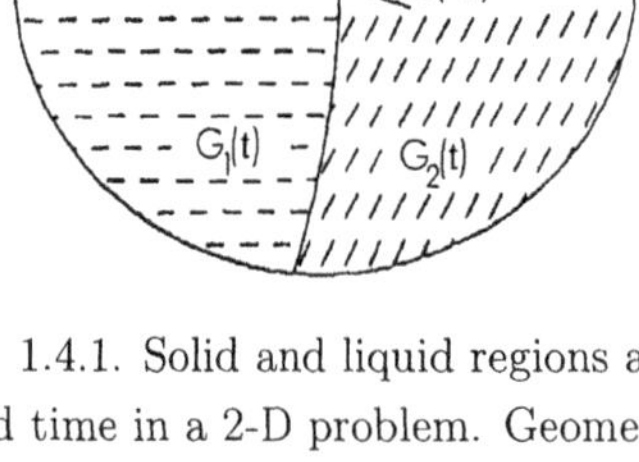

Fig. 1.4.1. Solid and liquid regions at a fixed time in a 2-D problem. Geometry number 1.

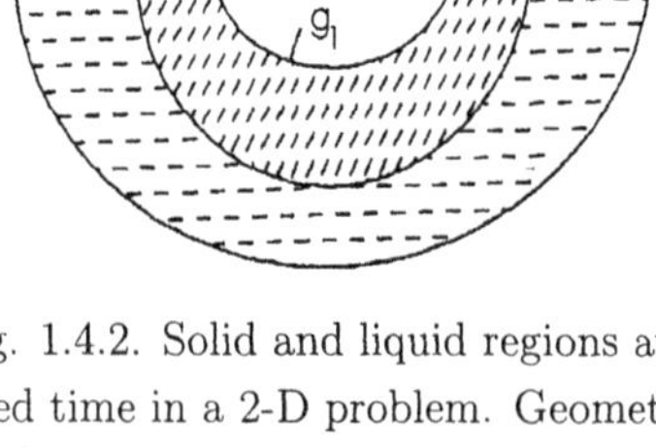

Fig. 1.4.2. Solid and liquid regions at a fixed time in a 2-D problem. Geometry number 2.

Boundary conditions on $\partial G_i(t)$

$$T_i(x,t) = g_i(x,t), \; x \in \partial G_i(t), \; i = 1, 2; \; 0 < t < t_*, \tag{1.4.5}$$

$$g_1 < T_m \text{ and } g_2 > T_m.$$

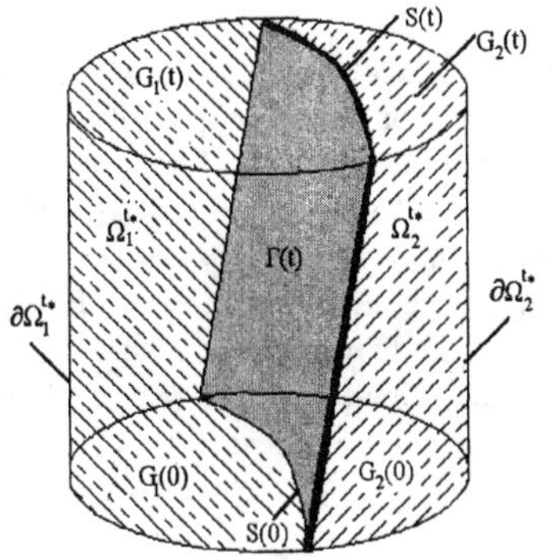

Fig. 1.4.3. Cylinder over $G(0)$ in Fig. 1.4.1.

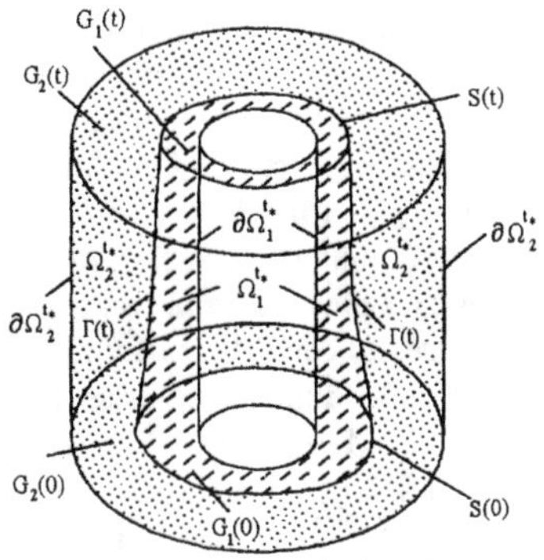

Fig. 1.4.4. Cylinder over $G(0)$ in Fig. 1.4.2.

Initial conditions

$$T_i(x,0) = f_i(x),\ x \in G_i(0),\ i = 1,\ 2, \tag{1.4.6}$$

$$f_1 < T_m \text{ and } f_2 > T_m.$$

Isotherm conditions on $\Phi(x,t) = 0$

$$T_i = T_m,\quad i = 1, 2. \tag{1.4.7}$$

Energy balance at $\Phi(x,t) = 0$

$$[-\vec{q}\,]_2^1 \cdot \vec{n} = \rho_1 \hat{l} \vec{V} \cdot \vec{n}, \tag{1.4.8}$$

where $\vec{q} = (q_1, q_2, q_3)$ is the heat flux vector for anisotropic bodies and is defined as

$$q_i = -\sum_{J=1}^{3} K_{ij} \frac{\partial T}{\partial x_J},\quad i = 1,2,3. \tag{1.4.9}$$

Here, K_{ij} are called thermal conductivity coefficients, $\vec{n}$ is the unit outward normal to the solid phase at $S(t)$, and $\vec{V}$ is the velocity of the free boundary. Equation (1.4.8) is a generalization of (1.3.10) to a multi-dimensional case.

If the heat equation (1.4.3) is written for anisotropic bodies, then $a_{jk}^i(x,t)$ are known functions which can be identified with $K_{ij}(x,t)$ in (1.4.9). In a more general case K_{ij} could be functions of temperature also and in that case the operator $\mathcal{P}_i$ will change (see (1.4.29)). The coefficients $b_{jk}^i(x,t)$ could be the space derivatives of $K_{ij}(x,t)$ and/or convective terms arising due to the motion of the ith phase. In the case of motion of phases, total derivative of T_i with respect to time should be considered. If so,

$$C_i\rho_i \frac{DT_i}{Dt} = C_i\rho_i \frac{\partial T_i}{\partial t} + C_i\rho_i \sum_{j=1}^{3} u_j^i \frac{\partial T_i}{\partial x_j},\quad i = 1, 2, \tag{1.4.10}$$

where u_j^i, $j = 1, 2, 3$ are components of the velocity of the ith phase. Motion in the liquid phase may arise due to natural convection and/or shrinkage or expansion of the volume, etc. In (1.4.4), $b_{jk}^i(x,t)$ could be taken to be either known or unknown quantities. r^i is the rate at which heat is generated/annihilated per unit volume. If $b_{jk}^i(x,t)$ are unknown and arise due to the liquid motion, then (1.4.4)–(1.4.8) have to be supplemented with other equations to determine them. The motion of the solid phase is generally neglected but the motion of the melt plays an important role in solidification/melting problems. In the one-dimensional case, u_x could be easily determined with the help of mass balance condition (1.3.9). In the multi-dimensional case, the velocity of the melt can be determined only by supplementing (1.4.4)–(1.4.8) with the equations of fluid dynamics such as Navier-Stokes equations discussed in § 1.4.7.

Instead of temperature-prescribed boundary conditions on $\partial G_i(t)$, $i = 1, 2$, other types of boundary conditions can also be prescribed provided solidification takes place in such a way that only one phase-change boundary separates the two phases. This restriction is inherent in the classical formulation. Solidification/melting problems could be much more complicated. For example, if ice pieces are put in a glass of water, then a free boundary exists at the boundary of each ice piece. Two free boundaries which may be disjoint at some instant of time may join at a later instant. A piece of ice on melting, may break into two. Such problems are extremely difficult to investigate. In (1.4.7) it has been assumed that there is thermodynamic equilibrium at the phase-change boundary.

Equation (1.4.8) can be expressed as

$$\sum_{j,k=1}^{3} a_{jk}^1 \frac{\partial T_1}{\partial x_j}\frac{\partial \Phi}{\partial x_k} - \sum_{j,k=1}^{3} a_{Jk}^2 \frac{\partial T_2}{\partial x_j}\frac{\partial \Phi}{\partial x_k} = -\rho \hat{l}\,\frac{\partial \Phi}{\partial t}. \tag{1.4.11}$$

Equation (1.4.11) can be obtained as follows. Let $\vec{q}^{\,i}$ be the heat flux vector in the ith phase given by (1.4.9) with K_{jk} replaced by a_{jk}. Then

$$\vec{q}^{\,i}\cdot\vec{n} = \vec{q}^{\,i}\cdot \text{grad}\Phi/|\text{grad}\Phi| = \left(q_1^i\frac{\partial \Phi}{\partial x} + q_2^i\frac{\partial \Phi}{\partial y} + q_3^i\frac{\partial \Phi}{\partial z}\right) / \,|\text{grad}\Phi|,\ i = 1, 2, \tag{1.4.12}$$

$$= \left(-\sum_{J,k=1}^{3} a_{Jk}^i \frac{\partial T^i}{\partial x_J}\frac{\partial \Phi}{\partial x_k}\right) / \,|\text{grad}\Phi|,\ i = 1, 2. \tag{1.4.13}$$

we have

$$\frac{d\Phi}{dt}(x,y,z,t) = \frac{\partial \Phi}{\partial t} + \frac{\partial \Phi}{\partial x}\frac{dx}{dt} + \frac{\partial \Phi}{\partial y}\frac{dy}{dt} + \frac{\partial \Phi}{\partial z}\frac{dz}{dt} = 0, \tag{1.4.14}$$

or

$$(\vec{V}\cdot\vec{n})|\text{grad }\Phi| = -\frac{\partial \Phi}{\partial t}. \tag{1.4.15}$$

Substitution of (1.4.13) and (1.4.15) into (1.4.8) results in equation (1.4.11). The negative sign on the r.h.s. of (1.4.11) will arise in (1.3.6) also if we take $\Phi(x,t) = x - S(t)$ in (1.3.6).

The energy balance condition (1.4.8) has been obtained on the basis of the principle of conservation of energy stated in § 1.4.7. The detailed derivation of (1.4.8) using energy conservation principle is given below.

Consider the three-dimensional problem of solidification of a melt in which we take densities of solid and liquid phases to be the same and consider an isotropic case for the sake of simplicity. Let P be a point on the interface $S(t)$ at some given time t, and Q be a point on $S(t+\Delta t)$. Q lies along the vector normal to $S(t)$ at P as shown in Fig. 1.4.5. Solidification progresses from P to Q. Take $\vec{PQ}$ to be the direction of the vector $\vec{n}_x$ normal to the surface $S(t)$ at a fixed time where $\vec{n}_x = (n_{x1}, n_{x2}, n_{x3})$, $n_{x_i} = \cos(\vec{n}, \vec{x}_i)$, $i = 1, 2, 3$, $\vec{n}$ is the unit vector normal to the interface $S(t)$ and $\vec{n} = (\vec{n}_x, n_t)$. $n_t = \cos(\vec{n}, \vec{t})$, $\vec{t}$ is the unit vector in the direction of t and $\vec{x}_i$, $i = 1, 2, 3$ are unit vectors in the x_1, x_2, x_3 directions, respectively. Here, x_1, x_2, and x_3 are Cartesian coordinates of a point in R^3. Let Δw be a small area on $S(t)$ enclosing the point P. Construct a cylinder with base Δw, height $|\vec{PQ}|$ and axis PQ. The latent heat released during the solidification of this volume element is $\rho l \Delta w |\vec{PQ}|$. The rate at which latent heat is released during solidification of this volume element is

$$\lim_{\Delta t \to 0} \rho l \left|\frac{\vec{PQ}}{\Delta t}\right| \Delta w = \rho l \vec{V} \cdot \vec{n}_x \Delta w. \tag{1.4.16}$$

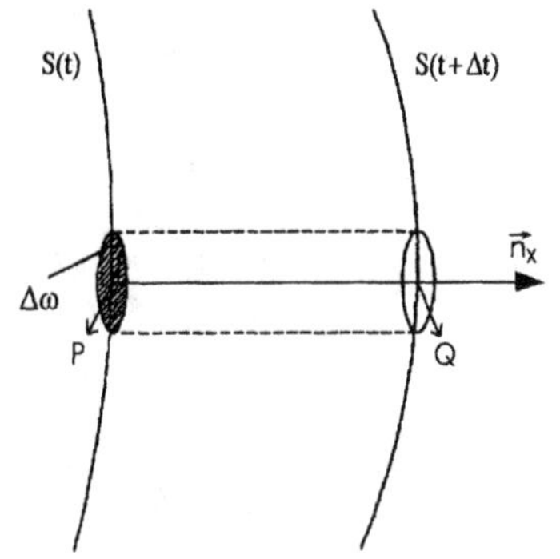

Fig. 1.4.5. Geometry in the derivation of the Stefan condition

Here, $\vec{V}$ is the velocity of the interface. On using the energy conservation principle in the cylinder of volume $\Delta w|\vec{PQ}|$ which requires calculating the difference in the rates at which heat flows into the cylinder across the left face and leaves through the face on the right and adding to it the rate at which heat is released and cancelling $\Delta\omega$, we obtain

$$K_S \frac{\partial T_S}{\partial n_1} + K_L \frac{\partial T_L}{\partial n} = -\rho l \vec{V} \cdot \vec{n}_x. \tag{1.4.17}$$

where $\vec{n}_1 = -\vec{n}$ and $\vec{n}_1$ is the outward normal to the base of the cylinder at P.

1.4.2 Alternative forms of the Stefan Condition

The Stefan condition (1.4.11) in multipledimensions can be expressed alternatively in a form more suitable for analytical and numerical treatment. It will be assumed that $\Phi(x,y,z,t)=0$ can be written as

$$\Phi(x,y,z,t)=z-S(x,y,t)=0. \tag{1.4.18}$$

Alternative expressions like $x-f(y,z,t)$ or $y-g(x,z,t)$ can also be used for Φ. Sometimes the physics of the problem requires the free boundary to be expressed in the form $t=w(x,y,z)$ (see (3.3.46)). Differentiation of (1.4.7) with respect to x and y gives

$$\frac{\partial T_i}{\partial x}=-\frac{\partial T_i}{\partial z}\frac{\partial z}{\partial x}\,,\quad \frac{\partial T_i}{\partial y}=-\frac{\partial T_i}{\partial z}\frac{\partial z}{\partial y},\ i=1,2. \tag{1.4.19}$$

From (1.4.18), we have

$$\frac{\partial \Phi}{\partial x}+\frac{\partial \Phi}{\partial z}\frac{\partial z}{\partial x}=0,\quad \frac{\partial \Phi}{\partial y}+\frac{\partial \Phi}{\partial z}\frac{\partial z}{\partial y}=0,\quad \frac{\partial \Phi}{\partial z}=1. \tag{1.4.20}$$

Equation (1.4.11) can be rewritten as

$$\sum_{J=1}^{3}\left(\sum_{k=1}^{3}a^1_{Jk}\frac{\partial \Phi}{\partial x_k}\right)\frac{\partial T_1}{\partial x_J}-\sum_{J=1}^{3}\left(\sum_{k=1}^{3}a^2_{Jk}\frac{\partial \Phi}{\partial x_k}\right)\frac{\partial T_2}{\partial x_J}=\rho\hat{l}\frac{\partial S}{\partial t}. \tag{1.4.21}$$

On using (1.4.19) and (1.4.20) in equation (1.4.21) the latter can be written as

$$\left\{\sum_{J=1}^{3}\left(\sum_{k=1}^{3}a^1_{Jk}\frac{\partial \Phi}{\partial x_k}\right)\frac{\partial \Phi}{\partial x_J}\right\}\frac{\partial T_1}{\partial z}-\left\{\sum_{J=1}^{3}\left(\sum_{k=1}^{3}a^2_{Jk}\frac{\partial \Phi}{\partial xk}\right)\frac{\partial \Phi}{\partial x_J}\right\}\frac{\partial T_2}{\partial z}=\rho\hat{l}\frac{\partial S}{\partial t}. \tag{1.4.22}$$

Note that in (1.4.22), the derivative of temperature is with respect to z but not with respect to x and y. Equation (1.4.22) looks like the Stefan condition in a one-dimensional problem.

In an isotropic case,

$$a^i_{Jk}=K_i\delta_{Jk},\ \text{ where } \delta_{Jk} \text{ is the 'Kronecker delta function'}. \tag{1.4.23}$$

Using (1.4.23) in (1.4.22), the Stefan condition can be expressed as

$$\left\{1+\left(\frac{\partial S}{\partial x}\right)^2+\left(\frac{\partial S}{\partial y}\right)^2\right\}\left\{K_1\frac{\partial T_1}{\partial z}-K_2\frac{\partial T_2}{\partial z}\right\}=\rho\hat{l}\frac{\partial S}{\partial t}. \tag{1.4.24}$$

On using (1.4.19) and (1.4.20), we have

$$\vec{n}=\frac{\nabla \Phi}{|\nabla \Phi|}=\frac{\nabla T_i}{|\nabla T_i|}. \tag{1.4.25}$$

Also on the free boundary

$$\vec{V}\cdot\vec{n}=\vec{V}\cdot\nabla T_i|\nabla T_i|=-\partial T_i\partial t/|\nabla T_i|,\ i=1,2. \tag{1.4.26}$$

Isotherm conditions (1.4.7) can be differentiated with respect to t to obtain the second relation in (1.4.26). Substituting (1.4.26) in (1.4.8), we obtain

$$|K_1\nabla T_1| - |K_2\nabla T_2| = -\rho\hat{l}\frac{\partial T_1}{\partial t}/|\nabla T_1| = -\rho\hat{l}\frac{\partial T_2}{\partial t}/|\nabla T_2|. \tag{1.4.27}$$

1.4.3 The Kirchhoff's transformation

Consider the case in which the thermal conductivity varies with the temperature, but is independent of position and time. The heat equation for isotropic material in this case can be written as

$$\begin{aligned}\rho C\frac{\partial T}{\partial t} &= \operatorname{div}(K(T)\operatorname{grad} T) + A(T,x,y,z,t),\\ &= K\nabla^2 T + \frac{\partial K}{\partial T}\left\{\left(\frac{\partial T}{\partial x}\right)^2 + \left(\frac{\partial T}{\partial y}\right)^2 + \left(\frac{\partial T}{\partial z}\right)^2\right\} + A,\end{aligned} \tag{1.4.28}$$

A is the rate at which heat is supplied per unit volume.

Kirchhoff's transformation is defined as [24],

$$\theta = \frac{1}{K_0}\int_0^T K(\alpha)d\alpha, \tag{1.4.29}$$

where K_0 is the value of $K(T)$ when $T = 0$. It can be seen that

$$\frac{\partial\theta}{\partial t} = \frac{K}{K_0}\frac{\partial T}{\partial t},\quad \frac{\partial\theta}{\partial x} = \frac{K}{K_0}\frac{\partial T}{\partial x},\quad \frac{\partial\theta}{\partial y} = \frac{K}{K_0}\frac{\partial T}{\partial y},\quad \frac{\partial\theta}{\partial z} = \frac{K}{K_0}\frac{\partial T}{\partial z}, \tag{1.4.30}$$

and

$$\nabla^2\theta - \frac{\rho C}{K}\frac{\partial\theta}{\partial t} = -\frac{A}{K_0}, \tag{1.4.31}$$

where K and C are now functions of θ. In many cases, for example, in metals near absolute zero temperature, both K and C are proportional to the temperature and $\rho C/K$ can be taken to be a constant.

If A is independent of the temperature, then there will be no difference between the structure of the Fourier's heat conduction equation and equation (1.4.31). However if we use Kirchhoff's transformation (1.4.29), the choice of boundary conditions will be limited. When T is prescribed as $T = f(x,y,z,t)$ on the boundary, the expression

$$\theta = \frac{1}{K}\int_0^{f(x,y,z,t)} K(\alpha)d\alpha, \tag{1.4.32}$$

gives the boundary condition for θ. Similarly, if the normal derivative of temperature T is prescribed, the normal derivative of θ can also be prescribed.

The extension of the formulation (1.4.4)–(1.4.8) to R^n, $n > 3$ is simple. Many generalizations of this formulation are possible and some of them will be discussed later.

1.4.4 Boundary conditions at the fixed boundary

(A) Standard boundary conditions

(AI) Type I boundary condition

In the boundary condition of this type temperature is prescribed. Let ∂G be the boundary of the region $G \subset R^3$. Type I boundary condition is of the form

$$T|_{\partial G} = f(x, y, z, t),\ t > 0,\ (x, y, z) \in \partial G. \tag{1.4.33}$$

If $f < T_m (> T_m)$ then solidification (melting) starts instantaneously, i.e., there is no waiting time for the phase-change to take place. Since energy can neither be produced nor disappear at infinity, if G is unbounded, we have

$$\lim_{|x| \to \infty} T(x, t) = T(x, 0),\ \ t > 0. \tag{1.4.34}$$

In order to obtain a physically meaningful solution to the diffusive heat equation for unbounded domains, it is required that

$$\lim_{|x| \to \infty} \frac{\partial T}{\partial n}(x, t) = 0,\ \ t > 0. \tag{1.4.35}$$

Temperature-prescribed boundary conditions are known as *Dirichlet type boundary conditions.*

Sometimes, while constructing approximate solutions more than the prescribed number of boundary conditions are required to match the number of unknowns. In such a case additional boundary conditions are generated. Let us consider a radially symmetric one-dimensional heat conduction problem in which the temperature at $r = a$ is prescribed as

$$T(r, t)|_{r=a} = f(t). \tag{1.4.36}$$

If (1.4.36) is differentiated with respect to time and the Fourier's heat equation is used, then

$$\left.\frac{\partial T}{\partial t}\right|_{r=a} = k \left(\frac{\partial^2 T}{\partial r^2} + \frac{1}{r}\frac{\partial T}{\partial r} \right)_{r=a} = f'(t) \tag{1.4.37}$$

Equation (1.4.37) provides one more boundary condition for the problem.

(AII) Boundary condition of radiative-convective type

This boundary condition can be written as

$$-K \left.\frac{\partial T}{\partial n}\right|_{\partial G} = \beta E (T^4 - T_0^4)|_{\partial G} + (C_1 T - C_2 T_0)|_{\partial G}. \tag{1.4.38}$$

Here, $\vec{n}$ stands for the unit outward normal, β is Stefan-Boltzmann constant, E is the emissivity of the surface which is defined as the ratio of the heat emitted by the body to the black body at the same temperature, T_0 is the ambient temperature, and C_1 and C_2 are functions of time. β is positive and has dimensions [watt]/[meter]$^2\times$ [Temperature]^4, $0 \leq E \leq 1$. From this boundary condition, three types of well known boundary conditions can be obtained as particular cases. If $\beta = 0$ and $C_1 = C_2 = h_t$ then we have a *convective type boundary condition.* $h_t\,(>0)$ is called 'heat transfer coefficient' and has dimensions [Power] / [Temperature] [length]^2. If $\beta = 0$, $C_1 = 0$ and $C_2T_0 = g(t)$, then we have a *flux-prescribed boundary condition.* Flux prescribed boundary conditions are known as *type II boundary condition.* Convective type boundary conditions are called *type III boundary conditions* and radiative-convective boundary conditions are called *type IV boundary conditions.*

More details about these boundary conditions can be found in [24]. These four types of boundary conditions are commonly prescribed in heat transfer problems.

(B) Non-standard boundary conditions

(BI) Non-local boundary condition

To explain the nature of a non-local boundary condition, we consider a simple one-dimensional problem of fluidized-bed coating. Our interest in this problem is more on pointing out the characteristic features of a non-local boundary condition than its rigorous formulation. A thin metallic plate of thickness $2b$, area A_p and the initial temperature T_0 ($T_0 > T_m$) is immersed vertically in a pool of plastic coating material [29] which is maintained at temperature T_∞ which is also the temperature away from the boundary layer. The softening temperature of the coating material is $T_m > T_\infty$. Because of the high thermal conductivity and finite heat capacity of the metallic plate, the plate can be treated as a lumped parameter system in which the temperature can be regarded as a function of time only. Let $T_p(t)$ be the temperature of the plate. In the one-dimensional problem considered here, let the metallic plate be situated at $x = 0$. The problem is that of finding the temperature $T(x,t)$ of the softened plastic, temperature $T_p(t)$ and the extent $x = \delta(t)$ upto which the plastic has softened. The unknown quantities can be obtained by solving the following system of equations.

$$\rho_f C_f \frac{\partial T}{\partial t} = K_f \frac{\partial^2 T}{\partial x^2}, \quad 0 < x < \delta(t),\ t > 0, \tag{1.4.39}$$

$$T(0,0) = T_0, \;\; T(0,t) = T_p(t), \;\; T(\delta,t) = T_m, \tag{1.4.40}$$

$$\frac{m_p C_p}{A_p}(T_0 - T_p(t)) = h_t t(T_m - T_\infty) + \rho_f C_f \int_0^{\delta(t)} (T - T_\infty)dx, \tag{1.4.41}$$

$$-K_f \frac{\partial T}{\partial x}\bigg|_{x=\delta} = h_t(T_m - T_\infty) + \rho_f C_f (T_m - T_\infty)\frac{d\delta}{dt}, \tag{1.4.42}$$

The subscripts f and p stand for fluid plastic and plate, respectively, m stands for mass, ρ for the density and h_t is the heat transfer coefficient. Equation (1.4.41) is the boundary condition at $x = 0$ which because of the integration over $0 \leq x \leq \delta$ in the second term on the r.h.s. is a non-local boundary condition. The l.h.s. in (1.4.41) is the heat given out by the plate per unit area over the time interval $[0, t]$ which is equal to the sum of the heat transported by convection into the coating material and the heat received by the fluid plastic. Equation (1.4.42) is the energy balance equation at $x = \delta(t)$.

(BII) Boundary condition of the fifth type

Boundary condition of the fifth kind arises in the following context. Suppose the boundary of a region is in contact with a thermostat. The continuity of the temperature and of the flux at the surface of contact, say $x = 1$, in the case of a one-dimensional slab $0 \leq x \leq 1$ can be expressed as

$$T(1,t) = \theta(1,t) \quad , \quad \bar{C}\frac{\partial \theta}{\partial t} = -K\frac{\partial T}{\partial x}, \text{ at } x = 1, \tag{1.4.43}$$

where T and θ are temperatures of the slab and of the thermostat, respectively, and $\bar{C}$ is the 'heat capacity' of the thermostat. The thermostat maintains a uniform temperature. On combining the two equations of (1.4.43), we obtain a boundary condition of the fifth type

$$\bar{C}\frac{\partial T}{\partial t} = -K\frac{\partial T}{\partial x}, \text{ at } x = 1. \tag{1.4.44}$$

(C) Boundary conditions with multivalued functions

In this boundary condition the prescribed quantities are multivalued functions. So it is not a classical boundary condition in which the prescribed functions are assumed to be sufficiently regular functions. Such boundary conditions are suitable for weak formulations and can be described with the help of 'subdifferentials' (see (4.3.44) for its definition). The boundary condition derived below is an extension of the fifth type of boundary condition (1.4.44) in the weak form.

Let Ω be a bounded open domain in R^3 which is occupied by both solid and liquid regions. The boundary $\partial\Omega$ of Ω is smooth, and is surrounded by a high conducting material of negligible thickness so that the temperature of the surrounding material is taken to be a function of time alone. Let the time interval be given by $0 \leq t \leq t_*$, for some positive number t_*, and $Q = \Omega \times (0, t_*)$, $\sum = \partial\Omega \times (0, t_*)$, $Q = Q_S \cup Q_L$ where Q_S and Q_L are the regions occupied by the solid and the liquid, respectively. $\sum = \sum_L \cup \sum_S \cup \sum_{LS}$ where $\sum_L$ and $\sum_S$ are the boundaries of solid and liquid regions and $\sum_{LS}$ is the portion of $\sum$ which is neither solid nor liquid. We assume that the measure of the set $\sum_{LS}$ is zero. Because of the perfect thermal contact between $\partial\Omega$ and the outside material, the temperature is taken to be continuous and so

$$T|_{\sum} = \theta|_{\sum}, \tag{1.4.45}$$

where $T(x,t)$ is the temperature of Ω and $\theta(t)$ is the temperature of the outside material.

The heat balance on $\sum_L$ gives (see [30])

$$\hat{C}\frac{\partial\theta}{\partial t} + K_L\frac{\partial T}{\partial n} + g(x,t,\theta) = 0, \tag{1.4.46}$$

where $\vec{n}$ is the outward normal to $\partial\Omega$, $\hat{C}$ is the heat capacity of the surrounding material and $g(\cdot)$ is the rate of heat flow from outside of Ω to $\partial\Omega$.

The heat balance on $\sum_S$ gives

$$\hat{C}\frac{\partial\theta}{\partial t} + K_S\frac{\partial T}{\partial n} + g(x,t,\theta) = 0. \tag{1.4.47}$$

Define a multivalued function ξ as

$$\xi(r) = \begin{cases} \frac{1}{K_L} & , \quad r > 0, \\ \left(\frac{1}{K_S}, \frac{1}{K_L}\right) & , \quad r = 0, \\ \frac{1}{K_S} & , \quad r < 0. \end{cases} \tag{1.4.48}$$

If $K_S > K_L$, then ξ is a monotone graph. Equations (1.4.46) and (1.4.47) can be combined into a single equation

$$\xi\hat{C}\frac{\partial\theta}{\partial t} + \frac{\partial T}{\partial n} + \xi g(x,t,\theta) = 0, \text{ on } \partial\Omega. \tag{1.4.49}$$

In (1.4.49) normal derivative of the temperature on $\partial\Omega$ is prescribed in terms of a multivalued function. $T < 0$ is solid, $T > 0$ is liquid and $T = 0$ corresponds to the mushy region. Conditions of the type (1.4.49) are also called *dynamical boundary conditions* [31]. Weak formulation is a continuum model and in this formulation, solid, liquid and mushy regions (both solid and liquid phases are present in the mush) are modelled with the help of a single equation which is valid in the region Q in the distributional sense. Therefore the boundary conditions over different portions of the boundary should be combined into a single equation which results in (1.4.49). Stefan problems with dynamical boundary conditions have been investigated by several authors (cf. [31, 32]).

1.4.5 Conditions at the free boundary

In the Stefan problem two boundary conditions are prescribed at the free boundary. One specifies the temperature and another accounts for the energy balance. The isotherm condition in (1.4.7) is the simplest type of temperature-prescribed boundary condition and is justified only if the free boundary is planar and is in thermodynamic equilibrium. In this case, isotherm temperature is the equilibrium temperature T_m. The equilibrium temperature is affected by the curvature and the surface tension in the free boundary.

This effect is known as *Gibbs-Thomson effect.* Similarly equilibrium temperature may depend on the velocity of the free boundary and in this case the free boundary condition is known as a *kinetic condition.* An explanation of curvature effects and kinetic condition requires thermodynamical and metallurgical concepts and so these effects will be discussed a little later in § 2.3.2.

One form of generalization of (1.4.8) is to consider the thermophysical parameters as general functions of space, time and temperature and include a heat flux acting at the free boundary. Such generalizations, at least in the one-dimensional problems, have been considered extensively and analyzed mathematically. Some of these extensions will be discussed later with respect to specific problems. Another type of generalization of free boundary condition is to consider the transport and accumulation of different forms of energies in the phase-change interface such as by heat conduction along the interface. This generalization of the Stefan condition, called *generalized Stefan condition*, will be discussed in § 2.4.

The notion of Stefan condition is not restricted to conditions of the form (1.4.8) which arise due to heat balance during phase changes. Stefan condition can be considered in a functional form, for example, the velocity of the free boundary can be prescribed as a function of temperature, temperature derivatives, time, and $S(t)$ (see (3.1.5)).

Implicit free boundary condition

In the Stefan condition (1.4.11), the velocity of the free boundary occurs explicitly. A boundary condition of this type is called an *explicit free boundary condition.* There are several problems in which the boundary conditions at the free boundary do not involve the normal velocity of the free boundary. For example the condition $\partial T/\partial x|_{x=S(t)} = 0$ does not involve the term $\partial S/\partial t$. Such boundary conditions are called *implicit free boundary conditions.* In some one-dimensional problems, by using suitable transformations, implicit boundary conditions can be easily converted to an equivalent explicit boundary condition. Implicit boundary conditions will be discussed further in § 3.3.

1.4.6 The classical solution

The definition of a classical solution of problem (1.4.4)–(1.4.8) is given below. Classical solutions of other problems to be discussed later can be defined on similar lines.

Definition: The classical solution of (1.4.4)–(1.4.8) is a quadruple (T_1, T_2, Φ, t_*) such that T_i, $\nabla_x T_i$ are continuous in $\cup_{0\le t<t_*} \overline{G_i(t)}$ and $\nabla_x^2 T_i$ and $\partial T_i/\partial t$ are continuous in $\cup_{0<t<t_*} G_i(t)$; $\Phi(x, y, z, t)$ has continuous first order partial derivatives in $\bar{G}^{t_*}$ and satisfies conditions imposed in (1.4.2) in § (1.4.2). The triple $(T_1, T_2, S(t))$ satisfies the equations (1.4.4)–(1.4.8) in which the coefficients satisfy the constraints in (1.4.3).

Here, $\nabla_x T_i = (\partial T_i/\partial x,\ \partial T_i/\partial y,\ \partial T_i/\partial z)$, $i = 1, 2$ and $\nabla_x^2 T_i$ can be similarly defined.

It may be noted that the existence of a classical solution is not guaranteed merely by prescribing sufficiently regular initial and boundary data satisfying the sign constraints, such as in (1.4.5) and (1.4.6). Some additional compatibility conditions discussed in later sections should also be satisfied. t_* cannot be arbitrarily taken and is related to the existence of the solution. In $R^n, n \geq 2$, existence and uniqueness of classical solutions have been mostly proved only locally-in-time, this too under suitable constraints.

1.4.7 Conservation laws and the motion of the melt

There is a clear and overwhelming experimental evidence [33] that except in very early times, convective heat transfer dominates over conductive heat transfer during solidification/melting. Motion of the liquid which may arise due to natural convection, forced convection, or shrinkage/ expansion of the volume becomes more significant when the phase-change is of relatively large volumes of material. Motion of the melt can be accounted for by solving Navier-Stokes equations [34] together with the energy equation.

Most equations of mathematical physics are derived on the application of the following conservation laws:

1. *Conservation of mass*: The time rate of increase of mass of a system is equal to the difference between the rate at which mass enters into the system, and the rate at which mass leaves the system (disregarding relativity effects).

2. *Conservation of momentum*: The time rate of change of linear momentum of the mass of a system is equal to the sum of all the forces acting on the system.

3. *Conservation of energy*: The time rate of change of energy (internal energy + kinetic energy + potential energy) stored in the system is equal to the sum of the time rate of work of the external forces, the rate at which energy is transported into the system (heat energy, electrical energy etc.) or leaving the system across its boundaries and the, rate at which energy is produced or consumed in the system.

The fourth conservation law which deals with the conservation of moment of momentum is not required in the context of this volume. When a conservation law is valid in any arbitrarily small neighbourhood of each material point, we say that the conservation law holds locally.

If the melt is in motion, then in order to determine the motion of the melt, energy equation should be solved in conjunction with Navier-Stokes equations. We derive below the complete system of equations briefly.

Control volume and applications of conservation laws

A 'control volume' refers to a fixed region in space which encloses a fixed volume V_0 of the fluid in space. Its surface A_0 is called control surface, which could be of any shape.

In the *Eulerian coordinate system*, the fluid is imagined to pass through this arbitrary control volume. At the same position we observe different particles at different times. In the *Lagrangian coordinate system*, the coordinate system is attached to the moving particle. Therefore the control volume consists of the same particles all the time. Let ϕ be some 'scalar specific property' (per unit mass) such as mass, momentum, energy of the fluid. We shall consider the conservation of a scalar property ϕ in the control volume. The time rate of change of the total property ϕ contained in V_0 is given by

$$\frac{\partial}{\partial t}\int_{V_0} \rho\phi dV_0. \tag{1.4.50}$$

The conservation equation for ϕ is

$$\frac{\partial}{\partial t}\int_{V_0} \rho\phi\, dV_0 = -\int_{A_0} (\rho\vec{V}\phi)\cdot\vec{n}\, dA_0 - \int_{A_0} \vec{J}\cdot\vec{n}\, dA_0 + \int_{V_0} \dot{f} dV_0, \tag{1.4.51}$$

where $\vec{V}$ is the velocity of the fluid, $\vec{n}$ is the unit outward drawn normal to A_0, $\vec{J}$ is the surface flux of ϕ by diffusion from inside to outside across the surface A_0, f is the production or annihilation of ϕ per unit volume in the control volume, ρ is the density and dot denotes the time derivative. dA_0 is the elementary surface area and dV_0 is the elementary volume. Using Gauss's theorem, surface integrals can be converted to volume integrals and the local conservation of ϕ implies

$$\frac{\partial}{\partial t}(\rho\phi) + \nabla\cdot(\rho\vec{V}\phi) = -\nabla\cdot\vec{J} + \dot{f}. \tag{1.4.52}$$

Conservation of mass : Equation of continuity

Take $\phi = 1,\ f = 0$ and $\vec{J} = 0$ in (1.4.52). We obtain the equation of continuity as

$$\frac{\partial \rho}{\partial t} + \text{div}\,(\rho\vec{V}) = 0. \tag{1.4.53}$$

Equation of conservation of linear momentum

In order to obtain the equation of conservation of linear momentum in the x-direction, take $\phi = V_1$, $\vec{J} = -\vec{\mathbf{p}}_x = -(p_{xx}\vec{i} + p_{xy}\vec{j} + p_{xz}\vec{k})$ and take F_1 instead of $\dot{f}$ in (1.4.52) where F_1 is the body force in the x-direction. V_1 is the x-component of the fluid velocity $\vec{V}$ and p_{ij} are stress components.

$$p_{ij} = -p\delta_{ij} + \tau_{ij},\ \ i = 1,2,3 \text{ and } j = 1,2,3, \tag{1.4.54}$$

$$p = -\frac{1}{3}(p_{xx} + p_{yy} + p_{zz}). \tag{1.4.55}$$

Here, τ_{ij} represent viscous stress components, and p is the fluid pressure. The r.h.s. of (1.4.52) in this present conservation law should represent the sum of the forces acting on the control volume which are surface forces and body forces. If g_x is the x-component of the body force per unit mass, then the equation of conservation of linear momentum in the x-direction for constant ρ is given by

$$\rho \frac{DV_1}{Dt} = -\frac{\partial p}{\partial x} + \frac{\partial \tau_{xx}}{\partial x} + \frac{\partial \tau_{yx}}{\partial y} + \frac{\partial \tau_{zx}}{\partial z} + \rho g_x. \tag{1.4.56}$$

DV_1/Dt is the material time derivative of V_1 in (1.4.56). Two more equations in y and z directions can be similarly obtained by taking $\phi = V_2$ and $\vec{J} = -\vec{\mathbf{p}}_y = -(p_{yx}\vec{i} + p_{yy}\vec{j} + p_{yz}\vec{k})$ for the second equation and $\phi = V_3$ and $\vec{J} = -\vec{\mathbf{p}}_z$ for the third equation. We have

$$\rho \frac{DV_2}{Dt} = -\frac{\partial p}{\partial y} + \frac{\partial \tau_{xy}}{\partial x} + \frac{\partial \tau_{yy}}{\partial y} + \frac{\partial \tau_{zy}}{\partial z} + \rho g_y, \tag{1.4.57}$$

$$\rho \frac{DV_3}{Dt} = -\frac{\partial p}{\partial z} + \frac{\partial \tau_{xz}}{\partial x} + \frac{\partial \tau_{yz}}{\partial y} + \frac{\partial \tau_{zz}}{\partial z} + \rho g_z, \tag{1.4.58}$$

where g_y and g_z are the y and z components of the body force per unit mass.

For Newtonian fluids [34]

$$\tau_{ij} = 2\mu e_{ij} + \mu^1 \delta_{ij}(e_{xx} + e_{yy} + e_{zz}), \;\; i = 1,2,3; \; j = 1,2,3, \tag{1.4.59}$$

where e_{ij} are the strain components. From (1.4.54) it is clear that

$$\tau_{xx} + \tau_{yy} + \tau_{zz} = 0, \tag{1.4.60}$$

which implies that $\mu^1 = \frac{-2}{3}\mu$, where μ is the 'coefficient of viscosity'.

Equations (1.4.56)–(1.4.58) can be written as a single equation in the vector notation.

$$\rho \frac{D\vec{V}}{Dt} = \rho\vec{F} - \operatorname{grad} p + \mu\nabla^2\vec{V} + \frac{\mu}{3}\operatorname{grad}\operatorname{div}\vec{V}, \;\; \vec{F} = (g_x, g_y, g_z), \;\; \vec{V} = (V_1, V_2, V_3). \tag{1.4.61}$$

Equation (1.4.61) is called *Navier-stokes equation* for Newtonian fluids. This equation is independent of the temperature.

The equation of conservation of energy

A general expression of specific (per unit mass) energy E is given by

$$E = e + \text{ mechanical energy } + \text{ potential energy } + \text{ chemical energy}. \tag{1.4.62}$$

Here, e is the specific internal energy or specific internal heat energy. Mechanical energy, potential energy and chemical energy are generally omitted in the formulation of Stefan problems and e is taken as specific enthalpy which is the sum of the latent heat and the sensible heat (see § 2.1.3 for their definitions). Let $E = e = h$, where h is the specific enthalpy. Taking $\phi = h$ and $\vec{J} = -K\nabla T$ in (1.4.52), where K is the thermal conductivity in the isotropic case, we obtain

$$\frac{\partial}{\partial t}(\rho h) + \nabla \cdot (\rho \vec{V} h) = \nabla \cdot (K \nabla T) + \dot{f}. \tag{1.4.63}$$

Here, f is the heat generated or absorbed per unit volume. $\vec{V}$ is to be obtained by solving Navier-Stokes equation and substituted in (1.4.63). To take into account mechanical and potential energies in (1.4.63), we take body force $\vec{F} = -\text{grad}\,\psi$, $\vec{J} = -K\nabla T - \vec{P}_V$, where $\vec{P}_V$ is given by

$$\vec{P}_V = (\sum_{k=1}^{3} p_{ik} V_k, \sum_{k=1}^{3} p_{2k} V_k, \sum_{k=1}^{3} p_{3k} V_k). \tag{1.4.64}$$

Here, p_{iJ} is the stress tensor defined in (1.4.54). We present below the final form of the energy equation for constant ρ case (cf. [34]).

$$\rho \frac{D}{Dt}(e + \frac{1}{2}|\vec{V}|^2 + \psi) = \nabla(K\nabla T) - \text{div}(p\vec{V}) + W_T + \dot{f}, \tag{1.4.65}$$

$$W_T = \frac{\partial}{\partial x}(\vec{V} \cdot \vec{\tau}_x) + \frac{\partial}{\partial y}(\vec{V} \cdot \vec{\tau}_y) + \frac{\partial}{\partial z}(\vec{V} \cdot \vec{\tau}_z), \tag{1.4.66}$$

$$\vec{\tau}_i = \vec{i}\tau_{ix} + \vec{j}\tau_{iy} + \vec{k}\tau_{iz}, \; i = x, y, z. \tag{1.4.67}$$

Here, $\frac{1}{2}\rho|\vec{V}|^2$ is the kinetic energy and for simplicity we take the body force to be $\vec{F} = -\,\text{grad}\,\psi$.

Chapter 2

Thermodynamical and Metallurgical Aspects of Stefan Problems

2.1 Thermodynamical Aspects

Several generalizations and extensions of the two-phase problem (1.4.4)–(1.4.8) are possible but before dealing with them, we give some definitions. Many terms like equilibrium temperature, thermal conductivity, latent heat, enthalpy, etc., have been used earlier without defining or explaining them for the reason that greater concern was shown for the mathematical formulation. Some thermodynamical aspects of Stefan problems will be discussed in this section.

2.1.1 Microscopic and macroscopic models

A system is called *microscopic* if it is of 'small dimensions', roughly of the size of an atom or a molecule, i.e., of the size of 10^{-10} meters. In a *microscopic* model, thermal, mechanical or chemical phenomenon is studied at an atomic or molecular level. In a *macroscopic* model, the system is 'large enough' of the order of one micron, be visible with a microscope using ordinary light. A macroscopic system consists of large numbers of atoms or molecules. There is an important difference between a microscopic model and a macroscopic model. In the microscopic model, the description of an individual particle motion/action, even if available would not disclose the gross behaviour/properties of the system. For example, it is a striking fact, and one which is difficult to understand in microscopic detail that simple atoms forming a gas can condense abruptly to form a liquid with very different properties.

It is well known that a transport phenomenon whether it is electrical, heat or mass transfer, occurs due to changes in the energy levels of atoms or molecules. Imagine applying Newtonian laws of motion to 10^{20} molecules and obtaining information for each molecule in a transport phenomenon such as heat transfer. It is an awesome task. All the mathematical formulations discussed earlier are based on macroscopic modelling. Does it meant that microscopic modelling is not required? However complicated, microscopic models cannot be totally discarded. Kinetic theory of matter (cf. [35, 36]) applies the laws of mechanics to individual molecules of a system and enables one to calculate, for example, the numerical values of heat capacities, heats of transformation, viscosity. These parameters can be explicitly defined in a macroscopic model, but their numerical values can be obtained only on the basis of a molecular model. As stated earlier some of the equations of mathematical physics, for example, the heat equation can be easily derived on the basis of a microscopic model.

The advantage of the macroscopic model or a continuum hypothesis, and in which we are interested is that the gross behavior of the system can be studied and laws of Newtonian mechanics can be applied to the bulk matter. An approach more general than macroscopic modelling is of statistical thermodynamics which, ignores the detailed consideration of molecules as individuals, and applies statistical methods to find the distribution of very large number of molecules that make up a macroscopic piece of matter over energy states of the matter. The equations of conservation of mass, momentum and energy arising in continuum mechanics can be obtained as particular cases of the Boltzmann equation [36].

Both kinetic theory and statistical thermodynamics were first developed on the assumption that the laws of mechanics deduced from the behavior of matter in bulk, could be applied without change to particles like molecules and electrons. As science progressed, it became evident that at least in some respects this assumption was not correct, that the conclusions drawn from it by logical methods did not agree with experimental facts. For example, experiments suggest that the specific heat of many solids at constant volume approach the Dulong-Petit [35] value of 3R (R is universal gas constant) at high temperatures, but decreases to zero at very low temperatures. This behavior of solids can be explained with the help of a quantum mechanics approach. The failure of small scale systems to obey the same laws as large scale systems led to the development of quantum theory and quantum mechanics. Statistical thermodynamics is best treated today from the view point of quantum mechanics. On a microscopic scale classical mechanics does not apply and must be replaced by quantum mechanics. For further details of quantum theory, the reader is referred to [35, 36, 37].

2.1.2 Laws of classical thermodynamics

We shall be dealing here mostly with equilibrium thermodynamics. Thermodynamics is the study of energy and its transformation. There are many different types of energy but most studies of thermodynamics are primarily concerned with two forms of energy: heat and work. Thermodynamics deals with the macroscopic properties of matter and is an empirical science. It is developed on a small number of principles which are generalizations made from experience.

Thermodynamic equilibrium: When a system is in thermal, mechanical and chemical equilibrium, it is said to be in *thermodynamic equilibrium*. In thermal equilibrium, the temperature will be the same at all points of the system. In mechanical equilibrium, all motions, expansions or contractions of the system are absent. Note that atoms are still in motion. When all the chemical reactions stop then the system is in chemical equilibrium.

Reversible and quasi-static process: A process is called reversible if the initial state of the system can be restored with no observable effects in the system and its surroundings. If a process is not reversible, it is called irreversible. If a process is carried out in such a way that at any time the system departs only infinitesimally from the equilibrium state it is called a quasi-static process.

First law of thermodynamics:

Internal energy (internal heat energy) of a system is the sum of all the individual kinetic energies of motion and energies of interaction (potential energies) of the particles in the system. Internal energy can be transformed to do work and produce heat. One form of the first law of equilibrium thermodynamics is

$$dU = d'Q - d'W. \tag{2.1.1}$$

Here, dU is the change in the internal energy of the system from equilibrium state a to another equilibrium state b, $d'Q$ is the heat flow into the system during the change of state and $d'W$ is the work done by the system when the system changes its equilibrium state from a to b. The dash indicates that the quantities are not exact differentials. Internal energy is a state property, i.e., internal energy in state b does not depend on the process by which the system has been brought from state a to state b. dU is an exact differential but not $d'Q$ and $d'W$. It can be easily shown that work is path-dependent and so is heat flow [35]. Equation (2.1.1) holds for both reversible and irreversible processes.

Second law of thermodynamics: Entropy

Some changes in a system can take place only in one direction. Consider an isolated system in which a body at temperature T_1 is in contact with a heat reservoir at temperature $T_2 > T_1$. Heat will flow from the reservoir to the body and raise its temperature to T_2. Is it possible for the body to cool down to temperature T_1 by releasing heat to the reservoir? The change in this direction is not possible. It may be noted that the total energy of the system consisting of the body and the reservoir is conserved even if a reverse change takes place. Therefore if we are looking for some property of the system whose change can tell us the direction in which the reverse change is possible, then it cannot be energy. This property of the system is called *entropy*, denoted by $\hat{S}$ and defined as

$$d\hat{S} = \hat{S}_2 - \hat{S}_1 = \int_1^2 \frac{d'Q}{T}, \text{(reversible changes)} . \tag{2.1.2}$$

Here, 1 and 2 are the two equilibrium states, of a system. Entropy is defined only for reversible processes. *The second law of thermodynamics*, states that processes in which the entropy of an isolated system (i.e., $dQ = 0$) would decrease do not exist. Specific entropy (entropy per unit mass) will be denoted by $\hat{s}$.

If hot water is mixed with cold water, then the entropy of the cold water will increase more than the decrease in the entropy of the hot water which can be checked from (2.1.2). So there will be an increase in the entropy of the system consisting of both cold and hot water. This tells us that heat cannot flow from cold water to hot water as decrease in the entropy of an isolated system is not possible. Unlike energy or momentum, entropy is not conserved. The first law of thermodynamics states that energy can neither be created nor destroyed and the second law states that entropy can not be destroyed but it can be created. The process is called isentropic if the entropy of the system does not change during the process.

2.1.3 Some thermodynamic variables and thermal parameters

Specific heat capacity or specific heat: If no phase-change takes place in a process, then the heat capacity $\hat{C}$ at any temperature is defined by the equation

$$d'Q = \hat{C}dT, \tag{2.1.3}$$

where $d'Q$ is the quantity of heat added to the system which changes its temperature by dT. The process could be reversible or irreversible. If the volume of the system does not change in the process, then $\hat{C}$ is denoted by $\hat{C}_V$ and is called heat capacity at constant volume. At constant volume, $dV = 0$. From (2.1.1), we obtain

$$d'Q = dU + PdV = \left(\frac{\partial U}{\partial T}\right)_V dT + \left[\left(\frac{\partial U}{\partial V}\right)_T + P\right] dV = \left(\frac{\partial U}{\partial T}\right)_V dT, \tag{2.1.4}$$

where the only work done is by the pressure P in changing the volume. From (2.1.3), we get

$$\hat{C}_V = \left(\frac{\partial U}{\partial T}\right)_V. \tag{2.1.5}$$

It may be noted that Q is not a function of temperature. Therefore $\hat{C}$ is not the derivative of Q with respect to T but is only the ratio of $d'Q$ and dT or the ratio of very small amount of heat supplied and the change in the temperature.

A *specific value* of an *extensive property* (a property such as energy which depends on the mass) is its value per unit mass. *Specific heat* is defined as the heat capacity per unit mass and will be denoted by C.

Enthalpy: If the pressure P does not change in the process, then from (2.1.1), we get

$$dU = d'Q - PdV. \tag{2.1.6}$$

Also

$$dU + PdV = dU + PdV + VdP = dU + d(PV) = d(U + PV). \tag{2.1.7}$$

Let $H = U + PV$, where P is constant. H is called *enthalpy* and *specific enthalpy* will be denoted by h. H is a function of state, or a *thermodynamical property*. In this volume, H has also been used for enthalpy per unit volume. If a thermodynamic quantity depends only on the state of the system and not on the process by which it has been brought to that state, then it is called a thermodynamical property. Temperature and pressure are examples of thermodynamical properties. From (2.1.6) and (2.1.7), we have

$$dH = dQ = \hat{C}_P dT, \tag{2.1.8}$$

or

$$\hat{C}_P = \left(\frac{\partial H}{\partial T}\right)_P. \tag{2.1.9}$$

Here, $\hat{C}_P$ is the heat capacity at constant pressure.

From the relation $dH = dQ$ (at constant pressure), it is clear that enthalpy can be regarded as the heat content of the system. A positive ΔH (change in H) means that heat is absorbed while a negative ΔH means that heat is released.

Gibbs free energy and Helmholtz free energy: We now ask 'Is it possible to describe the thermal equilibrium of a system in terms of some property of the system itself even if it is in contact with surroundings'? A system and its surroundings together is called a 'universe'. Universe is considered to be isolated and closed. For an isolated system, $d'Q = 0$ and so $d\hat{S} = 0$. Let $d\hat{S}$ and $d\hat{S}_*$ be the changes in the entropies of the system and its surroundings, respectively, then from the second law of thermodynamics

$$(d\hat{S} + d\hat{S}_*) \geq 0. \tag{2.1.10}$$

Let the system absorb some heat $d'Q$ from its surroundings so that the internal energy of the system changes from U to $U + dU$ and the system does some work $d'W = PdV$ in a 'PVT system'. If T is the temperature of the system and its surroundings, then

$$d\hat{S}_* = -d'Q/T \text{ and } (d\hat{S} - d'Q/T) \geq 0. \tag{2.1.11}$$

Substituting $d'Q$ from (2.1.1) in (2.1.11), we get

$$dU + PdV - Td\hat{S} \leq 0. \tag{2.1.12}$$

All the quantities in (2.1.12) belong to the system. In any reversible process between two equilibrium states, $d\hat{S} + d\hat{S}_* = 0$ and so

$$dU + PdV - Td\hat{S} = 0. \tag{2.1.13}$$

If P and T are constant, then

$$d(U + PV - T\hat{S}) = dU + PdV - Td\hat{S} = 0. \tag{2.1.14}$$

Therefore, the quantity $U + PV - T\hat{S}$ is a function of the state of the system. It is called *Gibbs free energy* denoted by G and

$$G = U + PV - T\hat{S} = H - T\hat{S}, \ H = U + PV. \tag{2.1.15}$$

Note that G is the free energy of the system and not the free energy per unit volume. Interestingly in deriving (2.1.15) we had assumed that the process is reversible. But G is a function of the state as all the quantities on the r.h.s. of (2.1.15) are functions of state. It does not matter whether the state of the system is changed by a reversible process or by an irreversible process.

Equation (2.1.15) is very general and applies to a system of any nature. The process may be a change of state, or a change of phase, or a chemical reaction. Suppose the temperature of the system rises by an amount dT under constant pressure. When the system changes from one equilibrium state to another then from (2.1.15), we get

$$dG = dH - Td\hat{S} - \hat{S}dT. \tag{2.1.16}$$

From (2.1.3), $dQ = \hat{C}_P dT$ (under constant pressure $d'Q = dQ$) and therefore from (2.1.2)

$$d\hat{S} = \hat{C}_P dT/T. \tag{2.1.17}$$

Since $dH = dQ = \hat{C}_P dT$, we have

$$dG = -\hat{S}dT. \tag{2.1.18}$$

The free energy decreases with the rise in temperature at constant pressure. The Gibbs free energy is the most useful of all thermodynamical properties and provides a practical criterion for a system to be in thermodynamical equilibrium. If a system is in thermodynamical equilibrium, then $dT = 0$ and

$$dG = 0. \tag{2.1.19}$$

Out of the infinitely many states in which a system can exist, the equilibrium state is the one whose free energy does not change in any process which is carried out under constant pressure and constant temperature. $dG = 0$ also implies that in an equilibrium state, free energy would be minimum.

Let us now consider a system at constant temperature and constant volume. It is easy to see that

$$d(U - T\hat{S}) = dU - Td\hat{S} = 0. \tag{2.1.20}$$

The quantity $F = U - T\hat{S}$ is called *Helmholtz free energy.* F is also a function of state and plays the same role at constant temperature and constant volume as Gibbs free energy plays at constant temperature and pressure. dF can be expressed as

$$dF = -\hat{S}dT - PdV. \tag{2.1.21}$$

Latent heat: When a solid metal piece is heated, the temperature rises and reaches an equilibrium temperature T_m (see the next section). When more heat is supplied, the heat is first absorbed without raising the temperature of the piece. The heat so absorbed is called latent heat of melting. It is measured in terms of per unit mass and is the ratio of the heat absorbed and the mass m undergoing the change of phase. Why is heat absorbed ? In solid metals, the free atoms are closely packed. Atomic arrangement is disordered in liquids in comparison to that in solids. So the latent heat of fusion (or melting) is the energy required to pull the atoms apart to the more openly packed structure of the liquid [38]. In general when a phase changes isothermally from solid to liquid or liquid to vapour or solid to vapour, the system absorbs heat. The heat absorbed in these three transformations is called latent heat of fusion, latent heat of vaporization and latent heat of sublimation, respectively. When a liquid solidifies latent heat is released. Latent heat released per unit mass is taken as positive, and is denoted by l. Latent heat of solidification is the negative of the latent heat of fusion.

Thermal conductivity: Thermal conductivity is a transport property and is not a function of state. Transport of heat energy takes place by molecular motion/vibration and is called heat conduction. Heat conduction in equilibrium thermodynamics takes place according to an empirical law called the law of Fourier's heat conduction. For isotropic bodies this law is given by (1.3.8) and for anisotropic bodies by (1.4.9).

2.1.4 Equilibrium temperature; Clapeyron's equation

Melting or freezing temperature: The free energy curves vs temperature (pressure constant) for solid and liquid phases of a metal are given in Fig. 2.1.1. Since free energies of the two phases are the same at the temperature T_m, the two phases can stay in equilibrium at T_m as no exchange of free energy can take place. T_m is called equilibrium temperature. T_m is also called melting or freezing temperature. It can be called an *ideal equilibrium temperature* as the phase-change takes place at T_m under ideal conditions which ensure that the phase-change boundary is planar. In alloys phase-change takes place over a temperature range. If the temperature of the liquid is less than T_m, the liquid is called *supercooled.* It is clear from Fig. 2.1.1. that for $T < T_m$, the free energy of the solid is less than the free energy of the liquid. Therefore solid phase is a stable phase for $T < T_m$ and supercooled liquid is not a stable phase.

For a solid, the specific enthalpy h is given by

$$h = C_S T, \;\; T < T_m. \tag{2.1.22}$$

For a liquid

$$h = C_L T + l, \;\; l > 0; \;\; T > T_m. \tag{2.1.23}$$

At $T = T_m$,

$$h \in [C_S T_m, C_L T_m + l]. \tag{2.1.24}$$

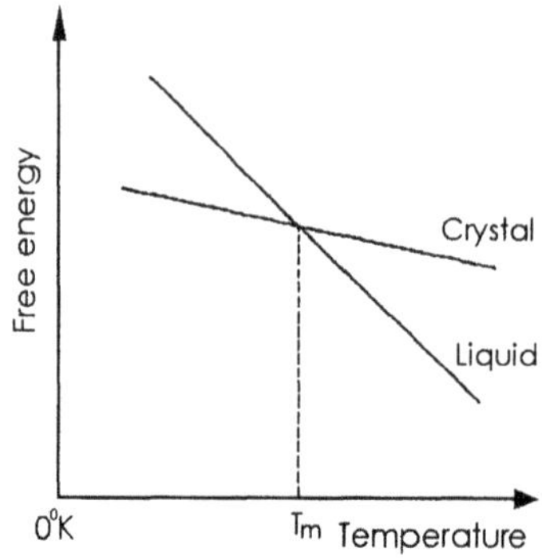

Fig. 2.1.1. Free energy vs temperature

From (2.1.15), $\Delta H = T_m \Delta \hat{S}$ as $\Delta G = 0$ at $T = T_m$, where ΔH and ΔG stand for the difference between their values in the two phases at $T = T_m$. Therefore

$$\Delta \hat{s} = \frac{l + (C_L - C_S)T_m}{T_m} = \frac{\hat{l}}{T_m}, \tag{2.1.25}$$

where $\Delta \hat{s}$ is the specific entropy of the liquid minus the specific entropy of the solid at $T = T_m$.

Clapeyron's equation: When two phases have different specific volumes their temperature of mutual equilibrium depends on the pressure as the phase-change causes work PdV to be done. Let $G_1(T, P)$ and $G_2(T, P)$ be the Gibbs free energies of the solid and liquid phases, respectively. Since the phases are in equilibrium

$$G_1(T, P) - G_2(T, P) = 0, \tag{2.1.26}$$

or

$$\left(\frac{\partial G_1}{\partial T} - \frac{\partial G_2}{\partial T}\right) dT + \left(\frac{\partial G_1}{\partial P} - \frac{\partial G_2}{\partial P}\right) dP = 0.$$

On using the definitions of G_1 and G_2 and (2.1.25), we obtain

$$\frac{dP}{dT} = \frac{l + (C_L - C_S)T_m}{(V_L - V_S)T_m}, \tag{2.1.27}$$

where V stands for the volume.

2.2 Some Metallurgical Aspects of Stefan Problems

2.2.1 Nucleation and supercooling

It is a common experience that when most liquids are cooled sufficiently, they become solid in course of time. A liquid metal normally solidifies by crystallization. In addition to the solid state, metals also exist in a glassy or an amorphous solid state, a state which is hard but not crystalline. The first step in the growth of a crystal is *nucleation*. A small nucleus appears in the liquid phase which then grows by the addition of more material from the liquid phase. The formation of a rain drop in a cloud is a familiar example.

The theory of 'nucleation' is important not only in crystal growth but also in understanding supercooling/superheating and supersaturation effects. Some authors use the term *undercooling* in the place of supercooling. A nucleus may form at the surface of the mold or inside the melt. When the probability of forming a nucleus is the same anywhere in the melt, then the system is in a state suitable for *homogeneous nucleation*. If some sites are more suitable for nucleation than others, then the system is in a state suitable for *heterogeneous nucleation*. In nature most nucleations occur heterogeneously. For heterogeneous nucleation we refer the reader to [39].

How does nucleation occur ? A phase transformation involves rearrangement of atoms and requires some irreversible departure from equilibrium. Phase transitions are driven by thermal fluctuations and in liquid metals, random fluctuations may create minute crystalline regions called 'clusters' or 'embryos' even at temperatures higher than T_m. On this already existing new phase which has grown randomly some more material of the new phase should grow for a crystal formation. Some sites such as rough surfaces of the mold, foreign bodies in the melt, and seeds of the new phase added to the melt help the process of nucleation. Further details of heterogeneous nucleation or even homogeneous nucleation are too complicated to be presented here.

For a nucleus to grow, some work has to be done by the system which results in a change in the free energy ΔG (per unit volume) which is the difference between the total free energy of the new phase (solid), and the total free energy of the old phase (liquid). Total free energy is the sum of the bulk energy and the surface energy, and is given by

$$\Delta G = r^3 \Delta G_V + \alpha r^2 \sigma, \tag{2.2.1}$$

where r is the linear dimension of the nucleus and r^3 its volume, ΔG_V is the change in the bulk free energy of the volume, αr^2 is the surface area of the nucleus and σ is the surface energy per unit area. σ is generally positive. ΔG_V is negative as the enthalpy of the liquid is greater than that of the solid. When the radius of the nucleus is very small, the second term on the r.h.s. of (2.2.1) dominates the first (σ is large when the nucleus is very small) term making ΔG positive. ΔG reaches its maximum when at $r = r_0$ (see Fig. 2.2.1.) and

$$\left.\frac{d(\Delta G)}{dr}\right|_{r=r_0} = 0. \tag{2.2.2}$$

Using (2.2.1) in (2.2.2), we obtain

$$r = r_0 = -2\alpha\sigma/(3\Delta G_V), \text{ and } A = \Delta G|_{r=r_0} = \frac{4\alpha^3\sigma^3}{27(\Delta G_V)^2}. \tag{2.2.3}$$

Since at equilibrium $\Delta G_V = 0$, ΔG becomes very large near $r = r_0$. The nucleus cannot grow beyond $r = r_0$ until this large energy barrier is crossed, for example by artificial seeding with the fragments of the new phase. Once this barrier of energy is crossed, ΔG starts decreasing and nuclei are formed when ΔG is negative. The critical size $r = r_0$ of the nucleus corresponds to an unstable equilibrium between the nucleus and its parent phase; ΔG makes larger one grow and smaller one shrink.

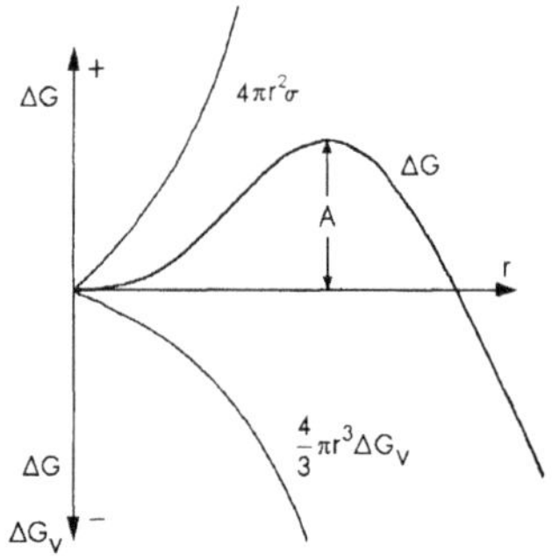

Fig. 2.2.1. ΔG, ΔG_V and surface energy vs r

Supercooled state: A state in which the temperature of a liquid is below the ideal equilibrium temperature T_m is called a supercooled state and the liquid is said to be supercooled. If the temperature of the solid is greater than T_m, the solid is said to be *superheated*. It may be mentioned that the term supercooling defined here is different from the *constitutional supercooling* which arises in alloy solidification and in which the freezing temperature may decrease with the rise in concentration.

Degree of supercooling

For simplicity the formation of a single nucleus in homogeneous nucleation will be considered. Let T be the nucleation temperature. From (2.1.18) and (2.1.25), we have

$$\frac{d(\Delta G_V)}{dT} = -\Delta\hat{S} = -\frac{\Delta H}{T} = \frac{l_m}{V_m T_m}. \tag{2.2.4}$$

Here, G_V is the free energy per unit volume. It has been assumed in (2.2.4) that $C_S = C_L$, a small change in the entropy due to a temperature change can be neglected and T_m can be used in the place of T. l_m is the latent heat per mole and V_m is the molar volume. The 'mole' or the 'mole number' of a substance is defined as the ratio of its mass and the molecular weight and the 'molar volume' is the ratio of volume and the mole number. Using Taylor's series expansion for ΔG_V, we obtain

$$(\Delta G_V)|_T = (\Delta G_V)|_{T_m} + (T - T_m)\frac{d(\Delta G_V)}{dT}|_{T_m} + \dots$$

$$= -(T_m - T)\frac{d(\Delta G_V)}{dT}|_{T_m} = -\frac{(T_m - T)l_m}{V_m T_m}. \qquad (2.2.5)$$

In deriving (2.2.5) we take $(\Delta G_V)|_{T_m} = 0$ as at $T = T_m$ the bulk free energies of solid and liquid phases are the same and higher order terms have been neglected. If $T < T_m$ in (2.2.5), then $(\Delta G_V)_T$ is negative. For large radius of the nucleus, ΔG can be taken to be equal to ΔG_V as the second term in (2.2.1) is very small. If the phase-change from liquid to solid takes place at a temperature lower than T_m, then the liquid is supercooled and $(T_m - T)$ gives the degree of supercooling. Note that ΔG is negative, which it should be for a crystal growth. If a system exists in a state of its lowest fee energy, then the system is in a stable state and it can remain in this state for a long time. However, systems often exist for long times in states not of lowest free energy, for example, a mixture of oxygen and hydrogen gases can stay unchanged at room temperature even though the stable state is water. Such systems are said to be *metastable.* Supercooled liquid and superheated solids are in metastable states and if they get some extra energy to rearrange a group of atoms into a next and more stable phase, the phase-change to a stable phase will take place. The stable state for a supercooled liquid is solid and for a superheated solid, it is liquid. The extra energy can be provided to the supercooled liquid, for example, by artificial seeding. During solidification, latent heat is released by the liquid which warms up the growing crystal and the solid-liquid interface attains the temperature T_m (see Fig. 2.3.7.).

Supercooling is generally small but can be as large as $0.2\,T_m$. For typical metals $T_m \sim 1000K$ to $2000K$. It may be noted that T_m is actually an ideal freezing temperature as the phase-change from liquid to solid generally takes place at a temperature lower than T_m. This is because of the effect of interface curvature and the kinetic undercooling which are discussed in the next section.

2.2.2 The effect of interface curvature

In the crystallization of a pure metal, at the interface between nucleated crystal and its melt, there is considerable activity amongst the atoms. This is because of the small amount of *activation energy* required to change the phase. In simple words, activation energy can be described as the least amount of energy required to change a metastable phase to a new and more stable phase. Some atoms break away from the crystal at the surface to join the liquid and some atoms in the liquid crystallize on the surface and become part of the crystal. If the rates of these two opposing processes are equal, then the surface is at the crystal-liquid equilibrium temperature. This equilibrium temperature need not be the ideal equilibrium temperature T_m. If the surface of the crystal is curved, and the center of curvature lies inside the crystal, then on an average the atoms at the surface are less surrounded by neighbouring atoms of the crystal than otherwise and their escape tendency into the liquid increases. Less energy is required for a phase-change and phase-change takes place at a lower temperature. The equilibrium temperature is lower than the ideal value T_m [40]. Conversely, if the center of curvature lies outside the crystal, then the equilibrium temperature will be higher than T_m. These effects are small unless radius of curvature is small. The deviation ΔT from the ideal equilibrium temperature T_m (which is the temperature of a planar phase-change interface) is given by

$$\Delta T = T_m^c - T_m = \bar{K}_c \Gamma_0 = \frac{\sigma \bar{K}_c T_m V_m}{l_m}, \tag{2.2.6}$$

where T_m^c is the new equilibrium temperature due to the curvature effect, $\bar{K}_c$ is the sum of the principal curvatures of the curved surface, σ is the surface energy per unit area and l_m is the latent heat per mole. Note that the curvature $\bar{K}_c$ and the constant Γ_0 are defined in such a way that a positive undercooling, i.e., a decrease in the equilibrium temperature is associated with a portion of solid-liquid interface which is convex towards the liquid phase (see Figs. 2.4.2. and 2.4.3.). $\bar{K}_c$ is negative in this case. Equation (2.2.6) is called *Gibbs-Thomson relation.* The relation (2.2.6) can be obtained as follows.

Consider a small solid particle (crystal) in the melt. The sum of the principle curvatures of the curved surface of this particle is denoted by $\bar{K}_c$. Let ΔG_r be the difference between the total free energies (per unit volume) of the new phase (very small crystal) and the old phase (liquid) when the curvature effects are included in the phase-change temperature T_m. Note that the total free energy in any phase is the sum of the free energy of the bulk and the surface energy.

If the volume of the liquid is much large than the volume of the crystal, and if $|T_m^c - T_m|$ is not very large, then it can be reasonably assumed that in solidification the free energy of the liquid does not change due to the curvature effect. In this case as discussed below ΔG_r can be regarded as the difference between the total free energies of a very small crystal (with curvature) forming at temperature T_m^c and a very small crystal forming at T_m without curvature.

$$\begin{aligned}\Delta G_r &= \text{total free energy of crystal at } T_m^c - \text{total free energy of liquid at } T_m^c \quad (2.2.7)\\ &= \text{total free energy of crystal at } T_m^c - \text{ total free energy of liquid at } T_m\\ &= \text{total free energy of crystal at } T_m^c - \text{ total free energy of crystal at } T_m. \quad (2.2.8)\end{aligned}$$

Since free energies of solid and liquid at T_m with no curvature are equal, we get (2.2.8). $|\Delta G_r|$ can also be calculated with the help of (2.1.15) and (2.2.8) and we get

$$|\Delta G_r| = |\Delta H - T_m^c \Delta \hat{S}| = |\Delta \hat{S}||(T_m - T_m^c)|. \tag{2.2.9}$$

Here, modulus of different quantities has been taken for convenience. $\Delta \hat{S}$ is the difference between the entropies (per unit volume) of solid and liquid at the temprature T_m. It can be shown [40] that if $C_S = C_L$ then

$$(\Delta H)_{T=T_m^c} = (\Delta H)_{T=T_m}. \tag{2.2.10}$$

The change in the free energy of the crystal due to curvature or the change in the free enthalpy of the crystal (per unit volume) due to curvature is due to the change in the internal pressure [40] and so

$$|\Delta G_r| = |\Delta P|. \tag{2.2.11}$$

The internal pressure ΔP is given by the relation [40]

$$|\Delta P| = \sigma|\bar{K}_c|. \tag{2.2.12}$$

When Gibbs-Thomson law is written as

$$T_m^c = T_m + \frac{\sigma \bar{K}_c T_m V_m}{l_m}, \tag{2.2.13}$$

then by convention $\bar{K}_c$ is taken as negative if the phase-change interface is convex towards the liquid, i.e., $T_m^c < T_m$ as σ, T_m and l_m are taken as positive. If $T_m > T_m^c$, then from (2.2.4) (2.2.9), (2.2.11), and (2.2.12) (modulus sign can be removed as we are dealing with positive quantities on both sides), we get

$$T_m - T_m^c = \sigma \bar{K}_c T_m V_m / l_m, \ \bar{K}_c > 0, \tag{2.2.14}$$

which is the relation (2.2.6).

2.2.3 Nucleation of melting, effect of interface kinetics, and glassy solids

Nucleation of melting: Melting of solid can begin without significant superheating in the solid. When nucleation of melting occurs over a solid surface, the nucleus consisting of the liquid phase will be enveloped basically by two different surfaces. On one side is a solid-liquid interface (see Fig. 2.2.2.) and on the other side is a gas-liquid interface. The surface energy of the gas-solid interface is generally greater than the combined gas-liquid and solid-liquid surface energies. This implies that the formation of the liquid film on the solid surface will not be opposed by the solid surface and the liquid film spreads over the solid surface without significant superheating. The equilibrium of surface energies is attained at the equilibrium temperature.

Effect of interface kinetics: It was suggested in [41], that the interface movement lowers the temperature of the solid-liquid interface below the equilibrium temperature T_m. The form of the growing phase adapts itself to satisfy the growth velocity-temperature requirement. The movement of the interface is dependent on a driving force which was called the viscous effect or the *interface kinetic effect.* The supercooling as a function of interface velocity can be expressed as

$$T(\vec{V}) = T_m - \vec{\phi} \cdot \vec{V}, \tag{2.2.15}$$

where $T(\vec{V})$ is the new melting temperature, $\vec{V}$ is the velocity of the interface and $\vec{\phi}$ is the viscous correction term. For small departures of $T(\vec{V})$ from T_m the rate at which solidification occurs is approximately proportional to the departure [42]. In high transition rates these effects appear on a macroscopic scale also. Kinetic undercooling is likely to be important only when $\vec{V} \cdot \vec{n}$ (normal velocity) is large, such as $\vec{V} \cdot \vec{n} \sim 10\text{m/s}$. The applicability of kinetic effect to a moving interface has been opposed by some authors [43] as the concept devolves from the thermodynamics of equilibrium systems which has been

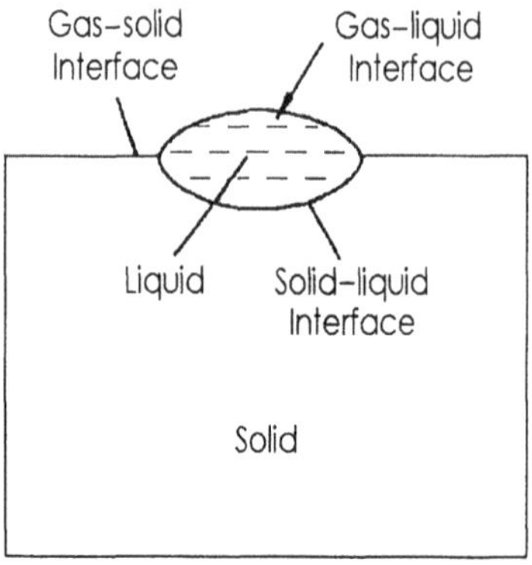

Fig. 2.2.2. Nucleation of melting

applied in (2.2.15) to transforming systems. The interfacial free energy of the solid-liquid interface is a thermodynamic quantity only in the limited context of stationary interfaces which are at equilibrium. See [42] and [44] for further information.

Glassy (amorphous) solids: At very high cooling rates, such as those encountered in rapid solidification of liquid metals, there may be insufficient time for the formation of nucleus and a *glassy solid* is formed. While still non-crystalline, it becomes a solid in its mechanical behaviour as its viscosity becomes very high. There is no phase-change boundary associated with glassy materials.

2.3 Morphological Instability of the Solid-Liquid Interface

A crucial assumption in the classical formulation of Stefan problems is that there exists a smooth surface which is the phase-change boundary such that on one side of this surface is a stable solid (liquid) region and the other side a stable liquid (solid) region. The solid region is identified only by its temperature which should be less than or equal to the equilibrium temperature T_m. Similarly liquid is identified only by its temperature which should be greater than or equal to T_m. The existence of such a planer solid-liquid interface is an ideal situation and is possible only in special cases such as *columnar solidification* of pure metals or *directional solidification* of alloys in 'Bridgman type furnace' [40] under a sufficiently high temperature gradient.

In unidirectional solidification, in its classical formulation, it has been established mathematically in several references such as in [45] that under certain conditions, such as, in the absence of volumetric heat sources, or non-existence a mushy region initially, a planar interface remains planar (mushy region does not develop) till the solidification is complete. In multi-dimensional problems, the existence of a sharp interface has been proved rigorously only for a short-time under suitable assumptions (see Chapter 10). Invariably, a sharp interface separating a stable solid region from a stable liquid region degenerates after a short time into a mushy region. The mushy region contains both solid and liquid phases and separates the stable solid region from the stable liquid region. The solid in the mushy region is present initially in the form of dendrites (Figs. 2.3.1. and 2.3.2.). Further growth of dendrites leads to a polycrystalline structure with one crystal from each nucleus (Fig. 2.3.3.). The diameter of a grain could be as small as 10^{-5}m and as large as 10^{-2}m. Mushy region may also develop during solidification if the liquid is supercooled or during melting if the solid is superheated. This degeneration of sharp interface into a mushy region is attributed to the morphological instability of the solid-liquid interface.

Fig. 2.3.1. Formation of dendrites

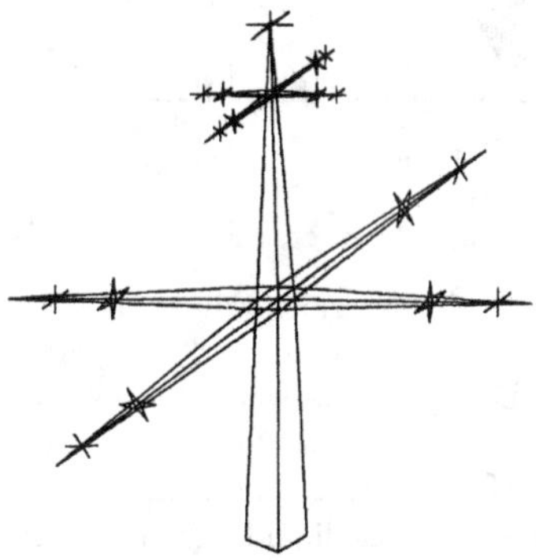

Fig. 2.3.2. Dendrite growth

Classical thermodynamic definitions of stability are inapplicable to the determination of the morphology of a growing interface, and current extensions of equilibrium thermodynamics have not yet furnished a fully acceptable alternative [46]. On the basis of a heuristics-based stability criterion, the interface is said to be morphologically unstable/stable if a small perturbation given to the interface grows/dies with time (see Fig. 2.3.4.). In Fig. 2.3.4. (a), the perturbations are growing, making the planar interface

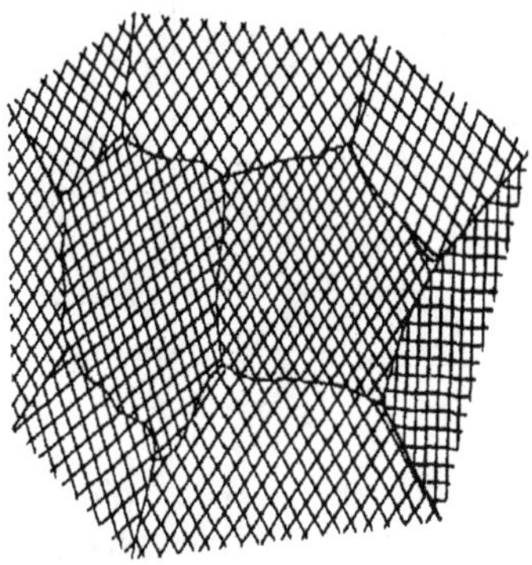

Fig. 2.3.3. Formation of polycrystalline structure.

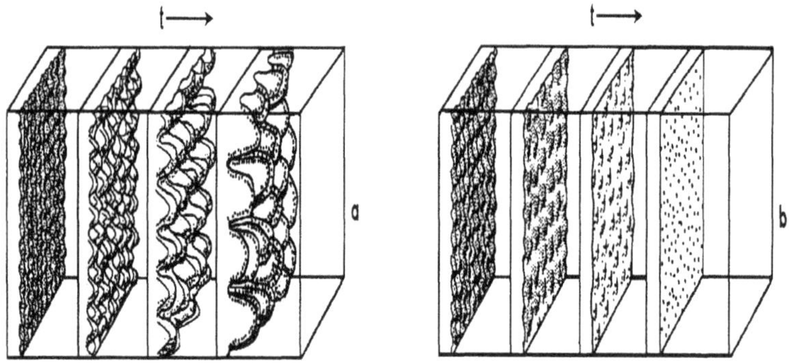

Fig. 2.3.4. (a) Small perturbations at the interface are growing

(b) Perturbations are dying.

unstable. Mushy region will develop in front of the interface. In Fig. 2.3.4. (b), the perturbations are dying and the planar interface remains planar. Figs. 2.3.5–2.3.7. show solid-liquid interface morphology and the temperature distributions in the solid and liquid regions. In Fig. 2.3.5., freezing temperature is equilibrium temperature T_m. The interface is planar and perpendicular to the direction of heat flow. The hottest part of the system is the liquid region. In Fig. 2.3.6. the growth of freezing front is columnar. The hottest part of the system is liquid. This growth is more common in alloys. In Fig. 2.3.7., the grains are equiaxed and have similar dimensions along all axes. The hottest part of the system is the crystal as the liquid is supercooled. The interface is stable in Fig. 2.3.5 and unstable in Figs. 2.3.6. and 2.3.7.

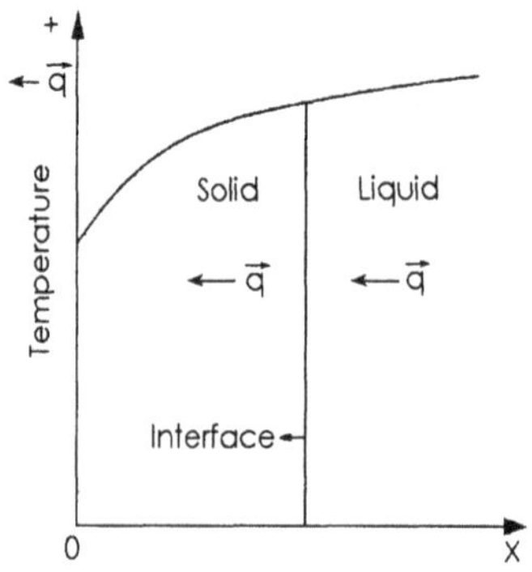

Fig. 2.3.5. Solid-Liquid interface morphology and the temperature distribution. Planar growth.

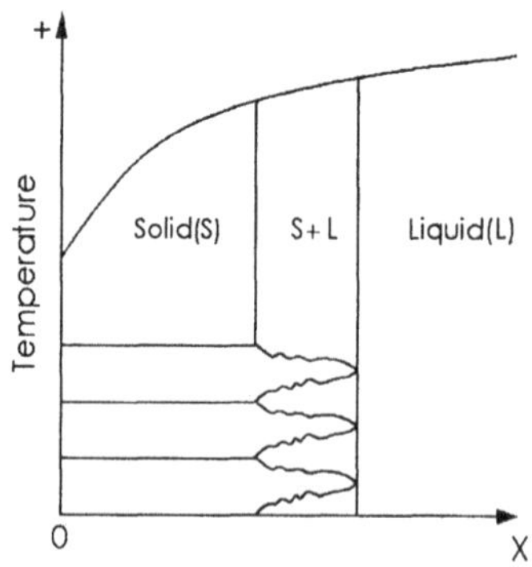

Fig. 2.3.6. Solid-Liquid interface morphology and the temperature distribution. Columnar growth.

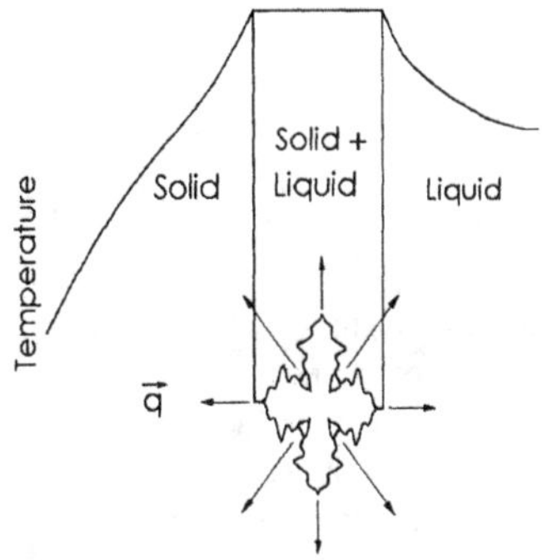

Fig. 2.3.7. Equiaxed dendrite growth

2.4 Non-material Singular Surface: Generalized Stefan Condition

A singular surface is a surface which does not have any inner structure so as to give rise to discontinuities in the bulk quantities. Solid-liquid interface and shock waves are examples of singular surfaces. The motion of a singular surface could be independent of the motion of the bulk phases.

A material surface is a surface which consists of the same material particles all the time. Fluid films and coatings are examples of a material surface. A surface which is not a material surface is called a 'non-material surface'.

It is well known that surfaces of bodies, and interfaces between pairs of bodies, exhibit properties quite different from those associated with their interiors. There have been several attempts (cf. [46, 47, 48, 49]) at formulating the equations of conservation of mass, motion and energy in a moving singular surface embedded in a three-dimensional continuum and dividing the continuum into two regions for which the singular surface is a common boundary. Some thermodynamic fields may have discontinuities across this singular surface. In the classical Stefan condition (1.4.8), the jump in the normal component of the heat flux vector across the phase-change boundary is given in terms of the amount of latent heat released or absorbed at the phase-change boundary. It is assumed that no mechanical work is done. The modern theory of surface thermodynamics and mechanics takes into account other aspects also such as the surface stress, in particular surface tension, and conduction of heat tangential to the surface while balancing energy along the singular surface. The Stefan condition derived below takes into account the surface tension and heat conduction while balancing energy in the phase-change boundary, and will be called *generalized Stefan condition* [51].

Consider a three-dimensional bounded region $\overline{R}$ which is occupied by solid and liquid phases. There exists a time dependent smooth singular non-material surface Γ which divides $\overline{R}$ into two parts R_S and R_L, $\overline{R} = R_S \cup R_L \cup \Gamma$, R_S and R_L are solid and liquid regions such that $R_S \cap R_L$ is empty. The subscripts S and L stand for the solid and the liquid phases, respectively. The external surfaces of R_S and R_L will be denoted by ∂R_S and ∂R_L, respectively (see Fig. 2.4.1.). Let $(R_S)_b = \partial R_S \cup \Gamma$ and $(R_L)_b = \partial R_L \cup \Gamma$. The external surface of R is $\partial R_S \cup \partial R_L \cup \partial\Gamma$ where the curve $\partial\Gamma$ is the boundary of Γ. The surface Γ is orientable and the unit normal $\vec{N}$ to Γ points into the liquid phase. Let $\vec{V}$ be the velocity of Γ in the direction of $\vec{N}$, i.e., $\vec{V} = |\vec{V}|\vec{N}$. It may be noted that only the normal component of the velocity of the interface can be determined as in (1.4.15). The tangential component of the interface motion for a massless interface cannot be derived a priori by recourse to principles of general validity [52].

We shall first balance the forces in $\overline{R}$. These results will be used in balancing energy later.

Conservation of forces in $\overline{R}$

$$\int_{\bar{R}} \rho\vec{B}\,dv + \int_{\partial R_S \cup \partial R_L} \boldsymbol{\tau}\vec{n}\,da + \int_{\partial\Gamma} \hat{\boldsymbol{\tau}}\vec{\nu}\,dl = 0. \tag{2.4.1}$$

Here, $\vec{B}$ is the body force per unit mass, $\vec{n}$ is the unit normal vector pointing outwards as the exterior boundary of a region is traversed in the anticlockwise direction, $\boldsymbol{\tau}$ is the bulk stress tensor and $\hat{\boldsymbol{\tau}}$ is the surface stress tensor, and $\vec{\nu}$ is the outward unit vector normal to $\partial\Gamma$ which is tangential to Γ. dv, da and dl are elementary volume, elementary area and elementary length, respectively. We take $\rho_S = \rho_L = \rho$ for simplicity.

$$\boldsymbol{\tau}\vec{n} = \left(\sum_{i=1}^{3} \tau_{ix}n_i,\ \sum_{i=1}^{3} \tau_{iy}n_i,\ \sum_{i=1}^{3} \tau_{iz}n_i\right),\quad \vec{n} = (n_1, n_2, n_3), \tag{2.4.2}$$

$\boldsymbol{\tau}_{ij}$, $i = 1, 2, 3$, $J = 1, 2, 3$ are components of bulk stress tensor $\boldsymbol{\tau}$.

If it is assumed that both solid and liquid regions are in equilibrium independently of each other, the balancing of forces in the solid region gives

$$\int_{R_S} \rho\vec{B}dv + \int_{\partial R_S} \boldsymbol{\tau}_S\vec{n}da + \int_{\Gamma} \boldsymbol{\tau}_S\vec{N}da = 0. \tag{2.4.3}$$

Similarly balancing of forces in the liquid region gives

$$\int_{R_L} \rho\vec{B}dv + \int_{\partial R_L} \boldsymbol{\tau}_L\vec{n}da - \int_{\Gamma} \boldsymbol{\tau}_L\vec{N}da = 0. \tag{2.4.4}$$

The minus sign in the last term of (2.4.4) arises because the outward normal to the liquid region at the boundary Γ points in the direction opposite to $\vec{N}$. Adding (2.4.3) and (2.4.4) and on subtracting the sum from (2.4.1), we obtain

$$\int_{\Gamma} \boldsymbol{\tau}_L\vec{N}da - \int_{\Gamma} \boldsymbol{\tau}_S\vec{N}da + \int_{\partial\Gamma} \hat{\boldsymbol{\tau}}\vec{\nu}dl = 0. \tag{2.4.5}$$

Applying surface divergence theorem to the last term of (2.4.5) gives integration over Γ and then assuming local equilibrium, we obtain

$$(\boldsymbol{\tau}_L - \boldsymbol{\tau}_S)\vec{N} + \mathrm{div}_\Gamma \hat{\boldsymbol{\tau}} = 0 \quad \text{in } \Gamma. \tag{2.4.6}$$

Here, $\mathrm{div}_\Gamma \hat{\boldsymbol{\tau}}$ is the surface divergence of $\hat{\boldsymbol{\tau}}$ (cf. [47]). Surface stress tensor generalizes the usual notion of surface tension and is consistent with the atomistic calculations, indicating the presence of compressive surface stresses in certain crystals [47]. If $\hat{\boldsymbol{\tau}} = \sigma \mathbf{1}$, where σ is a scalar valued function called surface tension and $\mathbf{1}$ is the inclusion mapping from the tangent space of Γ into the three-dimensional vector space [47], then

$$(\boldsymbol{\tau}_L - \boldsymbol{\tau}_S)\vec{N} + \bar{K}_c \sigma \vec{N} + \nabla_\Gamma \sigma = 0, \quad \text{in } \Gamma. \tag{2.4.7}$$

Here, $\bar{K}_c$ is the sum of principal curvatures of the singular surface.

The surface divergence of a smooth vector field $\vec{u} : \Gamma \to G$, where G is the translation space of a finite-dimensional Euclidean point space, is given by

$$\mathrm{div}_\Gamma \vec{u} = t_r D\vec{u}, \tag{2.4.8}$$

where the r.h.s. in (2.4.8) is the trace of the tangential derivative $D\vec{u}$ of $\vec{u}$ [47]. Note that $\vec{u}$ admits a unique decomposition. $\vec{u} = \vec{u}_S + u_n \vec{N}$, where $\vec{u}_S$ is a vector in a smooth tangential field and u_n is a scalar in a smooth scalar field. If $\boldsymbol{\tau}$ is a smooth tensor field, then for any scalar field ϕ

$$\mathrm{div}_\Gamma(\phi \boldsymbol{\tau}) = \phi \, \mathrm{div}_\Gamma \boldsymbol{\tau} + \boldsymbol{\tau} \nabla_\Gamma \phi. \tag{2.4.9}$$

Taking $\phi = \sigma$ and $\boldsymbol{\tau} = I$ (I is the identity tensor) in (2.4.9), we get

$$\mathrm{div}_\Gamma(\hat{\boldsymbol{\tau}}) = \mathrm{div}_\Gamma(\sigma I) = \sigma \mathrm{div}_\Gamma I + \nabla_\Gamma \sigma. \tag{2.4.10}$$

Using the relation $\mathrm{div}_\Gamma \mathbf{1} = \mathrm{div}_\Gamma I = \bar{K}_c \vec{N}$, we obtain the last two terms on the r.h.s. of (2.4.7).

Conservation of energy in the singular surface Γ

In order to obtain conservation of energy in Γ, we shall first derive the equation of conservation of energy in $\overline{R}$ followed by equations of conservation of energy in R_S and R_L. Subtracting the sum of energy equations for R_S and R_L from the energy equation for $\overline{R}$, we get the energy equation for Γ.

The equation of conservation of energy in $\overline{R}$ is

$$\begin{aligned} \frac{d}{dt} \int_{R_S \cup R_L} \rho e dv + \frac{d}{dt} \int_\Gamma \bar{\in} \, da &= - \int_{\partial R_S \cup \partial R_L} \vec{q} \cdot \vec{n} da - \int_{\partial \Gamma} \vec{\bar{q}} \cdot \vec{\nu} dl, \\ &\quad + \int_\Gamma (\boldsymbol{\tau}_L - \boldsymbol{\tau}_S)\vec{N} \cdot \vec{V} da + \int_{\partial \Gamma} \bar{\in} U dl, \end{aligned} \tag{2.4.11}$$

where e is the specific internal energy, $\bar{\in}$ is the surface energy of Γ per unit area, $\vec{q}$ is the heat flux vector of the bulk, $\vec{\bar{q}}$ is the tangential heat conduction vector in Γ, $(\boldsymbol{\tau}_L - \boldsymbol{\tau}_S)\vec{N}\cdot\vec{V}$ is the work done by the body forces in Γ and $U = -|\vec{V}|\vec{N}\cdot\vec{n}$ is the heat lost or gained due to geometry of $R_S \cup R_L$ as $\vec{N}$ may have a non-zero component along $\vec{n}$. Note that $(\boldsymbol{\tau}_L - \boldsymbol{\tau}_S)\vec{N}$ appears as a body force in (2.4.7) and the tangential work done by this force is zero. The angle which $\vec{N}$ makes with $\vec{n}$ is called *contact angle.* It is still being debated on how to prescribe the boundary conditions at the contact point [54]. By calculating

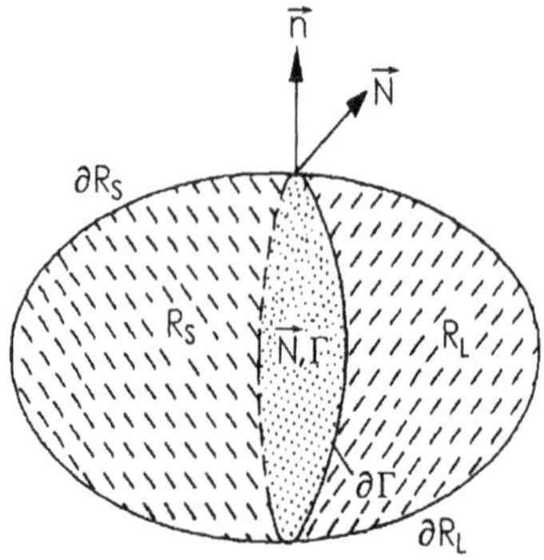

Fig. 2.4.1. Geometry in generalized Stefan condition.

the time derivative of the first term on the l.h.s. of (2.4.11) by first principles, it can be proved that

$$\frac{d}{dt}\int\limits_{R_S\cup R_L} \rho e dv = \int\limits_{R_S\cup R_L} \rho\frac{\partial e}{\partial t} + \int\limits_{\Gamma} \rho(e_S - e_L)\vec{V}\cdot\vec{N}da. \tag{2.4.12}$$

It may be noted that the region $\overline{R}$ is time independent but R_S and R_L vary with time in $\overline{R}$ and this point should be taken care in calculating the time derivative to get the last term in (2.4.12).

On balancing the energy separately in R_S and R_L, we get

$$\int\limits_{R_S} \rho\frac{\partial e_S}{\partial t}dv = -\int\limits_{\partial R_S} \vec{q}_S\cdot\vec{n}da - \int\limits_{\Gamma} \vec{q}_S\cdot\vec{N}da, \tag{2.4.13}$$

$$\int\limits_{R_L} \rho\frac{\partial e_L}{\partial t}dv = -\int\limits_{\partial R_L} \vec{q}_L\cdot\vec{n}da + \int\limits_{\Gamma} \vec{q}_L\cdot\vec{N}da. \tag{2.4.14}$$

Adding (2.4.13) and (2.4.14) and subtracting the sum from (2.4.11), we obtain

$$\int\limits_{\Gamma} \rho(e_S - e_L)|\vec{V}|\,da + \frac{d}{dt}\int\limits_{\Gamma} \bar{\in}da = -\int_{\Gamma}(\vec{q}_L - \vec{q}_S)\cdot\vec{N}da - \int_{\partial\Gamma} \vec{\bar{q}}\cdot\vec{\nu}dl$$
$$+\int\limits_{\Gamma}(\boldsymbol{\tau}_L - \boldsymbol{\tau}_S)\vec{N}\cdot\vec{V}da - \int\limits_{\partial\Gamma} \bar{\in}|\vec{V}|\vec{N}\cdot\vec{n}\,dl. \tag{2.4.15}$$

The second term and the last term on the r.h.s in (2.4.15) should be converted to a surface integral over Γ to obtain an equation for local energy balance in Γ. The conversion of the second term to the surface integral is straight forward. In [51], the contribution of the last term on the r.h.s. of (2.4.15) has been included as the third term in (2.4.16).

Local balance of energy in Γ gives

$$\rho(e_S - e_L)|\vec{V}| + \bar{\epsilon}' - \bar{K}_c\bar{\epsilon}\vec{V}\cdot\vec{N} + \operatorname{div}_\Gamma \vec{\hat{q}} + (\vec{q}_L - \vec{q}_S)\cdot\vec{N} - (\boldsymbol{\tau}_L - \boldsymbol{\tau}_S)\vec{N}\cdot\vec{V} = 0, \tag{2.4.16}$$

$$\bar{\epsilon}' = \frac{\partial\bar{\epsilon}}{\partial t} + \vec{V}\cdot\operatorname{grad}\bar{\epsilon}. \tag{2.4.17}$$

With the help of (2.4.6), the last term in (2.4.16) can be rewritten as

$$(\boldsymbol{\tau}_L - \boldsymbol{\tau}_S)\vec{N}\cdot\vec{V} = -(\operatorname{div}_\Gamma\hat{\boldsymbol{\tau}})\cdot\vec{V}. \tag{2.4.18}$$

Equation (2.4.16) is called a *generalized Stefan condition.*

Let us now briefly describe surface tension and its relation to $\bar{\epsilon}$. The effect of surface tension or capillarity can be explained on the basis of the assumption that on the outer surface of the liquid there exists a surface layer which could be a few molecules thick and whose properties differ from those of the bulk liquid. The surface film and the liquid can be considered as two phases of the liquid in equilibrium. The atoms in the free surface of a solid or a liquid have higher free energy than those inside because they have no neighbours and hence no cohesive bonds, on one side. Therefore, in equilibrium, a tension develops on the surface of the liquid. This tension is related to pressure by a relation of the form (2.2.12). σ is often assumed to be identical to the free energy of the surface. From the relation (2.1.15), we get

$$\sigma = \bar{\epsilon} - T\hat{S}, \tag{2.4.19}$$

where T is the phase-change temperature and $\hat{S}$ is the entropy.

The Stefan condition (1.4.8) can be easily derived as a particular case of (2.4.16). For the derivation of Gibbs-Thomson effect from (2.4.16), see [51]. Some other observations made in [51] are as follows. A simple criterion for the instability of the interface can be derived using (2.4.16). Let $\bar{\epsilon}' = 0, \bar{\epsilon} \neq 0,\ \vec{\hat{q}} = 0$ and $\hat{\boldsymbol{\tau}} = \sigma\mathbf{1}$ where the surface tension σ is constant. Under these assumptions, the relation given below can be obtained.

$$|\vec{V}| = \frac{(\vec{q}_L - \vec{q}_S)\cdot\vec{N}}{l\rho + \bar{K}_c(\bar{\epsilon} - \sigma)}. \tag{2.4.20}$$

Here, l is the latent heat per unit mass (for simplicity take $C_S = C_L$). Consider a very small solid growth protruding into the liquid (see Fig. 2.4.2.) at the solid-liquid interface. In this case $\bar{K}_c$ is negative and if $(\bar{\epsilon} - \sigma) > 0$, then $|\vec{V}|$ is increasing. This is an ideal situation for the dendritic growth as any solid protrusion in the liquid will grow. If $(\epsilon - \sigma) < 0$, then there is retardation. If $l\rho + \bar{K}_c(\bar{\epsilon} - \sigma) \to 0$, then $|\vec{V}| \to \infty$ and dendrites will grow indicating instability of the solid-liquid interface. If $l\rho + \bar{K}_c(\bar{\in} - \sigma) \to \infty$, then $|\vec{V}| \to 0$. This criterion may indicate branching or side protrusion in the dendrite. If the mean curvature is positive (see Fig. 2.4.3.) and $\bar{\in} - \sigma > 0$, then the surface Γ is locally retarded and for $\in -\sigma < 0$ it would be advanced.

In the derivation of (2.4.6) and (2.4.16), both the solid and liquid phases have been considered at rest. The motion of phases has been taken into account in [49] and a more general solidification theory has been presented which takes into account the coupling between thermal fields and kinematic fields. The liquid region is an ordinary Newtonian liquid and the solid phase an elastic material. The specific internal energy, specific entropy, stress tensor and the heat flux vector in the bulk are considered as functions of temperature, temperature gradient, deformation gradient and the rate of deformation gradient.

In Clapeyron's equation (2.1.27), it is assumed that Gibbs free energies of liquid and solid phases (two different phases of the same substance) are the same and that the system is in thermodynamic equilibrium. This means that neither phase can grow or decrease. In [50], a new Clapeyron's equation has been derived, which holds in the presence of nonuniform fields of pressure, temperature, density and velocity, and which does not require the equality of Gibbs functions across the interface as long as their difference is a function of time only, or in particular a constant along the interface. It was shown that even in the presence of constant pressure along the interface, the interface temperature can be changed by increasing or decreasing the velocity at which the process of solidification is taking place.

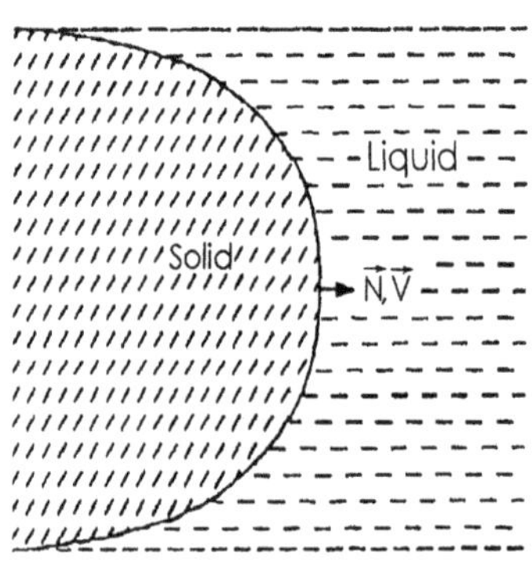

Fig. 2.4.2. Mean curvature $\bar{K}_c$ is negative

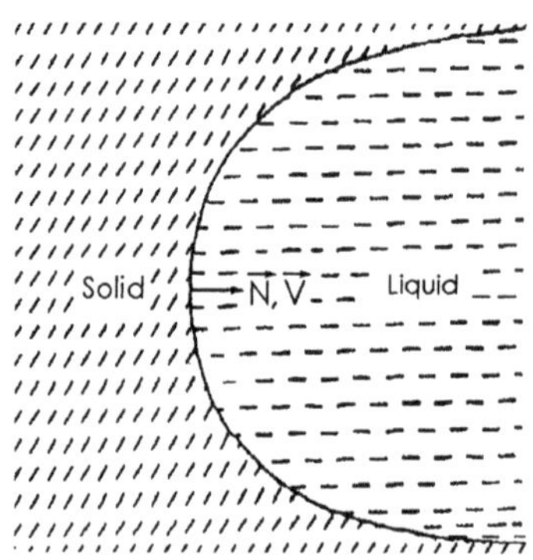

Fig. 2.4.3. Mean curvature $\bar{K}_c$ is positive.

Chapter 3

Extended Classical Formulations of n-phase Stefan Problems with $n \geq 1$

3.1 One-phase Problems

3.1.1 An extended formulation of one-dimensional one-phase problem

One-phase Stefan problems are mainly of three types. In the first type, one of the phases is considered at the melting temperature so that there is temperature gradient present only in one of the phases. Such problems could arise either in melting or solidification. Problems of the second kind are *ablation problems* in which a solid is melting and the liquid drains out instantaneously. In the third type, problems have to do with solidification of supercooled liquids, in which the temperature of the liquid is less than the equilibrium temperature leading to some changes in the energy balance equation at the interface. The melting problems in which the solid is superheated also belong to this class of problems.

One-dimensional one-phase problems with different formulations have been studied from the perspectives of mathematical analysis and computation. Instead of giving complete formulations of problems arising in different contexts, we give here a general formulation of the one-phase problem so that in later sections other one-phase problems can be treated as particular cases of this general problem. This will avoid unnecessary repetitions of some details common in different formulations.

A fairly general one-phase Stefan problem with nonlinear parabolic equation and nonlinear free boundary conditions has been discussed in [55] and is being presented here. Consider a time interval $(0, \hat{t}_*)$. For each $t \in (0, t_*)$, consider the set $\sum(t)$ of functions $p(\tau)$ which are continuously differentiable in $[0, t)$, continuous in [0,t], and are such that $p(\tau) \in (b_0, b_1)$ for $\tau \in (0, t)$, $p(0) = b$, $b_1 > b > b_0 > 0$. For $S(\tau) \in \sum(t_*)$

and $t \in (0, t_*)$ define the set $\Omega(t) \equiv \{(x, \tau) : 0 < x < s(\tau), 0 < \tau < t\}$ and let $C^{1,0}(\bar{\Omega}(t))$ be the set of functions $T(x,t)$ which are continuous in $\bar{\Omega}(t)$ together with their x-first derivatives.

Assume that $a(x, t, y_0, y_1, p_0, p_1)$ is a positive function for $0 \leq x < \infty$, $0 \leq t \leq t_*$, $-\infty < y_0$, $y_1 < \infty$, $p_0 \geq 0$, $-\infty < p_1 < \infty$. The function $q(x, t, y_0, y_1, p_0, p_1)$ is defined in the same domain as the function a, and $\tilde{h}(x)$, $f(t)$ are defined for $0 \leq x \leq b$, $0 \leq t \leq \hat{t}_*$, respectively, and $\psi(x,t)$ is defined for $x \geq 0$, $0 \leq t \leq \hat{t}_*$.

The problem is to find a triple $(t_*, S(t), T(x,t))$ such that

(i) $0 < t_* \leq \hat{t}_*$,

(ii) $S(\tau) \in \sum(t_*)$,

(iii) $T(x,t) \in C^{1,0}(\bar{\Omega}(t_*))$, T_{xx} and T_t are continuous in $\Omega(t_*)$,

(iv) the following equations are satisfied.

$$T_t - a^2(x, t, T, T_x, S, \dot{S})T_{xx} = q(x, t, T, T_x, S, \dot{S}),\ (x,t) \in \Omega(t_*), \tag{3.1.1}$$

$$T(x, 0) = \tilde{h}(x), \quad x \in [0, b], \tag{3.1.2}$$

$$T(0, t) = f(t), \quad t \in (0, t_*), \tag{3.1.3}$$

$$T(S(t), t) = \psi(S(t), t), \quad t \in (0, t_*), \tag{3.1.4}$$

$$\dot{S}(t) = \phi(S(t), t, T_x(S(t), t)), \quad t \in (0, t_*). \tag{3.1.5}$$

Here, $\phi(p, t, y_1)$ is a function defined for $p > 0$, $0 < t < t_*$, $-\infty < y_1 < \infty$, a^2 is the thermal diffusivity which has been considered as a function of independent and dependent variables, q is the heat source per unit time per unit volume. The temperature $T(x,t)$ can be made dimensionless by dividing it by T_m so that at the free boundary $T = 1$. By redefining the temperature as $T - 1$, the temperature at the free boundary can be taken to be zero. In the literature, both 0 and 1 are used as dimensionless isotherm temperatures. If we make the transformation $\hat{T} = T - \psi(x,t)$ in (3.1.1)–(3.1.5), then $\hat{T}$ becomes zero at the free boundary.

A mere formulation of the problem as given in (3.1.1)–(3.1.5) does not guarantee that its solution exists. Even if the solution exists, $T(x,t)$ and $S(t)$ may not satisfy the regularity conditions mentioned above. The appropriate conditions to be satisfied for the existence, uniqueness, well posedness, etc., of the solution of (3.1.1)–(3.1.5) and solutions of some other Stefan problems will be discussed in Chapter 10. Presently we are concerned with the formulation.

If the region $0 < x < S(t)$, $S(0) = b$ is taken as solid (could be identified as ice) and the region $x > S(t)$ $(S(t) < x \leq b_1)$ is taken as liquid at the equilibrium temperature

0 (or 1) (identified as water) and there is cooling at $x = 0$, then we have a one-phase solidification problem. The growth of the solid in the region $x > b$ depends on some appropriate conditions which the prescribed data has to satisfy. In particular the data has to satisfy some *compatibility conditions* at $x = b$ for $\dot{S}(t)$ to be positive. If $b = b_0 = 0$ then the data has to satisfy some compatibility conditions at $x = 0$. If the region $0 < x < S(t)$ is liquid (could be identified as warm water) and the region $x > S(t)$ is solid (identified as ice) at the equilibrium temperature and if there is heating at $x = 0$ then we have a one-phase melting problem. In the melting problem also some compatibility conditions are to be satisfied by the data for $\dot{S}(t)$ to be positive. These compatibility conditions have been discussed in Chapter 10 for several problems. Any of the boundary conditions (A) discussed in § 1.4.4 can be prescribed at $x = 0$ instead of (3.1.3).

Often the mathematical formulations of physical problems are presented in the form of a set of differential equations and boundary and initial conditions without defining the relevant function spaces in which a solution is sought, the spaces to which the known functions belong, ranges of dependent variables, and domains of independent variables. In a rigorous mathematical formulation all such details should be mentioned as done for problem (3.1.1)–(3.1.5). However, due to length constraints it is not possible to do so for every problem discussed in this volume. These details may differ for different problems.

A physical problem can be generalized to any extent but it may not be possible to throw any light on of its solution. We shall discuss only those generalizations of Stefan problems which have been rigorously explored. When dealing with the classical formulation of a Stefan problem, its solution should satisfy some regularity conditions given earlier in § 1.4.6 and as above.

In the place of (3.1.4) and (3.1.5) more general boundary conditions of the type

$$T(S(t),t) = Z\left(\partial T/\partial x, S(t), \dot{S}(t)\right), \tag{3.1.6}$$

and

$$W(T, \partial T/\partial x, S(t), \dot{S}(t)) = 0, \text{ on } x = S(t), \tag{3.1.7}$$

can be prescribed. Here, Z and W could be functionals and need not be pointwise functions of their arguments (cf. § (3.3.33)).

3.1.2 Solidification of supercooled liquid

Consider the following problem:

$$T_t = T_{xx}, \quad \text{in } D_{t_*} = \{(x,t) : 0 < x < S(t),\ 0 < t < t_*,\ S(0) = 1\}, \tag{3.1.8}$$

$$T(x,0) = T_0(x) \le 0\,,\ 0 \le x \le 1, \tag{3.1.9}$$

$$T_x(0,t) = g(t) \geq 0\,, \quad 0 < t < t_*, \tag{3.1.10}$$

$$T(S(t),t) = 0\,, \quad 0 < t < t_*, \tag{3.1.11}$$

$$T_x(S(t),t) = -\dot{S}(t),\ 0 < t < t_*. \tag{3.1.12}$$

By appropriate scaling of time and/or length, the various parameters have been taken to be unity in (3.1.8)–(3.1.12). This problem is concerned with the solidification of a supercooled liquid which initially occupies the region $0 \leq x \leq 1$ and ice at $T = 0$ occupies the region $1 < x < \infty$. It can be argued that the region $0 \leq x \leq 1$ cannot be solid. For if it were so then the region $x > 1$ would have to be liquid at the temperature zero and solidification would begin in the liquid. This would further imply that $T_x(S(t),t)$ and $\dot{S}(t)$ are positive, violating (3.1.12). This problem will be referred as *supercooled Stefan problem* or in short SSP. The analysis of such problems has been presented in § 4.4.1.

3.1.3 Multi-dimensional one-phase problems

The formulation given in (3.1.1)–(3.1.5) can be extended to multi-dimensional one-phase problems if the free boundary is defined by the equation $\Phi(x,t) = 0$, $x \in R^n$ as in §1.4.1. Multi-dimensional problems have not been investigated as exhaustively as one-dimensional problems and often the problems studied are not as general as described in the formulation (3.1.1)–(3.1.5). A typical multi-dimensional problem is the *ablation problem* described below.

A three-dimensional ablation problem

During melting of a solid, if the melt is removed as soon as it is formed, heat flux has to be prescribed at the phase-change interface for further melting to take place. Melting of a piece of ice when water formed is removed instantaneously and the melting of the surface of a spacecraft during re-entry into earth's atmosphere are examples of one-phase ablation problems.

Consider a half space $z \geq 0$ which, at time $t = 0$ is in a solid state. Heat input $Q(x,y,z,t) > 0$ is prescribed at the free boundary $z = S(x,y,t) = S(t)$ and it will be assumed that melting starts instantaneously at $t = 0$ (this is a minor assumption, see § (3.2.2)). The melt is drained out as soon as it is formed. The problem is to find the temperature $T(x,y,z,t)$ and the phase-change boundary which we shall denote by a short notation as $z = S(t)$. The following dimensionless equations are to be satisfied.

$$\frac{\partial T}{\partial t} = k\nabla^2 T\,, \quad z > S(t)\,, \quad t > 0;\ S(0) = 0, \tag{3.1.13}$$

$$T(x,y,z,t)|_{t=0} = f(x,y,z);\ \ f(x,y,0) = 1, \tag{3.1.14}$$

$$T(x,y,z,t)|_{z=S(t)}=1\,,\quad S(0)=0, \tag{3.1.15}$$

The melting temperature has been taken to be unity. The energy balance at $z=S(t)$ is given by the equation

$$Q(x,y,z,t)|_{z=S}+K\left\{1+\left(\frac{\partial S}{\partial x}\right)^2+\left(\frac{\partial S}{\partial y}\right)^2\right\}\frac{\partial T}{\partial z}\bigg|_{z=S}=l\rho\frac{\partial S}{\partial t}. \tag{3.1.16}$$

A Signorini-type boundary condition

Suppose in the above ablation problem melting starts at $t=t_0>0$, (instead of $t=0$) where t_0 is unknown. Then we have two types of boundary conditions on $z=S(t)$, $t>0$.

$$T(x,t,S(t),t)<1,\quad \left\{Q+K\frac{\partial T}{\partial n}\right\}_{z=S(t)}=0,\quad \frac{\partial S(t)}{\partial t}=0;\ t>0, \tag{3.1.17}$$

$$T(x,y,S(t),t)=1,\ \left\{Q+K\frac{\partial T}{\partial n}\right\}_{z=S(t)}=l\rho\frac{\partial S}{\partial t},\ \frac{\partial S}{\partial t}>0,\ t\geq t_0, \tag{3.1.18}$$

where $\vec{n}$ is the unit normal pointing into the solid. There is another way of expressing (3.1.17) and (3.1.18).

$$\left.\begin{aligned}&T(x,y,z,t)\leq 1,\\&\{Q+K\partial T/\partial n\}_{z=S(t)}\geq 0,\\&\{T(x,y,z,t)-1\}\{K\partial T/\partial n+Q\}=0.\end{aligned}\right\}\ t>0,\ z=S(t),\ S(0)=0 \tag{3.1.19}$$

$$Q+K\frac{\partial T}{\partial n}=\rho l\frac{\partial S}{\partial t}\,,\quad z=S(t),\ t>0;\quad \frac{\partial S}{\partial t}\geq 0. \tag{3.1.20}$$

The flux prescribed condition at the free boundary is now an inequality and not an equation. Such boundary conditions are called *Signorini-type boundary conditions* [56]. When melting or solidification does not start at $t=0$, then upto time $t=t_0>0$ or in other words till the time the temperature at the boundary $z=0$ becomes the melting temperature, a pure heat conduction problem without phase-change occurs. At $t=t_0$, the temperature of the half space is to be calculated, which serves as the initial temperature for the phase-change problem. In essence we are solving two heat conduction problems. If a Signorini-type boundary condition is prescribed, then we are solving one problem but the boundary condition is specified in terms of an inequality. This type of formulation is more suitable for mathematical analysis of ablation problems and their numerical computations. It is possible to obtain variational inequality formulations for some problems with Signorini type boundary conditions (cf. [22]). In problem (3.1.13)–(3.1.16), the melting starts over the whole boundary $z=0$. However, depending on the boundary conditions it is possible that the melting begins only over a portion of $z=0$. Such problems will be discussed in § 3.3.2.

3.2 Extended Classical Formulations of Two-phase Stefan Problems

3.2.1 An extended formulation of the one-dimensional two-phase problem

An extended formulation of problem (1.4.4)–(1.4.8) in the one-dimensional case and discussed in [57] is given below. Some results on the existence, uniqueness, and regularity of the solution to problem (3.2.1)–(3.2.10) will be presented in Chapter 10.

$$\mathcal{P}^{(1)}(T^{(1)}(x,t)) \equiv T^{(1)}_{xx} - \delta^{(1)} T^{(1)}_t = q^{(1)}(x,t), \quad \text{in } D^{(1)}, \tag{3.2.1}$$

$$D^{(1)} = \{(x,t) : 0 < x < S(t),\ 0 < t < t_*\},$$

$$T^{(1)}(x,0) = h^{(1)}(x),\ 0 \leq x \leq S(0),\ 0 < S(0) = b < 1, \tag{3.2.2}$$

$$T^{(1)}(0,t) = \phi^{(1)}(t), \quad 0 < t < t_*. \tag{3.2.3}$$

$$\mathcal{P}^{(2)}(T^{(2)}(x,t)) \equiv T^{(2)}_{xx} - \delta^{(2)} T^{(2)}_t = q^{(2)}(x,t), \text{ in } D^{(2)}, \tag{3.2.4}$$

$$D^{(2)} = \{(x,t) : S(t) < x < 1,\ 0 < t < t_*\},$$

$$T^{(2)}(x,0) = h^{(2)}(x),\ S(0) \leq x \leq 1, \tag{3.2.5}$$

$$T^{(2)}(1,t) = \phi^{(2)}(t), \quad 0 < t < t_*, \tag{3.2.6}$$

$$T^{(1)}(S(t),t) = T^{(2)}(S(t),t) = f(S(t),t),\ 0 < t < t_*, \tag{3.2.7}$$

$$\chi^{(1)}(S(t),t) T^{(1)}_x(S(t),t) - \chi^{(2)}(S(t),t) T^{(2)}_x(S(t),t)$$

$$= \dot{S}(t) + \mu(S(t),t),\ 0 < t < t_*. \tag{3.2.8}$$

Here, $\delta^{(1)}$ and $\delta^{(2)}$ are inverse of thermal diffusivities.

The boundary conditions (3.2.3) and (3.2.6) can be replaced by other boundary conditions such as

$$T^{(1)}_x(0,t) = g^{(1)}(T^{(1)}(0,t),t),\ 0 < t < t_*, \tag{3.2.9}$$

$$T^{(2)}_x(1,t) = g^{(2)}(T^{(2)}(1,t),t),\ 0 < t < t_*. \tag{3.2.10}$$

Boundary conditions of mixed type can also be prescribed.

The regularity conditions assumed in [57] for the analysis of the problem are: (1) $\phi^{(1)}(t)$, $\phi^{(2)}(t) \in C^1[0, t_*]$, (2) $h^{(1)}(x)$, $h^{(2)}(x)$ belong to $H^1(0,1)$ where $H^1(0,1)$ is the Sobolev space [58] endowed with the norm $||\bar{h}||_{H^1} = ||\bar{h}||_{L^2} + ||\bar{h}'||_{L^2}$; $\bar{h}(x) = h^{(1)}(x)$ for $0 \le x \le b$ and $\bar{h}(x) = h^{(2)}(x)$, $b \le x \le 1$, (3) the compatibility conditions for the initial and boundary data are satisfied, i.e., $\phi^{(1)}(0) = h^{(1)}(0)$ and $\phi^{(2)}(1) = h^{(2)}(1)$. $\delta^{(1)}$ and $\delta^{(2)}$ are constants, (4) The functions $q^{(1)}$ and $q^{(2)}$ are Hölder continuous in $D^{(1)}$ and $D^{(2)}$, respectively, with respect to x (or t) and $|q^{(i)}(x,t)| \le Q$ (constant), $(x,t) \in D^{(i)}$, $i = 1, 2$, (5) $f(x,t)$ and f_x are continuous and bounded in $\bar{D} = \bar{D}^{(1)} \cup \bar{D}^{(2)}$ and $f_{xx} - f_t$ is bounded and locally Hölder continuous in $\bar{D}$ with respect to x (or t). By redefining the temperature as $(T - f(x,t))$, (3.2.7) can be transformed to $T^{(i)} = 0$, $i = 1, 2$. This transformation can be used without redefining the assumptions about the changed data and the coefficients in (3.2.1)–(3.2.8). This transformation is the physical motivation for the term $\mu(S(t), t)$ in (3.2.8). Some of these regularity conditions can be relaxed, see Chapters 10 and 11.

The initial data and $f(x, 0)$ should also satisfy some compatibility conditions given below.

$$|h^{(1)}(x) - f(b,0)| \le \gamma_1 (b - x)^\alpha, \; 0 \le x \le b, \tag{3.2.11}$$

$$|h^{(2)}(x) - f(b,0)| \le \gamma_1 (b - x)^\alpha, \; b \le x \le 1, \tag{3.2.12}$$

where γ_1 and α satisfy conditions

$$\gamma_1 b^\alpha \ge d, \text{ and } \gamma_1 (1-b)^\alpha \ge d. \tag{3.2.13}$$

Here, d is such that $|\phi^{(i)}(t)| \le d$, $t \ge 0$. $\chi^{(i)}, \chi_x^{(i)}$, $\chi_{xx}^{(i)}$, $\chi_t^{(i)}$, $i = 1, 2$ are continuous in D and

$$|\chi^{(i)}(x,t)| \le \beta, \; (x,t) \in \bar{D}, \tag{3.2.14}$$

for some suitable $\beta \ge 0$. If the boundary conditions are of the type (3.2.9) and (3.2.10), then $g^{(i)}$, $i = 1, 2$ have to satisfy some conditions (cf. [59]).

For given $q^{(i)}$, $\phi^{(i)}$, $h^{(i)}$, $\chi^{(i)}$, $i = 1, 2$, and $f(x,t)$ and $\mu(x,t)$ satisfying the conditions imposed on each of them mentioned above; find t_*, $T^{(1)}(x,t), T^{(2)}(x,t)$ and $S(t)$ such that equations (3.2.1)–(3.2.8) are satisfied. In principle, the boundary conditions (3.2.7) and (3.2.8) on $x = S(t)$ can be replaced by conditions of the form

$$T^{(1)}(S(t),t) = T^{(2)}(S(t),t) = Z_1\left(\partial T^{(1)}/\partial x, \; \partial T^{(2)}/\partial x, \; S(t), \dot{S}(t), \right), \tag{3.2.15}$$

$$W_1(T^{(1)}, T^{(2)}, \partial T^{(1)}/\partial x, \partial T^{(2)}/\partial x, S(t), \dot{S}(t)) = 0, \text{ on } x = S(t). \tag{3.2.16}$$

Here, Z_1 and W_1 could be functionals and need not be pointwise functions of their arguments (cf. (3.3.33)).

3.2.2 Multi-dimensional Stefan problems of classes II and III

According to the change of phase initiating along the fixed boundary of the region originally occupied by the melt, classical Stefan problems can be divided into three classes [60]. In *class I problems* solidification or melting starts simultaneously at all points of a fixed surface of the region under consideration. In *class II problems* solidification begins at a 'portion' of the fixed surface. In *class III problems* it begins at a 'point' of the fixed surface. Class II and class III problems essentially differ from class I problems in that solidification (melting) in them both spreads along the surface and grows towards the interior. Under suitable assumptions, class I problems can be formulated as one-dimensional problems but class II and class III problems are necessarily multi-dimensional. The position of the interface in Fig. 1.4.2. is shown after solidification for a long time. It could be due to solidification at a point or over a portion of $\partial G_1(t)$ or over whole of $\partial G_1(t)$. After sufficiently long time, when the spread of solidification over $\partial G_1(t)$ is complete, then solidification growth will be only towards the interior. In Fig. 1.4.1, even for longer times, a class II problem persists. At least in the analytical solutions and the analysis of Stefan problems, this distinction between different classes affects both procedures and solutions.

A formulation of a multi-dimensional classical Stefan problem has already been given in § 1.4. Some more formulations will be discussed later. Because of their complexity, class II and III problems have not been studied widely. They can be formulated as one-phase or two-phase problems. A three-dimensional one-phase ablation problem in a semi-infinite region with melting due to a 'hot spot' was considered in [61] and a short-time asymptotic solution was obtained. We present below a class III two-phase Stefan problem discussed in [62] in which solidification initiates at a point due to a 'cold spot'.

Consider an axisymmetric cylindrical problem in which R and Z are cylindrical polar coordinates. A superheated melt at time $t = 0$ occupies the region $0 < R_0 \leq R < \infty$, $|Z| < \infty$. This melt is being cooled and a known flux $Q_S(Z,t)$ which is symmetric in Z is prescribed on the surface $R = R_0$ of the cylinder. The temperature $f_L(R,Z)$ of the melt at $t = 0$ is a known quantity which is taken to be a symmetric function in Z and is such that

$$\begin{aligned} f_L(R,Z) &= T_m\ , \quad Z = 0,\ R = R_0 \\ &> T_m\ , \quad Z \neq 0,\ R_0 \leq R < \infty, \end{aligned} \tag{3.2.17}$$

$$\frac{\partial f_L}{\partial Z} = 0, \quad Z = 0, R = R_0; \quad \frac{\partial^2 f_L}{\partial Z^2} \neq 0, \quad Z = 0,\ R = R_0. \tag{3.2.18}$$

The solidification will start instantaneously at the point $Z = 0$, $R = R_0$ and with time it will spread along the surface $R = R_0$ and also towards the interior $R > R_0$ of the cylinder. The equation of the solidification front can be written as

$$R = S(Z,t); \ |Z| \leq b(t), \ t > 0. \tag{3.2.19}$$

$|Z| = b(t)$ is the equation of the spread of solidification along $R = R_0$, and $R = S(Z,t) > R_0$ is the growth of the solidified layer towards the interior of the cylinder.

The dimensionless formulation of the problem is as follows:

In the liquid region

$$2\alpha^2 \frac{\partial T_L}{\partial V} = V \left(\frac{\partial^2 T_L}{\partial r^2} + \frac{1}{r}\frac{\partial T_L}{\partial r} + \frac{\partial^2 T_L}{\partial Z^2} \right), \; S(z,V) \le r < \infty, \; |z| < \infty, \; V > 0, \quad (3.2.20)$$

$$T_L(r,z,V)|_{V=0} = \hat{f}_L(r,z), \;\; 1 \le r < \infty, \; |z| < \infty, \quad (3.2.21)$$

$$\left. \frac{\partial T_L}{\partial r} \right|_{r=1} = \hat{Q}_L(z,V), \;\; |z| \ge B(V), \;\; V > 0. \quad (3.2.22)$$

In the solid region

$$2\frac{\partial T_S}{\partial V} = V \left(\frac{\partial^2 T_S}{\partial r^2} + \frac{1}{r}\frac{\partial T_S}{\partial r} + \frac{\partial^2 T_S}{\partial z^2} \right), \;\; 1 \le r \le S(z,V), \; |z| \le B(V), \; V > 0, \quad (3.2.23)$$

$$\left. \frac{\partial T_S}{\partial r} \right|_{r=1} = \beta \hat{Q}_L(z,V), \;\; |z| < B(V), \;\; V > 0. \quad (3.2.24)$$

At the solid-liquid interface

$$T_L(r,z,V)|_{r=S(z,V)} = 1, \;\;\; |z| \le B(V), \;\; V > 0, \quad (3.2.25)$$

$$T_S(r,z,V)|_{r=S(z,V)} = 1, \;\;\; |z| \le B(V), \;\; V > 0, \quad (3.2.26)$$

$$\left[1 + \left(\frac{\partial S}{\partial z} \right)^2 \right] \left(\frac{\partial T_S}{\partial r} - \beta \frac{\partial T_L}{\partial r} \right) \Bigg|_{r=S(z,V)} = \frac{2\lambda}{V}\frac{\partial S}{\partial V}, \; |z| \le B(V), \;\; V > 0, \quad (3.2.27)$$

$$S(z,V)|_{V=0} = 1, \;\; B(V)|_{V=0} = 0, \;\; S(z,V)|_{|z|=B(V)} = 1. \quad (3.2.28)$$

The following dimensionless variables have been used in (3.2.20)–(3.2.28).

$$z = Z/R_0, \; r = R/R_0, \; V = 2(k_S t/R_0^2)^{1/2}, \; \alpha^2 = k_S/k_L,$$

$$\lambda = l/C_S T_m, \; \beta = K_L/K_S, \; \hat{Q}_L(z,V) = Q_L(z,t) \cdot R_0/K_S T_m,$$

$$S(z,V) = S(Z,t)/R_0, \; \hat{f}_L(r,z) = f_L(R,Z)/T_m, \; B(V) = b(t)/R_0. \quad (3.2.29)$$

If the melt is superheated, i.e., if its temperature is greater than the melting temperature then for solidification to begin it is necessary (not sufficient), that the prescribed flux be infinite at $t = 0$. If at $t = 0$, $\hat{f}_L(1, 0) > 1$, then we shall have a heat conduction problem (without phase-change) up to time $t = t_m > 0$ such that at $t = t_m$, conditions (3.2.17) and (3.2.18) are satisfied. Time can be rescaled and this instant of time can be taken as $t = 0$ with the initial temperature changed to the temperature at $t = t_m$. A short-time analytical solution of a class II problem has been obtained in [63] in which solidification/melting starts at a portion of an edge in a plate. The prescribed flux need not be symmetrical. It is interesting to note that the growth of the free boundary for a short-time along $r = 1$ in the above cold spot problem could be of unusual type such as $0(t^{1/4})$ (similar results were obtained in [63]) but the growth the of free boundary along the interior is of well-known type such as of the order of $t^{1/2}$ or t or $t^{3/2}$.

Although class II and III problems appear to be interesting and distinct from class I problems there are only some short-time analytical solutions to them in the literature. There is no report available on the existence, uniqueness and regularity of the solutions of these problems.

3.2.3 Classical Stefan problems with n-phases, $n > 2$

Some one-dimensional problems with more than two phases

Classical Stefan problems with more than two phases are much more difficult to study than two-phase problems because of interactions among phases. Several types of generalizations of two-phase formulations to n-phase problems are possible. We shall report here some of the formulations studied in the literature. In principle, phase-change boundaries could be intersecting but such formulations have not be studied in detail. Some of these n-phase formulations are simple extensions of two-phase problems reported earlier in §§ 1.3 and 3.2.1. In an n-phase problem, it is difficult to prescribe the exact number of distinct phase-change boundaries without knowing the number of disjoint subregions and the physical situation. In several problems this has been taken to be $(n - 1)$. In each phase a suitable parabolic heat equation is satisfied and at each phase-change boundary, two boundary conditions are prescribed (cf. § 1.3)

The formulation of an n-phase one-dimensional problem in which the boundary conditions at the phase-change boundaries could be of type (1.3.5) and (1.3.6) has been considered in [64]. Odd and even numbered phases are identical so that there are only two different initial temperatures and two-different temperatures. The existence of a global solution to this problem which is nearly classical (the Stefan condition is satisfied in the integrated form) has been discussed in [64]. An analytical solution to a one-dimensional n-phase solidification problem in the region $0 \leq x < \infty$ has been discussed in [65]. At each free boundary, the temperature has a specified, constant, limiting value from each side. These limiting values may differ by a finite jump at different phase-change boundaries. Initially all the free boundaries coincide at $x = 0$ and the temperature takes the constant value T_m for $x > 0$ where T_m is the limiting value of the temperature as

$x \to \infty$. The Stefan condition is considered at the free boundaries. Densities of different phases could be different, giving rise to the movement of phases. Using local coordinates which are fixed in each phase, a similarity solution is presented.

The formulation of the one-dimensional multi-phase problem reported in [21] is a little different. At time $t = 0$, an unbounded plate of thickness $2d$ and temperature T_0 is dipped into a melt maintained at constant temperature $T_1 > T_0$. It is assumed that a known constant heat flux q is maintained from within the melt to the plate and in the 'boundary layer' the temperature changes linearly in the direction normal to the surface of the plate. Let T_k be the freezing temperature of the melt and T_n, the melting temperature of the plate. Depending on the magnitudes of different temperatures, the following three cases arise.

Case I

$$T_0 < T_k < T_1 < T_n. \tag{3.2.30}$$

Since $T_0 < T_k$, crust forms on the plate till some time $t_0 > 0$ after which the crust begins to melt as the temperature at the surface of the plate rises to T_k and at the free boundary heat flux q is acting. Melting will continue till time $t_1 > t_0 > 0$ when the crust has completely melted. After that, heating of the plate without a phase-change continues. The plate cannot melt because the melting temperature T_n of the plate is greater than T_1. For $0 \leq t \leq t_1$, there exists a two-phase problem with only one free boundary and for $t > t_1$ the problem is that of heat conduction in the plate without phase-change. Because of the assumption of symmetry it is sufficient to consider the region $0 \leq x \leq \infty$.

Case II

$$T_0 < T_k < T_n < T_1. \tag{3.2.31}$$

At any given time there will be only one phase-change boundary. The freezing temperature of the melt is lower than the melting temperature of the plate. First the crust forms and then the crust starts melting and when the plate attains temperature T_n, the plate starts melting.

Case III

$$T_0 < T_n < T_k < T_1. \tag{3.2.32}$$

There are three possibilities.

(P1) The plate melts and the melt crystallizes instantaneously with the immersion of the plate in the melt.

(P2) The melt crystallizes instantaneously but the melting of the plate starts late.

(P3) The plate begins to melt instantaneously but the crystallization of the melt is delayed.

Problem (P1) corresponding to the possibility (P1) can be formulated as (not made dimensionless) follows.

$$a^2(x)\frac{\partial^2 T_i}{\partial x^2} = \frac{\partial T_i}{\partial t}, \; i = 1, 2, 3, \; x \neq S_1(t), \; x \neq d; \; x \neq S_2(t), \; t > 0, \tag{3.2.33}$$

$$\begin{aligned} a^2(x) &= k_1^2, \; 0 < x < S_2(t), \\ &= k_2^2, \; S_2(t) < x < d, \\ &= k_3^2, \; d < x < S_1(t). \end{aligned} \tag{3.2.34}$$

Here, $x = S_1(t)$ is the equation of the freezing front in the solidification of the melt and $x = S_2(t)$ is the equation of the melting front in the melting of the plate. The subscripts 1, 2, and 3 refer to the regions $0 < x < S_2(t)$, $S_2(t) < x < d$ and $d < x < S_1(t)$, respectively.

$$\left.\frac{\partial T_1}{\partial x}\right|_{x=0} = 0; \; T_1|_{x=S_2(t)} = T_n = T_2|_{x=S_2(t)}; \; t > 0, \tag{3.2.35}$$

$$T_3|_{x=S_1(t)} = T_k; \; t > 0; \; T_1|_{t=0} = T_0, \; 0 < x < d, \tag{3.2.36}$$

$$T_2|_{x=d-0} = T_3|_{x=d+0}; \; K_2 \left.\frac{\partial T_2}{\partial x}\right|_{x=d-0} = K_3 \left.\frac{\partial T_3}{\partial x}\right|_{x=d+0}; \; t > 0, \tag{3.2.37}$$

$$K_1\frac{\partial T_1}{\partial x} - K_2\frac{\partial T_2}{\partial x} = \rho_1 l_1 \dot{S}_2(t), \; \text{for } x = S_2(t), \; t > 0; \; S_2(0) = d, \tag{3.2.38}$$

$$-q + K_3\frac{\partial T_3}{\partial x} = \rho_3 l_3 \dot{S}_1(t), \; \text{for } x = S_1(t), \; t > 0; \; S_1(0) = d. \tag{3.2.39}$$

The motion of the liquid due to the difference in densities has been neglected. Problems (P2) and (P3) can be formulated on the same lines as Problem (P1), but the time interval $t > 0$ has to be divided into several ones and in each interval an appropriate problem is to be formulated. For example in Problem (P2), if the melting of the plate begins at $t = t_0$ then for $0 \leq t < t_0$, we have a problem of pure heat conduction in the plate and crust formation in the melt. For $t_0 < t < t_1$, where t_1 is the time at which the temperature at $x = d$ becomes T_k, there will be both melting of the plate and freezing of the melt. For $t > t_1$, crust starts melting and melting of the plate continues or plate might have completely melted by that time. At each stage, temperature and the position of the free boundary / boundaries in the previous stage are to be ascertained.

In [21] the main interest in the study of Case III is to examine the possibilities of occurrences of three cases (P1), (P2) and (P3). This requires short-time ($t \to 0$)

analytical solution of Problem (P1) which was obtained by using fundamental solutions of the heat equation for a double layer and taking the limit as $t \to 0$. Both necessary and sufficient conditions for the occurrence of the case (P1) have been obtained in [21]. The solution to Problem (P3) will exist if $S_1(t) < d$ for $t > 0$ and therefore Problem (P3) does not have a solution. It has also been shown in [21] that the conditions under which both (P1) and (P2) are possible cannot coincide.

A three-phase problem with two free boundaries has been discussed in [66] which is concerned with both melting and evaporation. Consider a solid occupying the region $0 \leq x \leq a$ (one-dimensional problem). The boundary $x = 0$ is insulated and the solid is heated at $x = a$. First the solid melts at $x = a$ and for some time, solid and liquid regions separated by a free boundary occupy the region $0 \leq x \leq a$. With further heating, when vaporization temperature is attained at $x = a$, the liquid starts evaporating. There will now be two phase-change boundaries, viz., liquid-vapour and solid-liquid. If the solid is enclosed in a container, then heat transfer in the vapour is to be considered and there will be three-phases and two free boundaries. If the vapour is allowed to escape, then there are two phases and two free boundaries. Using finite-difference and finite element methods, numerical solution of this problem has been obtained out by several workers (cf. [66]).

3.2.4 Solidification with transition temperature range

A very pure metal has a fixed melting temperature, which is also its freezing temperature. In the case of alloys or metals with impurities, melting and freezing temperatures are not the same and phase-change takes place over a temperature range. Let freezing and melting temperatures be denoted by T_1 and T_2, respectively. For metals with impurities, both heat and mass transfer should be considered but if the concentration of impurity is small, then only heat transfer can be considered with phase-change taking place over a temperature range $T_1 \leq T \leq T_2$. The region whose temperature lies between T_1 and T_2 is called a mushy region. There are two phase-change boundaries. The solid-mush boundary separates the solid region from the mushy region and the liquid-mush boundary separates the liquid region from the mushy region. In [67] an analytical solution of a one-dimensional solidification problem in cylindrical symmetry with an extended freezing temperature range has been obtained. The finite-difference numerical solution of a one-dimensional solidification problem in a finite slab with an extended freezing temperature range has been presented in [68]. We present below a two-dimensional formulation of an extended freezing temperature range problem whose finite-difference numerical solution is reported in [69].

A two-dimensional region $0 \leq X \leq 1$, $-a \leq Y \leq a$ (X and Y are dimensionless coordinates of a point in a plane) at time $t = 0$ is occupied by a superheated melt. Solidification takes place over a temperature range $T_1 \leq T \leq T_2$. For $T < T_1$, the material is in a stable solid phase and for $T > T_2$, the material is in a stable liquid phase and for $T_1 < T < T_2$, a mushy region exists. Cooling is done at the boundary of the rectangular region in such a way that the mushy region is sandwiched between stable solid and stable liquid regions. It will be assumed that the solid-mush boundary can be expressed in the form $X = R_1(Y,t)$ and the liquid-mush boundary can be expressed in the form $X = R_2(Y,t)$. Without loss of generality it can be assumed that the solidification starts at $X = 0$ at time $t = 0$. Even if there is some delay in the starting of solidification, for example, if the solidification starts at $t = t_0 > 0$, then during the time $0 \leq t < t_0$, there exists only a heat conduction problem without phase change whose numerical solution is generally considered to be simple and so assumed to be known by using well known methods. By redefining the time scale it can be assumed that the solidification starts at $t = 0$.

The mathematical formulation of this three-phase problem in the dimensionless form as considered in [69] is as follows:

In the solid region

$$\frac{\partial T_S}{\partial t} = \alpha_S \left(\frac{\partial^2 T_S}{\partial X^2} + \frac{\partial^2 T_S}{\partial Y^2} \right), \quad 0 < X < R_1(Y,t), \ -a < Y < a; \ t > 0, \tag{3.2.40}$$

$$\left. \frac{\partial T_S}{\partial X} \right|_{X=0} = F_p(Y,t) \ \text{or} \ T_S|_{X=0} = T_p(Y,t); \ t > 0, \tag{3.2.41}$$

$$\left. \frac{\partial T_S}{\partial Y} \right|_{Y=a} = 0, \ 0 < X < R_1(a,t), \ t > 0, \tag{3.2.42}$$

$$\left. \frac{\partial T_S}{\partial Y} \right|_{Y=-a} = 0, \ 0 < X < R_1(-a,t), \ t > 0, \tag{3.2.43}$$

In the mushy region

$$\frac{\partial T_M}{\partial t} = \alpha_M \left(\frac{\partial T_M}{\partial X^2} + \frac{\partial T_M}{\partial Y^2} + \frac{\lambda}{\beta_1} \frac{\partial f_S}{\partial t} \right), \ R_1(Y,t) < X < R_2(Y,t), \ -a < Y < a; \ t > 0, \tag{3.2.44}$$

$$\left. \frac{\partial T_M}{\partial Y} \right|_{Y=a} = 0, \ R_1(a,t) < X < R_2(a,t), \ t > 0, \tag{3.2.45}$$

$$\left. \frac{\partial T_M}{\partial Y} \right|_{Y=-a} = 0, \ R_1(-a,t) < X < R_2(-a,t), \ t > 0, \tag{3.2.46}$$

$$T_M|_{t=0} = g(X,Y) \leq T_{ml}, \ 0 \leq X \leq R_2(Y,0), \ g(0,Y) = 1.0. \tag{3.2.47}$$

In the liquid region

$$\frac{\partial T_L}{\partial t} = \alpha_L \left(\frac{\partial^2 T_L}{\partial X^2} + \frac{\partial^2 T_L}{\partial Y^2} \right), \quad R_2(Y,t) < X < 1,\ t > 0, \tag{3.2.48}$$

$$T_L|_{t=0} = g(X,Y) \geq T_{ml}\,, \quad X \geq R_2(Y,0), \tag{3.2.49}$$

$$\left.\frac{\partial T_L}{\partial X}\right|_{X=1} = 0,\ \ t > 0, \tag{3.2.50}$$

$$\left.\frac{\partial T_L}{\partial Y}\right|_{Y=a} = 0,\ \ R_2(a,t) < X < 1,\ t > 0, \tag{3.2.51}$$

$$\left.\frac{\partial T_L}{\partial Y}\right|_{Y=-a} = 0,\ \ R_2(-a,t) < X < 1,\ \ t > 0. \tag{3.2.52}$$

At the solid-mush boundary $X = R_1(Y,t)$

$$T_S = T_M = 1.0, \tag{3.2.53}$$

$$\left\{1 + \left(\frac{\partial R_1}{\partial Y}\right)^2\right\}\left\{\frac{\partial T_S}{\partial X} - \beta_1 \frac{\partial T_M}{\partial X}\right\} = \lambda d_1 \frac{\partial R_1}{\partial t}. \tag{3.2.54}$$

At the liquid-mush boundary $X = R_2(Y,t)$

$$T_M = T_L = T_{ml}, \tag{3.2.55}$$

$$\left\{1 + \left(\frac{\partial R_2}{\partial Y}\right)^2\right\}\left\{\beta_1 \frac{\partial T_M}{\partial X} - \beta_2 \frac{\partial T_L}{\partial X}\right\} = \lambda d_2 \frac{\partial R_2}{\partial t}, \tag{3.2.56}$$

$$g(R_2(Y,0),Y) = T_{ml} \tag{3.2.57}$$

In the above formulation, d_1 and d_2 are solid fractions present at solid-mush and liquid-mush boundaries, respectively; f_S is the solid fraction in the mushy region, $g(X,Y)$ is the initial temperature/T_1, b is the length of the plate, a is the breadth of the plate/b, T is the temperature/T_1, T_{ml} is the temperature of the liquid-mush boundary/T_1. Other parameters are defined below.

$$\alpha = \text{thermal diffusivity}.\, t_m/b^2,\ \beta_1 = K_M/K_S,\ \beta_2 = K_L/K_S, \lambda = \rho l b^2/(K_S t_m T_1), \tag{3.2.58}$$

t_m is the time taken for the liquid to attain the temperature T_1 at $X = 0$. Densities of all the three regions have been taken to be equal and are denoted by ρ.

The constant d_1 in (3.2.54) is generally taken as unity. But in the case of 'eutectics' it can be taken to be less than unity. At the liquid-mush boundary, $d_2 = 0$. If in (3.2.56), $d_2 = 0$ is taken then we have an implicit free boundary condition which is not convenient for the numerical solution. By taking d_2 very small, an explicit free boundary condition can be generated. In § 3.3 some transformations to convert an implicit free boundary condition to an explicit free boundary condition are given but this treatment may give rise to some other difficulties in the numerical schemes. Although it is not a rigorous mathematical convergence proof, a reasonably accurate solution can be obtained if the numerical results converge as d_2 is gradually decreased. The accuracy of the numerical solution can be further checked by some methods such as integral heat balance calculation (cf. [69]).

The solid fraction in the mush depends on various physical parameters, such as the temperature of the mush, and the width of the mushy region. An exact mathematical expression for the solid fraction in the mush cannot be given and in its absence some approximate mathematical models are proposed. Two such models have been considered in [69] for the numerical solution of the problem.

Model I

$$f_S(X,Y,t) = \{d_1(T_{ml} - T_M(X,Y,t)) - d_2(1 - T_M(X,Y,t))\}/(T_{ml} - 1.0). \quad (3.2.59)$$

Model II

$$f_S(X,Y,t) = [d_1\{R_2(Y,t) - X\} - d_2\{R_1(Y,t) - X\}]/\{R_2(Y,t) - R_1(Y,t)\}. \quad (3.2.60)$$

In the first model, f_S is a linear function of the temperature of the mush and in the second model f_S is a linear function of the 'width' of the mushy region. Thermodynamically, the heat extracted from the system at the fixed boundaries should be equal to the heat given out by the system during solidification. If this balancing of heat is done in an integrated way over the whole region and over a period of time, then it is called *integral heat balance* verification. Integral heat balance is satisfied for model I but not for model II. In the first model we can calculate f_S from the temperature (calculated temperature) of the mush and this procedure is thermodynamically consistent but in the second model corresponding to the calculated f_S, the temperature is calculated in the numerical scheme and therefore integral heat balance is not satisfied.

3.3 Stefan Problems with Implicit Free Boundary Conditions

If in the place of (3.1.12), we have the condition

$$\left.\frac{\partial T}{\partial x}\right|_{x=S(t)} = 0, \quad (3.3.1)$$

then velocity of the free boundary is not explicitly prescribed. As stated earlier, free boundary conditions of the above type in which $\dot{S}(t)$ is not prescribed explicitly are known as implicit free boundary conditions. Free boundary condition (3.1.12) is an explicit boundary condition as $\dot{S}(t)$ is prescribed in it. Our main aim in this section is to present some transformations which convert a Stefan problem with an implicit free boundary condition to a Stefan problem with an explicit free boundary condition. An extensively studied problem with an implicit free boundary condition known as oxygen-diffusion problem will also be discussed later in detail.

3.3.1 Schatz transformations and implicit free boundary conditions

Problem 3.3.1. Find t_*, $T(x,t)$ and $S(t)$ satisfying the following system of dimensionless equations

$$T_{xx} - T_t = F(x,t), \quad a < x < S(t),\ 0 < t < t_*, \tag{3.3.2}$$

$$\alpha T_x(a,t) + \beta T(a,t) = f(t), \quad 0 < t < t_*, \tag{3.3.3}$$

$$T(x,0) = \phi(x), \quad -\infty < a \le x \le b = S(0), \tag{3.3.4}$$

$$T(S(t),t) = g(S(t),t),\ 0 < t < t_*, \tag{3.3.5}$$

$$\gamma(S(t),t)\dot{S}(t) = -T_x(S(t),t) + \hat{h}(S(t),t);\ \gamma \ne 0,\ 0 < t < t_*. \tag{3.3.6}$$

All the thermophysical parameters in this one-phase problem have been taken to be unity. This is possible by suitably choosing time and/ or length scales. In those Stefan problems in which the effect of thermophysical parameters is not to be investigated, it is convenient to take parameters to be unity. α and β are constants. Depending on the nature of the data, the problem (3.3.2)–(3.3.6) could be either a melting problem or a solidification problem. We report here some of the assumptions and for complete details see [70].

(i) $F(x,t)$, $g(x,t)$, $\hat{h}(x,t)$ and $F_x \in C$ for $a \le x < \infty$, $0 < t < t_*$.

(ii) $f(t) \in C$, $0 \le t < t_*$.

(iii) If $a < b$, then $\phi \in C^1$ for $a \le x \le b$ and $\phi(b) = g(b,0)$, $\phi'(b) = \hat{h}(b,0)$. These are compatibility conditions to be satisfied by the initial temperature and the functions g and $\hat{h}$ at the free boundary.

(iv) In addition to the regularity conditions to be satisfied by $S(t)$ and $T(x,t)$ in the classical solution of a Stefan problem mentioned in § 1.4.6 it will be assumed that

T_{xxx}, $T_{xt} \in C$ for $a < x < S(t)$, $0 < t < t_*$. The significance of this assumption will be made clear later.

Problem 3.3.2. Consider problem (3.3.2)–(3.3.6) with some changes. Let $\gamma(S(t), t) = 0$ in (3.3.6), so that

$$T_x(S(t), t) = \hat{h}(S(t), t). \tag{3.3.7}$$

Make some additional assumptions given below.

$$g_x(x, t) \neq \hat{h}(x, t), \quad a < x < \infty, \quad 0 < t < t_*, \tag{3.3.8}$$

and

$$\alpha = 1, \quad \beta = 0. \tag{3.3.9}$$

Problem 3.3.3. Let $\gamma(x, t) = g(x, t) = \hat{h}(x, t) = 0$ and $\alpha = 0$, $\beta = 1$ in Problem 3.3.1. In this case the free boundary conditions are given by

$$T(S(t), t) = 0, \quad 0 < t < t_*, \tag{3.3.10}$$

$$T_x(S(t), t) = 0, \quad 0 < t < t_* \tag{3.3.11}$$

The equivalence of Problems 3.3.1.–3.3.3. is established by the following propositions.

Proposition 3.3.1. If (S, T) is the solution of Problem 3.3.2., then (S, v) where $v = T_x$ and S is the same as in (S, T), is the solution of the following Stefan problem.

$$v_{xx} - v_t = F_x(x, t), \quad 0 < x < S(t), \quad 0 < t < t_*, \tag{3.3.12}$$

$$v(a, t) = f(t), \quad 0 < t < t_*, \tag{3.3.13}$$

$$v(x, 0) = \phi'(x), \quad a \leq x \leq b = S(0), \tag{3.3.14}$$

$$v(S(t), t) = \hat{h}(S(t), t), \quad 0 < t < t_*, \tag{3.3.15}$$

$$[\hat{h}(S(t), t) - g_x(S(t), t)]\dot{S}(t) = -v_x(S(t), t)) + F(S(t), t) + g_t(S(t), t), \; 0 < t < t_*. \tag{3.3.16}$$

Equation (3.3.12) can be easily derived if the partial derivative of (3.3.2) with respect to x is taken. If the material time derivative of (3.3.5) is taken, then we obtain

$$T_x(S(t), t)\dot{S}(t) + T_t = g_x(S(t), t)\dot{S} + g_t. \tag{3.3.17}$$

On using (3.3.7) and (3.3.2) in (3.3.17), the equation (3.3.16) can be obtained. Equation (3.3.17) suggests that we impose the condition $g \in C^1$, $b \leq x < \infty$. Other conditions in (3.3.13)–(3.3.15) can be easily derived.

Proposition 3.3.2. Conversely, if the pair (S, v) is the solution of (3.3.12)–(3.3.16) then the pair (S, T) where S is the same as in (S, v) and $T(x,t)$ is defined by

$$T(x,t) = \int_x^{S(t)} v(p,t)dp + g(S(t),t),\ a \le x \le S(t),\ 0 < t < t_*, \tag{3.3.18}$$

is the solution of Problem 3.3.2.

On repeated differentiations of (3.3.18) with respect to x, we get

$$T_x = v(x,t) \text{ and } T_{xx} = v_x. \tag{3.3.19}$$

Differentiation of (3.3.18), with respect to time gives

$$T_t = -\int_x^{S(t)} v_t(p,t)dp - v(S(t),t)\dot{S}(t) + g_x(S(t),t)\dot{S} + g_t(S(t),t). \tag{3.3.20}$$

Also

$$\int_x^{S(t)} v_t(p,t)dp = \int_x^{S(t)} (v_{xx} - F_x)dp = v_x(S(t),t) - v_x(x,t) - F(S(t),t) + F(x,t). \tag{3.3.21}$$

In order to obtain (3.3.2) from the solution of (3.3.12)–(3.3.16), we use (3.3.21) in (3.3.20) and then substitute v_x from (3.3.16). Derivation of other conditions in Problem 3.3.1. is straightforward. Even if $\alpha = 0$ and $\beta = 1$ in (3.3.3) the substitution $v = T_x$ works. Differentiation of (3.3.3) ($\alpha = 0$) with respect to t gives

$$T_t(a,t) = T_{xx}(a,t) - F(a,t) = f'(t) \text{ or } v_x(a,t) = f'(t) + F(a,t). \tag{3.3.22}$$

In this case it has to be assumed that T_t and T_{xx} are continuous at $x = a$ and $f(t) \in C^1$. We conclude that Problem 3.3.1. and the problem defined by equations (3.3.12)–(3.3.16) are equivalent. Equation (3.3.16) is an explicit free boundary condition.

Proposition 3.3.3. If (S, T) is the solution of Problem 3.3.3., then (S, v), where $v = T_t$, is the solution of the following problem.

$$v_{xx} - v_t = F_t(x,t),\ \ 0 < x < S(t),\ 0 < t < t_*, \tag{3.3.23}$$

$$v(a,t) = f'(t),\ \ 0 < t < t_*, \tag{3.3.24}$$

$$v(x,0) = \phi''(x) - F(x,0),\ \ a \le x \le b = S(0), \tag{3.3.25}$$

$$v(S(t),t) = 0,\ \ \ 0 < t < t_*, \tag{3.3.26}$$

$$F(S(t),t)\dot{S} = -v_x(S(t),t),\ 0 < t < t_*. \tag{3.3.27}$$

Equations (3.3.23)–(3.3.25) suggest that we impose the conditions that F_t and $f'(t) \in C$ and $\phi(x) \in C^2$.

The derivation of equations (3.3.23)–(3.3.26) is simple. To obtain (3.3.27), differentiate (3.3.11) with respect to t and use (3.3.2).

Proposition 3.3.4. Conversely, if (S, v) is the solution of (3.3.23)–(3.3.27), then (S, T) where S is the same as in (S, v), $\dot{S}$ is of one sign, and $T(x, t)$ is given by

$$T(x,t) = \int_x^{S(t)} \int_\eta^{S(t)} \{v(p,t) + F(p,t)\} dp d\xi, \ a \leq x \leq S(t), \ 0 < t < t_*, \tag{3.3.28}$$

is a solution of Problem 3.3.3.

To derive (3.3.2), differentiate (3.3.28) twice with respect to x so that

$$\frac{\partial T}{\partial x} = -\int_x^{S(t)} \{v(p,t) + F(p,t)\} dp \tag{3.3.29}$$

and

$$\frac{\partial^2 T}{\partial x^2} = v(x,t) + F(x,t) = \frac{\partial T}{\partial t} + F(x,t). \tag{3.3.30}$$

On substituting $x = S(t)$ in (3.3.28) and (3.3.29), we obtain (3.3.10) and (3.3.11). If it is assumed that $f(0) = \phi(a)$ and $\phi(b) = 0$ then (3.3.3) and (3.3.4) can be obtained.

Although our main concern in this section is to show how the transformations $v = T_x$ and $v = T_t$ transform Stefan problems with implicit free boundary conditions to Stefan problems with explicit free boundary conditions, some remarks on the analysis of Problems 3.3.1.–3.3.3. will be in order (see also Chapter 10). If T_x is continuous in $a \leq x \leq S(t)$, then the relation $\dot{S}(t) = -T_x(S(t), t)$ (for simplicity take $\gamma = 1$ and $\hat{h} = 0$ in (3.3.6)) implies that $\dot{S}(t)$ is continuous. This is true for the more general boundary condition (3.3.16) also if continuity assumptions are made for other functions involved. If the free boundary condition is of the form (3.3.7) or (3.3.11), then even if T_x is continuous, it cannot be directly concluded that $\dot{S}(t)$ is continuous.

Existence of unique solutions of Problems 3.3.2. and 3.3.3. has been discussed in [70] under suitable data assumptions. If $a < b$ and the data satisfy appropriate assumptions, then it can be proved that $0 \leq \dot{S}(t) \leq A$, $0 < t \leq t_0 < \infty$ for some constant A (see Proposition 10.1.20.). The method of proof is the same as in [71]. In the place of (3.3.2) a quasi-linear heat equation with some constraints can also be considered.

Conversion of a Stefan-type problem to a Stefan problem

We consider a one-dimensional Stefan-type problem with phases 1 and 2 in which phase 1 occupies the region $0 \leq x < S(t)$ and phase 2 occupies the region $S(t) < x \leq 1$. This problem differs from a Stefan problem only in the free boundary conditions. Let $T^{(1)}$

and $T^{(2)}$ be temperatures of phases 1 and 2, respectively. $T_x^{(1)}$ and $T_x^{(2)}$ are prescribed on $x = S(t)$ and another boundary condition on $S(t)$ is given in the form of a relation

$$\dot{S} = f(T^{(1)}(S(t),t),\ T^{(2)}(S(t),t)). \tag{3.3.31}$$

If the transformations $v^{(1)} = T_x^{(1)}$ and $v^{(2)} = T_x^{(2)}$ are used in the Stefan-type problem (equation (3.3.31) is not a Stefan condition) mentioned above and the problem is formulated in terms of $v^{(1)}(x,t)$ and $v^{(2)}(x,t)$, then at the free boundary $v^{(1)}$ and $v^{(2)}$ will be prescribed. If the heat equations in the two phases are of the form (3.3.2), then

$$T^{(i)}(S(t),t) = \int_0^t v_x^{(i)}(S(\tau),\tau)d\tau - \int_0^t F^{(i)}(S(\tau),\tau)d\tau,\ i = 1,2. \tag{3.3.32}$$

In view of (3.3.32), the equation (3.3.31) can be written as

$$\dot{S} = P_t(v_x^{(1)}, v_x^{(2)}, S(t)), \tag{3.3.33}$$

where for any $t \in [0,t_*]$, P_t is a functional (a real valued function) acting on functions $S(t)$, $v_x^{(1)}(x,t)$, $v_x^{(2)}(x,t)$, $x \in [0,S(t)]$ in $v_x^{(1)}(x,t)$ and $x \in [S(t),1]$ in $v_x^{(2)}(x,t)$ and $0 < t \le t_* < \infty$. As explained earlier, the Stefan condition (3.3.33) has been considered in the functional form and not as a heat balance condition.

3.3.2 Unconstrained and constrained oxygen-diffusion problem (ODP)

We shall first discuss a one-dimensional oxygen-diffusion problem, and use a shorter notation ODP for it. This problem was first formulated in [72] and studied later by several authors from various view points such as the existence and uniqueness, analytical and numerical solutions. Oxygen is fed to a tissue at the boundary $x = 0$ at which a constant concentration c_0 of oxygen is maintained. It is assumed that oxygen diffuses through the tissue and is absorbed at a constant rate α per unit volume. After some time, a steady state is reached. Suppose that in the steady state oxygen has penetrated upto a distance x_0 in the tissue. Then, at $x = x_0$, both the oxygen concentration and the flux are zero. Steady-state concentration and unknown x_0 can be easily obtained. If suitable dimensionalization is carried out as in [72], then the steady-state concentration can be obtained as $0.5(1-x)^2$, $0 \le x \le 1$. Once the steady state is reached, the boundary $x = 0$ is sealed. Oxygen diffusion and absorption starts again and the penetration depth of oxygen starts receding giving rise to a free boundary problem whose dimensionless formulation is given below.

$$\frac{\partial c}{\partial t} = \frac{\partial^2 c}{\partial x^2} - 1, \quad \text{in } D_{t_*} = \{(x,t) : 0 < x < S(t),\ 0 < t < t_* < \infty\}, \tag{3.3.34}$$

$$c(x,0) = 0.5(1-x)^2,\ 0 \le x \le 1, \tag{3.3.35}$$

$$\left.\frac{\partial c}{\partial x}\right|_{x=0} = 0, \tag{3.3.36}$$

$$c(S(t),t) = 0, \quad 0 < t < t_*; \;\; S(0) = 1, \tag{3.3.37}$$

$$\frac{\partial c}{\partial x}(S(t),t) = 0, \quad 0 < t < t_*. \tag{3.3.38}$$

Here, $c(x,t)$ is the concentration and $x = S(t)$ is the free boundary. It may be noted that this problem is similar to Problem 3.3.3. and with the help of the transformation $T = c_t$, it can be converted to a Stefan problem with an explicit free boundary condition. A typical feature of this problem is the non-compatibility of the initial and boundary data at $x = 0$. Since concentration is a physical quantity, $c(x,t)$ should be greater than or equal to zero. It can be proved rigorously that the classical solution (c, S) of (3.3.34)–(3.3.38) exists with $c(x,t) \geq 0$, $0 \leq x \leq S(t)$, $0 \leq t \leq t_*$. If the substitution $T = c_x(x,t)$ is used in (3.3.34)–(3.3.38), then by using the *maximum principle* [9] it can be easily concluded that the maximum value of $T(x,t)$ is zero and therefore $c_x(x,t) < 0$ for $0 < x < S(t)$. This implies $c > 0$ for $0 < x < S(t)$ or $c \geq 0$ for $0 \leq x \leq S(t)$. We shall discuss the existence of the solution of ODP a little later. It may be pointed out here that ODP is closely related to the supercooled Stefan problem (SSP) formulated in (3.1.8)–(3.1.12).

It can be easily checked that if the transformation $T = c_t$ is used in ODP then SSP formulation (see § 3.1.2) is obtained. If the concentration is defined by the relation

$$c(x,t) = \int_{S(t)}^{x} \int_{S(t)}^{\xi} (T(\eta,t) + 1) d\eta d\xi, \tag{3.3.39}$$

then ODP can also be obtained from SSP. Because of this connection between ODP and SSP, ODP can be regarded as a Stefan problem with an implicit free boundary condition even though ODP is concerned with the diffusion of mass and not with the heat transfer. The existence and uniqueness of ODP has been discussed in [73, 74, 75]. In [73], the existence and uniqueness of the solution of ODP was in essence proved by extending the domain of $c(x,t)$ to the interval $0 \leq x \leq 1$ and taking $c = 0$ for $S(t) < x \leq 1$. The solution of this extended domain problem was obtained as the limit of a one-parameter family of problems. If a constraint $c \geq 0$ is added in the extended domain ODP formulation in $0 \leq x \leq 1$, then it can be identified with an obstacle problem (see Chapter 7) whose variational inequality formulation exists. The existence and uniqueness of this obstacle problem has been proved in [74]. It has been proved that the obstacle problem never exhibits *blow-up*, in the sense that either (i) $c(x,t) \geq 0$, $c(x,t) \not\equiv 0$ for all $t > 0$ or, (ii) $c(x,t) \equiv 0$ for t greater than some finite time $\hat{t}$ (extinction time). In [75], the existence and uniqueness was proved using fixed point arguments. It may be noted that if the initial concentration is given by (3.3.35), then it is not necessary to add the constraint $c \geq 0$ to the problem formulation of ODP as in this case it has been proved that the unique solution exists and $c \geq 0$.

In the place of (3.3.35), let us take the initial concentration as

$$c(x,0) = g(x), \ \ 0 \leq x \leq 1, \ \ g \not\equiv 0. \tag{3.3.40}$$

We shall still call this changed problem as ODP.

If it is assumed that $g(x)$ is non-negative, non-increasing, sufficiently regular such as $g \in C^2$, g'' is Hölder-continuous at $x = 1$ and $g(x)$ satisfies compatibility conditions $g'(0) = 0$, $g(1) = g'(1) = 0$, $g''(1) = 1$, then ODP with (3.3.40) in the place of (3.3.35) possesses a very smooth solution with $c \geq 0$ (cf. [75]). It is interesting to note that if $g(x) = 0.5(1-x)^2$ then $g'(0) \neq 0$ but still the existence and uniqueness of ODP can be established [75].

Constrained and unconstrained ODP

We shall now consider ODP with initial concentration $\hat{g}(x)$ where $\hat{g}(x)$ can have any sign for $x \in [0,1]$. Add the constraint $c \geq 0$ in the formulation. This problem will be called *constrained oxygen diffusion problem* or in short CODP. If the constraint $c \geq 0$ is not imposed in the formulation, then we have an *unconstrained ODP* or in short UODP. CODP is equivalent to solving the following equation with a suitable initial and boundary data.

$$\frac{\partial c}{\partial t} = \frac{\partial^2 c}{\partial x^2} - H_v(c), \tag{3.3.41}$$

where H_v is the Heaviside function.

$$\left.\begin{aligned} H_v(c) &= 1, \quad c > 0 \\ &= 0, \quad c \leq 0. \end{aligned}\right\} \tag{3.3.42}$$

The solution of CODP and UODP coincide as long as $c \geq 0$. The solution of CODP generally exists, is unique, and remains well behaved even for a sufficiently long time. The global behaviour of UODP is a delicate matter as a *blow-up* may occur in which $\dot{S}$ becomes infinite in finite time. Blow-up will be dealt with further in § 4.4. The solution $T(x,t)$ of SSP is equal to c_t where $c(x,t)$ is the solution of UODP but in general $T(x,t)$ is not equal to $\hat{c}_t$ where $\hat{c}(x,t)$ is the solution of CODP. The non-consideration of the negativity set of $c(x,t)$ in CODP deprives its solution of many interesting features which the solutions of UODP have, for example, the approach of the free boundary to a negativity set [75].

ODP in a radially symmetric domain.

The existence of the solution of ODP in a cylindrical domain $r_0 \leq r \leq r_1$ was considered in [76]. ODP in cylindrical symmetry can be easily formulated and this formulation is not being given here. The steady-state concentration $c_0(r)$ in cylindrical symmetry can be obtained as

$$c_0(r) = \frac{r^2 - r_1^2}{4} - \frac{r_1^2}{2}\ln\frac{r}{r_1}, \tag{3.3.43}$$

where r_1 solves the equation

$$\frac{r_0^2 - r_1^2}{4} - \frac{r_1^2}{2}\ln\frac{r_0}{r_1} = c_1, \; c_1 = c_0(r_0). \tag{3.3.44}$$

Here, c_1 and r_0 are given, r_0 is the inner radius of the cylinder. A unique root $r_1 > r_0$ of (3.3.44) can be obtained. The existence of a global solution has been proved by extending the existence of the local-in-time solution. The proof of the results in the cylindrical geometry are not on the same lines as proofs developed for ODP given in (3.3.34)–(3.3.38). It has been proved in [76] that under suitable assumptions there exists a t_* such that $S(t_*) = r_0$.

Quasi-static two-dimensional ODP and the Hele-Shaw problem

A link between *quasi-static two-dimensional ODP* and the *Hele-Shaw problem* has been discussed in [77]. Let $p(x, y, t)$, where t is a parameter, be the pressure in the well-posed Hele-Shaw problem in which the free boundary is blown outward. The pressure in the liquid satisfies the equation

$$\nabla^2 p = 0 \text{ in } \Omega \subset R^2, \tag{3.3.45}$$

Let $\partial\Omega$, the boundary of the simply connected region Ω, be a free boundary.

On $\partial\Omega$, we have

$$p = 0, \; \frac{\partial p}{\partial n} = -V_n \text{ or } \nabla p \cdot \nabla w = -1, \text{ on } t = w(x, y), \tag{3.3.46}$$

where $t = w(x, y)$ is the equation of the free boundary and $\vec{n}$ is the unit outward normal to the free boundary.

Equation (3.3.45) can be obtained by taking $\vec{V} = -\text{grad} p$ in the equation of continuity of an incompressible fluid whose flow is considered in a narrow channel [78, 79]. Free boundary is the surface of an expanding fluid blob. The continuity of pressure at the free boundary after appropriate scaling gives $p = 0$. The steady-state diffusion equation for concentration and the boundary conditions at the free boundary in a two-dimensional ODP can be written as follows.

$$\nabla^2 c = 1, \text{ in } \Omega \subset R^2, \tag{3.3.47}$$

$$c = \frac{\partial c}{\partial n} = 0 \text{ on } t = w(x, y). \tag{3.3.48}$$

The Hele-Shaw problem (3.3.45)–(3.3.46) can be converted to a steady-state two-dimensional COD described in (3.3.47)–(3.3.48) by using the transformation given in (6.2.25).

The problem defined in (3.3.47)–(3.3.48) is generally well-posed and so is the Hele-Shaw problem in which the boundary blows outward [77]. However the Hele-Shaw 'suction problem' is an *ill-posed problem* and the quasi-static ODP has recently been found to be a very effective tool for revealing some unexpected regularity properties of the free boundary in the solution of unstable Hele-Shaw suction problem [77].

Chapter 4

Stefan Problem with Supercooling: Classical Formulation and Analysis

4.1 Introduction

In §§ 2.2.2 and 2.2.3, supercooling in the melt was taken into account by redefining the equilibrium phase-change temperature with the help of Gibbs-Thomson relation and the kinetic condition. When both these effects are considered together, we shall call the resulting equilibrium temperature relation as *modified Gibbs-Thomson relation.* In this case, the new equilibrium temperature denoted by T_e is given by

$$T_e(x,t) = T_m - (\sigma/[\hat{S}]_E)\bar{K}_c - \alpha\sigma|V_n|/[\hat{S}]_E\,. \tag{4.1.1}$$

The symbols T_m, σ, $\hat{S}$, $\vec{K}_c$ and V_n have the same meaning as explained in the earlier sections; the parameter α is related to the microscopic relaxation time, $[\hat{S}]_E$= difference in the entropies per unit volume of the liquid and the solid at the equilibrium temperature. The last two terms in (4.1.1) are the correction terms arising due to the curvature and the velocity of the interface, respectively. Equation (4.2.11) deals with the units associated with σ, $\hat{S}$ and α. Has the solidification of a supercooled liquid been correctly and satisfactorily modelled by the equilibrium temperature in equation (4.1.1)? To answer this question we have to go back to the formation of solid from a supercooled liquid in the solidification process. It may be noted that (4.1.1) was discussed earlier also with the help of some thermodynamical and metallurgical considerations. A rigorous mathematical justification will be provided now.

The origin of solidification of a supercooled liquid of a pure substance is not in the growth of a sharp interface separating solid and liquid phases but in the finite size effect of the interface lying between stable solid and stable liquid phases as shown in Fig. 4.1.1. In other words, it can be said that the solidification of a supercooled liquid cannot progress without the formation of a mushy region in front of the solidification front. The mushy region contains both solid and liquid phases and solid could be present in it in the form of dendrites of different types. Therefore it is clear that any mathematical modelling of the solidification of a supercooled liquid should take into account solid, liquid and mushy regions, i.e., the finite-size effect of the interface. Does it mean that the classical formulation of the Stefan problem with supercooling accounted for by the Gibbs -Thomson law is not justified as there is no mushy region in the formulation? The rigorous mathematical justification of the sharp freezing front model (classical formulation) with equilibrium temperature given by (4.1.1) comes from the fact that the modified Gibbs-Thomson relation can be obtained if appropriate limits of the *phase-field model* which considers the finite width of the interface, are taken. Some phase-field models are being discussed here with this objective.

What should be the starting point in the modelling of solidification of a supercooled liquid? Research workers have used different approaches giving rise to different models. The differences arose due to the incapability of some models to predict accurately some physical situations or non-acceptability of some models due to the so-called inconsistencies. The existing literature on supercooling is vast. Some selected approaches in the modelling of solidification of a supercooled liquid are being presented here. It is not possible to describe here even these selected models completely. What is presented below is a suitable compromise between length restrictions and understability of these models.

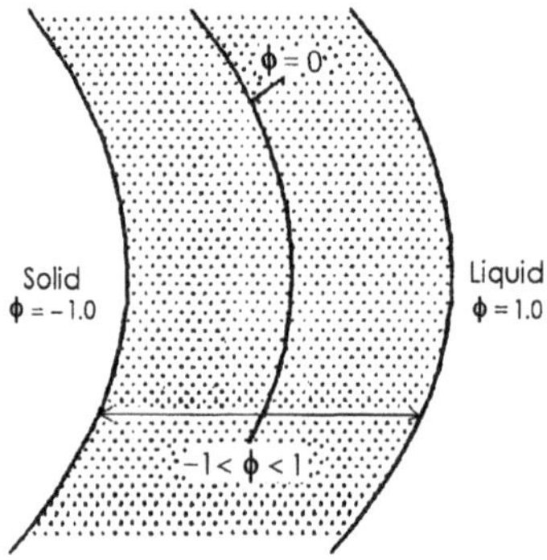

Fig. 4.1.1. Finite size of the interface

4.2 A Phase-field Model for Solidification using Landau-Ginzburg Free Energy Functional

If the interface in the solidification of a supercooled liquid is assumed to be of finite-width, then the modified Gibbs-Thomson relation would imply that there is a sharp interface which is continuously progressing through the finite-width interface. This in turn suggests exploring the enthalpy formulation given below in (4.2.1) in which a phase function $\phi(x,t)$ has been introduced which is continuously varying through the finite-width interface. Enthalpy $H(T)$ (per unit volume) is defined by

$$\begin{aligned} H(T) &= \rho_S C_S T_S + \rho_S \frac{l}{2}\phi, \quad T<0, \quad \phi=-1, \\ &= [-\rho_S l/2, \ \rho_L l/2], \quad T=0, \quad -1<\phi<1, \\ &= \rho_L C_L T_L + \rho_L \frac{l}{2}\phi, \quad T>0, \quad \phi=1. \end{aligned} \tag{4.2.1}$$

Here, ϕ is called a *phase function* or an *order parameter* which in the classical formulation has the value -1 in the liquid phase and 1 in the solid phase. Some authors take $\phi=1$ and $\phi=0$ in liquid and solid phases, respectively, and take l in the place of $l/2$. The phase function ϕ defines the state of the system. The phase-change temperature has been taken to be zero. $H(T)$ vs T has been plotted in Fig. 4.2.1. Enthalpy is a multi-valued function of temperature but the temperature is a continuous function of enthalpy. The equation for energy in terms of enthalpy is given by

$$\frac{\partial H}{\partial t} = \nabla^2(KT), \quad x \in \Omega \subset R^n, \ 0<t<t_*. \tag{4.2.2}$$

Equation (4.2.2) is valid only in the distributional sense as at $T=0$ the derivative of H is not defined (see § 5.2). K can have different values in different phases. It is clear from (4.2.1) that a sharp interface cannot be identified by $T=0$ as done in the classical formulation because at $T=0$ there exists a region and not a sharp interface. However a sharp interface can be identified by $\phi=0$ as ϕ is a continuous variable and $\phi=0$ corresponds to the center of the mushy region. An equation for $\phi(x,t)$ is needed. If ϕ is considered to be a continuous function of x and t, then it will look as in Fig. 4.2.2. When $\phi=0$, T is also zero. $T=0$ does not imply a sharp interface in the enthalpy model but $\phi=0$ does.

The starting point of a phase-field continuum model to be discussed below for the solid-liquid phase-change process has been taken here as the so-called Landau-Ginzburg free energy functional F defined as [80]

$$F\{\phi(x)\} = \int_\Omega \left\{ \frac{\xi^2}{2}(\nabla\phi)^2 + \frac{1}{8a} f(\phi(x,t)) - \frac{[\hat{S}]_E}{2}(T-T_m)\phi \right\} d\Omega. \tag{4.2.3}$$

Here, ξ is a length scale associated with the 'microscopic interaction strength' (ξ is the length such that two particles at a distance ξ apart will have the probability 1/2 of being in the same phase or state). a^{-1} is the measure of the depth of the *double-well* and is the indication of the barrier which must be crossed in the transition between the phases, $a^{1/2}\xi$ is the 'correlation length', $[\hat{S}]_E$ is the entropy difference between the phases at the equilibrium temperature per unit volume. All the terms in the integrand of (4.2.3) are in the units of energy per unit volume. $\Omega \subset R^n$, $n \geq 1$ is a fixed region in space in which the phase-change process is taking place. In the phase transition problems with two distinct phases such as solid and liquid, f is a symmetric double-well function of ϕ with two distinct minima, one for each phase. The last two terms in the integrand in (4.2.3) can be thought of as the 'local Helmholtz free energy density' and $(\nabla\phi)^2$ is the 'basic interaction term' arising due to the interaction of atoms with a mean field created by other atoms [81]. The minimum of $F(\phi)$ is attained at that value of ϕ which satisfies the following *Euler-Lagrange equation* derived under the condition that either ϕ is prescribed or $\nabla\phi \cdot \vec{n} = 0$ on the boundary of Ω.

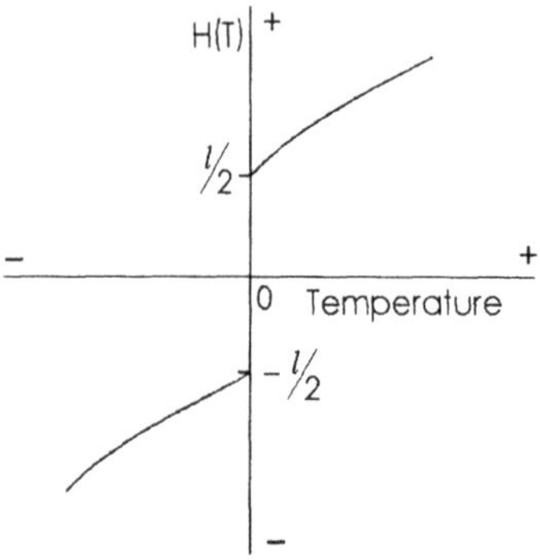

Fig. 4.2.1. Enthalpy vs temperature.

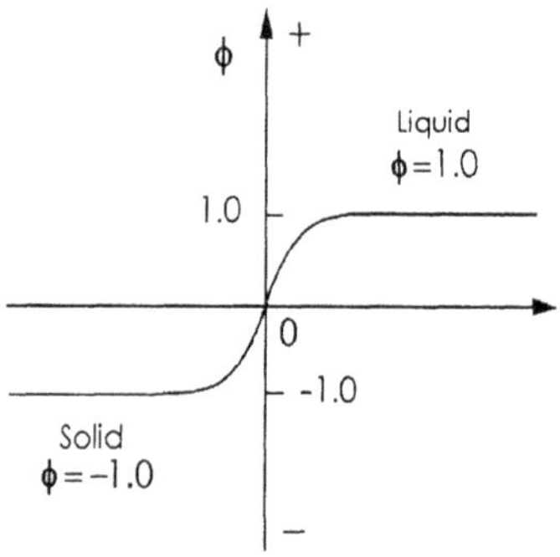

Fig. 4.2.2. Continuity of the phase function.

$$-\frac{\delta F(\phi)}{\delta\phi} = \xi^2\nabla^2\phi - \frac{1}{8a}f'(\phi) + \frac{[\hat{S}]_E}{2}(T - T_m) = 0, \tag{4.2.4}$$

Here, the dash denotes differentiation with respect to the argument. $\delta F/\delta\phi$ is the 'variational derivative' of ϕ which can be easily calculated (cf. [82]). Equation (4.2.4) holds for an equilibrium state and ϕ is a function only of x, $x \in \Omega$. Note that F is a function of ϕ and not of $T(x,t)$ and so T in (4.2.3) is fixed. For a transient or a non-equilibrium process, the free energy will not be minimum but will satisfy the following relation

$$\hat{\delta}\frac{\partial\phi}{\partial t}(x,t) = -\frac{\delta F}{\delta\phi}\{\phi(x,t)\}, \tag{4.2.5}$$

where $\hat{\delta} > 0$ is the relaxation time [83]. Equation (4.2.5) is valid for the case when ϕ is not conserved, i.e., the volume integral of $\phi(x,t)$ over Ω is not prescribed. The basis of (4.2.5) is the standard idea of 'non-equilibrium thermodynamics' which assumes that ϕ returns to equilibrium with a 'force' proportional to the extent to which it was out of equilibrium.

If ϕ is conserved, then

$$\frac{\partial\phi}{\partial t}(x,t) = \operatorname{div}\left[M \operatorname{grad}\left\{\frac{\delta F}{\delta\phi}(\phi(x,t))\right\}\right], \tag{4.2.6}$$

where M is a coefficient related to diffusion or an interdiffusion coefficient which may depend on $\phi(x,t)$ and on the temperature [83, 84]. The second law of thermodynamics requires that along the solution paths, the free energy should necessarily decrease with time, failing which the model will be 'thermodynamically inconsistent'. If F in (4.2.3) is a function of ϕ alone, then the second law of thermodynamics is satisfied but not when F is a function of the temperature and ϕ. The time derivative of $F\{\phi(x,t)\}$ gives

$$\frac{d}{dt}F\{\phi(x,t)\} = \int_{\Omega} \frac{\delta F}{\delta \phi}\{\phi(x,t)\}\frac{\partial \phi}{\partial t}d\Omega = -\frac{1}{\hat{\delta}}\int_{\Omega}\left(\frac{\delta F}{\delta \phi}\right)^2 d\Omega \leq 0. \tag{4.2.7}$$

In order to derive (4.2.7) take F as in (4.2.3) and proceed with the first principles to obtain the time derivative of F. In calculating dF/dt, at some stage of calculations Green's formula is to be applied to the volume integral of $\nabla\phi\cdot\nabla\partial\phi/\partial t$ and if it is assumed that on the boundary of Ω the normal derivative of ϕ is zero then we get (4.2.7). The second law of thermodynamics is satisfied even if ϕ is conserved, provided M grad $\delta F/\delta\phi$ is zero on the boundary of Ω.

Sometimes, since the order parameter ϕ alone is not sufficient to describe the local state of the system, additional fields are introduced. One such extra field is temperature. If the heat conduction is very 'fast' then it can be assumed that the material is at a uniform temperature but this assumption is not valid if heat conduction is 'slow'. A second equation for heat transport in the material should be considered along with equation (4.2.4) or (4.2.6). An example of this type of modelling is the phase-field model which consists of the following system of equations.

$$\alpha\varepsilon^2\frac{\partial\phi}{\partial t} = \varepsilon^2\,\nabla^2\phi + \frac{1}{2}(\phi-\phi^3) + \frac{\varepsilon[\hat{S}]_E}{3\sigma}(T-T_m), \tag{4.2.8}$$

$$\frac{\partial T}{\partial t} + \frac{l}{2}\frac{\partial\phi}{\partial t} = k\nabla^2 T. \tag{4.2.9}$$

In equations (4.2.8) and (4.2.9), ϕ and T are coupled but in deriving (4.2.8) no coupling between ϕ and T was considered. The double-well symmetric function f with distinct minima at $\phi=\pm1$ is taken as $f=(\phi^2-1)^2$. The parameter ε and α are defined as

$$\varepsilon = \xi a^{1/2}, \quad \alpha = \hat{\delta}/\xi^2. \tag{4.2.10}$$

ε is a measure of the interface thickness and is of the order of 10^{-8} cm, σ is the surface tension, $\hat{S}$ is the entropy per unit volume, k is the thermal diffusivity. The units of σ, $\hat{S}$, α are as follows:

$$\begin{aligned}\sigma :&= \text{unit of Energy}/(\text{length})^{(n-1)},\ \hat{S} := \text{unit of energy}/\{(\text{length})^n\cdot\text{degree}\}\\ \alpha :&= \text{unit of time}/(\text{length})^2,\ n = \text{spacial dimension of }\Omega.\end{aligned} \tag{4.2.11}$$

At a deeper level of physics there are significant differences between the Stefan problem and the phase-field model in that the phase-field model incorporates the microscopic physics in an average sense and also takes into account macroscopic parameters. Stefan problem deals only with macroscopic modelling.

Both the equations (4.2.8) and (4.2.9) hold good throughout Ω and the sharp interface separating solid and liquid phases has been replaced by a finite-width interface. Therefore there are no boundary conditions at the interface. The region corresponding to $T = 0$ is now given by $-1 < \phi < 1$, and $\phi = 0$ is the center of the interface. Initial and boundary conditions for both ϕ and T are to be prescribed. For example, on the boundary $\partial\Omega$ of Ω, we may have

$$\left.\frac{\partial \phi}{\partial n}\right|_{\partial\Omega} = 0; \quad \left.\frac{\partial T}{\partial n}\right|_{\partial\Omega} = 0, \tag{4.2.12}$$

or

$$\phi|_{\partial\Omega} = \phi(x),\ x \in \partial\Omega; \quad \left.\frac{\partial T}{\partial n}\right|_{\partial\Omega} = 0, \tag{4.2.13}$$

or any combination of these boundary conditions.

Note that for both $T < 0$ and $T > 0$, (4.2.9) is well defined and the region $T = 0$ is not defined by the enthalpy but is identified by the phase function ϕ, $-1 < \phi < 1$. The phase-field equations were first formulated in [85] on the basis of the model discussed in [80]. The last term in the integral in (4.2.3) can be taken in the form of a more general expression such as $-2u\gamma G(\phi)$ (cf. [86]), where u is the dimensionless temperature and

$$u = (T - T_m)/(l/C_P). \tag{4.2.14}$$

Here, γ is a microscopic parameter related to a macroscopic parameter and has the dimension of energy per unit volume, $G(\phi)$ is an increasing function related to the entropy distribution across the interface and

$$[\hat{S}]_E = 2\gamma \int_{-1}^{1} G'(\phi)d\phi. \tag{4.2.15}$$

If $G(\phi) = \phi$ and the units of energy and temperature are suitably chosen, then

$$[\hat{S}]_E = 4 \text{ units of energy per unit volume/degree}, \tag{4.2.16}$$

and

$$-2u\gamma G(\phi) = -2u\phi. \tag{4.2.17}$$

In Fig. 4.2.3. the last term on the r.h.s. in the equation to determine ϕ has been taken as $2u$. By defining ε and α through (4.2.10) and σ in units of free energy, the last term on the r.h.s. in (4.2.8) can be obtained.

The capillary length d_0 is defined as (cf. [87]).

$$d_0 = \sigma/\{[\hat{S}]_E\, l/C_P\}, \tag{4.2.18}$$

and it is easy to check that with the help of (4.2.18), (4.2.8) becomes

$$\alpha\varepsilon^2\frac{\partial\phi}{\partial t} = \varepsilon^2\nabla^2\phi + \frac{1}{2}(\phi - \phi^3) + \frac{\varepsilon}{3d_0}u. \tag{4.2.19}$$

Equation (4.2.19) has two length scales, viz., ε and d_0. Note that there is only one length scale in the modified Gibbs-Thomson relation (4.1.1). Using (4.2.18), (4.1.1) can be written in terms of u as

$$u = -d_0\bar{K}_c - \alpha d_0|V_n|. \tag{4.2.20}$$

In (4.1.1), specific heats, thermal conductivity and densities have been taken to be equal in both solid and liquid phases. This is generally done for convenience as the mathematics does not change significantly even if different thermophysical properties are taken in different phases. There are several advantages of phase-field models. A variety of phenomena ranging from motion by mean curvature to stable anisotropic crystal growth and single-needle dendrites are obtained by simply varying the parameters. Phase-field equations are amenable to numerical computations [88]. Computations involving self-interactions of the interface do not pose difficulties in the phase-field model. The phase-field equations are guaranteed to have a smooth solution (ϕ, T) if the initial and boundary data are sufficiently smooth (cf. [81, 89]).

Rigorous mathematical analysis for obtaining Gibbs-Thomson relation by taking appropriate limits of some parameters was first presented in [81]. The 'Inner' and 'outer' expansions of the order parameter $\phi(x,t)$ were obtained as $\xi \to 0$ and by matching these expansions, it was proved that under suitable assumptions

$$u(x) = -\sigma_0\bar{K}_c/4 + O(\xi),\ x \in \partial\Omega_0, \tag{4.2.21}$$

where $\partial\Omega_0$ is the center of the finite-width interface or where $\phi = 0$; $\bar{K}_c$ is the sum of the principal curvatures of the interface $\phi = 0$ and σ_0 is defined by the equation

$$\sigma_0 = \frac{2}{3}\xi. \tag{4.2.22}$$

It was shown that

$$\sigma = \sigma_0 + O(\xi^2). \tag{4.2.23}$$

Based on the mathematical procedure presented in [81], and by taking appropriate limits, modified Gibbs-Thomson relation was derived in [90] and the Gibbs-Thomson relation in [81].

In Figs. 4.2.3. and 4.2.4. (cf. [91, 90]), the appropriate limits of the parameters to be taken, and the corresponding Stefan and Stefan-like problems obtained have been shown. Although the limiting cases obtained in Fig. 4.2.3. are covered in Fig. 4.2.4., the limits are more explicitly defined in Fig. 4.2.3.

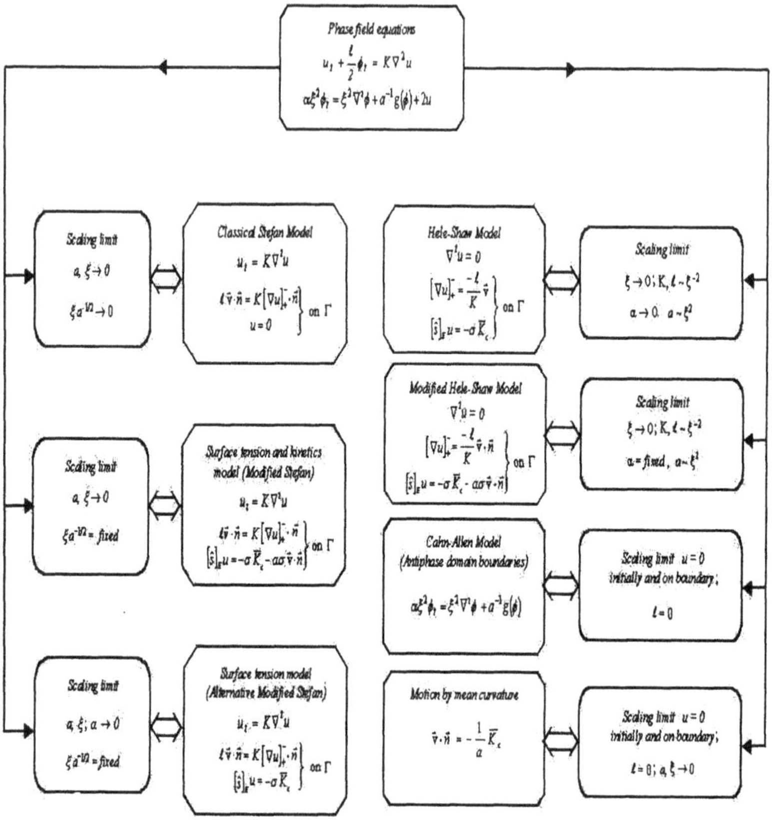

Fig. 4.2.3. Limits of phase-field equations I

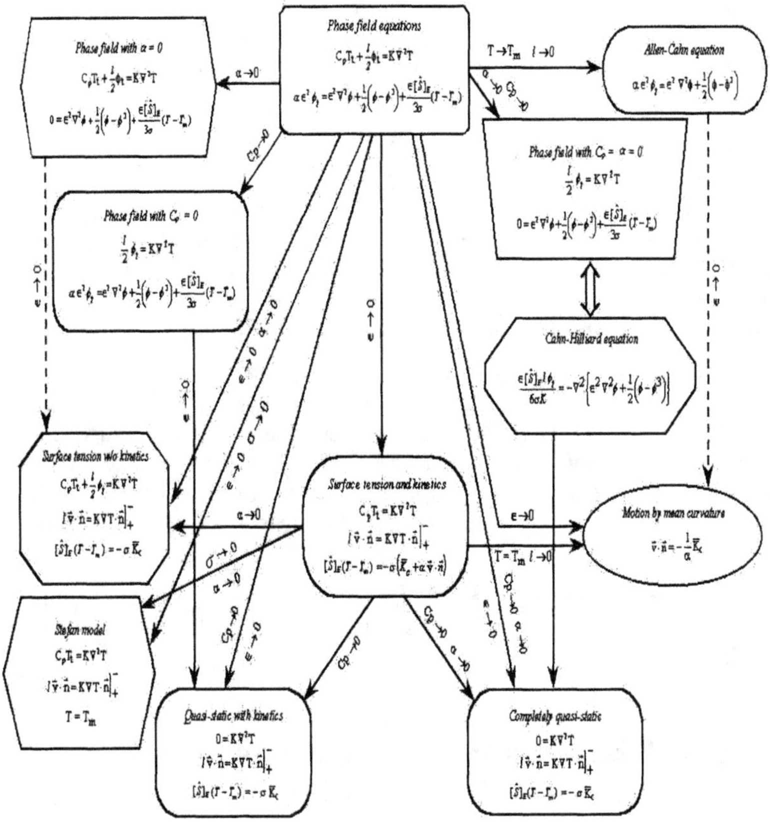

Fig. 4.2.4. Limits of phase-field equations II

The role of microscopic anisotropy in the macroscopic behaviour of the phase boundary was analysed in [92] using Landau-Ginzburg [80] free energy functional. The continuum limit for a lattice spin system with 'anisotropic interactions' J_i and spacing a_i leads to the following equation for the phase function ϕ.

$$\hat{\delta}\frac{\partial \phi}{\partial t} = \sum_{i=1}^{d} \xi_i^2 \frac{\partial^2 \phi}{\partial x_i^2} + \frac{1}{2}(\phi - \phi^3) + 2u, \tag{4.2.24}$$

where d is the spacial dimension of the space, $\hat{\delta}$ is the relaxation time and u is defined by (4.2.14). The temperature and ϕ are coupled through an equation of the type (4.2.9) and ξ_i is related to J_i and a_i. A modified Gibbs-Thomson law in polar coordinates (r, θ) was derived for the equilibrium temperature which has the form

$$[\hat{S}]_E u(r,\theta) = -[\sigma(\theta) + \sigma''(\theta)]\bar{K}_c - \hat{\delta}|V_n|\sigma(\theta)/\xi_A^2(\theta) + O(\xi^2). \tag{4.2.25}$$

Here, $\xi_A(\theta)$ is a measure of the thickness of the interface and $\sigma(\theta)$ is the angle dependent surface tension. In [93] the basic phase-field model was extended to physical problems in which fluid properties such as velocity, pressure and density variations were incorporated along with heat properties and a unified and consistent derivation of equations was obtained.

Since many phase-field models for solidification have been proposed in the literature, the phase-field model consisting of (4.2.8) and (4.2.9) will be called *Standard phase-field model* or SPF model.

4.3 Some Thermodynamically Consistent Phase-field and Phase Relaxation Models of Solidification

It was assumed in (4.2.3) that the energy functional F is a function of ϕ alone, and that T is a constant. If a more general energy functional of the form

$$F\{(T(x,t), \phi(x,t)\} = \int_\Omega \{f(T,\phi) + \frac{1}{2}B|\nabla\phi(x)|^2\}d\Omega, \tag{4.3.1}$$

is taken, in which B is either a positive constant or a function of T but not of ϕ, then F will not always decrease on the solution paths and the thermodynamical consistency condition (4.2.7) will not be satisfied. However either the SPF model can be obtained as a particular case of some thermodynamically consistent models or the modified Gibbs-Thomson relation can be derived independently of the SPF model. Several thermodynamically consistent models have been proposed in the literature (cf. [94, 95, 96, 97, 98]). Our interest here is only in briefly indicating the derivation of some thermodynamically consistent models from which the modified Gibbs-Thomson relation can be obtained by a rigorous limiting process.

Entropy Functional

The model proposed in [97] and described below is based on an entropy functional and is on the same lines as the model proposed earlier in [94]. The starting point of the formulation in [97] is to define an entropy functional $\hat{F}\{\phi(x,t), e(x,t)\}$ for a subregion Ωg of a closed and bounded domain $\Omega \subset R^3$ in which the phase-change is taking place.

$$\hat{F} = \int_{\Omega_g} \left\{ \hat{S} - \frac{1}{2}\varepsilon^2 (\nabla\phi)^2 \right\} d\Omega_g, \tag{4.3.2}$$

Here, $\hat{S}(e,\phi)$ is the entropy per unit volume, ϕ is the phase function as defined in § 4.2, $e(x,t)$ is the internal energy density and ε is a constant. $e(x,t)$ satisfies the equation

$$\dot{e} + \nabla \vec{q} = 0 \quad \text{in } \Omega g, \tag{4.3.3}$$

where $\vec{q}$ is the heat flux vector. The time derivative of $\hat{F}$ can be calculated as follows:

$$\frac{d}{dt}(\hat{F}) = \int_{\Omega_g} \left\{ \left(\frac{\partial \hat{S}}{\partial e}\right)_\phi \dot{e} + \left[\left(\frac{\partial \hat{S}}{\partial \phi}\right)_e + \varepsilon^2 \nabla^2 \phi \right] \dot{\phi} - \varepsilon^2 \nabla \cdot (\dot{\phi} \nabla \phi) \right\} d\Omega_g \tag{4.3.4}$$

$$= \int_{\Omega_g} \left\{ \vec{q} \cdot \nabla \left(\frac{\partial \hat{S}}{\partial e}\right)_\phi + \left[\left(\frac{\partial \hat{S}}{\partial \phi}\right)_e + \varepsilon^2 \nabla^2 \phi \right] \dot{\phi} \right\} d\Omega_g$$

$$- \int_{A_g} \left\{ \left(\frac{\partial \hat{S}}{\partial e}\right)_\phi \vec{q} + \varepsilon^2 \dot{\phi} \nabla \phi \right\} \cdot \vec{n} \, dAg. \tag{4.3.5}$$

Here, A_g is the surface area of Ω_g, dAg is the elementary surface area and $\vec{n}$ is the outward unit normal to A_g. The derivation of (4.3.5) from (4.3.4) requires a simple application of Gauss's theorem. The thermodynamical consistency of this entropy model requires that the local entropy production be non-negative. If the entropy flux through the surface of Ω_g is subtracted from $d\hat{F}/dt$, then the entropy produced can be obtained. The consistency condition requires that

$$\frac{d\hat{F}}{dt} + \int_{A_g} \left(\frac{\vec{q}}{T} \cdot \vec{n} + \varepsilon^2 \dot{\phi} \nabla \phi \cdot \vec{n} \right) dAg \geq 0, \tag{4.3.6}$$

as the rate at which entropy is produced cannot be negative. $\vec{q}/T$ in the first term in the surface integral in (4.3.6) is the entropy flux due to heat flow (see (2.1.2)). T is the absolute temperature. The quantity $\varepsilon^2 \dot{\phi} \nabla \phi$ is the entropy flux related to the changes in the phase function at the boundary of the subregion Ω_g. Both solid and liquid phases have been taken at rest. If $d\hat{F}/dt$ from (4.3.5) is substituted in (4.3.6) and the relation

$$\left(\frac{\partial \hat{S}}{\partial e}\right)_\phi = \frac{1}{T}, \tag{4.3.7}$$

is used, then we get the following consistency condition

$$\int_{\Omega_g} \left\{ \vec{q} \cdot \nabla \left(\frac{1}{T}\right) + \left[\left(\frac{\partial \hat{S}}{\partial \phi}\right)_e + \varepsilon^2 \nabla^2 \phi \right] \dot{\phi} \right\} d\Omega_g \geq 0. \tag{4.3.8}$$

The derivation of (4.3.7) deserves some discussion. In (2.1.15), the phase function ϕ has not been considered. Therefore, as suggested in [94], it will be assumed that even when the free energy density G and the entropy density $\hat{S}$ depend on the order parameter ϕ and the temperature T, they are concave in T and e, respectively (though not necessarily in ϕ) and they still obey 'Legendre transform relation' and its inversion [99].

$$G(T, \phi) = \inf_e [e(\hat{S}, \phi) - T\hat{S}(e, \phi)], \tag{4.3.9}$$

and

$$\hat{S}(e, \phi) = \inf_T \left(\frac{e(\hat{S}, \phi)}{T} - \frac{G(T, \phi)}{T} \right). \tag{4.3.10}$$

The differentiations of (4.3.10) with respect to e and ϕ give the following results.

$$\left.\frac{\partial \hat{S}}{\partial e}(e, \phi)\right|_\phi = \frac{1}{T(e, \phi)}, \tag{4.3.11}$$

and

$$\left.\frac{\partial \hat{S}}{\partial \phi}(e, \phi)\right|_e = -\frac{1}{T(e, \phi)} \left.\frac{\partial G}{\partial \phi}(T(e, \phi), \phi)\right|_T, \tag{4.3.12}$$

where $T(e, \phi)$ is the value of T at which the infimum in (4.3.10) occurs. For those values of T where no phase-change occurs, G is differentiable, and

$$\frac{\partial \{G(T, \phi)/T\}}{\partial T} = -e/T^2. \tag{4.3.13}$$

The consistency condition (4.3.8) gets satisfied if $\vec{q}$ and $\dot{\phi}$ are chosen such that

$$\vec{q} = M_T \nabla \left(\frac{1}{T}\right), \tag{4.3.14}$$

and

$$\hat{\delta}\dot{\phi} = \left(\frac{\partial \hat{S}}{\partial \phi}\right)_e + \varepsilon^2 \nabla^2 \phi. \tag{4.3.15}$$

Here, $\hat{\delta}$ is the relaxation time and is a positive constant, and M_T could be a positive constant or a function of temperature but not of ϕ. In both [94] and [97], the starting point for the derivation of phase-field equations is the same entropy functional (4.3.2) but the consistency condition $d\hat{F}/dt \geq 0$ is satisfied in different ways. We describe very briefly the procedure adopted in [94]. The transient equations for $\phi(x,t)$ and $e(x,t)$ have been obtained from (4.3.2) using the same arguments which lead to (4.2.5) and (4.2.6). While $e(x,t)$ is conserved, ϕ may or not be conserved. The quantities $\delta\hat{F}/\delta\phi$ and $\delta\hat{F}/\delta e$, which are required in deriving transient equations for ϕ and e can be easily calculated, see [94]. Having obtained the transient equations, the consistency condition can be established as in (4.2.7). Note that the time derivative of the entropy functional contains both $\delta\hat{F}/\delta\phi$ and $\delta\hat{F}/\delta e$ and it should be nondecreasing.

The quantity $(\partial\hat{S}/\partial\phi)_e$ in (4.3.15) is not convenient to handle. So it should be expressed in terms of $(\partial G/\partial\phi)_T$ as in (4.3.12). Integration of (4.3.13) gives

$$G(T,\phi) = T\left(-\int_{T_m}^{T} \frac{e(\xi,\phi)}{\xi^2}d\xi + f(\phi)\right), \tag{4.3.16}$$

where $f(\phi)$ is an unknown function. A suitable definition of $e(T,\phi)$ agreeing with the classical definition of enthalpy can be developed if we take

$$\left.\begin{aligned} e(T,\phi) &= e_S(T) + p(\phi)\hat{L}(T), \\ &= e_L(T) + [p(\phi)-1]\hat{L}(T), \end{aligned}\right\} \tag{4.3.17}$$

where $p(\phi)$ is an arbitrary function except that $p(0)=0$ and $p(1)=1$. In the present model, we take $\phi=0$ is the solid phase and $\phi=1$ is the liquid phase. $\hat{L}(T)=e_L(T)-e_S(T)$. When $T=T_m$, $\hat{L}(T_m)=\tilde{l}_m$, where $\tilde{l}_m$ is the latent heat of fusion per unit volume. Any of the expressions for e given in (4.3.17) can be used in (4.3.16). If the second definition in (4.3.17) is used in (4.3.16), then we obtain

$$G(T,\phi) = T\left(-\int_{T_m}^{T} \frac{e_L(\xi)}{\xi^2}d\xi - [p(\phi)-1]Q(T) + f(\phi)\right), \tag{4.3.18}$$

where

$$Q(T) = \int_{T_m}^{T} \frac{\hat{L}(\xi)}{\xi^2}d\xi. \tag{4.3.19}$$

By differentiating (4.3.18) w.r.t. ϕ, we obtain,

$$\left(\frac{\partial G}{\partial\phi}\right)_T = -TQ(T)p'(\phi) + Tf'(\phi). \tag{4.3.20}$$

Note that e_L and e_S are functions of T and not of ϕ. It is now a simple matter to derive the phase-field model. Substituting e and $\vec{q}$ from (4.3.17) and (4.3.14) in the energy equation (4.3.3), we obtain

$$\dot{e}_L(T) + \dot{p}(\phi)\hat{L}(T) + [p(\phi)-1]\frac{d\hat{L}}{dt}(T) = -\nabla\cdot[M_T\nabla\left(\frac{1}{T}\right)]. \tag{4.3.21}$$

If $M_T = KT^2$ and $\phi = 0, 1$ in (4.3.21), then we get heat conduction equations in solid and liquid phases, respectively. In order to obtain kinetic equation for ϕ, we substitute $\partial \hat{S}/\partial \phi$ from (4.3.12) in (4.3.15) and use (4.3.20). We obtain

$$\hat{\delta}\dot{\phi} = Q(T)p'(\phi) - f'(\phi) + \varepsilon \nabla^2 \phi \,. \tag{4.3.22}$$

Here, $f(\phi)$ is a double-well symmetric function with two distinct minima at $\phi = 0$ and $\phi = 1$, and is taken as

$$f(\phi) = \phi^2(1-\phi)^2. \tag{4.3.23}$$

It is not possible to determine $p(\phi)$ uniquely. The equations (4.3.21) and (4.3.22) can be presented in some more suitable dimensionalized forms (see [97]).

The SPF model was obtained as a particular case of the phase-field model constructed in [95] subject to the conditions that the energy density vary linearly with the order parameter and the entropy be a non-concave function of the order parameter. Since SPF model can be obtained as a particular case of a thermodynamically consistent model, it can also be regarded as a consistent model. In [97], two phase-field models have been presented. One of them partially resembles the model in [85] and the other the model in [98]. Although SPF model cannot be obtained as a particular case of any of the models constructed in [97], it has been reported that by taking appropriate limits of the dimensionless equations derived in [97] and following the procedure discussed in [81], modified Gibbs-Thomson law can be derived.

Some thermodynamically consistent phase relaxation models for supercooling

The Stefan condition (1.4.11) imposed in the classical Stefan problem formulation does not arise in the phase-field models discussed above. Like an enthalpy formulation, phase-field model is also a continuum formulation, and ϕ is defined as

$$\left.\begin{aligned} \phi &= -1, \ T < 0, \\ &= (-1, 1), \ T = 0, \\ &= 1\ , \ T > 0. \end{aligned}\right\} \tag{4.3.24}$$

The equations of the type (4.3.24) are valid throughout Ω, which is the region under consideration. As mentioned earlier, in an enthalpy formulation, $T = 0$ does not give the middle of the mushy region but in the phase-field model, $\phi = 0$, gives the middle of the finite-size interface. In (4.3.24), the sharp interface separating solid and liquid regions is given by $\phi = 0$ and not by $T = 0$. It is well known that in the case of supercooling or superheating, a mushy region develops, and the free boundary progresses through the mushy region. Therefore to obtain an equiaxed or columnar growth of the interface or the amount of supercooling or superheating, the phase function ϕ, should be in some ways associated with the structure of the mushy region. It should be pointed out here that $T = T_m$ (or $T = 0$) is not the actual temperature of the mushy region. It is an average value of the temperature. The actual situation is shown in Fig. 4.3.1.

Several mathematical models based on relaxation dynamics have been constructed in the literature in terms of ϕ and T, with which modified Gibbs-Thomson relation, mushy region, dendritic growth, etc. (cf. [100, 101, 102, 103]) can be described. Most of these models are based on the weak formulation which is not the emphasis in this volume. We give below a classical formulation concerned with a family of thermodynamically consistent models of phase transitions. Phase diffusion is neglected in this formulation by assuming that no interfacial energy occurs [100]. This means that the term $\xi^2(\nabla\phi)^2/2$ occurring in (4.2.3) is taken to be zero. Therefore, only a rate-type constitutive law governing the evolution of ϕ accounts for the *phase relaxation*.

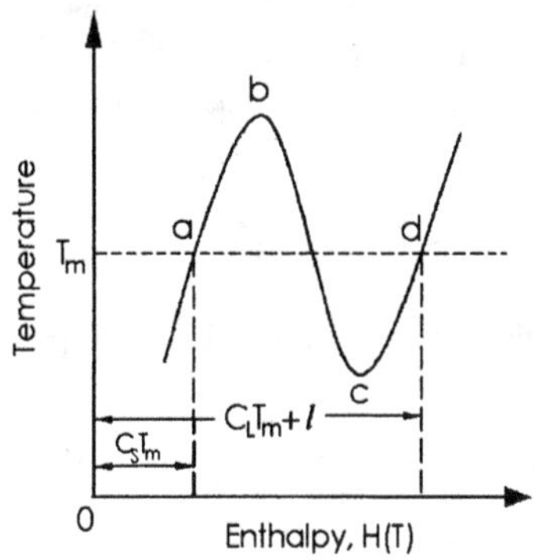

Fig. 4.3.1. ab-metastable state for the solid
cd-metastable state for the liquid.

The internal energy density $e(T, \phi)$ can be taken as

$$e(T, \phi) = f_0(T) + f_1(\phi), \tag{4.3.25}$$

where equation (4.3.25) is the generalization of the energy equation considered in the earlier phase transition models. If $C_S = C_L$, then the specific enthalpy (or internal energy) is given by

$$e = C_V T + l\phi, \ 0 \leq \phi \leq 1. \tag{4.3.26}$$

As $C_V = (\partial e/\partial T)_V > 0$, we should have $f_0'(T) > 0$ in (4.3.25). The dash denotes differentiation with respect to the argument. If $\phi = 0$ represents the solid phase, and $\phi = 1$ represents the liquid phase, which is a more energetic phase then we should have $f_1'(\phi) \geq 0$. For the sake of convenience, let

$$e(T, \phi) = T^2 F_0'(T) + T_m^c F_1(\phi), \tag{4.3.27}$$

where the functions F_0 and F_1 are assumed to be continuous with $F_1'(\phi) \geq 0$ and T_m^c is the phase-change temperature which could be T_m or be different from T_m. The energy balance equation has the standard form

$$\dot{e}(T,\phi) + \operatorname{div} \vec{q} = \dot{r}\ ;\ \vec{q} = -\hat{K}(T)\nabla T, \tag{4.3.28}$$

where the dot denotes time derivative, r is the heat source density function and e is given by (4.3.27). To derive a constitutive equation for ϕ it will be assumed that ϕ is not an explicit function of the space variables which means that $\nabla\phi = 0$ or that no interfacial diffusion occurs which was present in (4.2.4). Therefore the constitutive relation for ϕ can be assumed to be of the form

$$\dot{\phi} = \beta(T,\phi), \tag{4.3.29}$$

where β is an unknown continuous function. The functions F_0, F_1 and β are unknown but not completely arbitrary. For example, if the system of equations (4.3.28) and (4.3.29) is thermodynamically consistent, then the *Clausius-Duhem inequality* should be satisfied, which implies that

$$\dot{G} + \hat{S}\dot{T} + \frac{\vec{q}\cdot\nabla T}{T^2} \leq 0. \tag{4.3.30}$$

Here, G is the free energy density which is related to entropy $\hat{S}$, e and T by the following relation.

$$G(T,\phi) = \inf_T\{e(\hat{S},\phi) - T\hat{S}(e,\phi)\}. \tag{4.3.31}$$

The time derivative of G gives

$$\dot{G} = \frac{\partial G}{\partial T}\dot{T} + \frac{\partial G}{\partial \phi}\dot{\phi}. \tag{4.3.32}$$

Using (4.3.32) in (4.3.30), we obtain

$$\left(\frac{\partial G}{\partial T} + \hat{S}\right)\dot{T} + \frac{\partial G}{\partial \phi}\dot{\phi} - \frac{\hat{K}(T)|\nabla T|^2}{T^2} \leq 0. \tag{4.3.33}$$

The inequality (4.3.33) is satisfied if it is assumed that

$$\frac{\partial G}{\partial T} + \hat{S} = 0,\quad \frac{\partial G}{\partial \phi}\beta(T,\phi) \leq 0,\quad \hat{K} \geq 0. \tag{4.3.34}$$

Substituting $\hat{S} = -\partial G/\partial T$ in (4.3.31), and using (4.3.27), the following equation is obtained.

$$G - T\frac{\partial G}{\partial T} = T^2 F_0'(T) + T_m^c F_1(\phi). \tag{4.3.35}$$

It is easy to check that $G(T,\phi)$ given by

$$G(T,\phi) = -TF_0(T) + TB(\phi) - (T - T_m^c)F_1(\phi), \tag{4.3.36}$$

satisfies (4.3.35). $B(\phi)$ is an unknown arbitrary function. The second condition in (4.3.34) is satisfied if $\beta(T,\phi)$ is taken as

$$\beta(T,\phi) = -\frac{1}{g(T,\phi)}\frac{\partial G}{\partial \phi}\,; \; g(T,\phi) > 0. \tag{4.3.37}$$

For the sake of simplicity $g(T,\phi)$ is taken as

$$g(T,\phi) = F_2'(\phi)N(T). \tag{4.3.38}$$

If β is continuous, and $g > 0$ and continuous, then $N(T)$ should be continuous and $N(T) > 0$ for $\forall\, T$. Also F_2 should be a monotonically increasing C^1 function, i.e., $F_2'(\phi) > 0$ for $\forall\,\phi$. The constitutive equation for ϕ can now be written as

$$\frac{\partial G}{\partial \phi}(T,\phi) + N(T)F_2'(\phi)\dot{\phi} = 0. \tag{4.3.39}$$

With constitutive equations (4.3.28) and (4.3.39), a thermodynamically consistent phase relaxation model is complete. By suitable choice of unknown functions F_0, F_1, $F_2(\phi)$ and $N(T)$, several well known relaxation models can be derived as particular cases of this model (cf. [100]). Our interest here is only in analysing supercooling and superheating effects.

As discussed earlier, when a phase is in a stable equilibrium, its free energy is minimum. Therefore, for stable solid and stable liquid phases, i.e., for $\phi = 0$ and $\phi = 1$ at any fixed temperature, we have

$$\left.\frac{\partial G}{\partial \phi}\right|_{\phi=0} \geq 0 \text{ and } \left.\frac{\partial G}{\partial \phi}\right|_{\phi=1} \leq 0, \tag{4.3.40}$$

as $\phi = 0$ and $\phi = 1$ are the end points of the interval [0,1]. If ϕ_e is any internal equilibrium point, i.e., $\phi_e \in (0,1)$, then

$$\left.\frac{\partial G}{\partial \phi}\right|_{\phi=\phi_e} = 0. \tag{4.3.41}$$

If the mushy region is at equilibrium temperature T_m, i.e., $T_m^c = T_m$ then for all possible values of $\phi \in [0,1]$ there is equilibrium. From (4.3.36), at $T = T_m^c = T_m$, we get

$$G(T,\phi) = -TF_0(T) + TB(\phi). \tag{4.3.42}$$

From (4.3.41), $\partial B(\phi)/\partial\phi = 0$ for $\phi \in [0,1]$. Therefore $B(\phi)$ has a constant value in the interval [0,1]. In this case

$$\begin{aligned} B = I(\phi) &= 0, \text{ if } \phi \in [0,1] \\ &= \infty, \text{ otherwise.} \end{aligned} \tag{4.3.43}$$

$I(\phi)$ is called a *indicator function* which has no classical derivative in R, but since I is a proper *lower semi-continuous function*, its subdifferential ∂I exists. The subdifferential of the function $I(\phi)$ is defined as

$$\alpha \in \partial I(\phi) \text{ if and only if } \alpha(\phi - \xi) \geq I(\phi) - I(\xi) \text{ for } \forall\ \xi \in R,\ \alpha \in R. \qquad (4.3.44)$$

With the help of the above definition it is easy to check that $\partial I = H_g^{-1}$ where H_g^{-1} is the inverse of the *Heaviside graph* H_g defined below

$$\left.\begin{aligned} H_g(x) &= 0,\ x < 0 \\ &= (0,1), x = 0, \\ &= 1, x > 0. \end{aligned}\right\} \qquad (4.3.45)$$

It is easy to obtain H_g^{-1} or ∂I by using (4.3.44), and we have

$$\left.\begin{aligned} \partial I(x) &= \emptyset,\ x \notin [0,1], \\ &= \{0\},\ x \in (0,1), \\ &= R^+,\ x = 1,\ R^- \text{ if } x = 0. \end{aligned}\right\} \qquad (4.3.46)$$

To avoid ϕ taking values outside the interval [0,1], but to preserve strict minima at the ends $\phi = 0$ and $\phi = 1$, a double well function should be added to the indicator function. Therefore $B(\phi)$ should be of the form

$$B(\phi) = I(\phi) + B_1(\phi), \qquad (4.3.47)$$

where $B_1(\phi)$ is a smooth 'double-well function' forcing B to have only two strict minima at $\phi = 0$ and $\phi = 1$. Depending on the physical situation, $I(\phi)$ or $B_1(\phi)$ could be zero or non-zero. $B_1(\phi)$ is generally taken in the form

$$B_1(\phi) = \nu\phi^2(1-\phi)^2,\ \nu > 0. \qquad (4.3.48)$$

$F_1(\phi)$ in (4.3.36) has been assumed to be nondecreasing and it should have strict minima at $\phi = 0$ and $\phi = 1$, as G possesses these two minima. $F_1(\phi)$ satisfies these conditions if it is taken as [100]

$$F_1(\phi) = \frac{\lambda}{3}\phi^3(3 - 2\phi),\ \lambda > 0. \qquad (4.3.49)$$

Superheating and supercooling effects

When a solid is superheated, $\phi = 0$ even at those points of the region where $T > T_m$. Note that superheating or supercooling is assessed with respect to T_m and not with respect to T_m^c. Similarly, for a supercooled liquid, $\phi = 1$ even when $T < T_m$. Therefore the free energy G must exhibit minima at $\phi = 0$ and $\phi = 1$ for all temperatures T such that $|T - T_m|/T_m < \varepsilon$ where ε is a sufficiently small positive number. In particular, minima exists at $T = T_m$ also.

We first consider the case $T > T_m^c$. For $T > T_m^c$, the last term in (4.3.36) is non-positive if $F_1(\phi)$ is taken as in (4.3.49). If $B = B_1$ in (4.3.36), then G has an absolute minimum at $\phi = 1$. Fig. 2.1.1. suggests that the solid is superheated. For $T < T_m^c$, the last term in (4.3.36) is non-negative and G (with $B = B_1$) has an absolute minimum at $\phi = 0$. Fig. 2.2.1. suggests that the liquid is supercooled. Let $u(T) = \lambda(T - T_m^c)/(\nu T)$, $\lambda > \nu > 0$. If B_1 from (4.3.48) and F_1 from (4.3.49) are substituted in (4.3.36) and G is differentiated, then we get

$$\frac{\partial G}{\partial \phi} = 4\nu T\phi(\phi - 1)\left[\phi - (1 - u(T))2\right]. \tag{4.3.50}$$

If $u(T) \in (-1, 1)$, then G takes two strict minima at $\phi = 0$ and $\phi = 1$ (check $\partial^2 G/\partial\phi^2$ also) and superheating and supercooling can occur when T lies in the interval

$$\left(\frac{\lambda T_m}{\lambda + \nu} < T < \frac{\lambda T_m}{\lambda - \nu}\right). \tag{4.3.51}$$

In (4.3.51), T_m^c has been replaced by T_m as supercooling and superheating are assessed with respect to T_m. It is clear from (4.3.51) that if the mushy region exists at $T = T_m$, then supercooling and superheating effects do not occur.

In essence, the mathematical model described above consists of the following coupled differential equations.

$$\frac{\partial}{\partial t}(T^2 F_0'(T) + T_m^c F_1(\phi)) \; - \; \operatorname{div}(\hat{K}(T)\operatorname{grad} T) = \dot{r} \tag{4.3.52}$$

and

$$\frac{\partial}{\partial t}(F_2(\phi)) + \partial I(\phi) + \frac{T}{N(T)}B_1'(\phi) \ni \frac{T - T_m^c}{N(T)}F_1'(\phi). \tag{4.3.53}$$

Equation (4.3.52) can be obtained from (4.3.27) and (4.3.28) while (4.3.53) can be obtained from (4.3.36) and (4.3.39). Due to the presence of the term $\partial I(\phi)$, the l.h.s. in (4.3.53) is a multivalued function and the relation (4.3.53) cannot be written as an equation. The system consisting of (4.3.52) and (4.3.53) should be supplemented with the boundary and initial conditions. For example the boundary condition could be

$$\vec{q}\cdot\vec{n} = -\hat{K}(T)\frac{\partial T}{\partial n} = 0 \text{ on } \Sigma = \partial\Omega \times (0, t_*), \tag{4.3.54}$$

where $\partial\Omega$ is the boundary of the region Ω under consideration and the initial conditions could be

$$\{T^2 F_0'(T) + T_m^c F_1(\phi)\}_{t=0} = e_0, \;\; (F_2(\phi))|_{t=0} = \phi_0, \tag{4.3.55}$$

where e_0 and ϕ_0 are known quantities. In Fourier's heat conduction law, $\hat{K}(T) = K(T) > 0$, where K is the thermal conductivity. In irreversible thermodynamics,

$\vec{q} = \hat{K}(T)\nabla(1/T)$. $N(T)$ is a prescribed quantity. For example, in [95], $N(T)$ has been taken to be T. If $I(\phi) = 0$, then (4.3.53) is a classical equation.

If $\hat{K}(T) = K$, a constant, and if suitable assumptions are made about various functions then the existence and uniqueness of the system (4.3.52) and (4.3.53) has been proved in [100].

4.4 Solidification of Supercooled Liquid Without Curvature Effect and Kinetic Undercooling : Analysis of the Solution

In §§ 4.2 and 4.3 some phase-field models were considered. With their help supercooling/superheating effects can be discussed if appropriate limits of parameters are taken. These models provided rigorous justification of the incorporation of the modified Gibbs-Thomson relation in the classical formulation when curvature effect and kinetic undercooling are included. Many physical situations can be modelled by suitable choices of arbitrary functions occurring in phase-field models and phase-relaxation models. In this section we want to discuss issues like: (i) existence, uniqueness and well-posedness of solutions and (ii) blow-up and regularization of Stefan problems with supercooling. It may be made clear that by 'Stefan problem with supercooling' we mean classical formulations of either solidification of a supercooled liquid or one or two-phase solidification problems with modified Gibbs-Thomson relation.

4.4.1 One-dimensional one-phase solidification of supercooled liquid (SSP)

The mathematical model of this problem has already been discussed in (3.1.8)–(3.1.12). Because of its simplicity this problem has been extensively studied. It should be remembered that neither the curvature effect nor the kinetic condition has been incorporated in this model. It was mentioned in § 3.1.2 that ODP (equations (3.3.34)–(3.3.38)) can be transformed to SSP and vice-versa and $c(x,t)$ and $T(x,t)$ are related through the equation (3.3.39). We shall now consider a more general ODP and take

$$c(x,0) = c_0(x) \ , \ 0 \leq x \leq 1, \tag{4.4.1}$$

where $c_0(x)$ could be different from $0.5(1-x)^2$. From (3.3.39), we have

$$c_0(x) = \int_1^x d\xi \int_1^\xi (T_0(\eta)+1)d\eta \ ; \ c_0''(x) = T_0(x)+1, \tag{4.4.2}$$

and $C_x(0,t)$ is given by (cf. [104])

$$c_x(0,t) = c_0'(0) + \int_0^t g(\tau)d\tau. \tag{4.4.3}$$

Here, $g(t)$ is the same as in (3.1.10). The second equation in (4.4.2) suggests that $c_0(x)$ should be sufficiently regular. For the existence and uniqueness of solutions some compatibility conditions should also be satisfied by the initial temperature and the boundary conditions. If the boundary condition (3.1.10) is replaced by the boundary condition $T(0,t) = F(t)$, then we obtain [104]

$$c(0,t) = c_0(0) + \int_0^t F(\tau)d\tau. \tag{4.4.4}$$

The essential difference between the classical Stefan problems concerning solidification of a liquid and the solidification of a supercooled liquid is in the sign requirements of the temperature difference $(T - T_m)$. The temperature of a supercooled liquid is less than T_m, whereas in the classical Stefan problem without supercooling, the temperature of the liquid is greater than or equal to T_m. As mentioned earlier, whenever the classical solution of a Stefan problem is discussed here, it is implied that the solution satisfies some conditions of the form (i) - (iv) given in § 3.1.1. Depending on the problem under consideration appropriate changes have to be made in these conditions. A global-in-time classical solution is known to exist for the one-dimensional Stefan problem under suitable data assumptions (cf. [57, 75]). If the liquid is supercooled or if the solid is superheated, a solution may still exist for a short-time but a *finite-time blow-up* (explained below) may occur, which prevents continuation of the solution.

Assuming that a solution of a Stefan problem with superheating/supercooling exists, some important questions to be investigated further are:

(A) Can a classical solution of the problem exist for an arbitrarily long time ?

(B) Does there exist a constant $t_b > 0$ such that $\lim_{t\to t_b-} S(t) = 0$? Here $S(t)$ is progressing towards $x = 0$.

(C) Does there exist a constant $t_c > 0$ such that $S(t) > 0$ for $t \in (0, t_c]$ and $\lim_{t\to t_c} \dot{S}(t) = -\infty$?

Question (B) pertains to *finite-time extinction* and (C) pertains to the blow-up of a solution defined below.

Finite-time blow-up: If $S(t)$ is the free boundary in a free boundary problem, and if at some time $t = t_* < \infty$, $S(t_*) > 0$ and $\liminf_{t\to t^*-} \dot{S}(t) = -\infty$, then it is said that a finite-time blow-up has occurred.

Essential blow-up: If the solution of the free boundary problem cannot be continued beyond the blow-up time, then the blow-up is called essential. It may be pointed out that blow-up is not a physically realistic phenomenon and it indicates defects in the mathematical model of the problem.

Non-essential blow-up: It may be possible to continue the solution of the free boundary problem beyond the blow-up time $t = t_*$ with $S(t_*)$ and $\lim_{t \to t_*-} T(x,t) = \hat{T}_0(x)$ as the initial data. In this case we have a non-essential blow-up.

Many studies are available on the analysis of SSP. Each study contains several results. Only a few of these can be presented here. In [59], [104] and [105] the boundary condition (3.1.10) in SSP, has been replaced with the following:

$$T_x(0,t) = 0, \quad 0 < t < t_*. \tag{4.4.5}$$

Let Q be a constant defined as

$$Q = \int_0^1 (T_0(x) + 1)dx. \tag{4.4.6}$$

The quantity $(1 + T_0(x))$ in which the latent heat has been suitably scaled and taken as unity, represents the sum of the latent heat and the sensible heat. Q can be interpreted as the global energy of the supercooled liquid at $t = 0$ and should be greater than or equal to zero. Q plays an important role in analysing questions (A), (B) and (C) mentioned above.

Let $T_0(x)$ be non-positive and continuous in [0,1]. If $T_0(1) = 0$ and $T_0(x)$ is Hölder-continuous at $x = 1$, then the supercooled problem possesses one solution valid for sufficiently small time $t < t_0$. The uniqueness and well-posedness of the solution can also be proved. If $T_0(x)$ is constant not exceeding -1, then SSP has no solution [75].

Some of the results reported in [75], [104] and [105] are summarized below in the form of propositions 4.4.1.–4.4.5. It should be remembered that in the discussion pertaining to these propositions, the term SSP means the mathematical formulation (3.1.8)–(3.1.12) in which (3.1.10) has been replaced by (4.4.5). Similarly ODP would mean the problem obtained by transforming the new SSP with the help of the transformation (3.3.39).

Proposition 4.4.1. If the closure of the set $\{T_0(x) < -1\}$ is either void, or if it is compact and contains the origin, then in the solution of SSP the conditions $Q > 0$, $Q = 0$ and $Q < 0$ are necessary and sufficient for global existence, finite-time extinction and blow-up, respectively. Blow-up always occurs when $Q < 0$ while $Q = 0$ is necessary for a finite-time extinction.

If the constraint $c \geq 0$ is not added in ODP with initial concentration given by (4.4.1), then the free boundary in both SSP and UODP (unconstrained ODP) coincides and both will have a finite-time blow-up at a time at which the negativity set of $c(x,t)$ reaches $x = S(t)$. For $t \in (0, t_*)$, define

$$N(t) = \{x : 0 \leq x < S(t), \ c(x,t) < 0\}. \tag{4.4.7}$$

In [104, 105], several results concerning blow-up in SSP and UODP have been reported. Some of them are given below.

Proposition 4.4.2. If $N(t) \neq \emptyset$, then

(i) for any $t_2 \in (t_1, t_*)$, $N(t_1) \subset\subset N(t_2)$, i.e., the negativity set expands,

(ii) if for some $\hat{t} > t_1$, the boundary $\partial N(\hat{t}\,)$ touches the free boundary then $\dot{S}(t)$ is singular as $t \to \hat{t}$,

(iii) the case above actually occurs.

Proposition 4.4.3. A necessary and sufficient condition for an essential blow-up at the point $(S(\hat{t}\,), \hat{t}\,)$ is that the point belongs to $\partial N(\hat{t}\,)$.

Proposition 4.4.4. A necessary and sufficient condition for having $\dot{S}(t)$ singular at a point $(S(\hat{t}\,), \hat{t}\,)$ is that the set $\{(x,t) : T(x,t) = -1\}$ reaches the free boundary at such a point.

Proposition 4.4.5. There are initial data such that ODP corresponding to SSP with (4.4.5) admits global solutions with $\lim_{t \to t_0} \dot{S}(t) = -\infty$ for some $t_0 > 0$.

The last proposition is a case of a non-essential blow-up and in [104] this case has been demonstrated with the help of an example.

4.4.2 Regularization of a blow-up in SSP by looking at CODP

The question which will be discussed now is, 'how to stop blow-up in SSP'? This is called regularization. Note that there are methods other than the one discussed below for *regularization*. Let us first examine which physical quantity is responsible for the blow-up. The integral

$$E(x,t) = \int_x^{S(t)} (T(\xi,t)+1)d\xi = -c_x, \tag{4.4.8}$$

represents the thermal energy in the layer $(x, S(t))$ (cf. (3.3.39)). $E(0,t) = E(0,0) = Q$ (cf. (4.4.6)). Here, $c(x,t)$ is the concentration in ODP.

If the mean energy $\bar{E}$ in the supercooled layer is defined as

$$\bar{E}(x,t) = (S(t)-x)^{-1} \int_x^{S(t)} d\xi \int_\xi^{S(t)} (T(\eta,t)+1)d\eta, \tag{4.4.9}$$

$$= (S(t)-x)^{-1} c(x,t), \tag{4.4.10}$$

then blow-up occurs if and only if $\bar{E}(x,t)$ is negative at some point (x,t). This implies that at the blow-up point $c(x,t) < 0$. Thus any regularization of SSP or UODP has to prevent $\bar{E}(x,t)$ from taking negative values by shifting UODP to CODP. It seems reasonable to conjecture that this argument holds for multi-dimensional problems also (cf. [105]). The absolute minimum of $\bar{E}(x,t)$ is $E(0,t)$ or Q. Blow-up occurs if and only if $Q < 0$. If $Q < 0$, then it implies that the system does not have enough energy in the form of latent heat to cancel out the 'cold' of the supercooled liquid. Another way to look at the blow-up points is to look at the roots of the equation $T(x,t) + 1 = 0$. Either the set of points (x,t) for which $T(x,t) + 1 = 0$ is empty, or it has a positive distance from the free boundary. In the latter case, if this set reaches the free boundary at $t = t_0$ then $\lim_{t \to t_0} \dot{S}(t) = -\infty$ (cf. [75]). For $t < t_0$, $\dot{S}(t)$ is bounded. This implies that there is too much 'cold' too near the free boundary, and blow-up occurs because the energy cannot reach the point of blow-up by diffusion.

To stop finite-time blow-up in SSP some mathematical techniques have been suggested in [104] which consist of modifying UODP to CODP provided CODP is formulated classically, with a new free boundary appearing each time $c(x,t)$ becomes zero. We can call it 'nucleation' of new free boundaries. In this case we have always a finite-time extinction. This 'nucleation' of new free boundaries can be achieved in SSP also. Let $Q = 0$, $c_0 \geq 0$. This implies that $c_x(0,t) = 0$. There is no change in the energy of the system in this 'nucleation' but the creation of the new phase boundaries rearranges the energy fluxes in such a way that the blow-up that would occur due to the deficit of energy near $x = S(t)$ is prevented by a counter-balancing surplus near $x = 0$.

There are techniques such as incorporation of curvature effects and kinetic undercooling for the regularization of blow-up during solidification of a supercooled liquid. Curvature effects cannot be included in SSP as it is a one-dimensional slab problem. If the kinetic condition is also excluded in the formulation, then we have to resort to other techniques for regularization such as nucleation described above.

Some of the results obtained in [105] have been extended to a two-phase problem considered in [106]. Let us assume that there is some solid region (finite or infinite) to the right of $x = S(t)$ in SSP whose temperature is $T_1(x,t) < 0$ for $t > 0$. The boundary and initial conditions are such that $T(x,t) < 0$, $t > 0$ is guaranteed till the solution exists. The following proposition, which is false for the one-phase SSP problem, is true for the two-phase problem.

Proposition 4.4.6. If $Q = 0$, an essential blow-up will occur, while, for finite-time extinction $Q > 0$ is necessary.

The above result has been proved by obtaining an integral equation for $S(t)$.

4.4.3 Analysis of problems with some changes in the initial and boundary conditions in SSP

Let us consider the formulation (3.1.8)–(3.1.12) with the boundary condition

$$T(0,t) = f(t), \;\; 0 < t < t_*. \tag{4.4.11}$$

By using the transformation (3.3.39) and using (4.4.4), a formulation in terms of $c(x,t)$ can be obtained. Let us assume that in the transformed problem

$$c(x,0) = c_0(x), \;\; 0 \leq x \leq 1. \tag{4.4.12}$$

The following results have been proved in [107].

Proposition 4.4.7. Assume that for some $x_0 > 0$ the following condition is satisfied.

$$c_0(x) \leq 0, \quad \text{for} \;\; x \in (1 - x_0, 1). \tag{4.4.13}$$

Then the solution of SSP with (3.1.10) replaced by (4.4.11) does not exist.

Proposition 4.4.8. Assume that there exists $x_0 > 0$ such that the initial temperature $T_0(x) \geq -1$, $T_0(x), \not\equiv -1$ in $(1 - x_0, 1)$. Then SSP with (4.4.11) has at most one solution in the local sense.

Proposition 4.4.9. Assume that there exists $x_0 > 0$ such that $T_0(x) > -1$ for x in $(1 - x_0, 1)$. Then SSP with (4.4.11) possesses a unique local-in-time classical solution. Existence can also be proved in a special case in which $\lim_{x \to 1-} \inf T_0(x) > -1$.

The convexity of the free boundary in SSP has been proved in [108] with no flux boundary condition imposed at $x = 0$, i.e., in (3.1.10), $g(t) = 0$. In [109] also, the convexity of the free boundary has been established but under special assumptions on $T_0(x)$. The initial data considered in [108] is of the form

$$T_0(x) \leq 0, \; x \in [0,1]; \; T_0(1) = 0; \; T_0(x) \in C^2[0,1]. \tag{4.4.14}$$

The main tool used in [109] in proving the results was the introduction of the function

$$v(x,t) = T_{xx}/T_x, \tag{4.4.15}$$

and the study of its level curves and this was done in [108] also. SSP can be formulated in terms of $v(x,t)$ which will not be discussed here. Let

$$M = \{x : x \in [0,1], \; T_0^{'}(x) \neq 0.\}. \tag{4.4.16}$$

If $T_0(x) \not\equiv 0$, then M is not void. Set

$$P(x) = T_0^{''}(x)/T_0^{'}(x), \;\; x \in M. \tag{4.4.17}$$

Out of the several results reported in [108], we present only one here.

Proposition 4.4.10. Assume that (4.4.14) is satisfied, and $P(x)$ is non-negative and non-increasing in M. If $P(1) > T_0'(1)$, then $\ddot{S}(t) > 0$ for any $t \in (0, t_*)$ and SSP with $g(t) = 0$ in (3.1.10) is solvable for any arbitrary $t_* > 0$.

Some results can be obtained for ODP also by following the approach adopted for SSP. In particular it can be proved that the free boundary in ODP with $c_0(x) = 0.5(1-x)^2$ is concave. See [108] for further details.

In [110], SSP with (3.1.10) replaced by the prescribed temperature condition $T(0,t) = f(t)$ has been considered. The questions (A), (B) and (C) raised earlier in this section have been analysed. Several results have been obtained in terms of a quantity $R(t)$ defined as

$$R(t) = \frac{1}{2} + \int_0^1 x T_0(x) dx + \int_0^t f(\tau) d\tau. \tag{4.4.18}$$

The motivation for considering $R(t)$ comes from the equation

$$\frac{(S(t))^2 - 1}{2} = \int_0^1 x T_0(x) dx + \int_0^t f(\tau) d\tau - \int_0^{S(t)} x T(x,t) dx. \tag{4.4.19}$$

When $S(t) = 0$, we have $R(t) = 0$. Equation (4.4.19) and several other equations of this kind can be derived by using the following Green's identity (cf. [57]).

$$\iint_{D_t} (vP(T) - TP^*(v)) dx d\tau = \int_{\partial D_t} \left[(T_x v - T v_x) d\tau + k^{-1} T v dx \right], \tag{4.4.20}$$

where P is the heat operator $k^{-1}T_{xx} - T_t$ and P^* is its adjoint operator, $D_t = \{(x,t) : 0 < x < S(t), 0 < t < t_*\}$. By choosing v and v_x suitably, several relations of the type (4.4.19) can be developed.

Proposition 4.4.11. Assume that the solution of SSP (with temperature prescribed in the place of (3.1.10)) exists, and $T_0(x)$ and $f(t)$ satisfy the following hypotheses:

(i) $T_0(x) \le 0$ is an increasing function in [0,1],

(ii) $f(t) \le 0$ is a decreasing function of t, $t \ge 0$.

If case (C) (§ 4.4.1) occurs, then it implies $R(t_c) \le 0$ (t_c is the same as defined in (C)).

The following convective boundary condition at $x = 0$ in SSP was considered in [111] in the place of (3.1.10).

$$K \frac{\partial T}{\partial x}(0,t) = \bar{h}(T(0,t) - g(t)), \; g(t) \le 0, \; 0 < t \le t_*. \tag{4.4.21}$$

The questions (A), (B) and (C) (§ 4.4.1) were investigated and in answering these questions, the key role is played by the quantity $Q_1(t)$ defined as

$$Q_1(t) = 1 + \frac{\beta}{2} + \int_0^1 (1 + \beta x) T_0(x) dx + \int_0^t \beta G(\tau) d\tau, \tag{4.4.22}$$

where $\beta = \bar{h}/Kb$ is a dimensionless parameter, b has the dimension of length with which the space coordinate can be dimensionalized to obtain the supercooled region as $0 \le x \le 1$ and

$$G(t) = \frac{C}{l} g\left(\frac{b^2 \rho C t}{K}\right). \tag{4.4.23}$$

We present here one of the results reported in [111].

Proposition 4.4.12. If the solution of this changed SSP with (4.4.21) exists and the initial and boundary data satisfy the following conditions

$$\left.\begin{aligned} &\text{(i) } T_0(x) \ge M(x-1), \text{ and } T_0'(x) \ge 0,\ 0 \le x \le 1,\ 0 < M < 1,\\ &\text{(ii) } G(t) \ge -M,\ t > 0,\\ &\text{(iii) } T_0'(0) = \beta(T(0) - g(0)), \end{aligned}\right\} \tag{4.4.24}$$

and there exists a time t_b such that $Q_1(t_b) = 0$, then $t = t_b$ is the time in the case (B).

The solidification of a supercooled liquid has been studied in geometries other than a slab. In a one-dimensional radially symmetric spherical geometry or in a one dimensional radially symmetric cylindrical geometry, in principle, curvature effects can be included. The curvature effect and the kinetic condition have in general a regularizing effect. If curvature effect is not included in cylindrical and spherical geometries, then the questions (A), (B) and (C) become more meaningful and should be investigated. This has been done in the next two problems. These problems can be easily formulated on the same lines as the formulation given in (3.1.8)–(3.1.12).

One-dimensional radially symmetric solidification of a supercooled liquid in the spherical region $r_0 \le r \le 1$, $r_0 > 0$ was considered in [112]. The quantity $Q_2(t)$ given below plays a key-role in the study of questions (A), (B) and (C).

$$Q_2(t) = 1 + 3\int_{r_0}^{1} r^2 T_0(r)dr - 3r_0^2 \int_0^{\tau} g(\tau)d\tau. \tag{4.4.25}$$

Here, $T_0(r)$ is the initial temperature and $g(t)$ is the prescribed flux at $r = r_0$ in the spherical geometry. A result concerning finite-time extinction is as follows.

Proposition 4.4.13. If $T_0(r) \le 0$, $r_0 < r < 1$, and continuous and $T_0(r) \ge -h_1(1-r)$, $h_1 \le 1$, $0 < g(t) \le h_1$ for $t > 0$, then case (B) occurs and $Q_2(t_b) = r_0^3$.

Solidification of supercooled liquid in a one-dimensional radially symmetric cylindrical region $r_0 \le r \le b$ was considered in [113] with temperature prescribed at $r = r_0$. A flux prescribed case can also be discussed in a similar way. One existence and uniqueness result is given below.

Proposition 4.4.14. Let the initial temperature $T_0(r)$, $r_0 \le r \le b$ be such that

$$T_0(r) > -1, \quad b - d_0 < r < b \tag{4.4.26}$$

for some $d_0 \in (0, b - r_0]$. Then the solution of the cylindrical problem exists in the interval $0 < t < t_*$ for some $t_* > 0$. The case $T_0(b) = -1$ is allowed as long as conditions in (4.4.26) are fulfilled. Uniqueness of the solution can be proved as in [107].

The SSP considered in [114] is ODP transformed into SSP. The initial concentration in ODP is $0.5(1-x)^2$ but at $x = 0$ non-zero flux is prescribed which is a monotonically increasing function of time with $c_x(0,0) = -1$. This ODP can be transformed to SSP with the initial temperature zero and the flux at $x = 0$ given by $T_x(0,t) = g(t),\ 0 < t < t_*$. It is assumed that $g(t)$ is a non-negative piecewise continuous function in $(0, \infty)$, bounded in every interval $(0, t)$, $t > 0$. The questions (A), (B) and (C) have been analysed for the transformed SSP. By using an appropriate Green's identity, the relation given below can be obtained.

$$S(t) = 1 - \int_0^t g(\tau)d\tau - \int_0^{S(t)} T(x,t)dx, \quad 0 < t < t_*. \tag{4.4.27}$$

A unique solution of the transformed SSP exists if and only if

$$\int_0^t g(\tau)d\tau < 1 \text{ for any } t > 0. \tag{4.4.28}$$

For other results see [114].

4.5 Analysis of Supercooled Stefan Problems with the Modified Gibbs-Thomson Relation

4.5.1 Introduction

A supercooled liquid is in a metastable state, and, if left to itself, will remain in that state for a long time. In the absence of curvature effect or kinetic undercooling in the equilibrium temperature, supercooled liquid has a destabilizing effect on the planar phase-change boundary. Imagine a small solid protrusion into the supercooled liquid at the planar interface. Suppose that the temperature of the supercooled liquid is decreasing as we move away from the planar front. Let the equilibrium temperature be T_m which is the ideal phase-change temperature. This solid protrusion will have a large temperature gradient. According to Stefan condition (1.4.22), its growth velocity will be greater than that of its neighbours on the planar front, and so this protrusion will grow and not dye. This indicates instability.

The effect of surface tension is to stabilize the phase-change boundary. Since the surface tension decreases the equilibrium temperature, any protrusion at the planar free boundary gets liquified. Without surface tension, supercooling could lead to a totally unstable system. In real physical systems when supercooling is present, and which is generally the case, local instabilities such as dendrites will be present. These are counter balanced by the nonlinear stabilizing effects of surface tension. In some sense, these arguments also hold for kinetic undercooling as well [86] that kinetic undercooling has a stabilizing effect in the sense that in the perturbation of the planar boundary an unstable mode remains unstable but with smaller amplitude. Large surface tension is incompatible with large curvature as large surface tension in supercooling implies that the deviation of the new equilibrium temperature from T_m is negative and large. This means that protrusions will be dissolved. This also means that the interface is nearly planar and the curvature is small. Therefore if the initial and boundary conditions are suitably chosen, then supercooling can be restricted. Surface tension is an equilibrium phenomenon, as it arises in equilibrium thermodynamics, but kinetic undercooling is a non-equilibrium phenomenon. It may be noted that in the Stefan problem, phase-change boundary is moving but it is assumed that there is an instantaneous equilibrium so that laws of equilibrium thermodynamics can be applied.

4.5.2 One-dimensional one-phase supercooled Stefan problems with the modified Gibbs-Thomson relation

Consider the formulation given in (3.1.8)–(3.1.12) with $g(t) = 0$ in (3.1.10) and (3.1.11) taken as

$$T(S(t),t) = \varepsilon\dot{S}(t), \;\; \varepsilon > 0, \; \dot{S}(t) < 0. \tag{4.5.1}$$

Equations (3.1.8), (3.1.9), (3.1.10), (3.1.12) and (4.5.1) constitute our new SSP. In equation (4.5.1) kinetic undercooling has been incorporated. Note that since $S(t)$ is moving towards $x = 0$, $\dot{S}(t)$ is negative. In the problem considered is [115], $S(0) = 0$ and $\dot{S}(t) > 0$ which is different from our formulation. This change in essence does not affect the results of [115]. It has been established in [115] that if the initial temperature $T_0(x)$ is such that $T_0'(x)$ is continuous in $0 \leq x \leq 1$ and $T_0'(0) = 0$ then the kinetic undercooling will have a regularizing effect and blow-up will be prevented. In physical terms, the kinetic undercooling term allows more energy to enter the liquid at the moving boundary than would be the case if this energy is supplied by the latent heat alone (as in SSP without kinetic undercooling). The rate of change of energy is given by (cf. equation (4.4.8) for $E(x,t)$)

$$\frac{dE}{dt} = \frac{d}{dt}\int_0^{S(t)} \{T(x,t)+1\}dx = \varepsilon\dot{S}^2. \tag{4.5.2}$$

As $\varepsilon > 0$, there is an additional energy $\varepsilon\dot{S}^2$ available which compensates any energy deficit in $T_0(x)$. The kinetic undercooling term in (4.5.1) imposes a limit on the speed of the interface. For some results concerning a one-dimensional two-phase problem and a semi-infinite problem, see [115].

A one-dimensional supercooled problem in a semi-infinite region $0 \leq x < \infty$ has been considered in [116]. The region under consideration is $S(t) \leq x < \infty$, $t > 0$ and

$S(0) = 0$. Let $T_0(x) \leq 0,\ 0 \leq x < \infty,\ T_0(x) \in C^1[0,\infty)$ be the initial temperature of the supercooled liquid and

$$T_0(x) \to -1 - \tilde{\delta} \text{ as } x \to \infty, \tag{4.5.3}$$

where $\tilde{\delta}$ is a constant. If kinetic undercooling is considered, then

$$T(S(t),t) = -\varepsilon\dot{S}(t),\ \varepsilon \geq 0,\ \dot{S}(t) > 0,\ t > 0. \tag{4.5.4}$$

No fixed boundary condition at $x = 0$ is required as the region $x < 0$ is ice at time $t = 0$. Several cases for $\tilde{\delta} > 0, = 0, < 0$ have been analysed when $\varepsilon > 0$ and $\varepsilon = 0$. When $\varepsilon = 0$, and $\tilde{\delta} > 0$, a finite-time blow-up occurs. When $\varepsilon > 0$ and $\tilde{\delta} < 0$, similarity solutions exist and $S(t) = \beta\sqrt{t}$ (for β see [116]). When $\varepsilon > 0$, and $\tilde{\delta} > 0$, travelling wave solutions exist with $S(t) = \tilde{\delta}t/\varepsilon$. If $\varepsilon > 0$, and $\tilde{\delta} = 0$; $S(t) \sim dt^\alpha$, $\frac{1}{2} < \alpha < 1$, where d is a constant depending on several parameters including ε.

A radially symmetric spherical problem in the region $S(0) \leq r < \infty$, $S(0) \neq 0$ was also considered in [116]. Both surface tension and kinetic condition have been considered so that

$$T(S(t),t) = -\varepsilon\dot{S} - 2\sigma/S,\ \dot{S} > 0,\ t > 0. \tag{4.5.5}$$

Note that in this case the curvature $\bar{K}_c = 1/r$ and the free boundary is $r = S(t)$. Initial temperature is of the same type as in (4.5.3). If $\varepsilon = 0$ and $\tilde{\delta} > 0$, then even with surface tension a finite-time blow-up occurs. However if $\varepsilon > 0$, and $\tilde{\delta} > 0, = 0$ or < 0, finite-time blow-up does not occur. For other results see [116].

In [117], the effect of both kinetic undercooling and latent heat on the solution has been analysed. Consider SSP described by (3.1.8)–(3.1.12) in which instead of (3.1.10), the boundary condition $T(0,t) = f(t)$ is prescribed. The interface conditions are

$$T(S(t),t) = \varepsilon\dot{S},\ \varepsilon > 0,\ \dot{S} < 0, \text{ and } T_x(S(t),t) = -l\dot{S}, \tag{4.5.6}$$

$T_0(x)$ and $f(t)$ are continuous non-positive functions such that $T_0(0) = f(0)$. When $\varepsilon = 0$, and $l < 0$, (l is the latent heat) we have a one-phase Stefan problem concerning solidification. When $\varepsilon = 0$ and $l > 0$ we have a supercooled Stefan problem. In the latter case, if $\varepsilon > 0$, then it will be a supercooled problem with a kinetic condition. Several cases have been considered for different signs of ε and l but our interest at present is in the case $\varepsilon > 0$ and $l > 0$ for which the following result has been proved in [117].

Proposition 4.5.1. If $T_0(x)$ and $f(t)$ satisfy the assumptions stated above, then the supercooled problem with (4.5.6) in which $\varepsilon > 0$, and $l > 0$ admits a unique solution and the free boundary $S(t)$ is a smooth function in $(0, t_*)$ where t_* is such that $S(t_*) = 0$ and $S(t) > 0$ for $t < t_*$.

4.5.3 One-dimensional two-phase Stefan problems with the modified Gibbs-Thomson relation

Although a two-phase Stefan problem with the modified Gibbs-Thomson relation can be easily formulated, for the sake of better understanding of the results, we present here the formulation of a two-phase Stefan problem with a kinetic condition.

$$\frac{\partial T_S}{\partial t} = k_S \frac{\partial^2 T_S}{\partial x^2}, \quad \text{in } Q_1 = \{(x,t) : 0 < x < S(t),\ 0 < t < t_*\}, \tag{4.5.7}$$

$$\frac{\partial T_L}{\partial t} = k_L \frac{\partial^2 T_L}{\partial x^2}, \quad \text{in } Q_2 = \{(x,t) : S(t) < x < 1,\ 0 < t < t_*\}, \tag{4.5.8}$$

$$T_S(S(t)-0,t) = T_L(S(t)+0,t) = T(S(t),t),\ 0 \le t < t_*, \tag{4.5.9}$$

$$\dot{S}(t) = g(T(S(t),t)), \tag{4.5.10}$$

$$S(0) = S_0,\ 0 < S_0 < 1, \tag{4.5.11}$$

$$\left\{ K_S \frac{\partial T_S}{\partial x} - K_L \frac{\partial T_L}{\partial x} \right\}_{x=S(t)} = l\dot{S}(t),\ l > 0,\ 0 \le \dot{S}(t) \le \text{constant}\ , \tag{4.5.12}$$

$$T(x,0) = T_0(x),\ 0 \le x \le 1, \tag{4.5.13}$$

$$T_L(1,t) = h_L(t),\ T_S(0,t) = h_S(t). \tag{4.5.14}$$

This solidification problem has been considered in [118]. The main interest in [118] is to establish the existence and uniqueness of the classical solution of the problem if $g(T)$ is of the form

$$g(T) \in C^1(R) \text{ and } g(T) = T^m,\ m \in N. \tag{4.5.15}$$

The condition (4.5.12) is satisfied at $x = S_0$ also. $h_i(t) \in C^{\alpha_1}[0,t_*]$, $i = S, L$, $\frac{1}{2} < \alpha_1 < 1$, $T_0(x) \in C^{\alpha_2}[0,S_0] \cap C^{\alpha_2}[S_0,1]$, $2 < \alpha_2 < 3$, $T_0(0) = h_S(0)$, $T_0(1) = h_L(0)$ and S_0 satisfies a constraint (cf. [118]) which involves the L^∞-norms of the initial and boundary temperatures. No sign constraints have been prescribed on the initial and boundary data. It has been proved that there exists a t_* such that $\lim_{t\to t_*} S(t) = 1$. Travelling wave solutions of the problem have also been obtained. A similar problem with $g(T) \in C^1(R)$ and $|g(T)| \le a(|T|^m + 1)$, where a and $m \ge 1$ are positive constants, was considered in [119] and local existence of the classical solution was proved under some assumptions some of which are the same as in [118].

A two-phase problem in $-\infty < x < \infty$ with kinetic undercooling condition (4.5.10) has been considered in [120]. It was proved in [120] that if $|g(T)| \leq Me^{\gamma|T|}$ for some $M > 0$ and $\gamma \in (0, 1/4)$, then a global solution to the problem (4.5.7)–(4.5.14) exists under suitable assumptions. If $g(T) \geq ae^{\gamma_1|T|}$ for some constant a and large γ_1, then blow-up in finite time may occur. A critical value of the exponent γ was obtained in [121] as $\gamma = \sqrt{2\pi e}$ such that for $\gamma \in (0, \sqrt{2\pi e})$, a global-in-time solution exists and blow-up in finite time occurs if $g(T) \geq ae^{\gamma|T|}$.

In [122], the problem (4.5.7)–(4.5.14) was considered and a single function $\psi(x,t)$ represents both initial and boundary conditions.

$$T(x,t) = \psi(x,t), \ \ (x,t) \in \partial^b Q_{t_*}, \tag{4.5.16}$$

where

$$\partial^b Q_{t_*} = \{(x,0) : x \in (0,1)\} \cup \{(0,t) : t \in (0,t_*)\} \cup \{(1,t) : t \in (0,t_*)\}.$$

It was proved that if $g(T) \in C^{\beta}(R)$, $0 < \beta < 1$, $\psi \in W_q^{2,1}(Q_{t_*})$, $q > 3$ and the following conditions are satisfied (any one out of (4.5.17) and (4.5.18)),

$$g(T) \leq 0, \ T \leq m_1, \ g(T) \geq 0, \ T \geq m_2 \text{ for some } m_1, m_2 \in R, \tag{4.5.17}$$

$$|g(T)| \leq a_1|T| + a_2, \ a_1 \text{ and } a_2 \text{ are constants}, \tag{4.5.18}$$

$$g(\psi|_{x=0}) = g(\psi|_{x=1}) = 0, \tag{4.5.19}$$

then a global classical solution exists such that $S(t) \in C^{1+\lambda/2}[0,t_*]$, $\lambda \in (0,1)$. $W_q^{2,1}(Q_{t_*})$, $q \geq 1$ is the Sobolev space of functions $u(x,t)$ in the domain $Q_{t_*}(\bar{Q}_1 \cup \bar{Q}_2 \times (0,t_*))$ with bounded norm

$$||u||_{W_q^{2,1}(Q_{t_*})} = \Big(\int_{Q_{t_*}} (|u|^q + |u_x|^q + |u_{xx}|^q + |u_t|^q) dx dt\Big)^{1/q}. \tag{4.5.20}$$

Replace the condition (4.5.10) in problem (4.5.7)–(4.5.14) by the following kinetic condition

$$T^{\varepsilon}(S^{\varepsilon}(t),t) = -\varepsilon \dot{S}(t), \ \varepsilon > 0, \ 0 < t < t_*, \tag{4.5.21}$$

where T^{ε} and S^{ε} are written in place of T and S to signify that both T and S now depend on ε. This problem has been considered in [123], and the results given in Proposition 4.5.2. have been established.

Proposition 4.5.2. Let $T_0(x)$, h_S and h_L satisfy the following smoothness assumptions.

$$T_0(x) \in C^1[0,S_0] \cap C^1[S_0,1] \cap C[0,1] \text{ and } h_L(h_S) \in C^1(R) \cap L_{\infty}(R). \tag{4.5.22}$$

The consistency conditions at $x = 0$, and $x = 1$ for the initial temperature and boundary conditions are satisfied and the functions h_S and h_L do not change sign for $t > 0$. Under the above assumptions a unique classical solution $(T^\varepsilon, S^\varepsilon)$ (in the sense of § 1.4.6)of the problem (4.5.7)–(4.5.14) with (4.5.21) taken in the place of of (4.5.10) exists for every $\varepsilon > 0$ and for some $t_*^\varepsilon > 0$. Either $t_*^\varepsilon = +\infty$ or $\min\{1 - S^\varepsilon(t_*^\varepsilon), S^\varepsilon(t_*^\varepsilon)\} = 0$.

Depending on the boundary conditions, $\dot{S}$ could be greater or less than zero. By considering a sequence of problems $\{T^{\varepsilon_n}, S^{\varepsilon_n}\}$, $\varepsilon_n > 0$, $n = 1, 2 \ldots$ and taking the limit $\varepsilon_n \to 0$, the existence of a weak solution has been established in [123] and it has been shown that $T(S(t), t) \to 0$ for a.a. $t \in (0, t_*)$.

Suppose the initial data is such that for some $S_1 \in [S_0, 1]$

$$T_0(x)(x - S_1) > 0, \; x \in [0, 1], \;\; x \neq S_1, \; S_0. \tag{4.5.23}$$

If (4.5.23) is satisfied, then a supercooled liquid phase exists between stable solid and stable liquid phases. It has been proved in [123] that the simply connected supercooled liquid region disappears in finite time, after which the solution becomes a classical one. In (4.5.23), the degree of supercooling is not prescribed. It was pointed out earlier for a one-phase supercooled problem (cf. Proposition 4.4.1.) that the negativity set $\{T_0(x) < -1\}$ is responsible for a blow-up (roughly speaking). The initial data given below gives rise to a discontinuity in the free boundary.

Let S_1^+ and $S_1^- \in (S_0, 1)$ be such that

$$T_0(x) < -l, \;\; x \in (S_1^-, S_1^+), \tag{4.5.24}$$

$$T_0(x) > -l, \;\; x \in (S_0, S_1^-) \cup (S_1^+, 1), \tag{4.5.25}$$

where l is the latent heat, and $l > 0$. Under suitable assumptions, it has been shown in [123] that the free boundary is smooth everywhere except at a point $t = \hat{t} < t_*$. At this point $S(t)$ may have a jump. $S(\hat{t} + 0) - S(\hat{t} - 0) > 0$ will be exactly equal to the length of the interval in which $T(x, \hat{t} - 0) \leq -l$.

In the Neumann solution (1.3.11)–(1.3.16), the velocity of the interface (cf. (1.3.13)) is infinite at $t = 0$ which seems to be inconsistent with the assumption of local thermodynamic equilibrium at a constant freezing temperature. If the kinetic condition (4.5.26) is incorporated in (1.3.5), $\rho_S = \rho_L$ and in the place of $\hat{l} = \{l + (C_L - C_S)T_m\}$, l is taken, then we get a problem considered in [123]. Let

$$T_S(S(t), t) = T_L(S(t), t) = -\varepsilon \dot{S}(t), \; \varepsilon > 0, \; \dot{S} > 0. \tag{4.5.26}$$

Under suitable assumptions, the existence and uniqueness of the Neumann solution with the condition (4.5.26) and other changes have been proved. It has been shown that $T(x, t) \in C^\infty(Q_1) \cap C^\infty(Q_2)$ and $S(t) \in C^\infty(0, t_*)$, where $Q_1(Q_2) = \{x > 0, x - S(t) > 0 (< 0), \; 0 < t < t_*\}$. $S(t) \geq \gamma t$ for $t \in (0, t_*)$, where γ is a positive constant. An analytic solution has been obtained which is validated by a numerical solution obtained by a finite-difference scheme [123]. An equation for the phase-change temperature has been obtained in [124] by considering the entropy production in the interface, pressure-induced undercooling, and viscosity-induced variations in the freezing temperature. A fairly general expression for the latent heat depending on the thermal expansion coefficient, temperature, pressure, etc., has also been obtained. In deriving phase-change temperature and latent heat expressions, non-equilibrium thermodynamical considerations have been used.

Dissolution of a spherical ice ball in supercooled water has been discussed in [125]. This one-dimensional radially-symmetric spherical problem is closely related to a Stefan problem with modified Gibbs-Thomson relation. In [125], a weak solution has also been discussed. We give below only the classical formulation of the problem of dissolution of a solid ice ball of radius $S(t)$ in a supercooled liquid. Consider the following equations.

$$\frac{1}{r}\frac{\partial^2}{\partial r^2}(rT_i) = \frac{\partial T_i}{\partial t},\ i = S, L;\ \text{in } Q^i_{t_*}, \tag{4.5.27}$$

$$T(r,0) = T_0(r) \le 0,\ 0 \le r \le r_0,\ T_0'(0) = 0;\ T_L(r_0,t) = T_1(t),\ t > 0, \tag{4.5.28}$$

$$T_S(S(t),t) = T_L(S(t),\,t) = -\frac{1}{S(t)} + \beta\frac{dS}{dt},\ r = S(t),\ t > 0; S(0) = 1, \tag{4.5.29}$$

$$\frac{\partial T_S}{\partial r} - \frac{\partial T_L}{\partial r} = -l\frac{dS}{dt} \quad \text{on } r = S(t),\ t > 0, \tag{4.5.30}$$

Here, $\beta > 0$ is the relaxation parameter arising in the kinetic condition, the surface tension in (4.5.29) is taken unity and

$$Q^S_{t_*} = \{(r,t) : 0 < r < S(t);\ 0 < t < t_*,\ S(0) = 1\},$$

$$Q^L_{t_*} = \{(r,t) : S(t) < r < r_0;\ 0 < t < t_*,\ r_0 > 1\}.$$

It will be assumed that equations (4.5.27)–(4.5.30) have been suitably dimensionalized. In (4.5.29) both curvature effect and kinetic undercooling have been incorporated. Except β and l, all other parameters have been taken to be unity.

To discuss melting of ice in a supercooled liquid, the transformation

$$\hat{T} = rT(r,t) + 1, \tag{4.5.31}$$

is made which changes the problem (4.5.27)–(4.5.30) to the following problem.

$$\frac{\partial \hat{T}_i}{\partial t} = \frac{\partial^2 \hat{T}_i}{\partial r^2},\ r \in Q^i_{t_*},\ i = S, L,\ t > 0, \tag{4.5.32}$$

$$\hat{T}(r,0) = \hat{T}_0(r),\ r \in Q^L_{t_*} \cup Q^S_{t_*} \cup S(t);\ \hat{T}_S(0,t) = 1,\ t > 0;\ \hat{T}_L(r_0,t) = \hat{T}_1(t), \tag{4.5.33}$$

$$\hat{T}_i(S(t),t) = -\beta S(t)\dot{S}(t),\ i = S, L, \tag{4.5.34}$$

$$\left[\frac{\partial \hat{T}}{\partial r}\right]_{S(t)+0}^{S(t)-0} = -lS(t)\dot{S}(t). \tag{4.5.35}$$

If $\beta = 0$ in (4.5.29), then we have only curvature effect in $T(S(t),t)$. Equation (4.5.35) suggests that we have a Stefan problem with latent heat $-lS(t)$. Now consider the initial and boundary data such that we have a one-phase melting problem for $\hat{T}$. We consider the following formulation.

$$\hat{T}_S(0,t) = 1,\ t > 0;\ \hat{T}_S(S(t),t) = 0,\ t > 0;\ \hat{T}_S \geq 0,\ r \in [0,1],\ t > 0, \tag{4.5.36}$$

$$\hat{T}_L(r_0,t) = 0,\ \hat{T}_L \equiv 0 \text{ in } r \in (1, r_0), t > 0. \tag{4.5.37}$$

In terms of $T_L(r,t)$, the initial temperature of the liquid is

$$T_L(r,0) = -1/r,\ 1 \leq r \leq r_0. \tag{4.5.38}$$

Note that the surface tension which was taken as unity in (4.5.29), is still present. As the initial temperature of the solid is greater than or equal to $T_L(r,0)$, the ice ball with center at $r = 0$ may dissolve partially in the supercooled water at least for a short time. To preserve the classical one-phase Stefan problem structure, we have $\hat{T}(S(t),t) = 0$ which requires that $\beta \to 0$. The key idea in the existence proof for problem (4.5.36)–(4.5.38) in which $\beta = 0$ is to consider a sequence of problems for $\beta_n > 0$, $n = 1, 2 \ldots$ and take the limit as $\beta_n \to 0$. This method of proof is generally employed to develop the proof for the case $\beta \to 0$. Under suitable assumptions about the data, a unique classical solution $(\hat{T}_S^\beta,\ S^\beta)$ exists with the monotonically decreasing free boundary S^β. Some interesting results of [125] are the following:

1. The classical solution of the problem (4.5.27)–(4.5.30) exists and is unique for $\beta > 0$ and all time $t > 0$.

2. If $\beta = 0$ in (4.5.29), then the classical solution exists and is unique only locally in time.

3. There exists at least one weak solution for $\beta = 0$ with a monotone free boundary $r = S(t)$ for all time $t > 0$.

4. Non-existence of the classical solution for the problem with $\beta = 0$ for all time $t > 0$ for certain types of initial data is possible in this one-phase problem. There exists at least one point $t = \hat{t}$ of the discontinuity of $S(t)$ such that

$$0 \leq S(\hat{t} - 0) < S(\hat{t} + 0),$$

or the continuous function $S(t)$ on a small time interval $(0, t_*)$ cannot be absolutely continuous on $(\hat{t}, t_*)$ and it maps some zero-measure set of $(\hat{t}, t_*)$ onto some set of $Q_{t_*}^S \cup Q_{t_*}^L$ with a strictly positive measure.

It turns out that the results for $\beta > 0$ are not the same as for $\beta = 0$. The kinetic condition acts as a regularizing factor in the dissolution of ice ball in the supercooled water whereas surface tension alone does not have the same effect.

The formulation given in (4.5.27)–(4.5.30) was considered in [126] also but with some minor changes in the region under consideration and some change in the boundary condition. We shall continue with the notations used in (4.5.27)–(4.5.35) with the change that $S(0) = S_0 < 1$ and $r_0 = 1$. The boundary condition at $r = 1$ could be any of the type given below in (4.5.39) and (4.5.40)

$$\hat{T}(1,t) = \Theta(t),\ t > 0, \tag{4.5.39}$$

$$\hat{C}\hat{T}(1,t) = -\frac{\partial \hat{T}}{\partial r}(1,t) + \hat{T}(1,t) - 1,\ t > 0. \tag{4.5.40}$$

Both (4.5.39) and (4.5.40) are in non-dimensional form. Equation (4.5.40) is a boundary condition of the fifth kind (cf. (1.4.44)) . $\Theta(t)$ is the temperature of the thermostat and $\hat{C}$ is the specific heat of the thermostat. We shall call this changed problem, Problem (GP). Equation (4.5.40) models the heat exchange between the thermostat and the body. Assume that

$$\hat{T}_0(r) \in C^1[0, S_0] \cap C^1[S_0, 1] \cap C[0,1], \tag{4.5.41}$$

and the consistency conditions

$$\hat{T}_0(0) = 1,\ \hat{T}(S_0) = 0,\ \hat{T}_0(1) = \Theta(0), \tag{4.5.42}$$

are satisfied. The result stated in Proposition 4.5.2. can also be established for Problem (GP) (cf. [126]).

Asymptotic stability of stationary solutions of Problem (GP) was also studied in [126] and it was shown that under suitable assumptions the problem has one or two or three stationary solutions. One of them corresponds to the case in which the solid phase has melted completely. This solution is always asymptotically stable as $t \to +\infty$. The other two stationary solutions have non-empty solid regions. The solution with the largest radius for the solid is stable, the other is unstable.

A two-phase one-dimensional Stefan problem in the region $0 \le x \le 1$ with a kinetic condition of the form

$$T^- = T^+ = \varepsilon \dot{S}(t),\ \text{ at } x = S(t), \tag{4.5.43}$$

was considered in [127]. In (4.5.43), ε is a constant without sign restrictions. Similarly $\dot{S}(t)$ has no sign restrictions. T^+ is the temperature of the region to the right of $S(t)$, and T^- is the temperature of the region on the left of $S(t)$. There are no sign restrictions on T^+ and T^- and so superheating or supercooling may exist. $S(0) = S_0,\ 0 < S_0 < 1$. There are no sign restrictions on the initial and boundary data. The main result of this study is that introducing a kinetic term at the interface can prevent finite-time blow-up even if supercooling (superheating) exists. Under suitable assumptions on the initial and boundary data, the existence of a global-in time solution has been proved using Schauder's fixed point theorem. Here global-in-time solution means that either $t_* = \infty$ and $0 < S(t) < 1$, $t < \infty$, $0 < t \leq t_*$ or $t_* < +\infty$ and $\lim_{t \to t_*-} S(t) = 0$ or 1. Uniqueness of the solution has also been proved. The solution of the classical Stefan problem can be retrieved by taking the limit $\varepsilon \to 0$ if correct sign restrictions are imposed on the initial and boundary data and the free boundary is assumed to be nondecreasing.

4.5.4 Multi-dimensional supercooled Stefan problems and problems with the modified Gibbs-Thomson relation

The one-phase supercooled Stefan problem in two and three dimensions without regularization by the Gibbs-Thomson condition has been studied in [128] with particular interest in the mechanism of cusp formation for interfaces that are initially smooth. The problem considered is as follows:

$$T_t = \nabla^2 T\,,\; x \in \Omega(t) \subset R^N,\; t > 0, \tag{4.5.44}$$

$$T(x,0) = T_0(x) \leq 0, \;\text{ for } x \in \Omega(0), \tag{4.5.45}$$

$$T(x,t) = 0, \;\text{ for } x \in \partial\Omega(t),\; t \geq 0, \tag{4.5.46}$$

$$\vec{V}\cdot\vec{n} = -\nabla T\cdot\vec{n},\; x \in \partial\Omega(t),\; t \geq 0. \tag{4.5.47}$$

Here, $N = 2, 3$, $\partial\Omega(t)$ is the boundary of $\Omega(t)$ which is a free boundary and $\vec{n}$ is the outward unit normal on this boundary. As discussed earlier in the one-dimensional case, the equation

$$c_t = \nabla^2 c - \chi_{\Omega(t)},\; x \in R^N,\; t > 0, \tag{4.5.48}$$

where c stands for the concentration, is more suitable to handle a supercooled problem. $\chi_{\Omega(t)}$ is the *characteristic function* of $\Omega(t)$ which for each t takes the value unity if $x \in \Omega(t)$ and zero elsewhere. With the help of Baiocchi transformation [129], $c(x,t)$ can be defined as

$$c(x,t) = -\int_t^{S(x)} T(x,\xi)d\xi, \tag{4.5.49}$$

where the free boundary has been written as $t = S(x)$, $t > 0$, $S(x) = 0$ for $t = 0$ and $T(x,t)$ satisfies (4.5.44)–(4.5.47). We have

$$\begin{aligned} \nabla\cdot(\nabla c) &= \int_{S(x)}^t \nabla.(\nabla T)d\xi - \nabla(S(x)).\nabla T(x,S(x)), \\ &= \int_{S(x)}^t \frac{\partial T}{\partial t}d\xi + 1 = T(x,t) + 1 = c_t + 1. \end{aligned} \tag{4.5.50}$$

In obtaining (4.5.50), the relation (4.5.51) has been used which can be derived as follows. Let $\Phi(x,t) = t - S(x) = 0$. On using (1.4.15) and (1.4.25) in (4.5.47), we obtain

$$\nabla(S(x)).\nabla T(x, S(x)) = -1. \tag{4.5.51}$$

By differentiating (4.5.49), we get $c_t = T(x,t)$. In the one-dimensional Stefan problems, to prove the existence and uniqueness of solutions, fixed point arguments combined with classical 'parabolic regularity theory' can be used. In the multi-dimensional Stefan problems, there is lack of regularity in space at the free boundary which does not allow the fixed point arguments to be used. Suppose the free boundary $\partial\Omega(t) \in C^{m,\alpha}$ for each $t \geq 0$, m (integer) ≥ 0, $0 < \alpha < 1$ then $T(x,t)$ would be at most $C^{m,\alpha}$ in space by the classical regularity theory and then

$$\frac{\partial T}{\partial n} \in C^{m-1,\alpha}, \tag{4.5.52}$$

in space. Solving (4.5.47) by 'characteristics' one would obtain atmost $\partial\Omega(t) \in C^{m-1,\alpha}$ in a hypothetical fixed point argument. One way to overcome this difficulty is to regularize the supercooled problem by introducing curvature effects at the free boundary and write

$$T(x,t) = \varepsilon \mathcal{P}(\partial\Omega(t)), \tag{4.5.53}$$

where $\mathcal{P}$ is a regular second order operator. By taking the limit $\varepsilon \to 0$, the problem (4.5.44)–(4.5.47) can be obtained provided enough 'a priori' estimates can be proved.

A constructive approach has been adopted in [128] and a method has been described to obtain a short-time solution of the problem transformed in terms of $c(x,t)$. The manner in which cusps develop is described. By using perturbation analysis and matched asymptotic expansions, a mechanism of cusp formation is presented for interfaces that are initially smooth. Even small amount of undercooling can give rise to singularities.

A one-phase two-dimensional Stefan problem with kinetic condition has been considered in [130] in the region $\Omega_t \subset R^2 \times R^+$ where $\Omega_t = \{(x,y) : 0 < y < g(x,t),\ x \in R, 0 < t < t_*\}$, $y = g(x,t)$ is the equation of the free boundary, $g(x,0) = g_0(x) > 0$, $x \in R$. The free boundary conditions are

$$\nabla T \cdot \vec{n} = T\ ;\ \text{and}\ \ T = \vec{V} \cdot \vec{n} = g_t/(1+g_x^2)^{1/2}. \tag{4.5.54}$$

The remaining part of the problem formulation consists of the heat equation and initial and boundary conditions and is given below.

$$T_t = \nabla^2 T,\ (x,y,t) \in \Omega_t, \tag{4.5.55}$$

$$T(x,y,0) = T_0(x,y)\ ;\ T(x,0,t) = b(x,t). \tag{4.5.56}$$

Under suitable assumptions on the compatibility of initial and boundary conditions at $y = 0$, and at the initial position of the free boundary and some regularity conditions, the local-in-time existence and uniqueness of the classical solution have been proved using Schauder's fixed point theorem. It was observed that the introduction of curvature term in the temperature at the free boundary provides more regularity to the temperature than the kinetic condition alone.

A classical two-phase Stefan problem with the modified Gibbs-Thomson relation has been studied in [131] in a region $G \subset R^n, n \geq 2$. Let $G(t) = G \times \{t\}$ for $\forall t \in [0, t_*], G(t) = G_1(t) \cup G_2(t) \cup S(t)$, where $G_1(t)$ is the solid region, $G_2(t)$ is the liquid region and $S(t)$ is the phase-change interface separating solid and liquid regions. Let $G^{t_*} = \cup_{0 \leq t \leq t_*} G(t)$, $i = 1, 2$, and $G_i^{t_*} = \cup_{0 \leq t \leq t_*} G_i(t)$ and $G^{t_*} = G_1^{t_*} \cup G_2^{t_*} \cup \Gamma$ where $\Gamma = \cup_{0 \leq t \leq t_*} S(t)$. On Γ and the parabolic boundary, we have

$$T = -\sigma \bar{K}_c - \beta \vec{V} \cdot \vec{n}, \; \sigma > 0, \; \beta > 0 \text{ on } \Gamma, \tag{4.5.57}$$

$$T(x,t) = g(x,t) \text{ on } \{G \times \{t = 0\}\} \cup \{\partial G \times [0, t_*]\}, \tag{4.5.58}$$

$$S(0) = \Gamma_0. \tag{4.5.59}$$

Here, $\vec{n}$ is the unit normal vector pointing from $G_1(t)$ to $G_2(t)$. Equations (4.5.57)–(4.5.59) together with heat equations in the solid and the liquid regions and the Stefan condition at the interface completes the formulation of the problem. Thermal conductivities of the solid and the liquid phase are taken unequal. The main result of [131] is given by the following proposition.

Proposition 4.5.3. Let

$$g \in C^0\left(\bar{G} \times [0, t_*]\right) \cap C^{2+\alpha}\left(\bar{G}_1(0) \times [0, t_*]\right) \cap C^{2+\alpha}\left(\bar{G}_2(0) \times [0, t_*]\right), \tag{4.5.60}$$

for some fixed $\alpha > 0$, $G(0) = G_1(0) \cup G_2(0) \cup \Gamma_0$. Here, Γ_0 is the boundary of an open set with

$$\text{dist}(\Gamma_0, \partial G) > 0, \text{ and } \Gamma_0 \in C^{3+\alpha}. \tag{4.5.61}$$

Assume that g and Γ_0 satisfy the compatibility conditions

$$g_t(x,0) - K_i \nabla^2 g(x,0) = 0, \; x \in \partial G \cap \partial G_i(0), \; i = 1, 2, \tag{4.5.62}$$

and

$$g(x,0) = -\sigma \bar{K}_c(x) - \beta l \left[K_1 \frac{\partial g_1}{\partial n} - K_2 \frac{\partial g_2}{\partial n} \right](x,0), \; x \in \Gamma_0, \tag{4.5.63}$$

where $g = g_i$ in $G_i(0)$, $i = 1, 2$, $\bar{K}_c(x)$ is the sum of the principal curvatures of Γ_0. If the above conditions are satisfied, then there exists a time $t_0 > 0$ (depending on Γ, σ and β) such that the Stefan problem with condition (4.5.57) admits a unique solution (T, Γ) for $t \in [0, t_0]$ and

$$T \in C^{2+\alpha,(2+\alpha)/2}(\bar{G}_1^{t_0}) \cap C^{2+\alpha,(2+\alpha)/2}(\bar{G}_2^{t_0}), \tag{4.5.64}$$

and

$$\Gamma \in C^{3+\alpha,(3+\alpha)/2}. \tag{4.5.65}$$

The key idea in proving the existence and uniqueness of the solution is that when translated to local coordinates, (4.5.57) is a quasi-linear parabolic equation on a manifold without boundary. For a given $T(x,t)$ solve (4.5.57) to obtain the velocity of the interface. This provides us with the interface Γ which can be used to solve the parabolic problem in the bulk with the Stefan condition on Γ. A new temperature function $\hat{T} = \mathcal{P}(T)$ is obtained. Using the regularizing character of the operator $\mathcal{P}$, it can be proved that $\mathcal{P}$ is a contraction on a small time interval. Neumann condition or mixed conditions can also be prescribed on the fixed boundary $\partial G \times [0, t_0]$.

The local-in-time existence of the solution of a more general two-phase Stefan problem with the modified Gibbs-Thomson relation has been proved in [132]. The parabolic operator in the heat equation is of the form

$$\mathcal{P}_j T_J = \left(\frac{\partial}{\partial t} - \nabla_x (A_J(x,t)\nabla_x)\right) T_j = f_J \text{ in } G_J(t) \text{ for } \forall\, t \in [0, t_0],\ J = 1, 2, \tag{4.5.66}$$

and $0 \le t \le t_0$ is the time interval in which the solution exists. The domain $G_1(t) \in G \subset R^n$, $n \ge 2$ and $\partial G_1(t) = \Gamma(t) \subset G, \forall t \in [0, t_0]$, is the free surface. $G_2(t) = G \backslash \bar{G}_1(t)$, $0 \le t \le t_0 \cdot G_1$ and G_2 are solid and liquid regions, respectively.

$$T_J = -\sigma(x,t)\bar{K}_c(x,t) - \alpha(x,t)\,\sigma(x,t)\vec{V}\cdot\vec{n}, \text{ on } \Gamma(t),\ J = 1, 2, \tag{4.5.67}$$

where $\alpha \ge 0$, and the unit normal $\vec{n}$ is pointing into the liquid. The problem formulation can be completed by adding Stefan condition and initial and boundary conditions to the equations (4.5.66) and (4.5.67). To prove the existence of the solution, the problem in time varying domains $G_J(t)$, $J = 1, 2$ is first converted into a problem in fixed domain $\Omega_J(0)$, $J = 1, 2$ by using 'Hanzawa transformation' [133]. Then a local coordinate system is introduced in a neighborhood of $\Gamma(0)$. Under suitable consistency conditions (see [132]) and the regularity of parameters and coefficients, an initial approximation of the solution of an auxiliary problem makes it possible to construct a classical solution of the given Stefan problem on $\Gamma(0)$. To apply Newton's method of successive approximations it is necessary to obtain conditions for the invertibility of the 'Frechet derivative' [22] on the initial approximation. Once a regularizer of the Frechet derivative is constructed, the existence of the solution in a very small time interval can be proved with the help of Newton's method.

Multi-dimensional one-phase and two-phase Stefan problems with Gibbs-Thomson relation have been discussed in [134] under the assumption that the surface tension ε is small. Although classical solutions have not been discussed, some of the observations made in [134] are interesting. Under the assumption that the classical Stefan problem with $\varepsilon = 0$ has a smooth free boundary and that its classical temperature solution $\hat{T}$ exists, an approximate solution of the Stefan problem with $\varepsilon \ne 0$ is taken as $\hat{T} + \varepsilon u$ (one-phase problem). The existence and uniqueness of the weak solution of the problem formulated for u has been proved. The free boundary of $\hat{T} + \varepsilon u$ has also been analysed. In the two-phase problem the perturbations are of the form $(\hat{T}_1 + \varepsilon u_1, \hat{T}_2 + \varepsilon u_2)$ where $\hat{T}_1 + \varepsilon u_1$ is the temperature of water region and $\hat{T}_2 + \varepsilon u_2$ is the temperature of ice surrounding the water region. Assuming that Gibbs-Thomson relation lowers the equilibrium temperature, the following conclusions have been drawn. In the one-phase problem, small surface tension decreases the water region for all small times and the small surface tension increases the water region for all large times provided the data are radial or close to radial. The results have been extended to general shapes. This observation is radically different in the two-phase problem in which the small surface tension decreases the water region for all times provided the data are radial or close to radial.

On the basis of physical arguments, it was mentioned earlier in § 4.5.1 that surface tension has stabilizing effect on the growth of planar phase-change boundary. Investigation of stability or instability of perturbations of the planar phase-change boundaries is a vast subject (cf. [135, 136]) and requires rigorous mathematical treatment. We present here some simple results obtained in [137] on the morphological instability of the similarity solution of a one-dimensional Stefan problem in an infinite region. Consider the following two-dimensional one-phase problem of the solidification of a supercooled liquid.

$$T_t = \nabla^2 T, \quad S(x,y,t) \neq 0, \quad t > t_0, \; (x,y) \in R^2. \tag{4.5.68}$$

On the free boundary $S(x, y, t) = 0$, we have

$$T = -\sigma \bar{K}_c; \text{ and } (\nabla T . \nabla S) = S_t, \; t > t_0, \tag{4.5.69}$$

$$\bar{K}_c = (|\nabla S|^2 \nabla^2 S - \frac{1}{2} \nabla(|\nabla S|^2).\nabla S)/(2|\nabla S|^3) \tag{4.5.70}$$

$$T(x,y,t_0) = T_0(x,y); \; S(x,y,t_0) = S_0(x,y), \tag{4.5.71}$$

$$T_0(x) \to T_\infty \in (-1,0) \text{ as } x \to \infty; \; T_0(x) \to 0 \text{ as } x \to -\infty. \tag{4.5.72}$$

Consider a one-dimensional one-phase problem in the region $-\infty < x < \infty$ in which the region $x < 0$ is ice at the phase-change temperature $T \equiv 0$ and the region $x > 0$ is occupied by the supercooled water at $T = T_\infty$. A similarity solution of this one-dimensional problem can be easily obtained by following the procedure used to obtain the Neumann solution (cf. (1.3.11)–(1.3.13)). Let this solution for $\sigma = 0$ be denoted by (T_1, S_1), where

$$\left.\begin{aligned} T_1 &= T_\infty + 2\alpha \exp(\alpha^2)(1 - \operatorname{erf}(x/(2\sqrt{t}))), \text{ for } x > 2\alpha\sqrt{t}, \\ &= 0, \quad x < 2\alpha\sqrt{t} \end{aligned}\right\} \tag{4.5.73}$$

$$S_1(x,t) = (x - 2\alpha\sqrt{t}) = 0, \tag{4.5.74}$$

Here, α is the root of the transcendental equation

$$2\alpha \exp(\alpha^2) \int_\alpha^\infty \exp(-y^2)dy = -T_\infty. \tag{4.5.75}$$

This similarity solution is perturbed so that

$$T_\varepsilon(x, y, t) = T_1 + \varepsilon u(x, y, t) + O(\varepsilon^2), \tag{4.5.76}$$

$$S_\varepsilon(x, y, t) = S_1(x, t) + \varepsilon R(x, y, t) + O(\varepsilon^2). \tag{4.5.77}$$

As the phase-change boundary is no more planar after perturbation, we shall have a two-dimensional problem with the effect of surface tension included in the temperature at the free boundary as in (4.5.69). For linear stability analysis, T_ε and S_ε are substituted in (4.5.68)–(4.5.71) and the problem is reformulated in terms of u and R. The Fourier transform with respect to y of the equations yields one-dimensional time dependent initial-value problems for different values of the transform parameter whose solutions are analysed for the stability analysis (cf. [137]). The linear stability analysis showed that if the surface tension is non-zero, then each perturbation mode with a non-zero wave number will be stable. However, the solution is unstable with respect to perturbations with a zero wave number limit point in their Fourier spectrum.

Concluding Remarks

The discussion in §§ 4.1–4.5, is focussed on some modelling and analysis aspects of supercooling. The phenomena of nucleation, conditions at the contact line of manifolds belonging to the fixed boundary of the domain and the free boundary, and the modelling of surface tension and kinetic condition, are quite complicated. For further information references given in [54] are suggested. Surface tension can be taken into account by incorporating surface energy in the energy functional without introducing a sharp phase-change interface [138]. In [139], ideas and methods of the theory of minimal surfaces have been introduced as well as new estimates for time derivatives of non-integer order. Ideas and methods of the geometry of manifolds have been used in [140]. Landau-Ginzburg theory, which is the basis of standard phase-field models, relies on general thermodynamic considerations. In the limit, Stefan problem with modified Gibbs-Thomson law is obtained. Assuming that the solution of phase-field equations exists, an asymptotic solution of the phase-field equations has been constructed in [141] and the existence of its smooth solution has been proved.

4.5.5 Weak formulation with supercooling and superheating effects

The enthalpy equation (4.2.2) can be written as

$$C_P\frac{\partial T}{\partial t} + l\frac{\partial \chi}{\partial t} = \nabla \cdot (K(T)\nabla T), \text{ in } \Omega_* = \Omega \times (0, t_*), \ \Omega \subset R^3, \tag{4.5.78}$$

$$\chi \in H_g(T - T_m). \tag{4.5.79}$$

Here, $K(T) = K_S T$, $T < T_m$ and $K(T) = K_L T$, $T > T_m$ and H_g is the Heaviside graph defined in (4.3.45). χ is the liquid fraction present in any phase. In the classical Stefan problem $\chi = 1$ in the liquid phase and $\chi = 0$ in the solid phase. If the mushy region is present, then $\chi \in [0, 1]$. Equation (4.5.78) holds in the distributional sense. We shall now include supercooling and superheating effects in the weak formulation given by (4.5.78) and (4.5.79). If the interface is not at the equilibrium temperature T_m, then the condition $T = T_m$ at the interface in the absence of curvature effects, is replaced by the kinetic condition

$$\dot{S}(t) = \beta(T(S(t), t)), \tag{4.5.80}$$

where $\beta : R \to R$ is continuous and increasing and $\beta(0) = 0$. This represents supercooling or superheating effects due to the phase-change velocity (kinetic condition). The standard Stefan problem is then obtained in the limit as $\beta'(0) \to +\infty$ (expand β in powers of T). For small departures from the equilibrium temperature, the melting or solidification rate is proportional to the departure [42]. When a mushy region is present, i.e., when $\chi \in [0, 1]$ and χ is considered a function of x and t, (4.5.79) should be replaced by a relaxation law for χ of the form [101]

$$\alpha \frac{\partial \chi}{\partial t} + H_g^{-1}(\chi) \ni \beta(T - T_m) \text{ in } \Omega_*, \tag{4.5.81}$$

where α is a positive constant, $\chi = \chi(x, t)$, $(x, t) \in \Omega_{t_*}$ and the term on the r.h.s. of (4.5.81) is an approximation of the term on the r.h.s. of (4.5.80). In (4.5.81), there is no diffusion of χ. It has been assumed that the phase transition is much faster than the heat diffusion. In the classical Stefan problem the phase-change is driven by the release or absorption of latent heat at the interface. In the relaxation model (4.5.81), the latent heat released or absorbed is diffused into the system. The relation (4.5.81) accounts for the movement of the interface through the mushy region in the weak sense.

As mentioned earlier, in the metallurgical literature concerning solidification of a pure metal, there are two basic modes of solidification, viz., *directional solidification* and *equiaxed solidification*. In the directional solidification, which is also called columnar growth, the kinetic law (4.5.80) is considered at the phase-change boundary and the classical formulation is used. In the equiaxed solidification or phase relaxation model, a weak formulation is considered using (4.5.78) and (4.5.79). In casting, at first an equiaxed zone is formed in contact with the wall of the mold. Then a columnar region moves towards the interior, while in the remainder of the liquid, nucleation occurs and an equiaxed solid phase grows, until the two solid phases impinge on and eventually occupy the whole volume (see [142] and a large number of references reported there).

In [101], the existence and uniqueness of weak solutions of several Stefan problems with supercooling/superheating effects have been discussed using the theory of non-linear semi-groups of contractions. These weak formulations correspond to: (i) problems of phase relaxation in homogeneous systems (generalized to heterogeneous systems) in which energy equation is written using non-equilibrium thermodynamics, (ii) Stefan problems with phase relaxation, (iii) wave propagation for heat. In [101] and [102] only kinetic condition has been considered.

Weak formulation of the Stefan problem with surface tension has been considered in [143]. By coupling the heat equation with the Euler's equation of a non-convex functional (which represents the Gibbs free energy), two weak formulations are given and existence of the solution is proved for each one.

Chapter 5

Superheating due to Volumetric Heat Sources: The Formulation and Analysis

Some amount of superheating of the solid occurs with the melting of pure metals. When superheating is present, a 'mushy region' may develop in front of the phase front, separating the stable solid phase from the stable liquid phase. Note that a superheated solid is in a solid state but with time degenerates into mush. The temperature of the mushy region is taken to be equal to T_m if the phase-change is taking place at T_m. The degree of superheating is small and it is generally neglected in the formulation but when melting takes place due to strong volumetric heat sources, there is clear evidence of the superheating of the solid. The theoretical analysis of a sharp melting front model in the presence of strong volumetric sources indicated 'superheated regions' in the solid just in front of the interface [144]. Superheating or supercooling indicates the instability of the sharp interface and requires corrective measures in the modelling of classical Stefan problems. Some of the corrective measures are: (1) consider an enthalpy model in the place of the classical model, (2) introduce modified Gibbs-Thomson relation in the temperature relation at the phase-change interface in the classical formulation, and (3) consider a different type of formulation such as a 'classical enthalpy formulation' (CEF).

5.1 The Classical Enthalpy Formulation of a One-dimensional Problem

We shall first discuss CEF in the context of one-dimensional melting due to volumetric heat sources in a slab ($-1 \leq x \leq 1$). This physical problem arises in the 'spot welding' of two large metal sheets of equal thickness of some heat and electric conducting material. Two circular electrodes are placed opposite to each other with the two sheets held tightly in between the electrodes [145]. A high electric current is passed through the system for a short time. The energy produced by the current raises the temperature of the sheets. It will be assumed that the heating takes place due to *Joule heating* throughout the volume of the conducting material. If the conditions are symmetrical about $x = 0$ then it is sufficient to consider heat transfer in only one sheet with no flux condition at $x = 0$. Under certain assumptions, a one-dimensional problem can be formulated in the region $0 \leq X \leq 1$ (X is dimensionless). At $X = 0$, the two sheets are joined and at $X = 1$, the electrodes are cooled. According to the maximum principle, the temperature will be maximum at $X = 0$. The temperature at $X = 0$ will rise only due to the heat received by it due to Joule heating and not by conduction as other parts are at a lower temperature and because $\partial T/\partial X = 0$ at $X = 0$ (because of symmetry). Liquid will form at $X = 0$ when the specific enthalpy at $X = 0$ becomes $C_L T_m + l$. It is assumed that $t = 0$ is the time when the temperature at $X = 0$ rises to the melting temperature T_m. For $t > 0$ the specific enthalpy at $X = 0$ will rise from $C_S T_m$ to $C_L T_m + l$ (we assume the specific heat of the mushy region also to be C_S). During this time the enthalpy of the points in some neighbourhood of $X = 0$ will have risen. By the time melting starts at $X = 0$, i.e., the enthalpy at $X = 0$ rises to $C_L T_m + l$, a mushy region will have developed in some neighbourhood of $X = 0$. This clearly suggests a three-region formulation, given first in [146], and then extended in [147]. The dimensionless formulation described in [147] is given below.

In the solid region

$$\frac{\partial T_S}{\partial t} = \alpha_S^2 \frac{\partial^2 T_S}{\partial X^2} + A_S + B_S T_S; \quad A_S > 0, \ A_S + B_S T_S > 0,$$
$$S_1(t) < X < 1, \ 0 < t < t_e < \infty, \tag{5.1.1}$$

$$T_S(X, 0) = f_S^{(1)}(X), \ 0 \leq X \leq 1, \tag{5.1.2}$$

$$f_S^{(1)}(0) = 1, \quad \left.\frac{\partial f_S^{(1)}}{\partial X}\right|_{X=0} = 0, \ \alpha_S^2 \left.\frac{\partial^2 f_S^{(1)}}{\partial X^2}\right|_{X=0} + A_S + B_S > 0, \tag{5.1.3}$$

$$\frac{\partial T_S}{\partial X} + d_1 T_S = d_2(t), \text{ on } X = 1, \tag{5.1.4}$$

$$\left.\begin{aligned} &T_S = 1, \\ &\partial T_S/\partial X = 0. \end{aligned}\right\} \text{ on } X = S_1(t); \ S_1(0) = 0, \ t = S_1^{-1}(X), \ S_1^{-1}(0) = 0. \tag{5.1.5}$$

In the dimensionless form $T_m = 1$ and A_S, B_S and d_1 are known constants. The equation of the solid-mush boundary is $X = S_1(t)$ or $t = S_1^{-1}(x)$, t_e is the time at which the mushy region becomes extinct, and α is the dimensionless diffusivity.

In the mushy region

$$\frac{\partial H_M}{\partial t} = Q(H_M); \ S_2(t) < X < S_1(t), \ 0 < t < t_e, \tag{5.1.6}$$

$$H_M = 1, \text{ at } t = S_1^{-1}(X), \tag{5.1.7}$$

$$\left.\begin{aligned} Q(H_M) &= A_S + B_S,\ H_M = 1\,, \\ &= D_1 + D_2 H_M,\ 1 < H_M < C_L/C_S + \lambda\,, \\ &= A_L + B_L,\ H_M = C_L/C_S + \lambda\,. \end{aligned}\right\} \tag{5.1.8}$$

Here, H_M is the enthalpy per unit volume of the mushy region, $\lambda = l/(C_S T_m)$. D_1 and D_2 are unknown constants. An exact form of Q cannot be guessed. A simple assumption that Q is a linear function of enthalpy in the mushy region has been made in (5.1.8).

In the liquid region:

$$\begin{aligned} \frac{\partial T_L}{\partial t} = \alpha_L^2 \frac{\partial^2 T_L}{\partial X^2} + A_L + B_L T_L;\ A_L > 0,\ A_L + B_L T_L > 0, \\ 0 < X < S_2(t) < S_1(t),\ t \geq t_*, \end{aligned} \tag{5.1.9}$$

$$T_L \Big|_{X=S_2(t)} = 1, \tag{5.1.10}$$

$$(C_L/C_S)\alpha_L^2 \frac{\partial T_L}{\partial X} = -\frac{dS_2}{dt}\left\{(C_L/C_S) + \lambda - H_M\right\}, \text{ on } X = S_2(t),\ t \geq t_*,\ S_2(t_*) = 0, \tag{5.1.11}$$

$$\left.\frac{\partial T_L}{\partial X}\right|_{X=0} = 0. \tag{5.1.12}$$

Here, $X = S_2(t)$ is the equation of the liquid-mush boundary and $t = t_*$ is the time at which $S_2(t)$ starts growing. $S_2(t_*) = 0$.

The above formulation requires some elaboration. The conditions in (5.1.3) have interesting physical interpretations. For example $f_S^{(1)}(0) = 1$ implies that the temperature at $X = 0$ is the melting temperature. It is generally understood that if more volumetric heat is supplied, then the enthalpy at $X = 0$ should rise. But this does not always happen. For example, if the last condition in (5.1.3) is not satisfied then the temperature at $X = 0$ for $t > 0$ first becomes less than unity. After some time $\partial T/\partial t$ becomes greater than zero so that the temperature starts rising, and again the temperature at $X = 0$ becomes unity and then the third condition in (5.1.3) is satisfied. $t = 0$ has been taken as the instant of time at which this happens. In view of the boundary condition (5.1.12), we should have the second condition in (5.1.3). As mentioned earlier, the melting does not start at $t = 0$ but at $t = t_*$. This is the time at which the dimensionless enthalpy at $X = 0$ becomes $C_L/C_S + \lambda$. The instant t^* can be calculated by obtaining the solution of (5.1.6) subject to (5.1.7) and (5.1.8) (cf. [147]). All the temperatures have been made dimensionless by dividing them by the melting temperature $T_m(^oK)$. The dimensionless time t is obtained by dividing the real time by t_d, which is an appropriate time with dimensions and $t = 0$ is time at which all the conditions in (5.1.3) are satisfied for the first time.

Till $t = t_*$, there exists only solid-mush boundary $X = S_1(t)$ but for $t > t_*$, we have liquid-mush boundary $X = S_2(t)$ also. The assumption of the continuity of enthalpy at $X = S_1(t)$ implies $C_M = C_S$ which for the sake of simplicity has been assumed to be so in the whole of the mushy region. The condition $\partial T_S/\partial X = 0$ at $X = S_1(t)$ in (5.1.5) can be derived independently of the assumption of continuity of the enthalpy.

The energy balance at $X = S_1(t)$ gives

$$\dot{S}_1[H]_{S_1(t)+0}^{S_1(t)-0} = -\left[K\frac{\partial T}{\partial X}\right]_S^M, \quad \dot{S}_1 > 0. \tag{5.1.13}$$

$H_M \geq 1$ as $X \uparrow S(t)-0$ and $H_S = 1$ as $X \downarrow S(t)+0$. Since $T_M = 1$ in the mushy region $\partial T_M/\partial X = 0$ as $X \uparrow S(t)-0$. It can be easily concluded from (5.1.13) that $\partial T_S/\partial X \geq 0$. In view of $T_S(S_1(t),t) = 1$ and $T_S(X,t) < 1$ for $X > S_1(t)$, $\partial T_S/\partial X \leq 0$ as $X \to S(t)+0$. Therefore $\partial T_S/\partial X = 0$ at $X = S_1(t)$. This also implies that enthalpy is continuous at $X = S_1(t)$ if $\dot{S}_1$ is not zero in (5.1.13).

The problem in the solid region is independent of liquid and mushy regions and therefore the solid region problem is a one-phase Stefan problem with an implicit boundary condition at $X = S_1(t)$. Under appropriate assumptions, the existence and uniqueness of the solution of this one-phase problem is known [108]. It can be proved that $S_1(t)$ is nondecreasing and sufficiently smooth and $S_1(t)$ exists till the time $S_1(t) = 1$. If, instead of a heat source, a heat sink is considered, then the problem in the solid region will be similar to ODP. In [147], by considering series expansions of both initial temperature distribution $f_S^{(1)}(X)$ and $S_1(t)$, short-time asymptotic behaviour of $S_1(t)$ has been investigated in terms of coefficients in the series expansion of $f_S^{(1)}(X)$ and some physical parameters. Numerical results were also obtained for $T_S(X,t)$, $S_1(t)$ and $S_2(t)$ in CES and ODP. It was not possible to carry out numerical computations till $X = 1$ in ODP. When $S_1(t)$ is almost nearing unity, $\dot{S}_1(t)$ becomes large and the numerical scheme which uses moving grid points failed. We are not aware of better results.

The problem for the liquid region is not the standard Stefan problem in which latent heat is non zero and maintains its sign. Because of the boundary condition (5.1.11), the quantity $C_L/C_S + \lambda - H_M$ (call it latent heat), is zero at $X = 0$. Also when $S_2(t)$ crosses $S_1(t)$, the sign of 'latent heat' changes. Therefore the usual methods of proving existence results do not work here directly. However adopting a different procedure in the form of constructing approximating problems [146], existence of the solution of (5.1.9)–(5.1.12) has been proved and it has been shown that $S_2(t) < S_1(t)$. If the heat source does not vanish in the liquid region, then the mushy region will become extinct in finite time. The mushy region may become extinct before $S_1(t)$ reaches $X = 1$. In this case $S_2(t)$ will meet $S_1(t)$ after some finite time, say, $t = t_e$. For $t > t_e$, we shall have a two-phase classical Stefan problem. Initially $S_1(t)$ moves very slowly as $S_1(t) \sim O(e^{-1/t})$ (cf. [147]).

For a short-time, asymptotic behaviour of $S_2(t)$ which depends on $S_1(t)$ can also be calculated. When $S_1(t) \approx E_1 t^{1/2}$ and $S_2(y) \approx E_2 y^{1/2}$, $y = t - t_*$, $y \geq 0$ then E_2 is given by

$$E_2 = \left\{ 2E_1^2 - \alpha_L^2 + \sqrt{(2E_1^2 - \alpha_L^2)^2 + 24E_1^2\alpha_2^2} \right\}/4, \tag{5.1.14}$$

where E_1 is the root of the equation

$$\sqrt{\pi}E_1 \exp\left(E_1^2/\alpha_S^2\right)(1 - \operatorname{erf}(E_1/\alpha_S)) - \alpha_S = \alpha_S^3 \left.\frac{\partial^2 f_S^{(1)}}{\partial X^2}\right|_{X=0} /(A_S + B_S). \tag{5.1.15}$$

By taking initial temperature of the solid as constant, viz., $-P$ several results concerning the solution of CEF were obtained in [148] which include the asymptotic behaviours of $S_1(t)$ and $S_2(t)$, finite-time extinction of the mushy region, and the result that $\dot{S}_1(t_e) < \dot{S}_2(t_e)$.

In CEF, superheating in the solid does not occur. CEF regularises the classical two-phase sharp interface problem in the presence of heat sources in which superheating would otherwise occur. Weak formulation also provides regularization. Is there any connection amongst the solution of CEF, weak solution and the classical solution? This question has been analysed in [149] and [150] for a short time by considering a one-dimensional problem in the region $-\infty < x < \infty$ with the free boundary starting at $x = 0$. Since superheating, if present, will be restricted to a small neighborhood of $x = 0$, it is sufficient to consider the behaviour of the initial temperature $f(x)$. To understand the connection between the classical solution and the weak solution, we should first define a 'weak solution'.

5.2 The Weak Solution

5.2.1 The Weak Solution and its Relation to Classical Solution

The definition of the weak solution of a Stefan problem is based on the classical formulation of the Stefan problem. For simplicity, first a weak formulation of a one-dimensional two-phase Stefan problem will be defined in the region $0 \leq x \leq 1$. For multi-dimensional Stefan problems, weak formulations can be defined similarly (see § 11.2.1). In terms of the enthalpy $H(T)$ (per unit volume), the two heat equations in the solid and liquid regions can be written as a single equation as follows:

$$\frac{\partial H}{\partial t} = \frac{\partial}{\partial x}\left(K(T)\frac{\partial T}{\partial x}\right) + Q(T)\,,\ 0 \leq x \leq 1,\ 0 < t < t_*, \tag{5.2.1}$$

where $K(T)$ and $Q(T)$ are defined as

$$K = K_1(T), \text{ in the solid },\ 0 < x < S(t),$$

$$= \quad K_2(T), \text{ in the liquid, } S(t) < x < 1, \tag{5.2.2}$$

$$Q(T) = A_S + B_S T, \text{ in the solid,}$$

$$= A_L + B_L T, \text{ in the liquid.} \tag{5.2.3}$$

More general expressions of thermal conductivities and heat generation terms can be considered. Let the phase-change temperature be denoted by T_m. Enthalpy $H(T)$ is defined by (4.2.1) (take $\phi = 0$ for $T < T_m$ and $\phi = 2$ for $T > T_m$). By inverting (4.2.1), we obtain

$$T = \beta(H), \tag{5.2.4}$$

where

$$\left.\begin{aligned} \beta(H) &= H/(\rho_S C_S),\ H < \rho_S C_S T_m, \\ &= 0,\ \rho_S C_S T_m < H < \rho_L C_L T_m + \rho_L l, \\ &= H/\rho_L C_L - l/C_L,\ H > \rho_L C_L T_m + \rho_L l. \end{aligned}\right\} \tag{5.2.5}$$

Equation (5.2.1) should be supplemented with the initial and boundary conditions such as

$$T(x,0) = T_0(x),\ 0 \le x \le 1;\ \ T(0,t) = g_1(t) < 0,\ T(1,t) = g_2(t) \ge 0, \tag{5.2.6}$$

Equation (5.2.1) is satisfied in the distributional sense as $\partial H/\partial t$ does not exist at $T = T_m$ in the classical sense. The classical formulation of the above two-phase Stefan problem is simple and will not be given here. Multiply (5.2.1) by a suitable test function $\psi(x,t) \in C^{2,1}([0,1] \times [0,t_*])$ such that

$$\psi(0,t) = \psi(1,t) = 0,\ 0 < t \le t_*,\ \text{and } \psi(x,t_*) = 0,\ \text{in } 0 \le x \le 1, \tag{5.2.7}$$

and integrate over the (x,t) region. We obtain

$$\int_0^1 \int_0^{t_*} \left[\psi \frac{\partial}{\partial x}\left(K(T)\frac{\partial T}{\partial x}\right) + \psi Q - \psi \frac{\partial H}{\partial t}\right] dx dt = 0. \tag{5.2.8}$$

On integrating by parts the first term in the integrand in (5.2.8), we obtain

$$\begin{aligned} &\int_0^{t_*} \left[\psi K \frac{\partial T}{\partial x}\right]_{x=0}^{x=1} dt - \int_0^1 \int_0^{t_*} K \frac{\partial T}{\partial x}\frac{\partial \psi}{\partial x}\, dx dt + \int_0^1 \int_0^{t_*} \psi Q\, dx dt \\ &\qquad - \int_0^1 [\psi H]_0^{t_*}\, dx + \int_0^1 \int_0^{t_*} H \frac{\partial \psi}{\partial t}\, dx dt = 0. \end{aligned} \tag{5.2.9}$$

The first term on the l.h.s. of (5.2.9) in zero in view of (5.2.7). Integrating the second term in (5.2.9) by parts, we obtain

$$\begin{aligned} &- \int_0^{t_*} \left[TK \frac{\partial \psi}{\partial x}\right]_0^1 dt + \int_0^1 \int_0^{t_*} T \frac{\partial}{\partial x}\left(K \frac{\partial \psi}{\partial x}\right) dx dt + \int_0^1 \int_0^{t_*} \psi Q dx dt \\ &+ \int_0^1 H(T_0)\psi(x,0) dx + \int_0^1 \int_0^{t_*} H \frac{\partial \psi}{\partial t} dx dt = 0, \end{aligned} \tag{5.2.10}$$

On using (5.2.6) in the first term in (5.2.10), we get

$$\int_0^1 \int_0^{t_*} \left(T \frac{\partial}{\partial x} \left(K \frac{\partial \psi}{\partial x} \right) + H \frac{\partial \psi}{\partial t} + \psi Q \right) dx dt = \int_0^{t_*} g_2(t) K(g_2(t)) \frac{\partial \psi}{\partial x}(1,t) dt$$

$$- \int_0^{t_*} g_1(t) K(g_1(t)) \frac{\partial \psi}{\partial x}(0,t) dt - \int_0^1 H(T_0) \psi(x,0) dx. \qquad (5.2.11)$$

If a pair of bounded measurable functions (T, H) exists such that (5.2.11) is satisfied for all test functions $\psi(x,t)$ satisfying (5.2.7), then the pair (T, H) is called a weak solution of the classical two-phase one-dimensional Stefan problem in the region $0 \le x \le 1$, $0 \le t \le t_*$. Note that in the definition of a weak solution, the boundary conditions at the interface $x = S(t)$ are not required. So they do not appear in (5.2.1)–(5.2.6). A weak solution for multi-dimensional two-phase classical Stefan problem can be defined similarly. In this case integration by parts is to be done with the help of Green's formula. Other types of boundary conditions can also be considered in (5.2.6) and in that case test functions have to satisfy appropriate boundary conditions. The boundary conditions for test functions are chosen in such a way that all those terms appearing in the integration by parts, which are to be evaluated at the boundary and which are not prescribed, are eliminated from the equation (5.2.11) through which the weak solution is defined. For example, in (5.2.9) the boundary conditions $\psi(0,t) = \psi(1,t) = 0$ eliminate the term $[\partial T/\partial x]_0^1$ as these boundary conditions are not prescribed in the present problem. Similarly weak formulations of more complicated problems and their weak solutions can be defined.

A classical solution is a weak solution. This proposition holds for a general Stefan problem but it will be discussed here for a simple one-dimensional problem given in (5.2.1)–(5.2.6). To avoid repetition, we give here only outlines (see (5.2.12)–(5.2.14) for details). If a classical solution exists, then we know that (5.2.1) holds separately in solid and liquid regions. T is continuous in $0 \le x \le 1$, and a bounded measurable function H defined in (5.2.1) also exists. We have to show that the pair (T, H) satisfies (5.2.11). To prove this, we start with (5.2.8) and write the double integral as the sum of two integrals, $I_1 + I_2$. In I_1 the limits of integration for the x-variable are taken from 0 to $S(t) - 0$ (solid region) and in I_2 the limits for x are taken from 0 to $S(t) + 0$ (liquid region) (the limits for t remain the same). The procedure given in (5.2.12)–(5.2.14) is to be followed for both the integrals. On adding the two integrals, we get (5.2.11) with an additional term which is the last term in (5.2.13). The line integral along $x = S(t)$ in (5.2.13) is zero because the classical solution exists and Stefan condition is satisfied. $I_1 + I_2 = 0$ as both I_1 and I_2 are zero, the remaining terms give (5.2.11).

In the weak formulation, the phases are 'pointwise' characterized by enthalpy, whereas, in the classical formulation the phases are 'globally' determined by the interface. Therefore a weak solution of the Stefan problem will be a classical solution provided:

(i) initially no mushy region exists,

(ii) distributed heat sources (sinks) are absent as in their presence mushy region may develop,

(iii) a smooth surface exists which divides the region under consideration into two disjoint regions, in each of them parabolic heat equation are satisfied in the sense of § 1.4.6. Temperature is less than or equal to the melting temperature in one of the regions (solid region) and in another region temperature is greater than or equal to the melting temperature (liquid region) with continuity of temperature across the interface,

(iv) the Stefan condition is satisfied at the interface.

These assumptions are not rigorous, but they serve our present purpose, which is to show that the Stefan condition can be recovered from the weak solution. We shall continue with the formulation given in (5.2.1)–(5.2.7). Assume that a weak solution satisfying the initial and boundary conditions (5.2.6) and equation (5.2.11) exists. A smooth interface $x = S(t)$ exists. Let the region $\Omega(x,t) = \{(x,t) : 0 \le x \le 1, 0 \le t \le t_*\}$ be divided by the curve $x = S(t)$ into two disjoint parts $\Omega_1(x,t)$ and $\Omega_2(x,t)$. Ω_1 lies to the left of $x = S(t)$ and Ω_2 lies to the right. $T = 0$ on $x = S(t)$ (under suitable assumptions on the data this and some other assumptions can be proved also, see § 11.1), $\Omega_1(x,t) = \{(x,t) : 0 \le x < S(t)-0,\ 0 \le t < t_*\}$, and $\Omega_2(x,t) = \{(x,t) : S(t)+0 < x \le 1, 0 \le t < t_*\}$. If equation (5.2.1) with $Q(T) = 0$ is multiplied by a test function $\psi(x,t)$ and integrated over $\Omega_1(t)$, then we obtain

$$\int_0^{t_*}\int_0^{S(t)-0} \left[\psi\frac{\partial}{\partial x}\left(K(T)\frac{\partial T}{\partial x}\right) - \psi\frac{\partial H}{\partial t}\right] dx dt = I_1. \tag{5.2.12}$$

If integrations are performed in (5.2.12) using integration by parts and the *Stoke's Theorem* is applied, then we arrive at the following equation.

$$\begin{aligned} I_1 = &\int_0^{t_*}\int_0^{S(t)-0} \left\{T\frac{\partial}{\partial x}\left(K\frac{\partial \psi}{\partial x}\right) + H(T)\frac{\partial \psi}{\partial t}\right\} dx dt \\ &+ \int_0^{S(t)-0} H(T_0)\psi(x,0)dx + \int_0^{t_*} g_1(t)\, K(g_1)\frac{\partial \psi}{\partial x}(0,t)\, dt \\ &+ \int_{x=S(t)-0}\left[\left(\psi K\frac{\partial T}{\partial x}\right) dt + H\psi\, dx\right]. \end{aligned} \tag{5.2.13}$$

On using Stoke's theorem [82], the area integral is converted to the boundary integral and for this we use the relation $\psi\partial f/\partial x = \partial(\psi f)/\partial x - f\ \partial\psi/\partial x$ in (5.2.12).

On integrating (5.2.1) with $Q(T) = 0$ again over the region $\Omega_2(t)$ (call this integral I_2) we obtain an equation similar to (5.2.13) (with appropriate changes) but the curve $x = S(t)$ is traversed in a direction opposite to that in (5.2.13). Adding the two equations of the type (5.2.13), one each for $\Omega_1(t)$ and $\Omega_2(t)$ and subtracting the sum from (5.2.11) $(Q(T) = 0)$ and remembering that $I_1 + I_2 = 0$ as (5.2.1) holds, we obtain

$$\int_{x=S(t)} \psi\left\{\left[K\frac{\partial T}{\partial x}\right] dt + [H]dx\right\} = 0\ \cdot \tag{5.2.14}$$

Here, [f] means the jump in the quantity f across $x = S(t)$ as we move in the positive x-direction. Since (5.2.14) holds for any arbitrary test function ψ and at an arbitrary time, we have

$$\frac{dS}{dt} = \left[K\frac{\partial T}{\partial x}\right] / [H]\,, \quad x = S(t). \tag{5.2.15}$$

Equation (5.2.15) is the Stefan condition. A similar procedure can be followed for multi-dimensional problems also.

For a solution of the Stefan problem to be a classical solution, in addition to the regularity of the temperature and the phase-change boundary, the temperature of the solid region should be less than the melting temperature and the temperature of the liquid region should be greater than the melting temperature.

5.2.2 Structure of the mushy region in the presence of heat sources

In the presence of heat sources or if melting temperature is a function of x and t, or if mushy region exists initially, a weak solution (WS) will not be a classical solution (CSS). However, if the solution of CEF denoted by CES exists and WS is sufficiently regular, then WS $\equiv$ CES but WS $\not\equiv$ CSS. This is because in CEF, a mushy region is present initially. By considering different cases in terms of the sign of the initial temperature $f(x)$, and the derivatives of $f(x)$ at $x = 0$, CES, WS and CSS have been compared in [149] and [150]. The phase-change starts at $x = 0$ in the region $-\infty < x < \infty$. The existence of solutions has also been discussed. We present only one result here which is valid only for a short time.

If $f(x) > 0$ for $x < 0$ and $f(x) < 0$ for $x > 0$, $h(0) = 0$ and

$$f'(0+) = 0, \text{ and } K_S f''(0+) + Q = 0, \begin{cases} f'''(0+) - \dfrac{\beta Q}{K_S^2} < 0, \text{ then } CSS \equiv WS \\ \\ f'''(0+) - \dfrac{\beta Q}{K_S^2} > 0, \text{ then } CSS \not\equiv WS \end{cases} \tag{5.2.16}$$

where Q is the constant heat source in $-\infty < x < \infty$, and

$$\beta = \dot{S}(0) = -K_L h'(0-)/l. \tag{5.2.17}$$

By considering a one-dimensional problem in $0 \le x \le 1$, it has been shown in [151] that if the melting temperature is space-dependent, then a mushy region can appear spontaneously even in the absence of volumetric heat sources. The following weak formulation has been considered in [152].

$$H_t = (\phi(H))_{xx} + f(H), \;\; (x,t) \in (0,1) \times R^+, \tag{5.2.18}$$

$$\phi(H)(0,t) = (\phi(H))_x(1,t) = 0,\ t \in R^+, \tag{5.2.19}$$

$$H(x,0) = H_0(x),\ x \in [0,1]. \tag{5.2.20}$$

Here, $\phi(H)$ can be identified with the temperature and H with the enthalpy per unit volume. f, ϕ and H_0 satisfy some assumptions (cf. [152]). The source term $f(H)$ and the initial enthalpy $H_0(x)$ are so chosen that enthalpy is increasing in x and t. Initially the region [0,1] is solid. At $x = 1$ it passes to mush in time and then to liquid. Interfaces appear first between solid and mush and then between mush and liquid. These recede monotonically towards $x = 0$. Under suitable assumptions on ϕ, f, H_0 it has been shown that solid-mush and liquid-mush interfaces are continuous, enthalpy is continuous across the solid-mush boundary but not across the mush-liquid interface and the mushy region disappears in finite time.

In equation (5.1.13) it has been assumed that the mush is progressing on the solid. If the solid is progressing on mush, then it cannot be proved that the enthalpy is continuous at the solid-mush boundary as the Stefan condition (5.1.13) with a positive sign on the r.h.s. is to be considered. If $\dot{S}$ changes sign at both solid-mush and mush-liquid boundaries, then the regularity of the free boundary and of the velocity satisfying the Stefan condition has been examined in [153]. The enthalpy has been assumed to be an increasing function of x. An example has been constructed in [154] in which the velocity of the free boundary changes sign infinitely many times.

Classical enthalpy formulation arises in thermal switching problems also (cf. [155, 156]). Certain materials such as some types of glasses and vanadium dioxide, change phase at relatively low temperature, such as, at 70^0C. In the liquid state the electrical conductivity of such materials increases by a factor of 10^4. For such materials, in Joule heating, the heat source is given by $\sigma(H)E^2$, where σ is the electrical conductivity, H is the enthalpy and E is the electrical field which depends on σ and some other parameters. In essence, heat source may depend on the free boundary, and the latent heat l depends on the history of the free boundary [156]. The existence of the solution to a one-dimensional classical enthalpy formulation of the thermal switching problem have been proved in [156]. The emphasis is on the solution for the solid region and the solid-mush boundary. The non-homogeneous term in the heat equation contains the free boundary due to the dependence of the heat source on it.

The behaviour of mushy regions in spacial dimension, $n \geq 1$, under the action of volumetric heat source depending on, x, t, and enthalpy, has been studied in [157]. Conditions on the heat source necessary for the appearance of a mushy region inside a purely liquid or a purely solid phase have been discussed. As far as regular solutions are concerned, the behaviour of the solution at the free boundary separating a mushy region from a pure phase is related to the behaviour of the source on the free boundary. Under some Lipschitz-continuity conditions on heat sources, a mushy region can expand into a pure phase only if the heat source is non-zero and is of a suitable sign at the interface. Hence, both energy and heat flux are continuous across the interface, unlike the case when a pure phase expands into a mushy region. An example has been given in which a mushy region in a pure phase disappears and again reappears after some time. Both weak and classical solutions have been discussed. For some results pertaining to weak solutions of this problem, see § 11.2.1.

By considering a one-dimensional two-phase classical Stefan melting problem in the region $0 \leq x \leq 1$ with volumetric heat sources and writing its solution in terms of Green's functions, it was shown in [144] that there exists a region $S(t) \leq x \leq S(t) + d$, for some $d > 0$ in which the solid is superheated. This happens provided the heat source is large and the prescribed heat flux on the solid at $x = 1$ is small. Linear stability analysis of the classical solution of this one-dimensional two-phase Stefan problem was also carried out for two-dimensional perturbations (three-dimensional perturbations can also be considered). If the heat flux towards the free boundary from the superheated solid side is greater than from the liquid side, then this situation suggests instability.

Superheating occurs generally for very pure metals. Its occurrence suggests that for such problems we should have a mathematical model different from the classical two-phase formulation as in the classical two-phase model, the solid cannot be superheated. One of the ways to present the superheating of the solid is to introduce the mushy region in the formulation as done in CEF. Another way is to consider enthalpy formulation which is a weak formulation. A grain or dendrite model has been proposed in [158] to describe the microstructure of the mushy region and discuss its stability. The solid is considered in the form of a sequence of uniform one-dimensional grains of very small length which are part solid and part liquid in the mushy region. An explicit solution for the one-dimensional version of the model is obtained. It has different forms near the pure solid boundary, in the mush, and near the pure liquid boundary. An appropriate average of this solution for the microstructure formulation, reduces to the weak solution proposed in [145].

The simple microscopic model considered in [158] for the mushy region has been further extended in [159] by including modified Gibbs-Thomson relation that results from the curvature of the interface and the kinetic condition (the normal velocity of the interface is taken away from the liquid). The fine structure of the mush consists of regularly spaced nucleation sites (grains) in one-dimension and a lattice of squares in two-dimension. A method of multiple scales is employed and a classical formulation of the free boundary problem has been used to model the evolution of the two-phase microstructure. Then a microscopic model for the mush is obtained by an averaging procedure. Emphasis is not put on volumetric heating in [159].

5.3 Blow-up and Regularization

As mentioned earlier, a blow-up may occur during the solidification of supercooled liquid if regularization is not done. In the superheating of the solid also, in the absence of regularization, i.e., non-inclusion of surface tension effect and/or kinetic condition at the interface or absence of mushy region in the formulation, we can raise the three questions

(A), (B) and (C) (§ 4.4.1) that arose in the context of supercooling.

A heat source term of the form $T^{1+\alpha}$, $\alpha > 0$, has been considered in [160] in the one-phase melting Stefan problem with the superheated solid occupying the region $0 \leq x \leq S(t)$, $S(0) = S_0 > 0$. The initial temperature of the solid is $T_0(x) > 0$, $x \in [0, S_0]$, 0 is the phase-change temperature, and $T(0,t) = 0$, $0 < t \leq t_*$. The temperature $T_0(x)$ satisfies some assumptions such as: (1) $T_0(0) = T_0(S_0) = 0$, (2) $T_{0,x} < 0$ on (x_0, S_0) for some $x_0 \in (0, S_0)$, (3) $T_{0,xx} + T_0^{1+\alpha} \geq 0$ for $x \in (0, S_0)$. The formulation of this problem can be easily written. Let $(t_*, T(x,t), S(t))$ be the classical solution. The main result of [160] can be stated as follows.

Proposition 5.3.1. If $t_* < \infty$, then one (and only one) of the following will always happen.

(I) $S(t) \to S_\infty \in (0, \infty)$ as $t \uparrow t_*$ and there exists only one blow-up point $x^* \in (0, S_\infty)$.

(II) $S(t) \to +\infty$ as $t \uparrow t_*$ and the set $\{T(x,t)\}$ is bounded on $\{(t,x) : 0 \leq t < t_*, 0 \leq x \leq \min(a, S(t))$ for each $a > 0$.

It was pointed out earlier that questions (A), (B) and (C) (see § 4.4.1) can be answered for the supercooled problem (3.1.8)–(3.1.12) with the help of the quantity Q defined by (4.4.6). A superheated one-dimensional problem can be formulated on the same lines as the formulation of the supercooled problem (3.1.8)–(3.1.12) and so complete formulation will not be given here (cf. [161]). But, for the sake of clarity, it may be mentioned here that the region $0 \leq x < S(t)$, $S(0) = 1$, is the superheated solid and $x > S(t)$ is the liquid region with $T_L \equiv 0 \cdot S(t)$ is progressing towards $x = 0$. For simplicity we take $\partial T/\partial x = 0$ at $x = 0$. The quantity P_0 given by

$$P_0 = \int_0^1 (T_0(x) - 1)\, dx, \tag{5.3.21}$$

plays an important role in answering the questions concerning blow-up in superheating. $T_0(x) \geq 0$ is the initial temperature of the solid, $0 \leq x \leq 1$, and the latent heat of fusion has been taken as unity in (5.3.21) after scaling. In melting, latent heat is negative and P_0 is the energy required for the complete melting of the superheated solid $0 \leq x \leq 1$. If $P_0 > 0$, then

$$\int_0^{S(t)} (T(x,t) - 1)\, dx \tag{5.3.22}$$

will remain positive. The statements (A1), (B1) and (C1) given below hold.

(A1) If the solution exists for all time, $S(t) \to S_\infty > 0$ and $T(x,t) \to 0$ as $t \to \infty$ then $-\int_0^{S_\infty} dx$ is negative. (B1) If there exists some finite time $t = t_*$ such that the solution exists for $0 \leq t \leq t_*$ and as $t \to t_*$, $S(t) \to 0$ and $dS/dt \to -\infty$, then $P_0 = 0$. (C1) If the solution exists only for $0 \leq t \leq t_*$ but as $t \to t_*$, $dS/dt \to -\infty$ and $S \to S_* > 0$ then the solution blows-up. In the supercooled problem, roughly speaking, when the free boundary touches the negativity set $\{T(x,t) < -1\}$ then blow-up occurs.

In the superheating problem, if the free boundary touches the positivity set $\{T(x,t) > 1\}$ then blow-up occurs. The heat in the solid near the free boundary is too much and the advancing free boundary cannot remove the heat and bring down the temperature to the equilibrium level which is taken as zero.

It is possible to give more general sufficient conditions for a blow-up. If there exists a function $f(x)$ defined for $0 \le x \le S(0)$ such that: $f(0) \ge 0,\ f'(0) \ge 0,\ f''(x) \ge 0$ for $0 < x < S(0)$ and

$$\int_0^1 (T_0(x) - 1) f dx > 0, \tag{5.3.23}$$

then blow-up occurs.

The questions (A), (B) and (C) (§ 4.4.1) were discussed in [161] for a two-dimensional one-phase melting of a superheated solid also. If there exists some $f(x)$, $x \in R^2$ with $f \ge 0$ and $\nabla^2 f \ge 0$ in $\Omega(0) = \Omega_0 \subset R^2$, $\Omega(t) = \Omega \times \{t\}$, $\Omega \subset R^2$, $0 \le t \le t_*$, such that

$$\int_{\Omega_0} f(T_0 - 1)\, dx > 0, \tag{5.3.24}$$

then case (C) must occur. If case (B) is to occur, then the integral in (5.3.24) should be zero for all the harmonic functions f. If $\int\limits_{\Omega_0} (T_0 - 1)dx = 0$, but there is some harmonic function f such that $\int\limits_{\partial\Omega_0} f(T_0 - 1)\, dx \ne 0$, then blow-up occurs.

Chapter 6

Steady-State and Degenerate Classical Stefan Problems

6.1 Some Steady-State Stefan Problems

Steady-state free boundary problems occur more frequently in Stefan-like problems such as seepage through dam and free surface flows (cf. (1.1.5)–(1.1.10)). Only a few studies on steady-state Stefan problems have been reported. These problems can be studied from the point of view of the conditions of their origin, and the existence, uniqueness, and regularity of their solutions. In [162] some conditions which lead to steady-state solutions have been discussed.

Consider a bounded region $\Omega \subset R^n$, $n \geq 1$, with a sufficiently regular boundary $\partial\Omega = \Gamma_1 \cup \Gamma_2 \cup \Gamma_3$, $\Omega = \Omega_1 \cup \Omega_2 \cup S$ where Ω_1 is the solid region, Ω_2 is the liquid region and S is the sharp interface separating these regions. The portion Γ_1 of the boundary $\partial\Omega$ is kept at temperature $T = b > 0$, heat flux is prescribed on Γ_2 and the boundary Γ_3 is insulated. A volumetric heat sink $g(x)$, $x \in \Omega$, per unit volume is acting in Ω. A steady-state will be reached if the outflow of heat through Γ_2 is large, and g is small. The temperature $T(x)$ in Ω is defined in the following way.

$$\left.\begin{aligned} T(x,t) &= T_1(x) < 0, \quad \text{if } x \in \Omega_1, \\ &= 0, \quad \text{if } x \in S, \\ &= T_2(x) > 0, \quad \text{if } x \in \Omega_2. \end{aligned}\right\} \tag{6.1.1}$$

The formulation of the steady-state problem consists of the following equations.

$$\nabla^2 T_i = -g, \quad \text{in } \Omega_i, \quad i = 1,2; \quad 0 \leq g < \infty, \tag{6.1.2}$$

$$T_1 = T_2 = 0, \quad K_1 \frac{\partial T_1}{\partial n} = K_2 \frac{\partial T_2}{\partial n}; \text{ on } x \in S, \tag{6.1.3}$$

$$T_2|_{\Gamma_1} = b > 0; \text{ and } \left.\frac{\partial T}{\partial n}\right|_{\Gamma_3} = 0, \tag{6.1.4}$$

$$\begin{aligned} -K_2 \left.\frac{\partial T_2}{\partial n}\right|_{\Gamma_2} &= q \text{ if } T_2 > 0; q > 0, \\ -K_1 \left.\frac{\partial T_1}{\partial n}\right|_{\Gamma_2} &= q \text{ if } T_1 > 0, \end{aligned} \tag{6.1.5}$$

where, $\vec{n}$ is the unit normal vector on S pointing into the liquid.

If

$$\theta = K_2 T^+ - K_1 T^-, \tag{6.1.6}$$

where T^+ is the positive part of T, and T^- is the negative part of T, then we have the following problem to be solved.

$$\nabla^2 \theta = -g \text{ in } \Omega,\ \theta|_{\Gamma_1} = B,\ -\left.\frac{\partial \theta}{\partial n}\right|_{\Gamma_2} = q,\ \left.\frac{\partial \theta}{\partial n}\right|_{\Gamma_3} = 0, \tag{6.1.7}$$

where $B = K_2 b > 0$.

The main result in [162] is concerned with the existence of a critical flux $q = q_c(B, g)$ such that:

(i) (q, g) with $q \le q_c(B, g)$ implies $T > 0$ in Ω. This means that only one-phase will be present in the steady-state,

(ii) (q, g) with $q > q_c(B, g)$ implies T has both negative and positive values in Ω. This in turn implies that both solid and liquid phases will be present in the steady-state.

First, (6.1.7) is formulated as a variational problem. Then with the help of the properties of the solution of the variational problem, q_c has been obtained. Some steady-state problems have been considered whose explicit solutions verify the theoretical results. The problem considered in [162] is a generalization of the problem in [163] in which $g \equiv 0$. It has been proved in [163] that there exists a $q_1 > 0$ such that for all $q > q_1$, we have a steady-state two-phase problem in Ω, and

$$q_1 = \frac{K_2 b}{a_0} \text{ meas } (\Gamma_2), \tag{6.1.8}$$

where a_0 is some constant having dimensions of $(\text{length})^n$.

6.2 Degenerate Stefan Problems

The term *degenerate Stefan problem* is used for a Stefan problem in which the mathematical nature of the differential equation or the boundary condition changes if some parameter associated with the problem varies within its admissible range of values. Some of the commonly studied degenerate Stefan problems are:

(i) *Quasi steady-state Stefan problems* or quasi-static Stefan problems in which the heat equation is elliptic but the free boundary is time dependent. Such problems may arise if the the specific heat $C \equiv 0$ or the temperature has attained a steady-state. The Hele-Shaw problem described in (3.3.45)–(3.3.46) is a quasi steady-state degenerate Stefan problem.

(ii) *Parabolic-elliptic Stefan problems* in which the specific heat $C = C(T)$ is such that

$$\begin{aligned} C(T) &= >0, \quad \text{if } T > 0, \\ &= 0, \quad \text{if } T = 0. \end{aligned} \tag{6.2.9}$$

Quasi steady-state Stefan problems

Using the theory of conformal mappings, a two-dimensional quasi-static moving boundary problem can be described by a non-linear Löwner-Kufarev equation [164] and a functional relation $\mathcal{F}$ between the shape of the free boundary and its velocity can be established. Together with the initial data this leads to an initial-value problem. Assuming that $\mathcal{F}$ satisfies certain conditions, the existence of a local-in-time solution of this initial value problem has been proved in [165]. This method is mainly applicable to those free boundary problems in which the domain is shrinking. Continuity conditions on $\mathcal{F}$ are also not easy to satisfy. The proof is based on the convergence of Picard iterative method.

In the Stefan problem considered in [166], the free boundary conditions are

$$T = 0, \text{ and } \tilde{R}(x)V_n = (n_t,\, A(x,t)\,\nabla_x T); \text{ on } \Gamma(t). \tag{6.2.10}$$

Here, $\Gamma(t)$ is the free boundary, $\Gamma(t) \cap \partial\Omega = \emptyset$ and $\partial\Omega(t) = \Gamma(t) \cup \partial\Omega$, $\forall t \in [0, t_*]$, $\vec{n}$ is the unit normal outward to $\Gamma(t)$, $\Omega(t) = \Omega \times \{t\}$, $0 \le t < t_*$, $\Omega \subset R^n$, $n \ge 1$. $A(x,t)$ is a uniformly elliptic matrix associated with the parabolic operator $\mathcal{P}$ defined as

$$\mathcal{P}T \equiv \left(\frac{\partial}{\partial t} - \nabla_x(A(x,t)\nabla_x)\right)T = F \text{ in } \Omega(t), \; 0 < t < t_*. \tag{6.2.11}$$

Suppose

$$\tilde{R}(x) = 0, \; x \in \Gamma(0), \; \tilde{R}(x) > 0 \text{ in } \Omega(0), \tag{6.2.12}$$

and $\partial\tilde{R}/\partial\lambda_n < 0$ in some neighbourhood N_0 of $\Gamma(0)$ in Ω. Let $(\lambda_n(x),\, \omega(x))$ be the local coordinates in N_0, with $\omega(x)$ the local coordinates of the projection $P(x)$ of a point $x \in N_0$ on $\Gamma(0)$ and $\lambda_n(x)$ the distance from x to $P(x)$. For the conditions in (6.2.12), we have a degenerate Stefan problem. If $\tilde{R}(x) \equiv 0$ then we have the oxygen-diffusion problem in R^n. Generally the existence, uniqueness and regularity of solution of degenerate Stefan problems are discussed in the context of their weak solutions [167, 168, 169, 170, 171, 172], but the weak solutions are not sufficiently regular to be called classical solutions. In some cases the existence of the classical solutions of degenerate Stefan problems has been investigated which are discussed below.

Using Hanzawa transformation [133] so as to consider a problem in a fixed domain and an analogue of Moser-Nash theorem [173] and assuming some compatibility and other conditions, the existence of the unique solution of the degenerate problem (6.2.10)–(6.2.12) and the solution of oxygen-diffusion problem for $R(x) \equiv 0$ has been established in [166] on a sufficiently small time scale.

Degenerate parabolic-elliptic problems

The parabolic-elliptic Stefan problems have been analysed mostly for their weak solutions. Only few studies have been reported on the analysis of the classical solutions of these problems. The regularity of the classical solution of the two-phase one-dimensional degenerate Stefan problem described in (6.2.13)–(6.2.17) has been discussed in [174]. Let the regions $-1 < x < S(t)$ and $S(t) < x < 1$ be denoted in the following equations by superscripts 1 and 2, respectively.

$$\beta^1(T)T_t^1 - T_{xx}^1 = 0, \quad -1 < x < S(t), \; 0 < t < t_*, \tag{6.2.13}$$

$$\beta^2(T)T_t^2 - T_{xx}^2 = 0, \quad S(t) < x < 1, \; 0 < t < t_*, \tag{6.2.14}$$

$$T^{1,2}(x,0) = T_0^{1,2}(x,0), \quad S(0) = 0, \tag{6.2.15}$$

$$T^1(-1,t) = g^1(t) < 0, \;\; T^2(1,t) = g^2(t) > 0, \;\; 0 < t < t_*, \tag{6.2.16}$$

$$T^{1,2}(S(t),t) = 0; \;\; S(t) = T_x^1(S(t)-0,t) - T_x^2(S(t)+0,t). \tag{6.2.17}$$

The basic assumptions are:

(P_1) $\beta^{1,2} \in C^\infty(\bar{R}_\mp)$, $\beta^{1,2}(T) \geq 0$ and $\beta^{1,2}(T) = 0$ if and only if $T = 0$,

(P_2) $\partial_x T_0^{1,2}(0) > 0$, $\mp g^{1,2}(t) \geq \delta > 0$ and $T_0^{1,2}(x)\,x \geq 0$ where the inequality holds if and only if $x = 0$.

The main result of [174] is the following proposition.

Proposition 6.2.1. Under the assumptions (P_1) and (P_2), the following results hold good.

(1) If the data $T_0^{1,2}(x) \in C^4[\mp 1, 0]$ and $g^{1,2}(t) \in C^2[0,t_*]$ satisfy second order compatibility conditions, then the unique weak solution has the following regularity

$$S(t) \in C^\infty(0,t_*]; \; T^{1,2}(x,t) \in C^\infty\left((\mp 1, S(t)] \times [0,t_*]\right). \tag{6.2.18}$$

(2) If the data $T_0^{1,2}(x) \in C^\infty[\mp 1, 0]$ and $g^{1,2}(t) \in C^\infty[0,t_*]$ satisfy the C^∞ compatibility conditions, then the unique weak solution is also C^∞-smooth up to the boundaries $t = 0$ and $x = \pm 1$.

Definition 6.2.1. : C^∞-compatibility conditions

The data $T_0^{1,2}(x)$ and $g^{1,2}(t)$ are called *m-order compatible* at $(x,t) = (0,0)$ and $(\mp 1, 0)$ if $T_0^{1,2}(x) \in C^{2m}[\mp 1, 0]$, $g^{1,2}(t) \in C^m[0, t_*]$ and there exist functions $\tilde{S}(t) \in C^{2m}[0, t_*]$ and $\tilde{T}^{1,2}(x,t) \in C^{2m,m}[\tilde{S}(t), \mp 1] \times [0, t_*]$ such that

$$\tilde{T}^{1,2}(\mp 1, t) = g^{1,2}(t), \ \tilde{T}^{1,2}(\tilde{S}(t), t) = 0, \tag{6.2.19}$$

$$\tilde{f}(x,t) \ = \ \beta^{1,2}(\tilde{T}^{1,2})\tilde{T}_t^{1,2} - \tilde{T}_{xx}^{1,2} = \mathrm{O}(t^m), \tag{6.2.20}$$

$$\tilde{S}(0) = 0, \ \tilde{g}(t) \equiv \tilde{S}'(t) - \tilde{T}_x^1(\tilde{S}(t) - 0, t) + \tilde{T}_x^2(\tilde{S}(t) + 0, t) = \mathrm{O}(t^m). \tag{6.2.21}$$

If $T_0^{1,2}(x) \in C^\infty[\mp 1, 0]$, $g^{1,2}(t) \in C^\infty[0, t_*]$ and (6.2.19)–(6.2.21) are satisfied for any m, then the data are called *C^∞-compatible.*

A weak variational formulation of the multi-dimensional degenerate parabolic-elliptic Stefan problem has been presented in [167]. By applying the parabolic regularization technique, the existence, uniqueness and stability of the solution with respect to the data, have been analysed. Boundary control aspects have also been discussed. By considering the weak formulation of a two-phase degenerate Stefan problem, the Lipschitz continuity of the free boundary (under suitable assumptions) has been proved in [169] in some small time interval. These results have been further extended in [170] and under suitable assumptions the free boundary is Lipschitz continuous and temperatures satisfy (6.2.13) and (6.2.14) classically.

6.2.1 A Quasi-static Stefan problem and its relation to the Hele-Shaw problem

A one-phase quasi-static Stefan problem (QSSP) in the region $\Omega(t) = \Omega \times \{t\}$, $\Omega \subset R^n$, $n \geq 2$ can be formulated by taking the specific heat $C \equiv 0$ in the heat equation in the one-phase Stefan problem. Temperature is static but the free boundary is time dependent. Let $\partial\Omega(t)$ be the free boundary which is also the outer boundary of $\Omega(t)$. $\Omega(t)$ could be expanding or shrinking for $t > 0$. Shrinking region problems or suction problems, are generally ill-posed. The Hele-Shaw problem (HSP) in R^2 given in (3.3.45)–(3.3.46) can be identified with a Stefan problem if the pressure p of the fluid is regarded as the temperature of the liquid in QSSP. If $p > 0$, then p is the temperature of the warm liquid which is expanding on ice. If $p < 0$, then p is the temperature of the supercooled liquid, and ice is expanding on this supercooled liquid. The pressure of the liquid can be increased (decreased) by injecting (withdrawing) fluid though the inner fixed boundary of $\Omega(0)$ if $\Omega(0)$ is considered to be a doubly-connected region. In this case the fluid is surrounding a region $G \subset R^n$, $n \geq 2$ and the inner static boundary of $\Omega(0)$ is ∂G. The boundary conditions on ∂G could be

$$\frac{\partial p}{\partial n} = -Q, \ \ Q > 0 \text{ (pressure increases)}, \tag{6.2.22}$$

$$\frac{\partial p}{\partial n} = Q, \;\; Q > 0 \text{ (pressure decreases)}, \tag{6.2.23}$$

where, $\vec{n}$ is the unit outward normal to ∂G.

Fluid can also be injected (withdrawn) though point sources (sinks) or distributed sources (sinks) situated in $\Omega(0)$. $\Omega(t)$ could be a simply connected region with outer boundary as the free boundary. When sources or sinks are present, the governing differential equation will have singularities. If the two-dimensional Hele - Shaw cell has porous plates though which a uniform suction is applied, then the formulation for pressure can be written as

$$\nabla^2 p + F(t) = 0, \;\; F > 0, \;\; x \in \Omega(t); \;\; p = 0 \text{ and } \partial p/\partial n = V, \text{ on } \partial\Omega(t). \tag{6.2.24}$$

Here, V is the outward normal component of the velocity of the free boundary.

By using the transformation

$$\hat{p} = p + F(t)\, U, \tag{6.2.25}$$

where U is the solution of the problem

$$\nabla^2 U = 1 \text{ in } \Omega(t); \;\; U = 0 \text{ and } \partial U/\partial n = 0 \text{ on } \partial\Omega(t), \tag{6.2.26}$$

this uniform suction problem can be converted to a standard Hele-Shaw problem except at the singularities of F.

For a two-dimensional HSP, complex variable methods can be used to find some exact analytical solutions. For example, it has been shown in [161] and [175] that due to suction from a point sink, a limacon can become a cardioid with a cusp. For Hele-Shaw problems with shrinking regions a variety of cusp may occur.

For both expanding and contracting one-phase Stefan problems, the free boundary has been shown in [176] to depend continuously and monotonically on the specific heat C taken to be greater than or equal to zero. In some cases, temperature has also continuous dependence on C (specific heat). In particular, taking $C \to 0$, the free boundary in the Stefan problem approaches that of the Hele-Shaw problem and it follows that even well-posed Stefan problems can have free boundaries which can get arbitrarily close to forming cusps. If the free boundary in an expanding Hele-Shaw problem has cusps as $t \to \infty$, then Stefan problem in the limit $t \to \infty$, also develops cusps [177].

For $n \geq 2$, Hele-Shaw suction problem with boundary conditions (6.2.22) and (6.2.23) and one-phase supercooled water problem were discussed in [177]. Existence of weak solutions for both the problems can be proved if and only if the initial domain belongs to a certain class of domains. Uniqueness does not hold in general. In the ill-posed Hele-shaw suction problem, fingering configuration can arise from a suitable initial domain whose boundary is smooth and nearly spherical. In the supercooled water problem the initial domain should belong to a certain class which depends on the initial temperature. Regularity of the free boundary has also been discussed.

Chapter 7

Elliptic and Parabolic Variational Inequalities

7.1 Introduction

Rigorous definitions of elliptic and parabolic variational inequalities will be given a little later. Before that we ask a question, What are the essential features of a variational inequality formulation? Let us consider a simple problem of finding the point $x_0 \in (a, b)$ at which the unique minimum of a real valued function $f(x) \in C^1[a, b]$ exists. If $f(x)$ is a convex function, then x_0 can be obtained by solving the equation $f'(x_0) = 0$. In general for $x_0 \in [a, b]$, three cases arise:

1. If minimum is attained at $x_0 = a$, then $f'(x_0) \geq 0$.

2. If minimum is attained at $x_0 = b$, then $f'(x_0) \leq 0$.

3. If x_0 is an interior point, then $f'(x_0) = 0$.

These three conditions can be expressed in terms of a single inequality

$$f'(x_0)(x - x_0) \geq 0, \; \forall \, x \in [a, b]. \tag{7.1.1}$$

Equation (7.1.1) is an example of a *variational inequality* whose characteristic features are that it is an inequality and it is satisfied for all x varying over the interval $[a, b]$.

We shall now generalize this notion and consider a space of functions, say, $H^1(\Omega), \Omega \subset R^n$, $n \geq 1$ is an open bounded set and let f be a functional on $H^1(\Omega)$. Find a function $u = u_0 \in H^1(\Omega)$ or belonging to a subset of $H^1(\Omega)$ such that $f(u_0)$ is minimum as u varies over $H^1(\Omega)$ or over a subset of $H^1(\Omega)$. Immediately several questions arise such as: What sort of function spaces should be considered so that a solution can be found? What should be the form of the functional defined on these spaces? Are there equivalent formulations? Answers to these require sophisticated functional analysis tools whose detailed description is beyond the scope of this book. In the next few sections an attempt will be made to answer some of the above questions in simple mathematical terms. After discussing the theoretical background of elliptic and parabolic variational inequalities, the formulations of some classical Stefan problems as variational inequalities have been given. It may be noted that for any given classical Stefan problem it may not be possible to formulate it as a variational inequality problem. Weak formulations of Stefan problems, which are continuum models, are easily amenable to variational inequality formulations as variational inequalities are themselves continuum models but this is not the case with classical formulations. To make this volume self-contained, some relevant definitions and theorems are given in the Appendices A–D.

The elliptic variational inequalities will be discussed first. Whenever it is possible to formulate transient Stefan problems as variational inequalities, they are formulated as parabolic variational inequalities. Elliptic variational inequalities, which are concerned with elliptic or steady-state free boundary problems, serve as a good starting point for discussing parabolic variational inequalities. This is because many of the ideas involved and approaches used in proving results in the analysis of elliptic variational inequalities can be extended, with appropriate changes, to the analysis of parabolic variational inequalities. This does not mean that every elliptic problem can be extended to a time dependent parabolic problem.

7.2 The Elliptic Variational Inequality

7.2.1 Definition and the basic function spaces

Let Ω be a bounded open subset of R^n with smooth boundary $\partial\Omega$. An inequality of the form

$$a(u, v-u) \geq (q, v-u),\ \forall\, v \in \mathcal{M};\ u \in \mathcal{M}, \tag{7.2.1}$$

where $a(u,v)$ is a *quadratic bilinear form* (cf. Appendix A), $a(u,v) : H^1(\Omega)\times H^1(\Omega) \to R$, and

$$(q, v) = \int_\Omega qv dx,\ q \in L^2(\Omega), \tag{7.2.2}$$

$$\mathcal{M} = \left\{v \in H^1(\Omega);\ v - d \in H_0^1(\Omega);\ v \geq \psi,\ \psi \in C(\bar{\Omega})\right\}, \tag{7.2.3}$$

is an example of an *elliptic variational inequality*. q, d and ψ, are known functions. $v = d$ on $\partial\Omega$ is the trace sense. ψ is called an *obstacle* and the problem is called an *obstacle problem* or an *elliptic variational inequality with obstacle.* If there exists a function $u_0 \in \mathcal{M}$ which satisfies (7.2.1) for all $v \in \mathcal{M}$, then u_0 is called a solution of the elliptic variational inequality. It can be proved that $\mathcal{M}$ is a non-empty closed convex set (cf. § 7.2.4 for the proof). Problems with more than one obstacle can also be studied (cf. [178]) but such problems will not be discussed here. It is well known in the calculus of variations that the problem formulated in a proposed class of functions may not possess a solution in that class. This difficulty can be overcome by broadening the class of functions in which the problem is formulated and therefore the admissible functions are considered in 'Sobolev spaces' or the 'space of distributions'. A 'Hilbert space' $H^1(\Omega)$ has been considered in (7.2.1) as we would like that at least the first order *weak derivatives* of functions belonging to $\mathcal{M}$ exist. The choice of an appropriate Sobolev space depends on the physical problem under consideration. For example, if in a given physical problem, a solution of the form $y = |x|$, $x \in R$, makes sense then $H^1(R)$ is an appropriate space for the admissible functions. Several other questions concerning variational formulations arise which will be explained after an adequate mathematical exposition of the concepts and notions of variational inequalities.

Let $A : H^1(\Omega) \to H^{-1}(\Omega)$ (H^{-1} is the *dual space* of the Hilbert space H^1, i.e., the space of continuous linear real valued functions on $H^1(\Omega)$) be an operator such that for a fixed $u \in H^1(\Omega)$, $A_u(v) = a(u,v)$, $v \in H^1(\Omega)$. A_u defines the mapping $v \to a(u,v)$. It can be easily proved that A_u is linear and if $a(u,v)$ is continuous, i.e., $|a(u,v)| \leq M\|u\|\|v\|$ for some $M \in R$ then

$$\|A_u\|_{H^{-1}(\Omega)} = \sup_{\|v\|_H = 1} |A_u(v)| \leq M\|u\|_{H^1(\Omega)}, \tag{7.2.4}$$

and hence

$$\|A\|_{\mathcal{L}(H^1, H^{-1})} \leq M_1 \text{ (constant)}. \tag{7.2.5}$$

Therefore, A is continuous and belongs to $H^{-1}(\Omega)$ as A is a linear continuous real valued function on $H^1(\Omega)$. By *Riesz representation theorem*, we have

$$_{H^{-1}(\Omega)}\langle A_u, v\rangle_{H^1(\Omega)} = (u^*, v), \ \forall\, v \in H^1(\Omega), \tag{7.2.6}$$

where u^* is some fixed element of $H^1(\Omega)$ and depends on A_u.

Conversely, if A is linear and satisfies (7.2.4), then $a(u,v)$ is a continuous bilinear form. It is clear that

$$a(u,v) = {}_{H^{-1}(\Omega)}\langle A_u, v\rangle_{H^1(\Omega)}. \tag{7.2.7}$$

With each $q \in L^2(\Omega)$, we can associate a continuous linear functional $B_q \in H^{-1}(\Omega)$ which is defined as

$$B_q : v \to (q,v) = \int_\Omega q v dx, \ v \in H^1(\Omega). \tag{7.2.8}$$

For a rigorous proof of this statement see [22]. A sketchy proof can be given using the following arguments. With each $q \in L^2(\Omega)$, an element of the dual space of $L^2(\Omega)$ can be associated. If the dual space of $L^2(\Omega)$ is denoted by $L^{2*}(\Omega)$ then $L^{2*}(\Omega) \hookrightarrow H^{-1}(\Omega)$.

In view of these arguments, we can write

$$_{L^2(\Omega)}\langle q, v\rangle_{H^1(\Omega)} = {}_{H^{-1}(\Omega)}\langle B_q, v\rangle_{H^1(\Omega)} = \int_\Omega q v dx = (q,v), \ v \in H^1(\Omega). \tag{7.2.9}$$

The pairing $\langle q, v\rangle$ is the value of the functional B_q at v in the sense of (7.2.9). We shall not justify at other places the use of (q, v) in the place of $\langle q, v\rangle)$. The above discussion suggests that a true variational inequality (or equality) in the context of (7.2.1) should be of the form

$$a(u, v-u) = {}_{H^{-1}(\Omega)}\langle A_u, v-u\rangle_{H^1(\Omega)} \geq {}_{H^{-1}(\Omega)}\ \langle q, v-u\rangle_{H^1(\Omega)},$$

$$\forall\, v \in H^1(\Omega),\ u \in H^1(\Omega). \qquad (7.2.10)$$

The space $H^k,\ k \geq 1$ can also be considered and weaker or stronger conditions can also be imposed on the functions involved in the variational inequality formulation.

The form of $a(u, v)$ depends on the elliptic operator considered in a given physical problem. For example, if the elliptic operator $A : H^1(\Omega) \to H^{-1}(\Omega)$ is defined as

$$Au = \nabla^2 u,\ u \in H^1(\Omega), \qquad (7.2.11)$$

then

$$a(u, v) = \int_\Omega \nabla u \cdot \nabla v dx. \qquad (7.2.12)$$

If the elliptic operator A is defined as

$$Au = -\sum_{i,J}^{n} a_{iJ}(x)\frac{\partial^2 u}{\partial x_i \partial x_J} + \sum_{i=1}^{n} b_i(x)\frac{\partial u}{\partial x_i} + e(x)u,\ x \in \Omega, \qquad (7.2.13)$$

then the bilinear form in (7.2.1) is defined by the relation

$$a(u, v) = \int_\Omega \left\{\sum_{i,J}^{n} a_{iJ}\frac{\partial u}{\partial x_i}\frac{\partial v}{\partial x_J} + \sum_{i}^{n}\left(b_i + \sum_J \frac{\partial a_{iJ}}{\partial x_J}\right)\frac{\partial u}{\partial x_i}v + euv\right\} dx. \qquad (7.2.14)$$

The relationship between an elliptic operator and the bilinear form which occurs in the inequality will be discussed later (cf. (7.2.24) and (7.2.37)). The coefficients $a_{iJ},\ b_i$ and e in the elliptic operator in (7.2.13) should belong to appropriate function spaces so that the variational inequality formulation makes sense. We shall see later that for the existence and uniqueness of solutions, it will be required that $a(u, v)$ satisfies some conditions such as continuity, coercivity and these conditions also put some restrictions on the coefficients in the elliptic operator. The non-homogeneous Dirichlet problem corresponding to the elliptic operator A in (7.2.13) can be stated as the problem of finding a function $u(x),\ x \in \Omega$ such that

$$Au = q, \text{ in } \Omega; \quad u = d, \text{ on } \partial\Omega. \qquad (7.2.15)$$

If continuous derivatives of u exist in Ω and d is continuous, then (7.2.15) is satisfied in the classical sense or pointwise sense. If q is square integrable in Ω, then $Au = q$ is satisfied in the sense of distributions. If only the weak derivatives of u exist, then $Au = q$ is satisfied in the distributional sense and $u = d$ is satisfied on $\partial\Omega$ in the trace sense.

7.2.2 Minimization of a functional

A minimization problem and its equivalent elliptic variational inequality

The variational inequality (7.2.1) can be expressed in some other forms also. We shall first show that the variational inequality problem

$$(z, v-z) + a(z, v-z) \geq (q, v-z), \ \forall\, v \in \mathcal{M}^*, \ z \in \mathcal{M}^*, \ q \in L^2(\Omega), \tag{7.2.16}$$

where $\mathcal{M}^*$ is a convex set of an inner product space and Ω is as in (7.2.1), is equivalent to the problem of minimization of a functional $P(v)$, where

$$P(v) = (v, v) + a(v, v) - 2(q, v), \ v \in \mathcal{M}^*, \ q \in L^2(\Omega). \tag{7.2.17}$$

Let $a(u, v)$ be symmetric and z be a solution of (7.2.16). By definition $a(u, v)$ is linear in both the arguments but it need not be symmetric. We shall show that $P(z) \leq P(v), \forall\, v \in \mathcal{M}^*$ so that z is the minimum of $P(v)$.

$$\begin{aligned} P(v) - P(z) &= (v, v) - (z, z) + a(v, v) - a(z, z) - 2(q, v) + 2(q, z) \\ &= (z-v, z-v) + 2(z, v-z) + a(v, v) - a(z, z) - 2(q, v-z). \end{aligned} \tag{7.2.18}$$

Also

$$a(u, u) = a(u-w, u-w) - a(w, w) + 2a(u, w). \tag{7.2.19}$$

Take $u = z$ and $w = z - v$ in (7.2.19). It can be seen that

$$a(v, v) - a(z, z) \geq 2a(z, v-z). \tag{7.2.20}$$

On using (7.2.20) in (7.2.18), we get

$$P(v) - P(z) \geq (z-v, z-v) + 2(z, v-z) + 2a(z, v-z) - 2(q, v-z). \tag{7.2.21}$$

If (7.2.16) holds, then from (7.2.21), $P(v) \geq P(z), \ \forall\, v \in \mathcal{M}^*$.

To prove the converse, suppose that $P(z) \leq P(v), \ \forall\, v \in \mathcal{M}^*$. Since $\mathcal{M}^*$ is a convex set, if $v \in \mathcal{M}^*$ and $z \in \mathcal{M}^*$ then $(1-\alpha)z + \alpha v \in \mathcal{M}^*, \ 0 < \alpha < 1$. Therefore,

$$P(z + \alpha(v-z)) - P(z) \geq 0,$$

or

$$\alpha(v-z, v-z) + \alpha\, a(v-z, v-z) + 2\{(z, v-z) + a(z, v-z) - (q, v-z)\} \geq 0. \tag{7.2.22}$$

For (7.2.22) to hold for an arbitrarily small α, (7.2.16) should hold.

7.2.3 The complementarity problem

By considering the minimization of the functional (a real valued function)

$$f(v) = a(v, v) - (q, v), \tag{7.2.23}$$

we shall now obtain other forms of elliptic variational inequalities. Although $a(u, v)$ given in (7.2.14) can also be considered, a simple bilinear form given by

$$a(u, v) = \int_\Omega \nabla u \cdot \nabla v \, dx, \tag{7.2.24}$$

will be considered for illustration. It will be assumed that a minimizing function $u(x)$ exists such that

$$f(u) \leq f(v), \ \forall \, v \in \mathcal{M}, \ u \in \mathcal{M}, \tag{7.2.25}$$

where the set $\mathcal{M}$ is defined by (7.2.3). Since $\mathcal{M}$ is a convex set, for $0 \leq \alpha \leq 1$, we have

$$f(u + \alpha(v - u)) - f(u) \geq 0, \ \forall \, v \in \mathcal{M}, \tag{7.2.26}$$

or

$$\int_\Omega \{\nabla(u + \alpha(v - u)) \cdot \nabla(u + \alpha(v - u)) - \nabla u \cdot \nabla u\} dx - \\ 2p(u + \alpha(v - u)) + 2p(u) \geq 0. \tag{7.2.27}$$

Here, p is a continuous linear functional on $H^1(\Omega)$ and is defined by

$$p(v) = \int_\Omega qv dx. \tag{7.2.28}$$

On using the linearity of both p and the gradient operator, from (7.2.27) we obtain

$$\alpha^2 \int_\Omega \nabla(v - u) \cdot \nabla(v - u) dx + \alpha \int_\Omega \{\nabla u \cdot \nabla(v - u) - q(v - u)\} dx \geq 0. \tag{7.2.29}$$

For (7.2.29) to hold for an arbitrarily small α, we should have

$$\int_\Omega \{\nabla u \cdot \nabla(v - u) - q(v - u)\} dx \geq 0. \tag{7.2.30}$$

If it is assumed that $u \in H^2(\Omega) \cap C(\Omega)$, the first term in (7.2.30) can be integrated by parts (application of Green's formula). Since v satisfies the same boundary conditions as u, we have $v - u = 0$ on $\partial\Omega$. On using this condition on $\partial\Omega$ and doing the integration by parts, we obtain

$$\int_\Omega (\nabla^2 u + q)(v - u) dx \leq 0, \ \forall \, v \in \mathcal{M}, \ u \in \mathcal{M}. \tag{7.2.31}$$

Let $v = u + \zeta$, $\zeta \geq 0$, $\zeta \in C_0^\infty(\Omega)$. If $u \geq \psi$, then $v \geq \psi$. Substituting $\zeta = (v - u)$ in (7.2.31) and remembering that $\zeta \geq 0$ and $\zeta \in C_0^\infty(\Omega)$, we conclude that

$$\nabla^2 u + q \leq 0, \quad \text{a.e. in } \Omega, \ u \geq \psi. \tag{7.2.32}$$

Next, consider a subset Ω_1 of Ω, where

$$\Omega_1 = \{x \in \Omega;\ u(x) > \psi(x)\}, \ \Omega_1 \subset \Omega. \tag{7.2.33}$$

If $u(x)$ and $\psi(x)$ both are continuous in Ω, then for every $x \in \Omega_1$ there exists a neighbourhood of x in which $u(x) - \psi(x) > 0$. Therefore Ω_1 cannot be a closed set. For any $\zeta \in C_0^\infty(\Omega_1)$, $v = u - \varepsilon\zeta$ is in $\mathcal{M}$ provided $|\varepsilon|$ is sufficiently small. On substituting $v = u - \varepsilon\zeta$ in (7.2.31), we get

$$-\varepsilon \int_\Omega (\nabla^2 u + q)\zeta \, dx \leq 0. \tag{7.2.34}$$

If (7.2.34) holds for ε positive or negative and ζ is arbitrary, then it can be easily concluded that

$$\nabla^2 u + q = 0, \ \text{a.e. in } \Omega_1, \ u > \psi. \tag{7.2.35}$$

The set Ω_1 in which $u(x) > \psi(x)$ is called a *non-coincidence set* and the set $\Omega_2 = \{x \in \Omega : u(x) = \psi(x)\}$ is called a *coincidence set.* The boundary of the non-coincidence set is called a free boundary. If Γ is the free boundary, then

$$\Gamma = \partial\Omega_1 \cap \Omega. \tag{7.2.36}$$

On combining the results obtained in (7.2.32) and (7.2.35), we get the following variational inequality problem.

Find $u \in H^2(\Omega) \cap C(\Omega)$ such that

$$\left.\begin{aligned} &\nabla^2 u + q \leq 0, \\ &u \geq \psi, \\ &(\nabla^2 u + q)(u - \psi) = 0, \end{aligned}\right\} \text{a.e. in } \Omega, \tag{7.2.37}$$

$$u - d \in H_0^1(\Omega).$$

The boundary condition $u = d$ is satisfied on $\partial\Omega$ in the trace sense. The free boundary problem to be studied in Ω_1 is to find a pair (u, Γ) such that

$$\nabla^2 u + q = 0, \ \text{in } \Omega_1, \tag{7.2.38}$$

$$\left. u = \psi;\ \frac{\partial u}{\partial x_i} = \frac{\partial \psi}{\partial x_i}, \ 1 \leq i \leq n \right\} \text{on } \Gamma, \tag{7.2.39}$$

$$u = d \quad \text{on } \partial\Omega_1 \cap \partial\Omega. \tag{7.2.40}$$

The second condition on Γ in (7.2.39) arises due to the fact that $u - \psi$ takes its minimum on Γ (assuming u and ψ to be smooth). The problem stated in (7.2.37) is called a *complementarity problem* and it is a standard problem in 'quadratic programming'. Its 'finite-difference discretization' will have the following form.

$$\left.\begin{aligned} BU + D &\leq 0, \\ U &\geq \Psi, \\ (BU + D)(U - \Psi)^T &= 0, \end{aligned}\right\} \tag{7.2.41}$$

where B, U, D and Ψ are appropriate matrices obtained after discretization of (7.2.37). For example, B could be a $n \times n$ matrix and U, D and Ψ could be $n \times 1$ matrices (n stands for the number of nodal points). On introducing the following substitutions

$$U - \Psi = E \text{ and } Y = -(BU + D), \tag{7.2.42}$$

in (7.2.41), we have a problem of finding E such that

$$BE = -Y - (B\Psi + D), \; E^T Y = 0; \; E \geq 0, \; Y \geq 0. \tag{7.2.43}$$

Matrices B, D and Ψ are known. The problem (7.2.43) is equivalent to the following programming problem provided B is symmetric and positive definite [179]. Minimize

$$(B\Psi + D)^T E + \frac{1}{2} E^T B E, \text{ for } E \geq 0. \tag{7.2.44}$$

If a more general form of $a(u, v)$ is considered such as the one considered in (7.2.14), then by following the procedure indicated in (7.2.26)–(7.2.35), one can easily obtain the complementarity problem of the form (7.2.37) in which in the place of $\nabla^2 u$ we shall have Au given in (7.2.13).

The variational inequality (7.2.37) can be transformed into an inequality of the form (7.2.1). Let $\zeta \in C_0^\infty(\Omega)$, $\zeta \geq 0$ and u be the solution of (7.2.37). If $v = u + \zeta$, then $v \geq \psi$ and v belongs to $\mathcal{M}$. Multiplying the first equation in (7.2.37) by ζ and integrating over Ω, we get

$$\int_\Omega (\nabla^2 u + q)\zeta \leq 0. \tag{7.2.45}$$

Integrating (7.2.45) by parts and remembering that $\zeta = v - u = 0$ on $\partial\Omega$, we get an inequality of the form (7.2.1). The arguments used in obtaining (7.2.45) hold good even if $\nabla^2 u$ is replaced by Au given in (7.2.13). The inequality (7.2.1) also implies the problem in (7.2.37). To show this, we first consider the case $u \geq \psi$. Take $\zeta \in C_0^\infty(\Omega)$, $\zeta \geq 0$. If $v = u + \zeta$, then we have $v \geq \psi$ and therefore v belongs to $\mathcal{M}$. Substituting ζ in place of $(v - u)$ in (7.2.1) and integrating by parts it can be easily seen that $\nabla^2 u + q \leq 0$, a.e., in Ω provided $u \in H^2(\Omega) \cap C(\Omega)$. Next consider the set $\Omega_1 = \{x \in \Omega : u(x) > \psi(x)\} \subset \mathcal{M}$. Follow the procedure which was used to arrive at (7.2.35). It is then easy to show that $(\nabla^2 u + q)(u - \psi) = 0$ for $u \geq \psi$.

7.2.4 Some existence and uniqueness results concerning elliptic inequalities

The minimization problem associated with the functional in (7.2.23) and the complementarity problem in (7.2.37) are also called variational inequality problems. We shall see later that variational inequalities can be expressed in some other forms as well. The equivalence of some different formulations of variational inequalities will be discussed in § 7.2.5 in the context of Problem 1.1.12. Questions pertaining to the existence, uniqueness, and stability of the solutions of variational inequalities arise at this point. As mentioned earlier, sophisticated functional analytic tools are required to answer them (cf. [178], [180], [181]). To understand the basic concepts we shall first discuss some results on the existence and uniqueness of the solutions of elliptic variational inequalities. These results will be helpful in parabolic variational inequalities also. It may be noted that for the existence of the minimum of a function (functional) it is not necessary for the function to be continuous. For example if $f(x) = |x|$, $x \neq 0$ and $f(0) = -2$ then f has a minimum value -2. In this case $f(x)$ is not continuous but is *lower semi continuous* (l.s.c.). The conditions under which the unique minimum of some of the functionals exist are discussed below in the form of theorems which provide answers to some problems.

Problem 7.2.1. Given a real vector space X, a function $f : X \to R$ and a set $Y \subset X$. Find the minimum of f in Y, i.e., find $y_0 \in Y$ such that $f(y_0) = \inf_{y \in Y} f(y)$.

The answer to this problem is contained in the following theorem.

Theorem 7.2.1. If X is a reflexive Banach space, $f : X \to R$ is a convex and l.s.c. function, $Y \neq \emptyset$ is a closed convex subset of X, and either, Y is bounded or f is coercive, then Problem 7.2.1. has a solution. This solution is unique if f is strictly convex. This theorem is called 'theorem of minimization of convex functionals'.

Problem 7.2.2. Let W be a Hilbert space and $\mathcal{P} \subset W$ be a non-empty closed convex set, $g \in W'$ (W' is the dual space of W and the elements of W' are linear, continuous real valued functions) and $f : W \to R$ is defined by

$$f(v) = \frac{1}{2}\|v\|_W^2 - g(v). \tag{7.2.46}$$

Find $u_0 \in \mathcal{P}$ such that $f(u_0) = \inf_{v \in \mathcal{P}} f(v)$.

Theorem 7.2.2. The solution to Problem 7.2.2. exists and is unique.

Proof. We shall show that f is both strictly convex and coercive and so Theorem 7.2.1. can be applied. Since W is a Hilbert space, it is reflexive. Let λ_1 and λ_2 be any two scalars such that $\lambda_1, \lambda_2 \geq 0$, $\lambda_1 + \lambda_2 = 1$ and v_1 and v_2 ($v_1 \neq v_2$) be any two vectors belonging to W. We have

$$\lambda_1 \|v_1\|^2 + (1-\lambda_1)\|v_2\|^2 - \left\{\lambda_1^2 \|v_1\|^2 + 2\lambda_1(1-\lambda_1)(v_1, v_2) + (1-\lambda_1)^2 \|v_2\|^2\right\}$$

$$= \lambda_1(1-\lambda_1)\left\{\|v_1\|^2 - 2(v_1, v_2) + \|v_2\|^2\right\}, \tag{7.2.47}$$

$$\geq \lambda_1(1-\lambda_1)\left\{\|v_1\|^2 - 2\|v_1\|\,\|v_2\| + \|v_2\|^2\right\}, \tag{7.2.48}$$

$$\geq \lambda_1(1-\lambda_1)\{\|v_1\| - \|v_2\|\}^2 > 0, \text{ if } v_1 \neq v_2 \text{ and } \lambda_1 \neq 0. \tag{7.2.49}$$

Also

$$\begin{aligned} f(\lambda_1 v_1 + \lambda_2 v_2) &= \tfrac{1}{2}\|\lambda_1 v_1 + \lambda_2 v_2\|_W^2 - g(\lambda_1 v_1 + \lambda_2 v_2) \\ &= \tfrac{1}{2}\left\{\lambda_1^2 \|v_1\|^2 + \lambda_2^2 \|v_2\|^2 + 2\lambda_1\lambda_2(v_1, v_2)\right\} - \lambda_1 g(v_1) - \lambda_2 g(v_2) \\ &< \tfrac{1}{2}\left\{\lambda_1 \|v_1\|^2 + \lambda_2 \|v_2\|^2\right\} - \lambda_1 g(v_1) - \lambda_2 g(v_2) \\ &< \lambda_1 f(v_1) + \lambda_2 f(v_2). \end{aligned}$$

Therefore f is strictly convex. From the Riesz representation theorem, we have

$$g(v) \leq \|g\|_{W'} \|v\|_W\,. \tag{7.2.50}$$

On using (7.2.50) in (7.2.46), we get

$$\begin{aligned} f(v) &\geq \frac{1}{2}\|v\|_W^2 - \|g\|_{W'}\|v\|_W, \\ &\geq \gamma \|v\|_W^2\,, \quad \text{for some positive constant } \gamma \text{ as } \|v\| \to \infty,\ \forall\, v, \end{aligned} \tag{7.2.51}$$

and so f is coercive.

Problem 7.2.3. Given a Hilbert space W, $g \in W'$ and $\mathcal{P} \neq \emptyset$, a closed convex set of W, find $u_0 \in \mathcal{P}$ such that

$$(u_0, u_0 - v)_W \leq g(u_0 - v), \ \forall\, v \in \mathcal{P}. \tag{7.2.52}$$

Theorem 7.2.3. u_0 is the solution of Problem 7.2.2. if and only if u_0 is a solution of Problem 7.2.3. or in other words Problems 7.2.2. and 7.2.3. are equivalent.

Proof (in outlines). First, we prove that Problem 7.2.2. implies Problem 7.2.3. If $u_0 \in \mathcal{P}$ and $v \in \mathcal{P}$, then $u_0 + \alpha(v - u_0) \in \mathcal{P}$ for $0 \leq \alpha \leq 1$ as $\mathcal{P}$ is a convex set. If the minimum is attained at u_0, then

$$F(\alpha) = f(u_0 + \alpha(v - u_0)) \geq f(u_0), \text{ and } F'(0) \geq 0. \tag{7.2.53}$$

Let $G(\alpha) = u_0 + \alpha(v - u_0)$, then

$$\frac{d}{d\alpha}F(G(\alpha))|_{\alpha=0} = \frac{1}{2}\left(\frac{d}{d\alpha}\|G(\alpha)\|_W^2\right)_{\alpha=0} - \left(\frac{d}{d\alpha}g(G(\alpha))\right)_{\alpha=0} \geq 0. \tag{7.2.54}$$

The first term on the r.h.s. of (7.2.54) can be easily calculated from the first principles and we obtain

$$\frac{1}{2}\left(\frac{d}{d\alpha}\|G(\alpha)\|_W^2\right)_{\alpha=0} = (u_0, v-u_0)_W. \tag{7.2.55}$$

On using the linearity of g

$$\frac{d}{d\alpha}g(G(\alpha))\bigg|_{\alpha=0} = \frac{d}{d\alpha}(g(u_0)+\alpha g(v-u_0))\bigg|_{\alpha=0} = g(v-u_0) \tag{7.2.56}$$

On combining the results in (7.2.55) and (7.2.56), we obtain (7.2.52). To prove that Problem 7.2.3. implies Problem 7.2.2., it will be assumed that the solution of Problem 7.2.3. exists. For the existence proof, see [22]. It can be easily shown that if the solution exists then it is unique. Let u_1 and u_2 be two solutions of Problem 7.2.3.. We have

$$(u_1, u_1-u_2)_W \le g(u_1-u_2),\ u_2 \in \mathcal{P}, \tag{7.2.57}$$

and

$$(u_2, u_2-u_1)_W \ \le\ g(u_2-u_1),\ u_1 \in \mathcal{P}. \tag{7.2.58}$$

Adding (7.2.57) and (7.2.58) and using the definition of 'scalar product' and the linearity of g, we get

$$(u_1-u_2, u_1-u_2) \le 0, \tag{7.2.59}$$

which implies $u_1 = u_2$. In view of the fact that Problem 7.2.2. implies Problem 7.2.3. the unique solution of the Problem 7.2.3. is also the unique solution of Problem 7.2.2. The inequality (7.2.52) is also called a variational inequality and here the inequality is defined with the help of a scalar product, whereas, in (7.2.1) it has been defined with the help of a bilinear form.

The functional (7.2.46) is a special case of the functional

$$f(v) = \frac{1}{2}a(v,v) - g(v), \tag{7.2.60}$$

where $a : W \times W \to R$ is a bilinear form. If with each pair $(u,v) \in W \times W$, $a(u,v)$ associates the scalar product $(u,v)_W$ or in other words $a(u,v) = (u,v)_W$ then $|a(u,v)| = |(u,v)_W| \le \gamma\|u\|_W\|v\|_W$ $(\gamma = 1)$ and a is continuous. $a(u,v) = (u,v)_W = (v,u)_W = a(v,u)$ and a is symmetric. a has also the coercivity property on W, i.e., $a(u,u) \ge \alpha\|u\|_W^2$ for $\alpha = 1$ as $(u,v) = \|u\|^2 \ge \|u\|^2$. On the other hand if a is coercive, symmetric and continuous then a scalar product $((u,v))$ can be defined with the help of a bilinear form as

$$((u,v)) = a(u,v),\ \forall\ u,\ v \in W. \tag{7.2.61}$$

Let $|||\cdot|||_W$ be the norm associated with the inner product defined in (7.2.61). We shall show that $\|\cdot\|_W$ and $|||\cdot|||_W$ are equivalent and therefore the continuous linear functionals defined on W for the two norms are the same. $\alpha\|u\|_W^2 \leq a(u,u) = ((u,u))_W = |||u|||_W^2 = ((u,u))_W = a(u,u) \leq \gamma\|u\|_W^2$. Therefore the two norms are equivalent and $a(u,v)$ is coercive and continuous with respect to the norm $|||\cdot|||_W$. Equivalence of norms implies that if $a(u,v)$ in (7.2.60) is symmetric, coercive and continuous with the respect to $\|.\|_W$, then the functional f in (7.2.46) is no more general then the functional in (7.2.60) as the functional in (7.2.60) can be written as

$$f(v) = \frac{1}{2}\ |||v|||_W^2 - g(v), \tag{7.2.62}$$

where $|||v|||_W = ((v,v)) = a(v,v)$, $v \in W$. Next we ask whether the minimum of the functional (7.2.60) can be obtained as the solution of a variational inequality.

Problem 7.2.4. Let W be a Hilbert space, $\mathcal{P} \subset W$, a non-empty closed convex set and $f : W \to R$ a functional defined by (7.2.60) in which the bilinear form $a(u,v)$ is continuous. Find $u_0 \in \mathcal{P}$ such that

$$f(u_0) \leq f(v), \ \ \forall\, v \in \mathcal{P}. \tag{7.2.63}$$

Problem 7.2.5. Let W, $\mathcal{P}$, f and $a(u,v)$ be the same as in Problem 7.2.4.. Find $u_0 \in \mathcal{P}$ such that

$$a(u_0, u_0 - v) \leq g(u_0 - v), \ \forall\, v \in \mathcal{P}. \tag{7.2.64}$$

If u_0 is the minimum of $f(v)$ in (7.2.60), then

$$\frac{d}{d\alpha} f(u_0 + \alpha(v - u_0))|_{\alpha=0} \geq 0, \ \forall\, v \in \mathcal{P}. \tag{7.2.65}$$

The derivative in (7.2.65) can be easily calculated by using the bilinearity of a and the linearity of g and it can be shown that (7.2.63) implies (7.2.64) only if $a(u,v)$ is symmetric. If a solution of Problem 7.2.5. exists, then it can be proved that it is unique [22]. If it can be proved that a unique solution of Problem 7.2.4. exists, then the equivalence of Problems 7.2.5. and 7.2.4. can be established. If a is coercive on $\mathcal{P} - \mathcal{P}$, i.e.,

$$\exists\, \alpha,\ \alpha > 0,\ e.g.,\ a(u-v, u-v) \geq \alpha\,\|u - v\|_W^2,\ \forall\, u, v \in \mathcal{P}, \tag{7.2.66}$$

then the solution of Problem 7.2.5. exists [22]. $\mathcal{P} - \mathcal{P} = \{x - y :\ x \in \mathcal{P},\ y \in \mathcal{P}\}$. It can be proved that if a is symmetric and, continuous on $\mathcal{P}$ and coercive on $\mathcal{P} - \mathcal{P}$ then it is coercive on $\mathcal{P}$.

$$\begin{aligned} a(u,u) &= a(u-\zeta, u-\zeta) - a(\zeta,\zeta) + 2a(u,\zeta),\ u \in \mathcal{P},\ \zeta \text{ fixed in } \mathcal{P} \\ &\geq \alpha\,\|u-\zeta\|_W^2 - \gamma\,\|\zeta\|_W^2 - 2\gamma\,\|u\|\|\zeta\|_W,\ \alpha > 0,\ \gamma > 0 \\ &\geq \alpha\,\|u-\zeta\|_W^2 - \gamma\,\|\zeta\|_W\ (3\|\zeta\|_W + 2\,\|u-\zeta\|_W). \end{aligned} \tag{7.2.67}$$

It is now easy to prove that $a\,(u,u) \to +\infty$ as $\|u\| \to \infty$.

Lions - Stampacchia Theorem

Theorem 7.2.4. Let W be a Hilbert space. $\mathcal{P} \subset W$ be a non-empty closed convex set, $g \in W'$ and $a : W \times W \to R$, a continuous bilinear form on W and coercive on $\mathcal{P} - \mathcal{P}$. There is one and only one $u_0 \in \mathcal{P}$ such that

$$a(u_0, v - u_0) \geq g(v - u_0), \ \forall \ v \in \mathcal{P}, \tag{7.2.68}$$

and further the application which associates u_0 to every g is continuous, i.e., the Problem 7.2.5. is well-posed.

Let $\mathcal{P} = \mathcal{M}$ (for $\mathcal{M}$ see (7.2.3)) in Theorem 7.2.4. so that we can discuss stability of the solution of the variational inequality (7.2.1). A comparison of (7.2.68) and (7.2.1) suggests that the linear functional g is defined as

$$g(v) = (q, v) = \int_\Omega qv dx, \ \forall \ v \in \mathcal{M}. \tag{7.2.69}$$

Note that a test function v satisfies the boundary condition. Therefore, to prove continuous dependence of the solution on the data, it is to be proved that if u_1 and u_2 are two different solutions of (7.2.68) corresponding to the data q_1 and q_2, then

$$\alpha \|u_1 - u_2\|_W \leq \|q_1 - q_2\|_{W'(\Omega)}, \ \alpha > 0. \tag{7.2.70}$$

Let g_1 and g_2 be associated with q_1 and q_2 through (7.2.69). From (7.2.68), we have

$$a(u_1, u_2 - u_1) \geq g_1(u_2 - u_1), \ (\text{take } v = u_2) \tag{7.2.71}$$

and

$$a(u_2, u_1 - u_2) \geq g_2(u_1 - u_2), \ (\text{take } v = u_1). \tag{7.2.72}$$

Adding (7.2.71) and (7.2.72) and changing the sign, we obtain

$$\begin{aligned} a(u_1 - u_2, u_1 - u_2) &\leq g_1(u_1 - u_2) - g_2(u_1 - u_2) \\ &\leq \|g_1 - g_2\|_{W'} \|u_1 - u_2\|_W. \end{aligned} \tag{7.2.73}$$

Using the coercivity of $a(u, v)$ on $\mathcal{P}$, (7.2.73) can be written as

$$\alpha \|u_1 - u_2\|_{W(\Omega)} \leq \|q_1 - q_2\|_{W'(\Omega)}, \ \alpha > 0. \tag{7.2.74}$$

This proves continuous dependence of the solution on the data . The Problem 7.2.5. is a generalization of Problem 7.2.3. as it imposes weaker conditions on the form of $a(u, v)$. There is an interesting geometrical interpretation of the minimization of the functional (7.2.46). Using Riesz representation theorem, $f(v)$ in (7.2.46) can be written as

$$f(v) = \frac{1}{2}\|v\|_W^2 - (u^*, v), \ v \in \mathcal{P}, \ u^* \in W. \tag{7.2.75}$$

Define a functional $F(v)$ as

$$\begin{aligned} F(v) &= f(v) + \frac{1}{2}\|u^*\|_W^2 \\ &= \frac{1}{2}\|v\|_W^2 - (u^*, v)_W + \frac{1}{2}\|u^*\|_W^2 \\ &= \frac{1}{2}\|v - u^*\|_W^2. \end{aligned} \tag{7.2.76}$$

If there is an element of $\mathcal{P}$ which minimizes F, then it also minimizes f and minimizing F amounts to finding a $v \in \mathcal{P}$ whose distance from a fixed $u^* \in W$ is minimum, i.e., $\|u_0 - u^*\|$ is minimum. If $u_0 \in \mathcal{P}$ is such that $\|u_0 - u^*\|$ is minimum then

$$\|u_0 - u^*\| \leq \|v - u^*\|, \ \forall\, v \in \mathcal{P}. \tag{7.2.77}$$

It can be proved that the inequality (7.2.77) is equivalent to the following inequality

$$(u_0 - u^*, v - u_0) \geq 0, \ \forall\, v \in \mathcal{P}. \tag{7.2.78}$$

Hint: let u_0, u^* and v be the vertices of a triangle. Use the proposition that the sum of the two sides of the triangle is greater than the third side. Problems 7.2.2. and 7.2.3. are both equivalent to the problem of finding $u_0 \in \mathcal{P}$ for a given u^* (determined by g) satisfying (7.2.78) for $\forall\, v \in \mathcal{P}$. u_0 is called the projection of u^* on $\mathcal{P}$. The inequality (7.2.78) is yet another way of writing the variational inequality associated with functional (7.2.46).

Some of the results discussed above can be proved under weaker conditions by considering a reflexive Banach space in the place of a Hilbert space but in a Hilbert space more interesting results can be obtained which in some cases have interesting interpretations.

In (7.2.64) (we consider this inequality as it is more general than (7.2.46)), g is linear. Therefore, (7.2.64) can be written as

$$a(u_0, u_0 - v) - g(u_0) \leq -g(v), \ \forall\, v \in \mathcal{P}. \tag{7.2.79}$$

This suggests that we can consider variational inequalities of the type

$$a(u, v - u) + p(u) \leq p(v), \ \forall\, v \in \mathcal{P}, \tag{7.2.80}$$

where $p : W \to R$ is not necessarily linear. In view of (7.2.80), we want to study now the minimization problem associated with the functional

$$f(v) = m(v) + p(v); \ f : W \to R \cup \{\infty\}, \tag{7.2.81}$$

where p is not linear and $m(v)$ could be a functional whose *Gâteaux derivative* denoted by $\nabla m(v)$ exists. The motivation for considering Gâteaux derivative comes from the fact that in seeking the minimum of a function defined from $R \to R$ we look for those points at which the classical derivative of the function is zero. On using the linearity of

$a(u, v)$ it was possible to calculate the derivative in (7.2.65) and therefore the Gâteaux derivative of the bilinear form $a(u, v)$ exists. If the continuity of $a(u, v)$ is assumed, then it can be proved that the mapping $v \rightarrow a(u, v)$ is continuous (see (7.2.4)). What type of functional $m(v)$ should be? Note that $p(v)$ can be easily handled as it is enough if it is l.s.c. and a *proper convex functional.* A question similar to the one raised about the minimum of the functional in Problem 7.2.4 can be asked for the functional (7.2.81).

Problem 7.2.6. Given a Hilbert space W, $\mathcal{P} \subset W$, a non-empty closed convex set and $f : W \rightarrow R \cup \{+\infty\}$, a functional of the form $f(u) = m(u) + p(u)$; find $u_0 \in \mathcal{P}$ such that

$$f(u_0) \leq f(v), \ \forall\, v \in \mathcal{P}. \tag{7.2.82}$$

The conditions under which u_0 is the solution of (7.2.82) are given in the following theorem.

Theorem 7.2.5. If $m(u)$ is finite, convex and *G-differentiable* (*Gâteaux differentiable*) on $\mathcal{P}$ and p is convex and a proper functional on $\mathcal{P}$, then $u_0 \in \mathcal{P}$ is the minimum of $f(u)$ in Problem 7.2.6. if and only if u_0 satisfies the inequality

$$_{W'}\langle \nabla m(u_0), u_0 - v \rangle_W + p(u_0) \leq p(v), \ \forall\, v \in \mathcal{P}. \tag{7.2.83}$$

The inequality (7.2.83) presents yet another form of variational inequality. We shall not pursue further the generalizations of the functionals whose minimum could be obtained.

Variational equation

If $\mathcal{P} = W$ in Problem 7.2.4. and we seek the minimum of f for $\forall\, v \in W$, then in (7.2.65), α could be any real number and not restricted to $0 \leq \alpha \leq 1$. In this situation, in the place of (7.2.65), we shall have an equation $f'(0) = 0$ and we get a *variational equation* to determine the solution u_0 and we have

$$a(u_0, z) = g(z), \ z \in W. \tag{7.2.84}$$

However, this is not the case always. For example, if f is of the form (7.2.81) in which $m(v)$ is G-differentiable but not $p(v)$, then even if $\mathcal{P} = W$, we have an inequality of the form (7.2.83) (cf. [22]).

The bilinear form $a(u, v)$ in (7.2.14) is not symmetric and therefore the scalar product which by definition is symmetric cannot be defined in terms of $a(u, v)$. The bilinear form (7.2.24) is symmetric and let us examine what happens if a scalar product is formally defined in terms of $a(u, v)$ by the equation

$$(u, v) = a(u, v) = \int_\Omega \nabla u \cdot \nabla v dx, \ u, v \in H^1(\Omega). \tag{7.2.85}$$

All other properties of the scalar product are satisfied by (u, v) if it is defined by (7.2.85), except the property $(u, u) = 0 \Rightarrow u = 0$. We have already seen that if a bilinear form can be equated with a scalar product then the bilinear form is continuous and coercive and both these properties are very useful in analysing the solutions of the variational inequalities. It is possible to define a scalar product with the help of (7.2.85) after some modifications in the formulation of the variational inequality. Let $\Omega \subset R^n$ be an open bounded set with $\partial\Omega$ of class C^0, i.e., every point of $\partial\Omega$ has a neighborhood U such that $\partial\Omega \cap U$ is the graph of a continuous function (cf. [182]), $u \in H^k(\Omega)$, $k \in N$ and $D^\alpha u$ (α-th order weak derivative, see Appendix D) $\in L^2(\Omega)$, $|\alpha| \leq k$. H^k can be provided with the scalar product

$$(u, v)_{H^k(\Omega)} = \sum_{|\alpha|\leq k} (D^\alpha u, D^\alpha v)_{L^2(\Omega)}, \tag{7.2.86}$$

and

$$(u, u)_{H^k(\Omega)} = \sum_{|\alpha|\leq k} \int_\Omega |D^\alpha u|^2 d\Omega. \tag{7.2.87}$$

The norm $\|u\|_{H^k(\Omega)} = (u, u)^{1/2}$. The scalar product in $H_0^k(\Omega)$ ($H_0^k(\Omega)$ is the closure of the space $C_0^\infty(\Omega)$ in $W^{k,2}(\Omega)$, see Appendix D) can be defined as

$$(u, v)_{H_0^k(\Omega)} = \sum_{|\alpha|=k} (D^\alpha u, D^\alpha v)_{L^2(\Omega)}. \tag{7.2.88}$$

In the space $H_0^k(\Omega)$, the equivalence of the norms generated by the scalar products defined in (7.2.86) and (7.2.88) can be asserted with the help of the following theorem.

Theorem 7.2.6. Let $\Omega \subset R^n$ be an open bounded set and $k \in N$ (N is the set of positive integers). There exist two constants β_1 and β_2 such that for every $u \in H_0^k(\Omega)$ the following inequality holds.

$$\begin{aligned} \beta_1 \left(\sum_{|\alpha|=k} \int_\Omega |D^\alpha u|^2 dx \right)^{\frac{1}{2}} &\leq \left(\sum_{|\alpha|\leq k} \int_\Omega |D^\alpha u|^2 dx \right)^{\frac{1}{2}} \\ &\leq \beta_2 \left(\sum_{|\alpha|=k} \int_\Omega |D^\alpha u|^2 dx \right)^{\frac{1}{2}}. \end{aligned} \tag{7.2.89}$$

The inequality on the l.h.s. of (7.2.89) can be easily proved and we have $\beta_1 = 1$. To prove the inequality on the r.h.s., the *Poincare's inequality* [22] can be invoked which states that if $\psi \in C_0^\infty(\Omega)$ then

$$\int_\Omega |\psi|^2 \, dx \leq \beta \sum_{i=1}^{n} \int_\Omega \left| \frac{\partial \psi}{\partial x_i} \right|^2 dx, \tag{7.2.90}$$

for some constant $\beta > 0$. Since $H_0^k(\Omega)$ is the closure of the space $C_0^\infty(\Omega)$, (7.2.90) holds even if $\psi \in H_0^k(\Omega)$. If the r.h.s. in (7.2.90) tends to zero, then the l.h.s. also tends to zero which implies $\psi \equiv 0$. The above discussion suggests that the scalar product can be defined with the help of (7.2.88) provided the space $H_0^1(\Omega)$ is considered in the place of $H^1(\Omega)$.

Next we show that the existence and uniqueness results are not affected if the space $H_0^1(\Omega)$ is considered in the place of $H^1(\Omega)$. Let u be the solution to the variational inequality

$$\int_\Omega \nabla u \cdot \nabla(v-u)dx \geq \int_\Omega q(v-u)dx,\ \forall\, v \in \mathcal{M},\ u \in \mathcal{M}, \tag{7.2.91}$$

where $\mathcal{M}$ is defined by (7.2.3). If $w = u - d$, then $w = 0$ on $\partial\Omega$ and $w \in H_0^1(\Omega)$. u is the solution to (7.2.91) if and only if $w = u - d$ is the solution to

$$\int_\Omega \nabla w \cdot \nabla(v-u)dx \geq \langle q, v-u\rangle - \int_\Omega \nabla d \cdot \nabla(v-u)dx. \tag{7.2.92}$$

We have, $v - u = v - d - w = \zeta - w$, and both ζ and w belong to $H_0^1(\Omega) \subset H^1(\Omega)$. The inequality (7.2.92) can be written as

$$\int_\Omega \nabla w \cdot \nabla(\zeta - w)dx \geq \langle F, \zeta - w\rangle,\ \forall\, \zeta \in M_1, \tag{7.2.93}$$

where

$$\langle F, \zeta\rangle = \langle q, \zeta\rangle - \int_\Omega \nabla d \cdot \nabla\zeta dx,\ d_{x_i} \in L^2(\Omega),\ 1 \leq i \leq n, \tag{7.2.94}$$

and

$$\mathcal{M}_1 = \{\zeta \in H_0^1(\Omega) : \zeta \geq \psi - d\}. \tag{7.2.95}$$

It can be seen that if $d \in H^1(\Omega)$, and $q \in H^{-1}(\Omega)$ then F is linear and continuous on $H^1(\Omega)$, i.e., $F \in H^{-1}(\Omega)$.

It will now be shown that $\mathcal{M}_1$ is a non-empty closed convex set, $a(w,\zeta)$ is continuous and coercive on $\mathcal{M}_1$, so that Theorem 7.2.4. can be applied. If $\zeta_1,\ \zeta_2 \in \mathcal{M}_1$, then $\alpha\zeta_1 + (1-\alpha)\zeta_2 \geq \alpha(\psi - d) + (1-\alpha)(\psi - d) \geq \psi - d$, for $0 < \alpha < 1$. Therefore $\mathcal{M}_1$ is a convex set. The functions of $H_0^1(\Omega)$ are absolutely continuous functions if Ω is an open subset of R^1. Let ψ be continuous on $\Omega = (0, b)$. Consider the function $[\psi - d]^+ = [\psi - d + |\psi - d|]/2$. $[\psi - d]^+$ is the positive part of $\psi - d$, therefore $[\psi - d]^+ \geq \psi - d$. As $\zeta = v - d = 0$ on $\partial\Omega$, $\psi - d \leq 0$ on $\partial\Omega$ and so $[\psi - d]^+ = 0$ on $\partial\Omega$. $\mathcal{M}_1$ is non-empty as it contains $[\psi - d]^+$. We shall now show that the space $\mathcal{M}_1$ is closed (complete in the norm defined through (7.2.88)). Let $\{v_n\}$ be a convergent sequence in $\mathcal{M}_1$ whose limit is $\hat{v}$. It is to be proved that $\hat{v} \in \mathcal{M}_1$. The sequence $\{v_n\}$ can be thought of as a subsequence of a sequence $\{\hat{v}_n\}$ in $H_0^1(\Omega)$ and $\{v_n\}$ converges to $\hat{v} \in H_0^1(\Omega)$ in the a.e. sense. Since $v_n \in \mathcal{M}_1$, $v_n \geq \psi$ for $\forall\, n$. $\{v_n\}$ is a subsequence of $\{\hat{v}_n\}$ therefore $\hat{v} \geq \psi$ and so $\hat{v} \in \mathcal{M}_1$. In the space $H_0^1(\Omega)$, scalar product can be defined as

$$(w, \zeta) = a(w, \zeta) = \int \nabla w . \nabla\zeta\ dx. \tag{7.2.96}$$

We have $a(w,w) = (w,w) \geq \|w\|^2$ and therefore $a(w,\zeta)$ is coercive and continuous on $H_0^1(\Omega)$ or coercive on $\mathcal{M}_1 - \mathcal{M}_1$. The conditions of Theorem 7.2.4. are satisfied and a unique solution w_0 of (7.2.92) exists in $\mathcal{M}_1$. The unique solution u_0 of (7.2.91) is then given by $u_0 = w_0 + d$.

Till now only Dirichlet problem has been considered for the elliptic operator of the form ∇^2. We now consider a problem of the form

$$-\nabla^2 u + \lambda u = q, \text{ a.e. in } \Omega,\ u \in H^2(\Omega) \cap C(\Omega),\ q \in L^2(\Omega), \tag{7.2.97}$$

$$u = d, \text{ a.e. on } \partial\Omega,\ d \in L^2(\Omega). \tag{7.2.98}$$

The differential equation (7.2.97) can be obtained as an *Euler equation* of an appropriate minimization problem, for example, in (7.2.60) take $a(u,v)$ given by (7.2.100), g defined by (7.2.69) and $v \in H^2(\Omega) \cap C(\Omega)$. If an obstacle is introduced such as $v \geq \psi$ then the obstacle problem for (7.2.97)–(7.2.98) can be formulated as a variational inequality of the form (see the derivation of the complementarity problem in (7.2.37) obtained from the minimization problem) given below in (7.2.99).

$$a(w, \zeta - w) \geq \langle q, \zeta - w\rangle,\ \forall\, \zeta \in H_0^1(\Omega),\ w \in H_0^1(\Omega), \tag{7.2.99}$$

where

$$a(w,\zeta) = \int_\Omega \nabla w . \nabla \zeta dx + \lambda \int_\Omega w\zeta\, dx,\ \forall\, w,\ \zeta \in H_0^1(\Omega). \tag{7.2.100}$$

It can be proved that if $\lambda > -1/\beta$, where β is the same as in the Poincare's inequality (7.2.90) then $a(w,\zeta)$ is coercive on $H_0^1(\Omega)$ (cf. [22]).

If the boundary condition is of the Neumann type such as

$$\frac{\partial u}{\partial n} = r, \text{ a.e. on } \partial\Omega,\ r \in L^2(\Omega), \tag{7.2.101}$$

where $\vec{n}$ is the unit outward normal to $\partial\Omega$, then the obstacle problem results in a variational inequality of the form

$$a(u, v-u) \geq \int_\Omega q(v-u)dx + \int_{\partial\Omega} r(v-u)dx,\ \forall\, v \in H^1(\Omega),\ u \in H^1(\Omega). \tag{7.2.102}$$

In this case $a(u,v)$ has the same form as in (7.2.100) but $u, v \in H^1(\Omega)$ (boundary condition is different). By choosing $\lambda > 0$ and $\nu = \min(1,\lambda)$, it can be shown that (cf. [22])

$$a(v,v) \geq \nu \|v\|^2_{H^1(\Omega)}, \tag{7.2.103}$$

where the norm in $H^1(\Omega)$ is defined by (7.2.86). Note that the Neumann boundary condition (7.2.101) has already been incorporated in (7.2.102). The test function need not satisfy the Neumann boundary condition as it occurs naturally in the inequality. When $-\nabla^2 u + \lambda u \le q$ is multiplied by $(v - u)$ and integrated over Ω and integration by parts is done we get (7.2.102). The integration by parts will give an integral over $\partial\Omega$, and the Neumann boundary condition is incorporated. If $\lambda = 0$, the form $a(u, v)$ is not coercive, and in this case the problem with Neumann condition does not always have a solution; and when it does, it is not unique.

The boundary conditions could be of mixed type, for example, let $\partial\Omega = \partial\Omega_1 \cup \partial\Omega_2$, $\partial\Omega_1 \cap \partial\Omega_2 = \emptyset$ and

$$-\nabla^2 u + \lambda = q, \text{ a.e. in } \Omega, \tag{7.2.104}$$

$$u = d, \text{ a.e. on } \partial\Omega_1, \tag{7.2.105}$$

$$\frac{\partial u}{\partial n} = r, \text{ a.e. on } \partial\Omega_2. \tag{7.2.106}$$

Then the variational inequality formulation is still given by (7.2.102) with $\partial\Omega$ in the integral on the r.h.s. of (7.2.101) replaced by $\partial\Omega_2$. Since $a(u, v)$ remains unchanged, in this case also, (7.2.103) holds good and $a(u, v)$ is coercive. The test functions have to satisfy the boundary condition (7.2.105).

Operators such as the one considered in (7.2.13) or even more general operators can be considered and the existence and uniqueness of solutions of the variational inequalities so obtained can be discussed under suitable assumptions (cf. [22]). It is not always possible to formulate a given physical problem as a variational inequality problem or obtain a variational equation for the problem.

Regularity of the solution is an important aspect of the study of variational inequalities. Some results on the regularity of solutions will be given in the context of parabolic variational inequalities. With the help of the obstacle problem of the string discussed earlier in Problem 1.1.12., some other aspects of variational inequalities such as the choice of appropriate spaces in which the solutions are sought, and the restrictions on the obstacle so that the set $\mathcal{M}$ in (7.2.3) is non-empty, will be discussed in the next section. In general the space of unknown functions should be large enough so that the existence of solutions can be discussed, but small enough so that a unique solution can be obtained. In physical problems, smoothness of solutions cannot be ignored. In principle, the data spaces should be general so as to accommodate various types of data but the continuous dependence of the solution on the data is required for well-posedness.

7.2.5 Equivalence of different inequality formulations of an obstacle problem of the string

We restate briefly an obstacle problem of the string as a problem of finding a continuous function $v(x)$ which minimizes the energy functional $f(v)$, where

$$f(v) = \frac{1}{2}\int_0^b \left(v'\right)^2 dx, \; v \in \mathcal{M}_2, \tag{7.2.107}$$

$$\begin{aligned}\mathcal{M}_2 &= \{v \in H^1(0,b) : v(0) = v(b) = 0,\ v(x) \geq \psi(x),\ \forall\, x \in (0,b)\} \\ &= \{v \in H_0^1(0,b) :\ v(x) \geq \psi(x), \forall\, x \in (0,b)\} \qquad (7.2.108)\end{aligned}$$

Since we are dealing with a physical problem, v and ψ should be continuous functions. It is clear from Fig. 7.2.1. that the first order derivatives of $v(x)$ and $\psi(x)$ need not be continuous but they should belong to $L^2(\Omega)$ which is also suggested by the integral in (7.2.107).

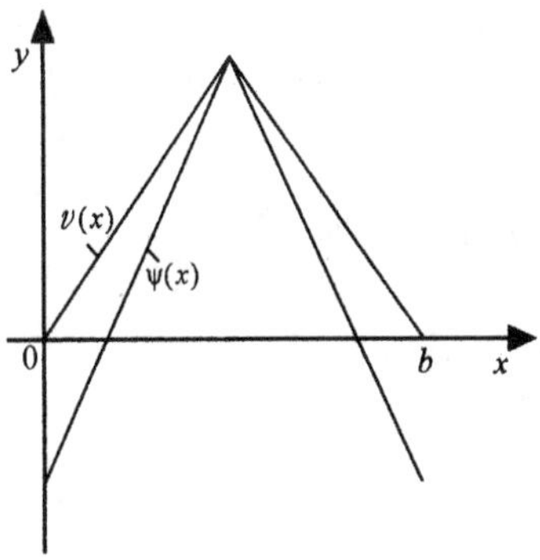

Fig. 7.2.1. Differentiability of $v(x)$ is not required.

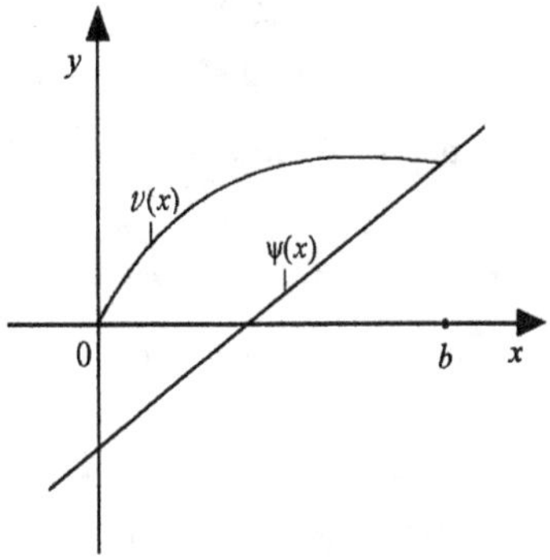

Fig. 7.2.2. The set $\mathcal{M}_2$ is empty.

In Fig. 7.2.2., although $v(x) \geq \psi(x)$, v does not satisfy the boundary conditions and so the set $\mathcal{M}_2$ is empty. For $\mathcal{M}_2$ to be non-empty, ψ must satisfy one of the following two conditions.

$$(\psi 1) :\ \psi\ \in C^0(\Omega),\ \psi(0+) < 0 \ \text{ and } \ \psi(b-) < 0. \qquad (7.2.109)$$

$$(\psi 2) :\ \psi\ \in H^1(\Omega),\ \psi(0) \leq\ 0 \ \text{ and } \ \psi(b) \leq\ 0. \qquad (7.2.110)$$

Note that by increasing the smoothness of ψ, the smoothness of the solution is not increased. For example, if ψ is taken as a parabola then $\psi \in C^\infty(\Omega)$, $\Omega = (0,b)$, but still the second derivative of the solution $u'' \notin C^2(0,b)$ if the parabola does not pass through the points 0 and b. In showing the equivalence of different forms of variational inequality formulations given below for the obstacle problem, the treatment is not strictly rigorous. A rigorous treatment would require that the condition $u'' \leq 0$ is considered in the sense of distributions in the formulation (I) given below. If $u \in H^1(0,b)$, then it is not necessary that u'' belongs to $L^1(0,b)$. It will be assumed for the present discussion that $u, v \in H^2(\Omega) \cap C(\Omega)$ and u'' belongs to $L^1(\Omega)$. The obstacle ψ satisfies one of the two conditions ($\psi 1$) or ($\psi 2$). For clarity, the three different formulations are restated below.

(I) Find $u \in \mathcal{M}_2$ ($\mathcal{M}_2$ as in (7.2.108)) such that $u(0) = u(b) = 0$, $u'' \leq 0$, $u \geq \psi$ and $(u - \psi)u'' = 0$.

(II) Find u such that

$$\int_0^b u'(x)(v'(x) - u'(x))dx \geq 0, \ \forall\, v \in \mathcal{M}_2, \ u \in \mathcal{M}_2. \tag{7.2.111}$$

(III) Minimize

$$\int_\Omega (\partial v/\partial x)^2 dx, \ \forall\, v \in \mathcal{M}_2.$$

If $u', v' \in L^2(\Omega)$, then u'^2 and $u'v' \in L^1(\Omega)$. Further let $u'' \in L^1(\Omega)$. It will be assumed in the following that all the equations hold good in the a.e. sense.

Equivalence of formulations (I), (II) and (III)

(III) $\Rightarrow$ (II). Assume that there exists a $u \in \mathcal{M}_2$ such that

$$\int_0^b u'^2 dx \leq \int_0^b w'^2 dx, \ \forall\, w \in \mathcal{M}_2. \tag{7.2.112}$$

Since $\mathcal{M}_2$ is a convex set, if $u, v \in \mathcal{M}_2$ then $w = u + \lambda(v-u) \in \mathcal{M}_2$, $0 < \lambda < 1$. From (7.2.112), we get

$$\int_0^b u'^2 dx \leq \int_0^b \{u'^2 + \lambda^2(v'-u')^2 + 2\lambda u'(v'-u')\}dx, \tag{7.2.113}$$

or

$$0 \leq 2\lambda \int_0^b u'(v'-u')dx + O(\lambda^2). \tag{7.2.114}$$

For (7.2.114) to be true for $\forall\, \lambda > 0$, we should have

$$\int_0^b u'(v'-u')dx \geq 0.$$

(II) $\Rightarrow$ (III). If $a(u,v)$ is defined by (7.2.24), then $a(u-v, u-v) \geq 0$ implies

$$\int_0^b 2u'(u'-v')dx - \int_0^b u'^2 dx + \int_0^b v'^2 dx \geq 0, \tag{7.2.115}$$

or

$$\int_0^b u'^2 dx \leq \int_0^b v'^2 dx - 2\int_0^b u'(v'-u')dx \leq \int_0^b v'^2 dx. \tag{7.2.116}$$

As (7.2.111) holds, the relation $\int_0^b u'(v'-u')dx \geq 0$ can be used in (7.2.116). The inequality (7.2.116) implies that u gives the minimum in (III).

(I) $\Rightarrow$ (II).

$$\int_0^b u'(v'-u')dx = u'(v-u)_0^b - \int_0^b u''(v-u)dx,$$

$$= -\int_0^b u''(v-u)dx. \tag{7.2.117}$$

Let $\Omega = (0,b) = I_1 \cup I_2$, where

$$I_1 = \{x \in \Omega : u(x) = \psi(x)\};\ u'' \le 0 \text{ in } I_1 \text{ if formulation (I) holds,} \tag{7.2.118}$$

$$I_2 = \{x \in \Omega :\ u(x) > \psi(x)\};\ u'' = 0 \text{ in } I_2 \text{ if formulation (I) holds.} \tag{7.2.119}$$

Using (7.2.118) and (7.2.119) in (7.2.117), we get

$$\int_0^b u'(v'-u')dx = -\int_{I_1} u''(v-\psi(x))dx \ge 0. \tag{7.2.120}$$

The integration by parts in (7.2.117) requires that u'' should belong to $L^1(\Omega)$. Therefore (I) $\Rightarrow$ (II) provided the solution u of (I) is such that $u'' \in L^1(\Omega)$. From Fig. 7.2.1 it is clear that the solution may exist even if $u'' \notin L^1(\Omega)$. It has been shown in [183] that if ψ be such that ψ'' is a *Radon measure* and its positive part can be represented by a function belonging to $L^1(0,b)$ then the condition $u'' \notin L^1(\Omega)$ is not required.

(II) $\Rightarrow$ (I). Consider (7.2.117) and the case $u \ge \psi$. Let $v = u+\zeta$, $\zeta \ge 0$, $\zeta \in C_0^\infty(\Omega)$, and $v \in \mathcal{M}_2$. If (II) holds, then

$$-\int_0^b u''\zeta dx \ge 0. \tag{7.2.121}$$

The relation (7.2.21) implies $u'' \le 0$ in Ω.

Next consider the case $u > \psi$. Let $v = u + \varepsilon\zeta$, $\zeta \in C_0^\infty(I_2)$. Extend ζ to Ω trivially, i.e., $\zeta \equiv 0$ in I_1. If ε is sufficiently small, v will be greater then ψ and we have

$$\int_0^b u'(v'-u')\ dx = -\int_0^b u''(v-u)\ dx = -\varepsilon \int_{I_2} u''\zeta\ dx \ \ge 0. \tag{7.2.122}$$

Since (7.2.122) holds for all sufficiently small ε which could be positive or negative, it can be concluded that

$$\int_{I_2} u''\zeta\ dx = 0. \tag{7.2.123}$$

As ζ is arbitrary, $u'' = 0$ on I_2.

The equivalence of different variational inequality formulations of problems involving more general elliptic operators can also be discussed. For example, it can be proved that the variational inequality (7.2.1) with $a(u,v)$ defined by (7.2.14) is equivalent to the variational inequality (7.2.37) in which the operator A is given by (7.2.13) provided u, the coefficients in the elliptic operator, and the obstacle ψ satisfy some conditions [22].

7.3 The Parabolic Variational Inequality

7.3.1 Formulation in appropriate spaces

Let Ω be an open bounded domain in R^n and $\Omega_{t_*} = \Omega \times \{0 < t < t_*\}$. One of the problems of mathematical physics is to solve the initial-boundary value problem stated below in (7.3.1)–(7.3.3).

$$T_t + AT = f, \quad \text{in } \Omega_{t_*} \tag{7.3.1}$$

$$T = g, \quad \text{on } \partial\Omega_{t_*}, \tag{7.3.2}$$

$$T = T_0, \quad \text{in } \Omega \times \{0\}, \tag{7.3.3}$$

where

$$A = -\sum_{i,J=1}^{n} a_{iJ}(x,t)\frac{\partial^2 T(x,t)}{\partial x_i \partial x_J} + \sum_{i=1}^{n} b_i(x,t)\frac{\partial T(x,t)}{\partial x_i} + d(x,t)T(x,t), \tag{7.3.4}$$

$$\partial\Omega_{t_*} = \partial\Omega \times (0, t_*). \tag{7.3.5}$$

$\partial\Omega_{t_*} \cup \Omega \times \{0\}$ is called a *parabolic boundary*. The weak form of problem (7.3.1)–(7.3.4) can be stated in terms of finding a function $T(x,t) \in H^1(\Omega_{t_*})$ satisfying the equation

$$(T_t, v - T) + a(t; T, v - T) = (f, v - T), \tag{7.3.6}$$

for a.a. $t \in (0, t_*)$ and all $v \in H^1(\Omega_{t_*})$. v satisfies the given initial and boundary conditions,

$$a\,(t; T, v) = \int_{\Omega} \left(\sum_{i,J} a_{iJ} \frac{\partial T}{\partial x_i}\frac{\partial v}{\partial x_J} + \sum_i \hat{b}_i \frac{\partial T}{\partial x_i} v + d\,Tv \right) dx, \tag{7.3.7}$$

$$\hat{b}_i = b_i + \sum \frac{\partial a_{iJ}}{\partial x_J}, \tag{7.3.8}$$

$$(v, w) = \int_{\Omega} vw\,dx. \tag{7.3.9}$$

It has been assumed in (7.3.4) that the data, the temperature derivatives with respect to x and t, and coefficients a_{iJ} are sufficiently smooth functions (cf. [1] and proposition 7.3.1.). There does not exist a minimization principle for parabolic problems. Often Biot's variational statement [184] is taken as a variational principle. Every elliptic problem cannot be generalized to a transient problem, for example, the obstacle problem for the string (Problem 1.1.12.) does not have a time dependent generalization. If an obstacle is introduced in the formulation (7.3.1)–(7.3.4), then we may obtain an inequality.

Parabolic variational inequality for the problem in (7.3.1)–(7.3.4) can be defined in any of the following two equivalent forms.

(I) For a.a. $t \in (0, t_*)$, let

$$E(t) = \{v \in H^1(\Omega);\ v \geq \psi(x,t) \text{ a.e. in } \Omega\}. \tag{7.3.10}$$

$E(t)$ is a closed convex subset of $H^1(\Omega)$. To make $E(t)$ a non-empty set, we shall take $\psi(x,t) \leq g(x,t)$ a.e. on $\partial\Omega$ for $t \in (0, t_*)$. A function $T(x,t) \in L^2(0, t_*; H^1(\Omega)), T(t) \in E(t)$, for $a.a.$ $t \in (0, t_*)$, i.e, for each fixed t, $T(x,t) = T(x) \in H^1(\Omega)$ is called a solution of the parabolic variational inequality (7.3.11) if

$$(T_t, v - T) + a(t; T, v - T) \geq (f, v - T), \text{ for } a.a\ t \in (0, t_*), \tag{7.3.11}$$

is satisfied for $\forall\ v \in L^2(0, t_*; H^1(\Omega))$ with $v(t) \in E(t)$ for a.a. $t \in (0, t_*)$, $v(x,0) = T_0$ and v satisfying (7.3.2). It is required that $\partial T/\partial t \in L^2(0, t_*; L^2(\Omega))$. $\psi(x,t)$ is called an obstacle. $f(x,t)$, $\psi(x,t)$ and $\partial\Omega$ should satisfy some smoothness conditions. Some of these conditions will be discussed for some specific problems to be discussed later.

(II) The second variational inequality formulation is the integrated form of (7.3.11), i.e., in the place of (7.3.11), we have

$$\int_0^{t_*} \{(\frac{\partial T}{\partial t}, v - T) + a(t; T, v - T) - (f, v - T)\} dt \geq 0. \tag{7.3.12}$$

Let

$$E = \{v \in L^2(0, t_*; H^1(\Omega));\ v \geq \psi(x,t) \text{ a.e. in } \Omega_{t_*},\ v(x,0) = T_0\}. \tag{7.3.13}$$

$v(x,t)$ satisfies the boundary condition (7.3.2). If there exists a function $T(x,t) \in E$ with $\partial T/\partial t \in L^2(0, t_*; L^2(\Omega))$ such that (7.3.12) is satisfied for all $v \in E$, then $T(x,t)$ is called a solution of the variational inequality (7.3.12).

Any solution of (7.3.11) will satisfy (7.3.12) and so the formulation (I) implies formulation (II). We assume that a solution $T(x,t)$ of (7.3.12) exists. For some $\varepsilon > 0$ such that $(t_0 - \varepsilon,\ t_0 + \varepsilon) \in (0, t_*)$, consider the following function.

$$\left.\begin{aligned} v &= T(x,t),\ t \notin (t_0 - \varepsilon, t_0 + \varepsilon), \\ &= \hat{v}, \qquad t \in (t_0 - \varepsilon, t_0 + \varepsilon). \end{aligned}\right\} \tag{7.3.14}$$

If v defined in (7.3.14) is used in (7.3.12), then

$$\int_{t_0-\varepsilon}^{t_0+\varepsilon} \left\{(\frac{\partial T}{\partial t}, \hat{v} - T) + a(t; T, \hat{v} - T) - (f, \hat{v} - T)\right\} dt \geq 0. \tag{7.3.15}$$

Since the interval $(t_0 - \varepsilon, t_0 + \varepsilon) \in (0, t_*)$ is arbitrary, we obtain (7.3.11). The complementarity problem corresponding to the parabolic variational inequality (7.3.11) in which the operator A is given by (7.3.4) consists of finding $T(x,t)$ satisfying the following system of equations.

$$\left.\begin{aligned} T_t + AT &\geq f, \\ T &\geq \psi(x,t), \\ (T_t + AT - f)(T - \psi(x,t)) &= 0, \end{aligned}\right\} \textit{ in } \Omega_{t_*}. \tag{7.3.16}$$

The relations in (7.3.16) are satisfied in the a.e. sense and $T \in L^2(0,t_*;H^2(\Omega))$. The equivalence of the variational inequality (7.3.12) and the complementarity problem (7.3.16) can be established. The proof is similar to the one discussed in the elliptic case (see § 7.2.3). Consider first the case $T \geq \psi$, where T is the solution of (7.3.12). If $\phi \in C_0^\infty(\Omega_{t_*})$, $\phi \geq 0$ and $\phi(x,0) = 0$ then $v = T + \phi$ belongs to the set E.

$$\int\limits_{\Omega_{t_*}} (T_t + AT - f)\phi\, dxdt = \int\limits_0^{t_*}\left\{\left(\frac{\partial T}{\partial t}, v - T\right) + a(t;T,v-T) - (f, v-T)\right\}dt \geq 0. \tag{7.3.17}$$

Since ϕ is arbitrary, we conclude that $\partial T/\partial t + AT - f \geq 0$ in Ω_{t_*}. Next, we consider the case $T > \psi$ and use the procedure followed in the elliptic case (equations (7.2.32)–(7.2.36)) with a difference that now $v = T - \varepsilon\phi$, $|\varepsilon| > 0$, ε very small, $\phi \in C_0^\infty(\Omega_{t_*})$ and $\phi(x,0) = 0$. It can be seen that $\partial T/\partial t + AT - f = 0$. Combining the two cases $T \geq \psi$ and $T > \psi$, (7.3.16) is obtained. To prove that (7.3.16) implies (7.3.12), consider the first integral on the l.h.s. of (7.3.17). On replacing ϕ by $(v - T)$, writing $v - T = v - \psi - (T - \psi)$ and using the third relation is (7.3.16) and the fact $v \geq \psi$, we obtain

$$\int\limits_{\Omega_{t_*}} (T_t + AT - f)(v - T)\, dxdt \geq 0, \ \forall\, v \in E. \tag{7.3.18}$$

Since (7.3.18) holds for all $v \in E$, T is the solution of (7.3.12).

In the elliptic case, the coercivity of the bilinear form played an important role in proving well-posedness and uniqueness of the solution. If the transformation

$$\hat{T} = e^{-\alpha t}T, \ \alpha > 0, \tag{7.3.19}$$

is used then $T_t + AT = f$ is transformed into

$$\hat{T}_t + (A + \alpha)\hat{T} = e^{-\alpha t}f, \tag{7.3.20}$$

and $a(t;T,v)$ is transformed into

$$\hat{a}(t;\hat{T},\hat{v}) = a(t;\hat{T},\hat{v}) + \alpha(\hat{T},\hat{v}). \tag{7.3.21}$$

If α is sufficiently large, then irrespective of the fact that $a(t;T,v)$ is coercive or not $\hat{a}(t;\hat{T},\hat{v})$ is coercive. Without any loss of generality it can be assumed from the start that $a(t;T,v)$ is coercive and if for a.a. t, $T \in H^1(\Omega)$ then

$$a(t;T,T) \geq \lambda \|T\|_{H^1_\Omega} \geq \lambda \int\limits_\Omega \left(|\nabla T|^2 + T^2\right) dx, \ \ \lambda > 0. \tag{7.3.22}$$

Uniqueness and stability of the solution of (7.3.12) can be proved if coercivity condition is imposed on $a(t;T,v)$ in (7.3.12). Let T_1 and T_2 be two different solutions of (7.3.12) satisfying the same initial and boundary conditions. Taking $v = T_2$ in (7.3.12) and $T = T_1$, we obtain

$$\int_0^{t_*} \left\{ \left(\frac{\partial T_1}{\partial t}, T_2 - T_1 \right) + a(t; T_1, T_2 - T_1) - (f, T_2 - T_1) \right\} dt \geq 0. \tag{7.3.23}$$

Similarly by taking $T = T_2$ and $v = T_1$, a second relation is obtained. Adding the two relations so obtained, it is easy to obtain the relation

$$\int_0^{t_*} \left\{ \frac{1}{2}\frac{d}{dt}(T_1 - T_2, T_1 - T_2) + a(t; T_1 - T_2, T_1 - T_2) \right\} dt \leq 0. \tag{7.3.24}$$

On integrating the first term under the integral in (7.3.24), remembering that $T_1 - T_2 = 0$ at $t = 0$ and using the coercivity condition (7.3.22), we obtain.

$$\frac{1}{2} \|T_1 - T_2\|^2_{t=t^*} + \lambda \int_0^{t_*} \|T_1 - T_2\|^2_{H^1(\Omega)}\, dt \leq 0. \tag{7.3.25}$$

Since both the terms in (7.3.25) are positive, each should be zero and this implies $T_1 = T_2$. The stability of the solution can be proved by following the proof given earlier in the elliptic case (cf. § 7.2.3 equation (7.2.74)). For proving the existence of the solution *penalty method* is generally used. The conditions under which a unique solution of the variational inequality (7.3.11) exists for the parabolic operator given in (7.3.4) have been discussed in [1] and are briefly given below in Proposition 7.3.1.

Proposition 7.3.1. Assume that the following conditions are satisfied.

(1) The operator $T_t + AT$, where A is given by (7.3.4) with coefficients defined in Ω_{t_*} is of parabolic type at (x,t), i.e.,

$$\sum_{i,J=1}^{n} a_{iJ}(x,t)\xi_i\xi_J \geq \lambda |\xi|^2, \; \forall\, \xi \in R^n, \; \lambda > 0. \tag{7.3.26}$$

It is said to be *uniformly parabolic* in Ω_{t_*} if (7.3.26) holds for all (x,t) in Ω_{t_*} and λ is independent of (x,t).

(2) $$\sum \|a_{iJ}\|_\alpha + \sum \|b_i\|_\alpha + \|d\|_\alpha \leq M(\text{Constant}), \; 0 < \alpha < 1, \tag{7.3.27}$$

where

$$\|T(x,t)\|_\alpha = \|T(x,t)\|_0 + H_\alpha(T(x,t)), \tag{7.3.28}$$

$$\|T\|_0 = \sup_{(x,t)\in\Omega_{t_*}} |T(x,t)|, \tag{7.3.29}$$

$$H_\alpha(u) = \sup_{(x,t),(x',t')} \frac{|T(x,t) - T(x',t')|}{(p_d((x,t),(x',t')))^\alpha}, \tag{7.3.30}$$

$$p_d((x,t),(x',t')) = |x - x'| + |t - t'|^{1/2}. \tag{7.3.31}$$

$p_d((x,t),(x',t'))$ is called *parabolic distance*.

(3) $\partial\Omega$ is in $C^{2+\alpha}$ (cf. Appendix D).

(4) $d(x,t) \geq 0$ and $f,\ g,\ D_x g,\ D_x^2 g,\ D_t g$ belong to $C^\alpha(\bar{\Omega}_{t_*})$. $C^\alpha(\bar{\Omega}_{t_*})$ is the space of functions $u(x,t)$ which are Hölder continuous with exponent α, $H_\alpha(u) < \infty$.

(5) $|D_x a_{iJ}| \leq M_1$ (constant) so that $a(t; u, v)$ can be defined. (7.3.32)

If the above conditions are satisfied, then there exists a unique solution of the obstacle problem (7.3.11) ((7.3.12)) and

$$D_x T,\ D_x^2 T,\ D_t T \text{ belongs to } L^p(\Omega_{t_*}),\ \forall p,\ 1 < p < \infty. \tag{7.3.33}$$

7.4 Some Variational Inequality Formulations of Classical Stefan Problems

7.4.1 One-phase Stefan problems

The variational inequality formulation is a fixed domain formulation and therefore weak enthalpy formulations of the Stefan problems are more suitable for variational inequality formulations than the classical formulations of Stefan problems. Even in the classical Stefan problems, variational inequality formulations of one-phase Stefan problems can be handled easily than the two-phase Stefan problems. This is because a one-phase Stefan problem can be formulated in terms of the 'freezing index' which together with its gradient is continuous throughout the region. Also the constraint that the temperature is greater than or equal to (less than or equal to) the melting temperature holds throughout the phase-change region. We describe below some variational inequality formulations of one-phase Stefan problems. Some results concerning analysis of solutions of these problems are also given.

Consider an open bounded domain Ω in R^3. At time $\mathrm{t} = 0$, Ω is filled with ice cold water at temperature $\mathrm{T} = 0$ where 0 is the dimensionless freezing temperature of water. Let $\partial\Omega = \partial\Omega_f(t) \cup \partial\Omega_e(t) \cup \partial\Omega_r(t)$ and on these three disjoint portions of the boundary, three different types of boundary conditions are prescribed. On the portion $\partial\Omega_f$ cooling is done so that ice formation takes place. For $t > 0$, $\Omega = \Omega_0(t) \cup \Omega_1(t) \cup \Gamma(t)$. $\Omega_0(t)$ is the ice region at time t, $0 \leq t \leq t_*$, $\Omega_1(t)$ contains water at temperature $\mathrm{T} = 0$ and $\Gamma(t)$ is the ice-water phase-change boundary whose equation is taken as $t = \phi(x),\ x \in \Omega$. $\Omega_0(t)$ and $\Omega_1(t)$ are disjoint regions. The mathematical formulation of this problem is as follows:

$$\frac{\partial T}{\partial t} = \nabla.(k\nabla T),\ \text{in } \Omega_0(t),\ 0 < t < t_*, \tag{7.4.1}$$

$$T = 0,\ \text{in } \Omega_1(t),\quad 0 < t < t_*, \tag{7.4.2}$$

$$T(x,t) = g(t) < 0,\ \text{on } \partial\Omega_f,\ 0 < t < t_*, \tag{7.4.3}$$

$$T(x,t) = 0,\ \text{on } \partial\Omega_r,\ 0 < t < t_*, \tag{7.4.4}$$

$$\alpha T = k\frac{\partial T}{\partial n},\ \text{on } \partial\Omega_e,\ 0 < t < t_*, \tag{7.4.5}$$

$$k\ \nabla T \cdot \nabla\phi(x) = l_c\ ;\ \text{and } T(x,\phi(x)) = 0, \tag{7.4.6}$$

$$T(x,0) = 0\ \text{ in } \Omega. \tag{7.4.7}$$

Here, $\vec{n}$ is the unit outward normal on $\partial\Omega_e$, $l_c = l/C > 0$, k is the thermal diffusivity and α is the constant for heat radiation. It has been assumed that $\Gamma(t)$ is monotone and ice is progressing on water.

Although the constraint $T \leq 0$ holds throughout Ω, the variational inequality formulation of this problem presents two difficulties. The above formulation is not a continuum model (because of the free boundary conditions) and secondly ∇T is discontinuous across $t = \phi(x)$ and so integration by parts which is required to obtain the bilinear form cannot be done in Ω unless we write the integral as the union of two integrals (see equation (7.3.7)).

Duvait's transformation

Using *Duvait's transformation* [185], temperature T(x,t) is transformed into another dependent variable u(x,t), called *freezing index* [186], such that u and ∇u are continuous throughout Ω. $u(x,t)$ is defined as

$$u(x,t) = \int_{\phi(x)}^{t} T(x,\tau)d\tau,\ \text{in } \Omega_0(t);\ u(x,t) = 0\ \text{ in } \Omega_1. \tag{7.4.8}$$

Note that $u(x,t)$ is continuous in $\Omega \times (0, t_*)$.

From (7.4.8), we can easily obtain the equation

$$\nabla u(x,t) = \int_{\phi(x)}^{t} \nabla T(x,\tau)d\tau - T(x,\phi(x)).\nabla\phi(x),\ \text{in } \Omega_0$$

or

$$\nabla u(x,t) = \int_{\phi(x)}^{t} \nabla T(x,\tau)d\tau, \quad \text{in } \Omega_0(t), \tag{7.4.9}$$

$$\nabla u(x,t) = 0, \text{ in } \Omega_1(t). \tag{7.4.10}$$

Also

$$\begin{aligned} \nabla \cdot (k\nabla u) &= \int_{\phi(x)}^{t} \nabla \cdot (k\nabla T)d\tau - k\nabla(T(x,\phi(x))).\nabla\phi(x), \\ &= \int_{\phi(x)}^{t} \frac{\partial T}{\partial \tau}d\tau - l_c, \\ &= \frac{\partial}{\partial t}\int_{\phi(x)}^{t} T d\tau - l_c, \\ &= \frac{\partial u}{\partial t} - l_c, \text{ in } \Omega_0(t), \end{aligned} \tag{7.4.11}$$

or

$$\frac{\partial u}{\partial t} - \nabla \cdot (k\nabla u) - l_c \;=\; 0, \quad \text{in } \Omega_0(t). \tag{7.4.12}$$

If u = 0, then $\partial u/\partial t - \nabla.(k\nabla u) - l_c = -l_c \le 0$. Therefore

$$\frac{\partial u}{\partial t} - \nabla \cdot (k\nabla u) - l_c \le 0, \quad \text{in } \Omega, \; 0 < t < t_*. \tag{7.4.13}$$

The initial and boundary conditions for u(x,t) can be obtained from those given for T(x,t) and the problem for the dependent variable u(x,t) can be stated as follows:

$$\left.\begin{aligned} &\partial u/\partial t - \nabla \cdot (k\nabla u) - l_c \le 0, \text{ in } \Omega_{t_*} = \Omega \times (0,t_*), \\ &u \le 0, \text{ in } \Omega_{t_*}, \\ &u(\partial u/\partial t - \nabla \cdot (k\nabla u) - l_c) = 0, \text{ in } \Omega_{t_*}. \end{aligned}\right\} \tag{7.4.14}$$

$$u(x,t) = \int_{0}^{t} g(\tau)d\tau = \hat{g}(t) < 0, \text{ and } \hat{g}(0) = 0; \text{ on } \partial\Omega_f(t), \; 0 < t < t_*, \tag{7.4.15}$$

$$u(x,t) = 0, \quad \text{on } \partial\Omega_r(t), \; 0 < t < t_*, \tag{7.4.16}$$

$$\alpha u(x,t) = k\frac{\partial u}{\partial n}, \quad \text{on } \partial\Omega_e(t), \; 0 < t < t_*, \tag{7.4.17}$$

$$u(x,0) = 0, \text{ in } \Omega. \tag{7.4.18}$$

The above equations are to be understood in the a.e. sense.

A variational inequality of the form (7.3.12) will now be obtained for equations (7.4.14)–(7.4.18). Define a set D as

$$D = \{v \in L^2(0,t_*;H^1(\Omega)),\ \dot{v} \in L^2(0,t_*;L^2(\Omega))\},$$

where

$$\begin{aligned} v &\leq 0 \text{ a.e. in } \Omega_{t_*},\ v(x,0)=0 \text{ in } \Omega, \\ v &= 0, \text{ on } \partial\Omega_r,\ 0<t<t_*, \\ v &= \hat{g}(t), \text{ on } \partial\Omega_f,\ 0<t<t_*. \end{aligned} \tag{7.4.19}$$

The test function $v(x,t)$ need not satisfy (7.4.17) as it will appear as a natural boundary condition in the formulation.

It can be seen that

$$\{(u_t, v-u) + (k\nabla u, \nabla(v-u)) - (l_c, v-u)\}_\Omega$$

$$= (u_t - \nabla\cdot(k\nabla u) - l_c,\ v-u)_\Omega + (k\nabla u\cdot\vec{n},\ v-u)_{\partial\Omega_e},\ 0\leq t\leq t_*. \tag{7.4.20}$$

The last term on the r.h.s. of (7.4.20) arises on doing the integration by parts of the term $(k\nabla u, \nabla(v-u))$. On $\partial\Omega_r$ and $\partial\Omega$, $v-u=0$. In view of the third condition in (7.4.14), we have

$$(u_t - \nabla.(k\nabla u) - l_c,\ v-u)_\Omega = (u_t - \nabla.(k\nabla u) - l_c,\ v)_\Omega \geq 0. \tag{7.4.21}$$

It may be noted that in Ω_{t_*}, $u_t - \nabla\cdot(k\nabla u) - l_c \leq 0$ and $v\leq 0$. On using (7.4.21) and (7.4.17) in (7.4.20), it can be written as

$$\{(u_t, v-u) + (k\nabla u, \nabla(v-u)) - (l_c, v-u)\}_\Omega - (\alpha u, v-u)_{\partial\Omega_e} \geq 0,$$

$$\text{for a.a. } t\in(0,t_*). \tag{7.4.22}$$

In (7.4.22), the boundary conditions (7.4.17) and (7.4.16) have been incorporated. Neumann boundary conditions are called *natural boundary conditions* because they are automatically taken care in the variational formulation and the test functions are not required to satisfy them. Dirichlet boundary conditions are to be satisfied by the test functions. On integrating (7.4.22) with respect to time, we get

$$\int_0^{t_*} \left[\{(u_t, v-u) + (k\nabla u, \nabla(v-u)) - (l_c, v-u)\}_\Omega - (\alpha u, v-u)_{\partial\Omega_e}\right] dt \geq 0. \tag{7.4.23}$$

The inequality (7.4.23) is the parabolic variational inequality associated with the one - phase Stefan problem (7.4.1)–(7.4.7). If there exists a function $u\in D$ (D as above) such that (7.4.23) is satisfied for all $v \in D$, then u is called a solution of the variational inequality (7.4.23). If $u \in L^2(0, t_*; H^1(\Omega))$ and the norm of u is defined as

$$\int_0^{t_*} \|u\|^2_{H^1(\Omega)}\, dt, \tag{7.4.24}$$

then $L^2(0, t_*; H^1(\Omega))$ is a Banach space.

The formulation given in (7.4.14)–(7.4.18) is also a variational inequality formulation and in this case we are looking for $u \in L^2(0, t_*; H^2(\Omega))$. It has been proved in [187, 188] that if the meas $(\partial\Omega_f) > 0$, then there exists a unique solution u(x,t) of the variational inequality (7.4.23) subject to the initial and boundary conditions (7.4.15)–(7.4.18) such that

$$\dot{u} \in L^2(0, t_*; H^1(\Omega)) \cap L^\infty(0, t_*; L^2(\Omega)); \text{ and } \dot{u} - \nabla \cdot (k\nabla u) \in L^2(\Omega_{t_*}). \tag{7.4.25}$$

The numerical solution of the variational inequality

$$(\dot{u}, v - u)_\Omega + (k\nabla u, \nabla(v - u))_\Omega + (\alpha u, v - u)_{\partial\Omega_e} \geq (l_c, v - u)_\Omega, \quad \text{for a.a. } t \in (0, t_*), \tag{7.4.26}$$

has been discussed in [188] by 'Galerkin approximation' in space and discretization by finite-difference in time. A variational formulation of a two-phase Stefan problem has been given in [189] and its numerical solution has been obtained by finite-difference methods.

The initial condition (7.4.18) implies that at t = 0, Ω is occupied by ice cold water. We shall now consider the problem in which at t = 0 ice occupies a region G_0 with temperature $\tilde{g}(x) \leq 0$, $G_0 \subset \Omega$. $\Omega - G_0$ is filled with water at temperature T = 0. Let $\Omega_0(t)$ be the region occupied by ice at any time t and $\Omega_1(t)$ be the water region so that $\Omega = \Omega_0(t) \cup \Omega_1(t) \cup \Gamma(t)$, where, Ω_0 and Ω_1 are disjoint regions and $\Gamma(t)$ is the ice-water interface whose equation is given by $t = \phi(x)$. If $x \in G_0$, then $\phi(x) = 0$. The initial and boundary conditions are given by (7.4.16)–(7.4.18). We assume that ice is progressing on water. The freezing index u(x,t) in this case is defined as

$$u(x,t) = \int_{\phi(x)}^{t} T(x,\tau)d\tau, \quad \text{in } \Omega_0(t) - G_0, \ 0 < t < t_*, \tag{7.4.27}$$

$$u(x,t) = \int_0^t T(x,\tau)d\tau, \text{ in } G_0, \ 0 < t < t_*, \tag{7.4.28}$$

$$u(x,t) = 0, \quad \text{in } \Omega_1(t), \ 0 < t < t_*; \text{ and } u(x,0) = 0, \text{ in } \Omega_1(0). \tag{7.4.29}$$

The above transformation was suggested in [190]. It is easy to check (see (7.4.11)) that

$$\nabla \cdot (k\nabla u) = u_t - l_c, \text{ in } \Omega_0(t) - G_0, \ 0 < t < t_*, \ l_c = l/C, \tag{7.4.30}$$

$$\nabla \cdot (k\nabla u) = u_t - \tilde{g}(x), \text{ in } G_0; \text{ and } u = 0, \text{ in } \Omega_1(t),\ 0 < t < t_*. \tag{7.4.31}$$

Define a function $f(x)$ such that

$$\left.\begin{aligned} f(x) &= \tilde{g}(x),\ \tilde{g}(x) \le 0,\ x \in G_0(x), \\ f(x) &= l_c > 0,\ x \in \Omega - G_0. \end{aligned}\right\} \tag{7.4.32}$$

In view of (7.4.30)–(7.4.32), we have

$$u_t - \nabla.(k\nabla u) - f = 0,\ u < 0,\ \text{in } \Omega_0(t),\ 0 < t < t_*, \tag{7.4.33}$$

and when $u = 0$, we have

$$u = 0;\ u_t - \nabla.(k\nabla u) - f = -f = -l_c,\ x \in \Omega - \Omega_0(t),\ 0 < t < t_*. \tag{7.4.34}$$

Combining (7.4.33) and (7.4.34), we get

$$u\,(u_t - \nabla.(k\nabla u) - f) = 0,\ \text{in } \Omega,\ 0 < t < t_*. \tag{7.4.35}$$

For $(x, t) \in \Omega_{t_*}$,

$$\begin{aligned} (u_t - \nabla.(k\nabla u) - f)\,(v - u) &= -(u_t - \nabla.(k\nabla u) - f)\,u + (u_t - \nabla.(k\nabla u) - f)\,v, \\ &\ge 0,\ \text{a.e. in } \Omega_{t_*},\ v \le 0,\ \text{in } \Omega_{t_*}. \end{aligned} \tag{7.4.36}$$

Therefore, the variational inequality problem corresponding to (7.4.30)–(7.4.32) can be stated as the problem of finding $u \in \Omega$ such that

$$\begin{aligned} (u_t, v - u)_\Omega + (k\nabla u, \nabla (v - u))_\Omega - (f, v - u)_\Omega - (\alpha u, v - u)_{\partial\Omega_e} \ge 0, \\ \text{for a.a. } t \in (0, t_*) \text{ and } \forall\ v \in D. \end{aligned} \tag{7.4.37}$$

Here, D is the same as defined in the context of problem (7.4.14)–(7.4.18).

By taking $\Omega = \{x : 0 \le x \le R_0 < \infty\}$, and $\Omega = \Omega_0(t) \cup \Omega_1(t) \cup \Gamma(t)$, where $\Omega_0(t)$ and $\Omega_1(t)$ are disjoint regions and $\Gamma(t)$ is the phase-change boundary, a one-dimensional one-phase problem has been considered in [178]. $\Omega_0(0) = \{x : 0 \le x < S_0,\ 0 < S_0 < R_0\}$ is the region initially occupied by hot water at the temperature $b(x) > 0$. $\Omega_1(0) = \{x : S_0 < x < R_0\}$ is occupied by ice at the melting temperature zero. A variational inequality with $u = \hat{g}(t) > 0$ (sufficiently smooth) prescribed at $x = 0$ has been studied in [178] when water is progressing on ice, i.e., $u \ge 0$ in $\Omega(t_*) = \Omega(0) \times (0, t_*)$. Using the 'penalty method' (cf. [1]), it has been shown that under suitable assumptions there exists a unique solution u to the variational inequality such that

$$u,\ u_x,\ u_t,\ u_{xx} \in L^\infty(\Omega(t_*)), \tag{7.4.38}$$

and $u \ge 0$, $u_t \ge 0$ in $\Omega_0(t_*)$. The region $\Omega_0(t) = \{x :\ u(x, t) > 0\}$, $0 \le t \le t_*$ is expanding continuously with time. Further, the water-ice boundary Γ admits the representation

$$\Gamma :\ x = \sigma(t),\ 0 \le t \le t_*,$$

where σ is a continuously increasing function of t with $S_0 = \sigma(0) < \sigma(t)$ for $t > 0$. For each $(x_0, t_0) \in \Gamma$, there exists a neighbourhood B_r of (x_0, t_0) such that

$$u_{xx},\ u_{xt}\ \in C(\bar{\Omega}_0 \cap B_r).$$

The one-phase Stefan problem has been studied by many authors with the help of the variational inequality formulation and references of several such studies can be found in [191]. In the water region, convection can be included in the formulation (cf. equation (7.4.43) given below). It is possible to obtain the complementarity problem in a constant-velocity case by applying Baiocchi transformation [192]

$$u(x,t) = \int_0^t T(x,\tau)\, d\tau,\ \ x \in \bar{\Omega} \times (0, t_*)\,, \tag{7.4.39}$$

to equation (7.4.43). Even in this simple situation, it is not possible to convert the complementarity problem to a variational inequality formulation due to difficulties arising in the transformation of boundary conditions and the geometry of the domain (cf. [191]). Several problems connected with the one-phase Stefan problem, such as, the exterior problem, the continuous casting model, and the degenerate case pertaining to a quasi-steady state model have been studied in [191]. In the interior problem, the melting of ice takes place due to a prescribed non-negative temperature on one part of the boundary of $\Omega \subset R^n$ and the Neumann boundary condition prescribed on the remaining part of the boundary of Ω. Some results of the regularity of the freezing index and its continuous dependence on the data in a strong sense can be found in [191]. The geometry of the exterior problem is similar to that of the expanding core model in the Hele-Shaw problem (cf. § 6.2.1). The variational inequality formulation of the exterior problem can be obtained in terms of the freezing index by using transformation (7.4.39). As discussed earlier, for large times, cusp-like singularities may develop on the free boundary. A strong geometric assumption about the data in the exterior problem leads to a star-shaped configuration without singular points [190].

A fundamental question which arises concerning the solution of a variational inequality is its relationship with the classical solution of the Stefan problem in which the temperature of the phase-change boundary is the melting temperature and the energy balance condition is satisfied pointwise on it. This question was essentially solved in [193, 194]. For the smoothness of the free boundary and the 'Caffarelli's criterion' on the local existence of the classical solutions, see references [193] and [195]. Under some assumptions it has been proved that for each $t > 0$, the n-dimensional Lebesgue measure of the free boundary $t = \phi(x)$ is equal to zero; $\phi(x)$ is a locally Lipschitz function and there exists a neighbourhood G_0 of the point (x,t) on the free boundary where the free boundary can be represented in suitable coordinates by the graph of a C^1 function in the form

$$x_n = S(x_1, x_2, \cdots\cdots x_{n-1}, t),\ \ S \in C^1, \tag{7.4.40}$$

and all the second derivatives of u are continuous upto the free boundary.

One-phase continuous casting model and its variational inequality formulation

The formulation of the problem and the analysis presented below is not classical. This problem is being discussed here to introduce a continuous casting model and because of some novelty in the expressions of bilinear form and the inner product used in the formulation.

The thermal energy conservation equation in a heat conducting body $\Omega \subset R^3$ is given by

$$\rho \frac{de}{dt} + \nabla \cdot \vec{q} = 0, \quad \text{in } \Omega_{t_*} = \Omega \times (0, t_*), \tag{7.4.41}$$

where e is the specific energy and $\vec{q}$ is the heat flux vector. If the phase-change is taking place in Ω at temperature T_m and the latent heat l is released, then e can be expressed as

$$e = CT + lH_v(T - T_m), \tag{7.4.42}$$

where $H_v(T)$ is the Heavyside function ($H_v(x) = 0$, if $x \leq 0$ and $H_v(x) = 1$, for $x < 1$). If the body is moving with a velocity $\vec{v}$, then the total derivative with respect to time is to be taken and (7.4.41) can be written as

$$\frac{\partial T}{\partial t} + \vec{v} \cdot \nabla T - \nabla.(\nabla T) = -l\left(\frac{\partial}{\partial t} + \vec{v} \cdot \nabla\right) H_v(T), \tag{7.4.43}$$

For simplicity, it has been assumed in (7.4.43) that C = 1.0, $\rho = 1.0$, $k = 1.0$, $T_m = 0$ and l is a constant. The equation (7.4.43) holds in Ω_{t_*} in the distributional sense and for its derivation see [196]. We consider ice-water system in which the temperature of water is zero and ice is progressing on water. If θ represents the difference between the solidification temperature and the actual temperature of ice, then (7.4.43) can be written as (cf. [196])

$$\frac{\partial \theta}{\partial t} + \vec{v} \cdot \nabla \theta - \nabla^2 \theta = -l \left(\frac{\partial}{\partial t} + \vec{v} \cdot \nabla\right) \chi_{(\theta > 0)}, \text{ a.e. in } \Omega_{t_*}, \tag{7.4.44}$$

where χ is the *characteristic function* of the set $\{\theta > 0\}(\chi = 1 \text{ for } \{\theta > 0\})$ which is now ice region and water region is the set $\{\theta \leq 0\}(\chi = 0 \text{ for } \{\theta \leq 0\})$. If the ice-water sharp boundary is written as $t = \phi(x)$, $x \in R^3$, $x = (x_1, x_2, x_3)$, then the classical Stefan condition and the isotherm condition can be written as

$$\{\nabla \theta \cdot \nabla \phi + l\,(1 + \vec{v} \cdot \nabla \phi)\}\,(x, \phi\,(x)) = 0; \text{ and } \theta = 0. \tag{7.4.45}$$

To complete the formulation, the initial and boundary conditions should be added to (7.4.44) and (7.4.45).

We shall now discuss variational inequality formulation of a problem arising in connection with the formation of the continuous ingot by the process of continuous casting.

A simple diagram depicting solidification of the melt in continuous casting is shown in Fig. 7.3.1. As soon as a sufficiently stable core is formed, the platform B begins to drop down with the velocity $\vec{v}$ in the direction x_1 and thus it draws out a continuous cast of cooling liquid. For appropriate assumptions under which a simple mathematical model of continuous cast is formulated, see [21]. The portion of the ingot taken into account in the present formulation occupies a cylindrical open bounded domain $\Omega = \Omega_1 \times \Omega_2 \subset R^n$, n = 2, 3 where $\Omega_1 = (0, b)$, b is the height of the lateral mold and $\Omega_2 = (0, a)$ if n = 2 because of the symmetry, or if n = 3, $\Omega_2 \subset R^2$ is a domain with Lipschitz boundary $\partial\Omega_2$ representing half the section of the ingot. Let $\Gamma_0 = \{0\} \times \Omega_2$, $\Gamma_1 = (0, b) \times \partial\Omega_2$, and $\Gamma_2 = \{b\} \times \Omega_2$. A point $x \in \bar{\Omega}$ has the coordinates $x = (x_1, x')$ where $x' = x_2$ if n = 2 and $x' = (x_2, x_3)$ if n = 3. In the notation $\nabla = (\partial_1, \nabla')$, $\nabla' = \partial_2$ if $n = 2$ and $\nabla' = (\partial_2, \partial_3)$ if n = 3. Here, ∂_1, ∂_2 and ∂_3 denote the partial derivatives with respect to x_1, x_2 and x_3, respectively. The initial temperature of the melt is given by

$$\theta(x_1, x', 0) = \theta_0(x_1, x') \leq 0, \ (x_1, x') \in \Omega. \tag{7.4.46}$$

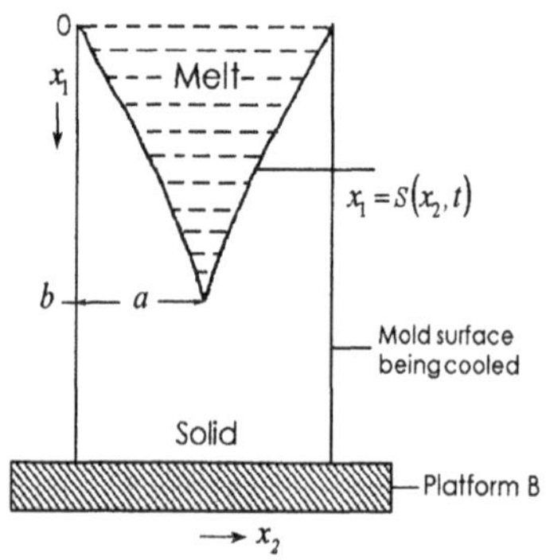

Fig. 7.4.1. Formation of continuous ingot

Boundary conditions which are prescribed on the lateral, bottom and upper surfaces of the cylinder are :

$$\theta(0, x', t) = 0, \ \text{on } \Gamma_0, \ t > 0, \tag{7.4.47}$$

$$-\nabla\theta \cdot \vec{n} = \alpha(\theta - \theta_1) \ \text{on } \Gamma_1, \ \theta_1 \geq 0, \ \alpha > 0, \tag{7.4.48}$$

$$\theta(b, x', t) = \theta_2(x', t), \ \text{on } \Gamma_2, \ t > 0. \tag{7.4.49}$$

The geometry considered in [191] is different from the geometry shown in Fig. 7.4.1. as only one branch of the free boundary has been considered in [191]. By symmetry consideration, one branch of the free boundary can be considered in Fig. 7.4.1. also but the boundary condition at $x_2 = 0$ will be different from that at $x_2 = a$ and not as in (7.4.48).

The boundary Γ_1 is being cooled. $\vec{n}$ is the unit outward normal on Γ_1. The energy equation in $\Omega_{t_*} = \Omega \times (0, t_*)$ is given by

$$\frac{\partial \theta}{\partial t} + v_0 \frac{\partial \theta}{\partial x_1} - \nabla^2 \theta = -l \left(\frac{\partial}{\partial t} + v_0 \frac{\partial}{\partial x_1} \right) \chi_{(\theta > 0)}, \text{ a.e. in } \Omega_{t_*}. \tag{7.4.50}$$

Here, $\bar{v} = (v_0, 0, 0)$ and v_0 is the constant casting velocity in the direction of x_1. Equation (7.4.50) holds in the sense of distributions. Let the free boundary be given by

$$x_1 = S(x', t), \; x' \in \Omega_2, \; t > 0, \tag{7.4.51}$$

where $S(x', t)$ is a smooth function, $S : \Omega_2 \times (0, t_*) \to [0, b)$. The conditions at the free boundary can be written as

$$\theta = 0; \;\; \text{and} \; \frac{\partial \theta}{\partial x_1} - \nabla' \theta . \nabla' S = l(v_0 - \frac{\partial S}{\partial t}). \tag{7.4.52}$$

The second condition in (7.4.52) can be obtained by using relations of the type (1.4.19) and (1.4.21) and remembering that $\partial(x_1 - S)/\partial t = v_0 - \partial S/\partial t$. The region $x_1 > S(x', t)$ is the solid region (see Fig. 7.4.1.) and the region $x_1 < S(x', t)$ is the melt region.

Using the maximum principle, it can be proved that under the boundary conditions considered, for $x_1 > S(x', t)$, $\theta > 0$. Only $\partial S/\partial t \leq v_0$ is physically admissible. In order to obtain variational inequality formulation of this casting problem, the following transformation has been used in [196] which regularizes the equation (7.4.50).

$$u(x, t) = v_0 \int_0^t \theta(x_1 + v_0(\tau - t), x', \tau) d\tau, \; x = (x_1, x') \in \bar{\Omega}, \; t > 0, \tag{7.4.53}$$

where θ is a non-negative function and is extended to $-\infty \leq x_1 \leq S(x', t)$ by taking it to be zero in this extension. If $\theta \in L^1_{loc}(\Omega_{t_*})$, then θ can be obtained in terms of u by the relation (cf. [196])

$$\frac{\partial u}{\partial t} + v_0 \frac{\partial u}{\partial x_1} = v_0 \theta, \text{ a.e. in } \Omega_{t_*}. \tag{7.4.54}$$

It can be shown that $u(x, t)$ formally satisfies the following equations.

$$\frac{\partial u}{\partial t} + v_0 \frac{\partial u}{\partial x_1} - \nabla^2 u = f \chi_{(u > 0)}, \text{ a.e. in } \Omega_{t_*}, \; u \geq 0, \tag{7.4.55}$$

$$u = 0, \text{ on } \Gamma_0; \text{ and } u = 0, \text{ at } t = 0, \tag{7.4.56}$$

$$\frac{\partial u}{\partial n} = \alpha (g - u), \text{ on } \Gamma_1, \tag{7.4.57}$$

$$\frac{\partial u}{\partial t} + v_0 \frac{\partial u}{\partial x_1} = v_0 \theta_2(x', t), \text{ on } \Gamma_2, \tag{7.4.58}$$

where f and g in (7.4.55) and (7.4.57) are given by (cf. [196])

$$f(x,t) = v_0\theta_0(x_1 - bt, x')\chi_{_I} - lv_0\left[1 - \chi_{_I}\right], \ x \in \Omega, \ t \geq 0, \tag{7.4.59}$$

$$g\left(x_1, x', t\right) = \int\limits_{[x_1 - bt]^+}^{x_1} \theta_1(\xi, t + \frac{\xi - x_1}{v_0}, x')d\xi, \ 0 \leq x_1 \leq b, \ t \geq 0. \tag{7.4.60}$$

Here, $[x_1 - bt]^+$ is the positive part of $x_1 - bt$ and $\chi_{_I}$ is the characteristic function of the set I related to the initial position of the free boundary $\Phi(0)$, and I is given by

$$I = \left\{(x_1, x', t) \in \Omega_{t_*};\ x_1 > S(x', 0) + bt\right\}, \tag{7.4.61}$$

$$I \subset \Omega^+ = \left\{(x_1, x', t) \in \Omega_{t_*};\ x_1 > S(x', t)\right\} = \{\theta > 0\},$$

$$\Phi(0) : x_1 = S(x', 0). \tag{7.4.62}$$

The equations (7.4.54) and (7.4.55) hold in the sense of distributions and the reader is referred to [196] for their derivation which is not straightforward.

For $t \geq b/v_0$, $\chi_{_I} = 0$ and $\chi_{(u>0)} \geq \chi_{_I}$ for $t > 0$ because $\partial S/\partial t \leq v_0$. To write the variational inequality formulation, we introduce the following notations.

$$a(u,v) = \int\limits_{\Omega} \nabla u \cdot \nabla v + v_0 \int\limits_{\Omega} v\frac{\partial u}{\partial x} + \alpha \int\limits_{\Gamma_1} uv, \ u(t), v(t) \in M_1(t), \tag{7.4.63}$$

$$(u,v) = \int\limits_{\Omega} uv + \frac{1}{v_0}\int\limits_{\Gamma_2} uv, \ u(t), v(t) \in M_1(t), \tag{7.4.64}$$

$$\langle G(t), v\rangle = \int\limits_{\Omega} f(t)v + \alpha \int\limits_{\Gamma_1} g(t)v + \int\limits_{\Gamma_2} \theta_2(t)v, \ v(t) \in M_1(t), \tag{7.4.65}$$

For a.a. $t \in (0.t_*)$

$$M(t) = \left\{v \in H^1(\Omega);\ v = 0, \ \text{ on } \Gamma_0\right\}, \ M_1(t) = \left\{v \in M(t);\ v \geq 0, \ \text{ on } \Omega\right\}. \tag{7.4.66}$$

The problem (7.4.55)–(7.4.58) has the following variational inequality formulation.

$$(\partial u/\partial t, v - u) + a(u, v - u) \geq \ \langle G(t), v - u\rangle, \ \ a.a.\ t \in (0, t_*),$$

$$\forall\ v(t) \in M_1, \ u(t) \in M_1. \tag{7.4.67}$$

If there exists a $u(t) \in M_1(t)$ such that (7.4.67) holds for all $v(t) \in M_1(t)$, then $u(t)$ is called a solution of the variational inequality (7.4.67). If may be noted that the boundary conditions (7.4.57) and (7.4.58) occur as natural boundary conditions in the formulation and so only the boundary condition on Γ_0 has been imposed on v.

Under suitable assumptions, existence of the unique solution of the variational inequality (7.4.67) has been discussed in [191, 196]. If has been shown that the free boundary has in fact the representation (7.4.51) such that $\partial S/\partial t \leq v_0$ and (θ, S) is the classical solution of the continuous casting problem.

Oxygen - diffusion problem

In the background of the previous formulations, variational inequality formulation of ODP discussed earlier in § 3.3.2 does not present much difficulty. We consider equations (3.3.34)–(3.3.38) and take $\Omega = [0,1]$, $\Omega_{t_*} = \Omega \times (0, t_*)$, $\Omega_0(t) = \{x;\ 0<x<S(t)\}$, $0 < t < t_*$. In the region $\Omega_0(t)$, the concentration $c > 0$ and in $\Omega \backslash \Omega_0(t)$, $c = 0$. Therefore in Ω, $c \geq 0$. Note that in ODP, the region under consideration is only $\Omega_0(t)$ and no constraint of the form $c \geq 0$ is imposed. The variational inequality problem is studied in the fixed domain Ω and the constraint $c \geq 0$ has been added to study an obstacle problem. It can be seen that

$$\int_0^{t_*}\int_\Omega \left(\frac{\partial c}{\partial t} - \frac{\partial^2 c}{\partial x^2} + 1\right)(v-c)\,dxdt = \int_0^{t_*}\int_0^1 \frac{\partial c}{\partial t}(v-c)\,dxdt$$

$$-\int_0^{t_*}\left[(v-c)\frac{\partial c}{\partial x}\right]_0^1 dx + \int_0^{t_*}\int_0^1 \frac{\partial c}{\partial x}\frac{\partial}{\partial x}(v-c)\,dxdt + \int_0^{t_*}\int_0^1 (v-c)\,dxdt. \qquad (7.4.68)$$

The test function $v(x,t)$ satisfies the same initial and boundary conditions as the concentration $c(x,t)$ and so the second term on the r.h.s. of (7.4.68) is zero. On rearranging equation (7.4.68), we obtain

$$\int_0^{t_*}\int_0^1 \frac{\partial c}{\partial t}(v-c)\,dxdt + \int_0^{t_*}\int_0^1 \frac{\partial c}{\partial x}\frac{\partial}{\partial x}(v-c)\,dxdt$$

$$+\int_0^{t_*}\int_0^1 (v-c)\,dxdt = \int_0^{t_*}\int_0^1 \left(\frac{\partial c}{\partial t} - \frac{\partial^2 c}{\partial x^2} + 1\right) v dxdt. \qquad (7.4.69)$$

We have $c_t - c_{xx} + 1 = 0$ if $c > 0$ and $c_t - c_{xx} + 1$ is equal to 1 if c $= 0$. Therefore, $c_t - c_{xx} + 1 \geq 0$ for $c \geq 0$ and $c\,(c_t - c_{xx} + 1) = 0$ in Ω. As $v \geq 0$, the last term on the r.h.s. of (7.4.69) is greater than or equal to zero and we get

$$\int_0^{t_*}\left\{(c_t, v-c) + \left(\frac{\partial c}{\partial x}, \frac{\partial}{\partial x}(v-c)\right) + (1, v-c)\right\} dt \geq 0. \qquad (7.4.70)$$

This completes the variational inequality formulation of ODP. If there exists a function $c\,(x,t) \in E$, where, $E = \{v \in L^2(0,t_*;\ H^1(\Omega))\ ;\ v \geq 0,$ a.e. in $\Omega_{t_*}, v(x,0) = c(x,0)\}$, and $c\,(x,t)$ satisfies (7.4.70) for all $v \in E$ then $c\,(x,t)$ is called a solution of the variational inequality (7.4.70). In the multi-dimensional case when $\Omega \subset R^n$, $n \geq 1$, the variational inequality formulation can be obtained following the procedure adopted in (7.4.68)–(7.4.70). The integration by parts will then require the use of Green's formula.

If oxygen-diffusion takes place in a subset of R^n, $n > 1$ then the initial concentration has to be suitably prescribed as in this case it is difficult to obtain it from the solution of a steady-state problem which was possible in the one-dimensional case. The existence, uniqueness and regularity of the solution of (7.4.70) was studied in [74]. Some regularity results for the one-phase Stefan problems have been reported earlier in this section and for further information, see [197, 198, 199].

Is it possible to obtain a one-phase Stefan problem from its variational inequality formulation? Consider the problem of finding a function $u(x,t)$ satisfying the following complementarity problem.

$$u_t - u_{xx} \geq f,\ u \geq \psi(x,t),\ \text{and } (u_t - u_{xx} - f)(u-\psi) = 0,\ a.e. \text{ in } \Omega_{t_*},$$

$$\Omega_{t_*} = \Omega \times (0 < t < t_*),\ \Omega = \{x : 0 \leq x \leq 1\}, \tag{7.4.71}$$

$$\begin{aligned} u(x,0) &= u_0(x) \geq \psi(x,0),\ x \in \Omega, \\ u(0,t) &= \hat{u}(t) \geq \psi(0,t),\ 0 < t < t_*. \end{aligned} \tag{7.4.72}$$

One can study the above variational inequality problem by employing standard methods for parabolic equations involving well known a priori L^p-estimates [9]. The solution is generally smooth in the sense that the distributional derivatives u_t, u_x, u_{xx} are in L^p, $p > 1$. u_t and u_{xx} are not continuous in general.

Now suppose that $u(x,t) > \psi(x,t)$ if and only if $0 < x < S(t)$, $0 < t < t_*$, $S(0) = 1$. In this case, we have $u_t - u_{xx} = f$ in $0 < x < S(t)$, $u(S(t),t) = \psi(S(t),t)$ (cf. § 7.2 equations (7.2.38)–(7.2.39)). If a new function $T(x,t)$ is defined as

$$T(x,t) = \frac{\partial}{\partial t}(u - \psi), \tag{7.4.73}$$

then $T(x,t)$ satisfies the following equations.

$$T_t - T_{xx} = F_t = (f - \psi_t + \psi_{xx})_t\,,\ 0 < x < S(t),\ t > 0, \tag{7.4.74}$$

$$T(x,0) = T_0(x) = u_0''(x) + (f - \psi_t)_{t=0},\ 0 \leq x \leq 1, \tag{7.4.75}$$

$$T(0,t) = \hat{T}(t) = \hat{u}_t - \psi_t|_{x=0},\ t > 0, \tag{7.4.76}$$

$$T(S(t),t) = 0,\ t > 0,\ S(0) = 1, \tag{7.4.77}$$

$$T_x\,(S(t),t) = \lambda\,(S(t),t)\,\dot{S}(t) = (f + \psi_{xx} - \psi_t)\,|_{x=S(t)}\dot{S}(t),\ t > 0 \tag{7.4.78}$$

By choosing F, $T_0(x)$, $\hat{T}(t)$ and $\lambda(S(t),t)$ suitably in (7.4.74)–(7.4.78), a one-phase Stefan problem or an ODP can be obtained. On inverting (7.4.73), we get

$$u\,(x,t) = \int_{\phi(x)}^{t} T(x,\eta)d\eta + \psi(x,t), \tag{7.4.79}$$

on the free boundary $t = S^{-1}(x) = \phi(x)$. From (7.4.79), $u\,(S(t),t) = \psi\,(S(t),t)$. In order to derive (7.4.78), calculate $u_t - u_{xx}$ from (7.4.79) at $x = S(t)$.

The equivalence of the weak formulation of a one-phase Stefan problem and its variational inequality formulation has been discussed in § 11.2.2.

7.4.2 A Stefan problem with a quasi-variational inequality formulation

Our interest in this section is only in introducing the *quasi-variational inequality*. For a rigorous discussion of the quasi-variational inequality, see [22] and the references mentioned there in this connection.

One of the forms of quasi-variational elliptic inequality considered by Bensoussan-Goursat-Lions (cf. [200]), and further discussed in [22], can be defined as follows: Let H_1 and H_2 be two Hilbert spaces and $v = (v_1, v_2)$ be the generic element of $H = H_1 \times H_2$. Define $a : H \times H \to R$ and $g : H \to R$ as follows:

$$a(w,z) = a_1(w_1,z_1) + a_2(w_2,z_2) + b_1(w_1,z_2) + b_2(z_2,w_1), \tag{7.4.80}$$

$$g(w) = g_1(w_1) + g_2(w_2). \tag{7.4.81}$$

Here, $a_1 : H_1 \times H_1 \to R$ and $a_2 : H_2 \times H_2 \to R$ are continuous bilinear forms, symmetric and non-negative along the diagonal, $b_1 : H_1 \times H_2 \to R$ and $b_2 : H_2 \times H_1 \to R$ are continuous bilinear forms, $g_1 : H_1 \to R$ and $g_2 : H_2 \to R$ are continuous linear functionals. Let $D \subset H = H_1 \times H_2$ be a non-empty closed convex set. If $v \in D$, then define $D(v) = D_1(v) \times D_2(v)$ where

$$D_1(v) = \{z_1 \in H_1 : (z_1, v_2) \in D\}, \text{ and } D_2(v) = \{z_2 \in H_2 : (v_1, z_2) \in D\}. \tag{7.4.82}$$

An inequality

$$a(u, u-v) \le g(u-v), \ \forall v \in D(u), \ u \in D \tag{7.4.83}$$

is an example of a quasi-variational inequality.

$D(v)$ is a non-empty closed convex set of $D_* = P_{H_1}(D) \times P_{H_2}(D)$, where P_{H_1} and P_{H_2} are 'projection operators'. If $u \in H_1$ and $W \subset H_1$ is a non-empty closed convex set, then $P_W(u)$ is that element of W which is nearest to u. The inequality (7.4.83) is a variational inequality when for $\forall\ u \in D,\ D(u) = Q = Q_1 \times Q_2$ with Q_1 being a non-empty closed convex set of H_1 and Q_2 being a non-empty closed convex set of H_2. $v \in D(u)$ suggests that quasi-variational inequality entails a convex set of competing functions which may depend on the possible solution whereas in the variational inequality formulation convex set is fixed.

A family of variational inequalities can be associated with the quasi-variational inequality (7.4.83). If u is fixed in D, then along u, we can consider a variational inequality concerned with finding a $w \in D(u)$ such that

$$a(w, w - v) \leq g(w - v),\ \forall\ v \in D(u). \tag{7.4.84}$$

A quasi-variational inequality formulation for the following one-phase Stefan problem has been studied in [201, 202]. Find a function $T(x,t)$ and the curve $\Gamma : x = S(t)$, $0 < t < t_*$ such that

$$T_t - T_{xx} = 0,\ 0 < x < S(t),\ 0 < t < t_*, \tag{7.4.85}$$

$$T = \lambda(x,t),\ x = S(t),\ 0 < t < t_*, \tag{7.4.86}$$

$$T_x = 0,\ x = S(t),\ 0 < t < t_*, \tag{7.4.87}$$

$$T(x,0) = T_0(x),\ 0 \leq x < S_0,\ S(0) = S_0, \tag{7.4.88}$$

$$T\,(0,t) = r\,(t)\,,\ 0 < t < t_*. \tag{7.4.89}$$

If the transformation $u = -T_x$ (see § 3.3) is used, then the boundary condition (7.4.87) becomes $u = 0$ and (7.4.86) becomes $u_x = -\lambda_x \dot{S} - \lambda_t$. The coefficient $-\lambda_x$ can be identified with the latent heat and λ_t with the displacement of the free boundary. If appropriate conditions are imposed on the initial and boundary data, then the problem (7.4.85)–(7.4.89) can be identified with a phase-change problem. Global existence of the classical solution of the transformed system has been discussed in [203] when $\lambda_x \equiv 1,\ \lambda_t \leq 0,\ -T_0'(x) \geq 0$ and $r'(t) \geq 0$. It turns out that in this case $\dot{S}(t) \geq 0$ and $u(x,t) > 0$ if $0 < x < S(t)$.

Let $\lambda(x,t),\ T_0(x)$ and $r(t)$ be given smooth functions for $0 \leq x \leq R_0,\ 0 \leq t \leq t_*$ and $S_0 \in (0, R_0)$. A quasi-variational inequality formulation of problem (7.4.85)–(7.4.89) will now be considered in the region $D = \{(x,t)\,,\ 0 < x < R_0,\ 0 < t < t_*\}$ in terms of a new variable $W(x,t)$ defined as

$$W(x,t) = \int_x^{R_0} (T\,(\xi,t) - \lambda\,(\xi,t))\,d\xi,\ (x,t) \in D. \tag{7.4.90}$$

It is understood that if $0 < t < t_*$, then $S(t) < R_0$. We extend $T(x,t)$ as $\lambda\,(x,t)$ in the region $S(t) \leq x \leq R_0,\ 0 \leq t \leq t_*$ and thus $W(x,t) = 0$ if $S(t) \leq x \leq R_0,\ 0 < t < t_*$.

Let $\Omega = \{(x,t) : 0 < x < S(t),\ 0 < t < t_*\}$. From (7.4.90), $\partial W/\partial t$ can be calculated as follows:

$$\begin{aligned} W_t &= \int_x^{R_0} \{T_t(\xi,t) - \lambda_t(\xi,t)\}\, d\xi,\ (x,t) \in D, \\ &= \int_x^{S(t)} \{T_{\xi\xi}(\xi,t) - \lambda_t(\xi,t)\}\, d\xi, \\ &= T_x(S(t),t) - T_x(x,t) - \int_x^{S(t)} \lambda_t(\xi,t) d\xi, \\ &= -T_x(x,t) - \int_x^{S(t)} \lambda_t(\xi,t)\, d\xi. \end{aligned} \tag{7.4.91}$$

For $(x,t) \in \Omega$, we have

$$W_x(x,t) = -(T(x,t) - \lambda(x,t)), \tag{7.4.92}$$

and

$$W_{xx}(x,t) = -T_x(x,t) + \lambda_x(x,t). \tag{7.4.93}$$

$W(x,t)$ satisfies the differential equation

$$-W_{xx}(x,t) + W_t(x,t) = f(x,t), \text{ for } (x,t) \in D, \tag{7.4.94}$$

where

$$\begin{aligned} f(x,t) &= -\lambda_x(x,t) - \int_x^{S(t)} \lambda_t(\xi,t)\, d\xi,\ (x,t) \in \Omega, \\ &= -\lambda_x(x,t) \text{ if } (x,t) \in D\backslash\Omega. \end{aligned} \tag{7.4.95}$$

$f(x,t)$ depends on $S(t)$, i.e., on the solution. It may be noted that $f(x,t)$ has been suitably extended to the region $D\backslash\Omega$ as it is required afterwards in the formulation. It will be assumed that $\lambda_x > 0,\ \lambda_t \leq 0$ and

$$T_0(x) > \lambda(x,0),\ 0 \leq x < S_0; \text{ and } T(S_0,0) = \lambda(S_0,0). \tag{7.4.96}$$

We define a function $g(x)$ as follows:

$$\begin{aligned} g(x) &= \int_x^{S_0} (T(\xi,0) - \lambda(\xi,0))\, d\xi, \text{ if } 0 \leq x \leq S_0, \\ &= 0, \text{ if } S_0 \leq x \leq R, \end{aligned} \tag{7.4.97}$$

so that $W(x,0) = g(x)$ and $g'(x) \leq 0$. Because of the assumption that $T_0(x) > \lambda(x,0)$ in $(0,S_0)$, the free boundary Γ starts at S_0. The boundary condition for $W(x,t)$ at $x = 0$ is given by

$$W_x(0,t) = \Psi(t), \tag{7.4.98}$$

where

$$\Psi(t) = -r\,(t) + \lambda\,(0,t)\,, \quad 0 \le t \le t_*.$$

Assume that $\Psi(t) \le 0$ for $0 < t < t_*$. It can be shown that W is greater than zero in Ω. On differentiating (7.4.94) with respect to x, we obtain

$$-W_{xxx} + W_{xt} = f_x = -\lambda_{xx} + \lambda_t, \text{ in } \Omega. \tag{7.4.99}$$

It will be assumed that

$$-\lambda_{xx} + \lambda_t \le 0 \text{ in } D. \tag{7.4.100}$$

Since $W_x = \lambda\,(x,t) - T\,(x,t) = 0$ for $(x,t) \in \Gamma$, and $\Psi(t) \le 0$, we obtain

$$-W_{xxx} + W_{xt} \le 0, \text{ in } \Omega, \tag{7.4.101}$$

and

$$W_x \le 0, \quad \text{on } \partial_p\Omega, \tag{7.4.102}$$

where $\partial_p\Omega$ is the parabolic boundary of Ω. Assuming W_x to be continuous in $\bar{\Omega}$, the maximum principle (see Appendix C) implies that $W_x < 0$ in Ω and on integrating W_x, it can be concluded that $W(x,t) > W(S(t),t) = 0$.

The variational inequality formulation of the problem (7.4.85)–(7.4.89) can now be obtained in terms of $W(x,t)$ as follows:

Find $W(x,t) \in M = \{v(x,t) \in H^1(D),\ v \ge 0\}$ such that

$$(-W_{xx} + W_t)\,(v - W) \ge f(v - W), \text{ a.e. in } D \text{ and } \forall\, v \in M,\ W \in M, \tag{7.4.103}$$

$$\begin{aligned} W(x,0) &= g\,(x)\,,\ 0 < x < R_0, \\ W_x(0,t) &= \Psi(t),\ 0 < t < t_*, \\ W(R_0,t) &= 0,\ 0 < t < t_*. \end{aligned}$$

For $W \ge 0$ and $v \ge 0$, we have

$$\begin{aligned} (-W_{xx} + W_t - f)\,(v - W) &= -\,(-W_{xx} + W_t - f)\,W + (-W_{xx} + W_t - f)\,v, \\ &= (-W_{xx} + W_t - f)\,v, \\ &= 0, \text{ if } W > 0 \text{ and } = -fv \text{ if } W = 0, \\ &\ge 0, \text{ for } \forall\, v \in M,\ W \in M. \end{aligned} \tag{7.4.104}$$

provided f is negative in $D\backslash\Omega$, which it is, under the assumption that $\lambda_x > 0$ (see (7.4.95)). It is clear from (7.4.95) that f is a function of $(x,t;\ S(t))$ and $S(t) =$

$\min\{x, W(x,t) = 0\}$. Therefore f is a discontinuous function of $W(x,t)$ and (7.4.104) is not a variational inequality in the usual sense. In [201], authors call it a quasi-variational inequality. The existence of the solution of (7.4.103) has been proved in [201] by considering a sequence of variational inequality problems in which the approximations of the free boundary are taken as known. Fixed point theorem for monotone mappings was used in [201] for the existence proof and in [202] existence of the solution of the above problem has been proved using finite-difference approximations and this resulted in a smoother solution. Under appropriate smoothness assumptions on λ, g, Ψ, etc. (cf. [202]) it has been shown that a solution of (7.4.103) exists and is such that: (1) W_t, W_{xx} belong to $L^\infty(D)$, (ii) W_x is Hölder continuous (exponent 1/2) in $\bar{D}$ (iii) $W_x(x,t) < 0$ for $0 < x < S(t), 0 < t < t_*$, (iv) $S(t)$ is Hölder continuous and monotone decreasing, and $S(0) = S_0$, $S(t) > 0$.

7.4.3 The Variational inequality formulation of a two-phase Stefan problem

In the one-phase Stefan problems, the freezing index $u(x,t)$ proved to be very useful as the gradient of u is continuous in Ω (the region under consideration). The temperature constraint $T \geq 0 (\leq 0)$ also holds throughout Ω. In the two-phase problem although freezing index can be defined, the constraint $T \geq 0 (\leq 0)$ does not hold throughout the two-phase region. An approach different from the one adopted for the one-phase problems is required for the variational inequality formulation. If the weak enthalpy formulation is considered for the two-phase Stefan problem, then the enthalpy, which is a multi-valued function of temperature can be written as (for simplicity take specific heats and densities of the two phases to be unity)

$$h(T) = h_0(T) + l\, sgn^+(T), \tag{7.4.105}$$

where $h_0(T)$ is the sensible heat and sgn^+ represents Heaviside graph (cf. (4.3.45)). In terms of the freezing index u,

$$u(x,t) = \int_0^t T(x,t)dt, \ (x,t) \in \Omega_{t_*} = \Omega \times (0,t_*), \ \Omega \subset R^3, \tag{7.4.106}$$

the heat energy equation (4.2.2) can be written as (see [204] for the derivation of (7.4.107))

$$-h_0(u_t) + \nabla^2 u + h_1(x) \in l\ \partial\Psi(u_t), \ \text{a.e. in } \Omega_{t_*} \tag{7.4.107}$$

Here, $h_1(x)$ is the initial enthalpy of the material, and $\partial\Psi$ is the subdifferential of a convex function $\Psi(z) = \max(0, z)$ which coincides with the multivalued mapping $sgn^+(T)$. Using the definition of the subdifferential of a function (see (4.3.44), the parabolic variational inequality for the enthalpy formulation can be obtained (cf. [204] and the references mentioned there). Our interest here is in the classical formulation.

A variational inequality formulation for a two-phase Stefan problem in the region $\Omega \subset R^3$ has been studied in [205]. The phase-change boundary $\Phi(t) : t = S(x)$, $x \in R^3$, divides Ω into two disjoint regions Ω_1 and Ω_2, representing the solid and the liquid region, respectively. The temperature in the interior $\widetilde{\Omega}_1$ of Ω_1 is negative and in the interior $\widetilde{\Omega}_2$ of Ω_2 is positive. The cases of regions of zero temperature within Ω_1 and Ω_2 can also be considered. Initially Ω is occupied by ice. On the portion $\partial\Omega_b$ of the boundary $\partial\Omega$ of Ω, temperature $\hat{b}(t)$ is prescribed and the remaining boundary of Ω is insulated. Let x be a point in Ω which is initially at $t_0 = 0$ in the solid phase, and $t_1(x)$ be the first time that x is in the liquid phase, and $t_2(x) > t_1(x)$ be the next time at which x is again in the solid phase. In this way a sequence $\{t_n\}$ can be defined such that $t_0(=0),\ t_1,\ t_2, \cdots,$ are such that in (t_i, t_{i+1}), x is in the liquid (solid) phase if i is odd (even). Since the normal derivative of ∇T is not continuous throughout Ω we define another dependent variable $V(x,t)$ by the equation

$$V(x,t) = K_1 \int_{t_0}^{t_1} T(x,\tau)d\tau + K_2 \int_{t_1}^{t_2} T(x,\tau)d\tau + \ldots\ K_J \int_{t_i}^{t} T(x,\tau)d\tau, \qquad (7.4.108)$$

where for $t \in (t_i, t_{i+1})$, $J = 1\,(2)$ if i is even (odd). K_1 and K_2 are the thermal conductivities of solid and liquid regions, respectively and are taken as constant. Densities of both phases are taken to be equal to unity. If $T^+ = \sup(T,0)$ and $T^- = \sup(-T,0)$ where 0 is the phase-change temperature then $V(x,t)$ can be written as

$$V(x,t) = \int_0^t \left\{-K_1 T^-(x,\tau) + K_2 T^+(x,\tau)\right\} d\tau. \qquad (7.4.109)$$

Since T is continuous throughout the interior of Ω,

$$\frac{\partial V}{\partial t} = K_J T(x,t), \qquad (7.4.110)$$

where $J = 1(2)$ if i is even (odd) for $t \in (t_i, t_{i+1})$.

We are now interested in obtaining a differential equation for $V(x,t)$ which holds throughout Ω. On differentiating (7.4.108), we obtain

$$\nabla V(x,t) = K_1 \left\{ \int_0^{t_1(x)} \nabla T d\tau + T(x, t_1(x)) \right\} +$$

$$K_2 \left\{ \int_{t_1(x)}^{t_2(x)} \nabla T d\tau + T(x, t_2(x)) - T(x, t_1(x)) \right\} + \ldots + K_J \left\{ \int_{t_i}^{t} \nabla T d\tau - T(x, t_i) \right\}. \qquad (7.4.111)$$

At $t_1(x),\ t_2(x), \ldots,\ t_i(x)$, phase-change takes place and, $T(x, t_i) = 0,\ i = 1, 2\ldots$ Therefore

$$\nabla V(x,t) = K_1 \int_0^{t_1} \nabla T d\tau + K_2 \int_{t_1}^{t_2} \nabla T d\tau + \ldots\ K_J \int_{t_i}^{t} \nabla T d\tau. \qquad (7.4.112)$$

Note that V, $\partial V/\partial t$ and ∇V are continuous in Ω. ∇T is discontinuous only across $t = t_i(x)$ and not within the interval (t_i, t_{i+1}). When (7.4.112) is differentiated, we get

$$\nabla^2 V(x,t) = K_1 \int_0^{t_1} \nabla^2 T d\tau + K_2 \int_{t_1}^{t_2} \nabla^2 T d\tau$$

$$+\{K_1\nabla T(x,t_1).\nabla t_1(x) - K_2\nabla T(x,t_1).\nabla t_1(x)\} + ...$$
$$+K_J\int_{t_i}^{t}\nabla^2 T d\tau - K_J\nabla T(x,t_i).\nabla t_i(x). \tag{7.4.113}$$

If the Stefan condition (1.4.17) is used in (7.4.113), then we obtain (take $\rho_1 = \rho_2 = 1$)

$$\nabla^2 V(x,t) = C_1\int_0^{t_1}\frac{\partial T}{\partial \tau}d\tau + C_2\int_{t_1}^{t_2}\frac{\partial T}{\partial \tau}d\tau + C_J\int_{t_i}^{t}\frac{\partial T}{\partial \tau}d\tau + l - l + ... \tag{7.4.114}$$

Performing the integration in (7.4.114) and remembering whether J is odd or even, we get

$$\begin{aligned}\nabla^2 V(x,t) &= -C_1T(x,0) + l + \frac{C_2}{K_2}\frac{\partial V}{\partial t}, \text{ if } T(x,t) > 0,\\ &= -C_1T(x,t) + \frac{C_1}{K_1}\frac{\partial V}{\partial t}, \text{ if } T(x,t) < 0.\end{aligned} \tag{7.4.115}$$

Since $T(x,t)$ has the same sign as $\partial V/\partial t$ (see (7.4.110)), (7.4.115) can be written as

$$\nabla^2 V(x,t) = -C_1T(x,0) + l\hat{H}(\frac{\partial V}{\partial t}) + \frac{C_2}{K_2}\left[\frac{\partial V}{\partial t}\right]^+ - \frac{C_1}{K_1}\left[\frac{\partial V}{\partial t}\right]^-, \tag{7.4.116}$$

where

$$\begin{aligned}\hat{H}(\partial V/\partial t) &= 1, \text{ if } \partial V/\partial t > 0,\\ \hat{H}(\partial V/\partial t) &= 0, \text{ if } \partial V/\partial t < 0.\end{aligned}$$

In (7.4.116), $[x]^\pm$ are the positive and negative parts of x.

What happens when $\partial V/\partial t = 0$ or $T(x,t) = 0$? If there exists a sharp boundary Γ separating solid and liquid phases at which $T = 0$, then the meas(Γ) $= 0$ in Ω and (7.4.116) holds a.e. in Ω. Two other cases may arise. For example there may exist regions of non-zero measures in $\Omega_1(t)$ or $\Omega_2(t)$ whose temperatures are zero. It has been argued in [205] that if such a region exists in the solid region then we define $\hat{H}(\partial V/\partial t) = 0$ when $\partial V/\partial t = 0$ and if such a region exists in the liquid region then we define $\hat{H}(\partial V/\partial t) = 1$ if $\partial V/\partial t = 0$. In all the three cases (7.4.116) holds in Ω.

To obtain a variational inequality for the above problem, the procedure adopted is similar to the one used for obtaining the weak enthalpy formulation in (7.4.107). We write

$$\hat{H}(\frac{\partial V}{\partial t}) \in \partial g_0(\frac{\partial V}{\partial t}), \tag{7.4.117}$$

where $g_0(x) = x^+$ defined as $\sup(0,x)$ and ∂g_0 is the subdifferential of g_0. Now our objective is to look for an appropriate convex function $g_1(x) : R \to R$ so that we can write

$$\nabla^2 V(x,t) + C_1 T(x,0) \in \partial g_1(\frac{\partial V}{\partial t}). \tag{7.4.118}$$

If $g_1(x)$ is defined as

$$g_1(x) = \frac{1}{2}(C_1/K_1)\left(x^-\right)^2 + \frac{1}{2}(C_2/K_2)\left(x^+\right)^2 + lx^+, \tag{7.4.119}$$

then using the definition of the subgradient of a convex function, i.e., $f(x) \in \partial g(x) \Leftrightarrow g(\xi) - g(x) \geq (\xi - x)f(x)$ for all $\xi \in R$, a relation of the form (7.4.120) can be obtained. The relation (7.4.118) can be expressed as

$$g_1(\xi) - g_1(\frac{\partial V}{\partial t}) \geq \left(\nabla^2 V + C_1 T(x,0)\right)\left(\xi - \frac{\partial V}{\partial t}\right), \ \forall\, \xi \in R. \tag{7.4.120}$$

The boundary condition to be satisfied by $V(x,t)$ on $\partial\Omega_b$ can be expressed as

$$\begin{aligned} V(x,t) &= K_1 \int_0^{t_1} \hat{b}(\tau)\, d\tau + .. \ + K_J \int_{t_i}^{t} \hat{b}(\tau)\, d\tau, \text{ on } \partial\Omega_b, \\ &= \int_0^t \left\{K_2\left(\hat{b}(\tau)\right)^+ - K_1\left(\hat{b}(\tau)\right)^-\right\} d\tau, \text{ on } \partial\Omega_b. \end{aligned} \tag{7.4.121}$$

On the remaining boundary, we take

$$\frac{\partial V}{\partial n} = 0, \text{ on } \partial\Omega \backslash \partial\Omega_b. \tag{7.4.122}$$

Let the initial condition for V be given by

$$V(x,0) = 0, \quad x \in \Omega. \tag{7.4.123}$$

On taking $\xi = \Psi(x)$, $x \in \Omega$ in (7.4.120) and integrating over Ω, we obtain

$$G(\Psi) - G(\frac{\partial V}{\partial t}) \geq \int_\Omega \nabla^2 V\left(\Psi - \frac{\partial V}{\partial t}\right) dx + C_1 \int_\Omega T(x,0)\left(\Psi - \frac{\partial V}{\partial t}\right) dx, \tag{7.4.124}$$

where

$$G(\Psi) = \int_\Omega g_1(\Psi(x))dx. \tag{7.4.125}$$

If the spaces to which Ψ, V and $\partial V/\partial t$ belong are defined suitably such as in (7.4.128), then (7.4.124) can be identified with a variational inequality formulation of the above two-phase problem. On integrating by parts the first term on the r.h.s. of (7.4.124) and assuming that Ψ and $\partial V/\partial t$ satisfy the same boundary conditions, we obtain

$$G(\Psi) - G\left(\frac{\partial V}{\partial t}\right) + a\left(V, \Psi - \frac{\partial V}{\partial t}\right) \geq C_1 \int_\Omega T(x,0)\left(\Psi - \frac{\partial V}{\partial t}\right) dx, \tag{7.4.126}$$

where

$$a(V,Z) = \int_\Omega \nabla V.\ \nabla Z\ dx. \tag{7.4.127}$$

Let

$$M(t) = \left\{Z \in H^1(\Omega), Z|_{\partial\Omega_b} = b(t)\right\}, \ a.a.\ t \in (0, t_*) \tag{7.4.128}$$

$$b(t) = K_2\left(\hat{b}(t)\right)^+ - K_1\left(\hat{b}(t)\right)^-. \tag{7.4.129}$$

If there exists a V and the following conditions hold:

1. V, $\partial V/\partial t \in L^2(0, t_*; H^1(\Omega))$,
2. $\partial V/\partial t \in M(t)$ (defined in (7.4.128),
3. The inequality (7.4.126) is satisfied for all $\Psi \in M(t)$ for a.a. $t \in (0, t_*)$,
4. $V = 0$ when $t = 0$,

then V is called a solution of the variational inequality (7.4.126). Under the assumptions that $T(x, 0) \in L^2(\Omega)$ and $b(t) \in L^2(0, t_*)$, a unique solution of (7.4.126) exists (cf. [205]). The numerical solution of the above variational inequality can be obtained with the help of a finite-difference scheme.

Chapter 8

The Hyperbolic Stefan Problem

8.1 Introduction

A *hyperbolic Stefan problem* is concerned with a phase-change problem in which the heat energy equation is hyperbolic. The change in the problem formulation introduced by the 'hyperbolic nature of the heat equation' is significant in many ways. The speed of heat propagation in the parabolic case is infinite which can be seen from the temperature solution given in (8.1.1) which is concerned with the problem of one-dimensional heat conduction in an unbounded medium with a constant initial temperature T_0 [24].

$$T(x,t) = \frac{T_0}{2(\pi kt)^{1/2}} \int_{-\infty}^{\infty} e^{-(x-x')^2/4kt} dx'. \tag{8.1.1}$$

For $t > 0$, howsoever small, and x howsoever large, $T(x,t)$ is different from T_0 which confirms the infinite speed of propagation of heat flow. In the parabolic case, heat flux is governed by the Fourier's law (cf. (1.3.8) and (1.4.9)). Thermal energy is transported by molecular motion which has finite speed and therefore Fourier's law seems to be a low order approximation to a more general constitutive law. Cattaneo [206] observed that the infinite speed of heat propagation in Fourier's law is due to diffusion and proposed that heat pulses ought to be transmitted by waves at finite but perhaps high speeds. This does not mean that Fourier's law leading to the diffusion equation should be discarded. *The relaxation time* τ in the hyperbolic model is generally very small in nearly all practical and exotic applications. As the heat flux equation (8.1.3) given below suggests that even on the shortest time scales of our daily lives, we get Fourier's law [207]. The use of hyperbolic heat equation in a mathematical model at very low temperatures is generally accepted, but its use at high temperatures has been debated in the literature and we refer the reader to an exhaustive review article [208] on the 'problem of second sound' (finite speed of heat propagation).

Our interest in the hyperbolic Stefan problems in this volume stems from the fact that models of these problems require many interesting physical concepts in addition to those associated with the Stefan problems. For example, in the hyperbolic Stefan problems, both temperature and flux can be taken to be discontinuous across the phase-change boundary. The delay in the release of latent heat and sensible heat can be considered. These, and several other variations in hyperbolic problems make these problems fairly interesting from the point of view of model and analysis. Many studies of hyperbolic Stefan problems have been conducted. There is still a need to collate the results of these studies in one source.

8.1.1 Relaxation time and relaxation models

In an idealized solid, for example, thermal energy is transported by two different mechanisms: (1) by quantized electronic excitations, which are called free electrons, and (2) by the quanta of lattice vibrations, which are called phonons. These quanta undergo collisions of a dissipative nature, giving rise to thermal resistance in the medium. A relaxation time τ is associated with the average communication time between these collisions for the commencement of the resistive flow. In essence τ is the relaxation parameter or the time which should elapse for the heat flow to take place after the temperature gradient is formed. There are different times of relaxation, so the mean relaxation time is generally not known. For such solids, it would be more important to know which modes carry the most heat and so we want the dominant rather than the mean mode of relaxation.

How does heat flux respond to the temperature gradient? This has been modelled in the literature in different ways. For example, the flux law in the hyperbolic model has been described as

$$\vec{q}(t+\tau) = -\nabla(KT),\ \tau > 0, \tag{8.1.2}$$

where, $\vec{q}$ is the heat flux vector and τ is the relaxation time [209]. In this model, if the temperature gradient is formed at time t then heat is released instantaneously at time $t+\tau$. The first order approximation of (8.1.2) gives

$$\tau\vec{q}_t + \vec{q} = -\nabla(KT),\ \tau > 0. \tag{8.1.3}$$

This model in (8.1.3) is commonly used in the place of Fourier's law if finite speed of heat propagation is considered. In the relaxation model (8.1.3) a delay of time τ in the heat flow is there but the heat release is not instantaneous at time $t+\tau$ and is distributed over a time interval. Equation (8.1.3) can be derived as a particular case of the following linearized constitutive equation for the heat flux vector $\vec{q}$ [210].

$$\vec{q}\,(x,t) = \int_{-\infty}^{t} K(t-r)\nabla T(x,r)dr = -\int_{0}^{\infty} K(p)\nabla T(t-p)dp, \tag{8.1.4}$$

where $K : (0,+\infty) \rightarrow R$ is a given kernel called the 'heat flux relaxation function' and accounts for the memory of the material. It can be seen that if $K(t) = (K_0/\tau)\exp(-t/\tau)$

in (8.1.4), then equation (8.1.3) is obtained. In (8.1.4), the temperature gradient induces heat flow after a delay of time τ and heat is released during a period of time, whereas, in the model (8.1.2), heat is released instantaneously after a lapse of time τ from the moment temperature gradient is formed.

We shall be dealing mostly with the heat flux model given in (8.1.3) which is called a *generalized Fourier's law* or a *non-Fourier's law*. Fourier's law cannot be derived from (8.1.4) unless ∇T is constant for all time and the thermal conductivity is defined as $\int_0^\infty K(p)dp$. It is clear that the flux law given in (8.1.3) is a particular case of a more general law stated in (8.1.2). Why do we not accept (8.1.2) as the generalized Fourier's law? This question has been discussed in [211] with the help of a simple experiment. Consider a piece of ice of unit volume at temperature $T < 0$. It is heated by a heat source of constant intensity $F > 0$. Let C_S and C_L be the specific heats of ice and water, respectively. The temperature of ice increases at the rate F/C_S until it reaches $T = 0$ (melting temperature). The temperature remains at zero for time $C_S l/F$ which is also the time taken for l units of heat to be supplied to ice. In the parabolic Stefan problem, the specific energy $e(t)$ (or specific enthalpy) at time t is given by

$$e(t) = \beta(T(t)) + l\xi(t). \tag{8.1.5}$$

Here, $\beta(T) = C_S T$ for $T \leq 0$, $\beta(T) = C_L T$ for $T > 0$, ξ is the water-fraction and $\xi \in H_g(T)$, where H_g is the Heaviside graph (cf. (4.3.45)). To be consistent with the assumption made in the constitutive law (8.1.2), the response of energy to the latent heat should also be delayed and the equation (8.1.5) in the hyperbolic Stefan problem may be written as

$$e(t) = \beta(T(t)) + l\xi(t - \tau). \tag{8.1.6}$$

Let us assume that there is no delay in the release of sensible heat so that the energy conservation law $\rho C T_t = -\nabla \vec{q}$ holds. In the classical Stefan problem if ice attains the temperature zero at time $t = 0$ and is receiving heat, then it starts storing latent heat for $t > 0$. The relation (8.1.6) suggests that storing of latent heat does not take place during the time internal $(t, t + \tau)$. Therefore ice does not melt during time $(t, t + \tau)$ and its temperature goes on rising. At time $t + \tau$, this 'superheated ice' which has sucked in enough energy suddenly changes to water. This is a very unstable and unrealistic situation.

The constitutive equations for the heat flux vector have been considered in much more general form than the one considered in (8.1.4). For example in [210], a general theory of heat conduction with finite wave speeds has been built, in which the heat flux vector is taken as a functional of temperature, temperature gradient and temperature summed histories. Heat flux, like the entropy, is determined by the functional for the free energy. In [212], temperature, temperature gradient and the time rate of temperature are taken as independent variables. When the constitutive equations include rate of temperature, there arises local entropy production in addition to entropy production due to conduction. Within the framework of rate-dependent theory, thermal waves can occur in the finite theory but not in the corresponding linearized theory.

8.2 Model I : Hyperbolic Stefan Problem with Temperature Continuity at the Interface

8.2.1 The mathematical formulation

Several physical and mathematical models have been proposed in the literature for the hyperbolic Stefan problem. In model I, the response of flux to the temperature gradient is delayed but the storage or release of energy as latent heat is not delayed. Temperature is assumed to be continuous and known across the phase-change boundary and heat flux is given by (8.1.3) which is taken as a 'local constitutive law'. The energy conservation equation is derived on the basis of energy conservation law (cf. § 1.4.7) and therefore the energy equation in a medium which is not undergoing phase-change is given by

$$\rho C\frac{\partial T}{\partial t} = -\text{div}\vec{q}, \tag{8.2.1}$$

where $\vec{q}$ is given by (8.1.3). On using (8.1.3), $\vec{q}$ can be eliminated from (8.2.1) and we obtain

$$\tau\rho CT_{tt} + \rho CT_t = K\nabla^2 T. \tag{8.2.2}$$

Equation (8.2.2) is a hyperbolic equation and is commonly known as *telegrapher's equation.* In the one-dimensional case the wave speed of this telegrapher's equation is $(k/\tau)^{1/2}$ because the characteristic curves are given by (cf. [12])

$$\frac{k}{\tau}\left(\frac{dt}{dx}\right)^2 = 1, \tag{8.2.3}$$

and the lines $x \pm (k/\tau)^{1/2}t =$ constant, are the characteristics.

A one-dimensional two-phase hyperbolic Stefan problem

For simplicity, we first consider a one-dimensional problem in the region $\Omega = \{x : 0 \le x \le 1\}$. $x = S(t)$, $0 < t < t_*$, is the phase-change boundary. Let the region $0 < x < S(t)$ be solid, and the region $S(t) < x < 1$, be liquid. It will be assumed that in both these regions equations of the type (8.2.2) hold good. The thermal properties and relaxation parameters of the two regions could be different but densities are taken to be equal.

Differential equations

$$\tau_S\rho C_S(T_S)_{tt} + \rho C_S(T_S)_t = K_S\nabla^2 T_S,\ 0 < x < S(t),\ 0 < t < t_*, \tag{8.2.4}$$

$$\tau_L\rho C_L(T_L)_{tt} + \rho C_L(T_L)_t = K_L\nabla^2 T_L,\ S(t) < x < 1,\ 0 < t < t_*. \tag{8.2.5}$$

Initial conditions

$$\text{(i)}\ \ T(x,0) = T_0(x),\ x \in \Omega. \tag{8.2.6}$$

$$\text{(ii)} \quad \frac{\partial T}{\partial t}(x,0) = g_0(x), \; x \in \Omega. \tag{8.2.7}$$

On what physical basis can $\partial T/\partial t$ at $t = 0$ be prescribed? There is no clear answer to this. If $q(x,0)$ is continuously differentiable in the whole of the region Ω, then (8.2.7) can be replaced by

$$\text{(iii)} \quad q(x,0) = q_0(x), \; x \in \Omega. \tag{8.2.8}$$

Boundary conditions at the fixed boundaries

If the temperature is prescribed on the fixed boundary, then the same boundary condition continues in the same form in the present model but if the boundary condition is of the type (1.4.38), i.e., flux is prescribed then we have the following boundary condition on the boundary $\partial\Omega$ of Ω.

$$q(x,t) = \beta E(T^4 - T_0^4) + (C_1 T - C_2 T_0), \text{ on } \partial\Omega. \tag{8.2.9}$$

Boundary conditions at the free boundary

(i) Isotherm condition : $T = T_m$ (constant) on $x = S(t)$. Isotherm condition also implies continuity of temperature. We shall see later that in some formulations temperature could be discontinuous across the free boundary. This is not surprising because hyperbolic equation admits discontinuous solutions.

(ii) Energy balance condition : If the phase-change is taking place from liquid to solid, then the energy balance at $x = S(t)$ will have the form

$$\rho \hat{l} \frac{dS}{dt} = q_L(S(t),t) - q_S(S(t),t), \; \hat{l} = l + (C_L - C_S)T_m. \tag{8.2.10}$$

Equation (8.2.10) is not suitable for obtaining analytical and numerical solutions as it is in terms of fluxes. So it will be expressed in a different form. On differentiating (8.2.10) with respect to time and using (8.2.1) and (8.1.3), we get

$$\rho \hat{l} \frac{d^2 S}{dt^2} = (q_L)_t + (q_L)_x \frac{dS}{dt} - (q_S)_t - (q_S)_x \frac{dS}{dt},$$

or

$$\begin{aligned}\rho \hat{l} \frac{d^2 S}{dt^2} = \rho\Big(C_S \frac{\partial T_S}{\partial t}\Big|_{S(t)} &- C_L \frac{\partial T_L}{\partial t}\Big|_{S(t)}\Big)\frac{dS}{dt} + \Big(-K_L \frac{\partial T_L}{\partial x} - q_L\Big)/\tau_L\Big|_{S(t)} \\ &+ \Big(K_S \frac{\partial T_S}{\partial x} + q_S\Big)/\tau_S\Big|_{S(t)}.\end{aligned} \tag{8.2.11}$$

If isotherm conditions are differentiated with respect to time, we obtain

$$(T_L)_t = -(T_L)_x \frac{dS}{dt}, \text{ and } (T_L)_t = -(T_L)_x \frac{dS}{dt}. \tag{8.2.12}$$

On substituting (8.2.12) in (8.2.11), we to obtain the following equation.

$$\rho \hat{l} \frac{d^2S}{dt^2} = \rho \left(C_L \frac{\partial T_L}{\partial x} - C_S \frac{\partial T_S}{\partial x} \right) \left(\frac{dS}{dt} \right)^2 + (K_S/\tau_S) \frac{\partial T_S}{\partial x} - (K_L/\tau_L) \frac{\partial T_L}{\partial x}$$
$$+ \; q_S/\tau_S - q_L/\tau_L, \text{ on } x = S(t). \tag{8.2.13}$$

If $\tau_S = \tau_L$, then the last term in (8.2.13) can be written with the help of (8.2.10).

For a one-phase problem, the condition (8.2.13) can be easily obtained, for example, take $T_L \equiv 0$, and $\tau_S = \tau_L = \tau$. In the above hyperbolic Stefan problem, there are three velocities, one that of the free boundary and two of the wave fronts in the two phases. Can the velocity of the free boundary exceed that of the wave front? This question has been studied in [213] with the help of an analytical solution of a one-phase problem. Consider a semi-infinite solid $0 \leq x < \infty$ which is initially in a solid state at the melting temperature T_m. This solid is heated at $x = 0$ so that we have a one-phase melting problem. It can be easily seen that the pair $(S(t), T_L)$, where

$$S(t) = bt, \; b > 0 \text{ and } b^2 \neq k/\tau, \tag{8.2.14}$$

and

$$T_L(x,t) = T_m + (l/C_L) \left[\exp\left\{ b(x - bt)/(\tau b^2 - k) \right\} - 1 \right], \tag{8.2.15}$$

is a solution of the hyperbolic Stefan problem (8.2.16)–(8.2.18).

$$\tau \rho C_L (T_L)_{tt} + \rho C_L (T_L)_t = K_L \nabla^2 T_L, \; 0 < x < S(t), \; 0 < t < t_*, \tag{8.2.16}$$

$$T_L = T_m, \text{ on } x = S(t), \tag{8.2.17}$$

$$\frac{d^2S}{dt^2} + (1/\tau) \frac{dS}{dt} = (C_L/l)(T_L)_x \left[\left(\frac{dS}{dt} \right)^2 - k/\tau \right], \text{ on } x = S(t). \tag{8.2.18}$$

The constant b in (8.2.14) can be determined, for example, from the temperature prescribed condition at $x = 0$, i.e.,

$$T_L(0,t) = T_m + (l/C_L) \left[\exp(b^2 t/(k - \tau b^2)) - 1 \right]. \tag{8.2.19}$$

We shall now examine three cases: (i) $b^2 > k/\tau$, (ii) $b^2 < k/\tau$ and (iii) $b^2 = k/\tau$.

Case 1. $b^2 < k/\tau$. It can be seen from (8.2.15) that $T_L > T_m$ for $0 < x < S(t)$ and this is a physically realistic case.

Case 2. $b^2 > k/\tau$. From (8.2.15), $T_L(x,t) < T_m$ for $0 < x < S(t)$ and we have a supercooled water progressing on ice. If $(T_L)_x$ is calculated from (8.2.15) and substituted in (8.1.3) and the differential equation so obtained is integrated, then we obtain

$$q(x,t) = \rho l b \; \exp\left[b(x - bt)/(\tau b^2 - k) \right], \; 0 \leq x \leq S(t). \tag{8.2.20}$$

From (8.2.20), flux is positive and so the heat is flowing from the supercooled water to ice which is not possible as it violates laws of thermodynamics. Therefore the solution in which $b^2 > k/\tau$ occurs is not physically realistic.

Case 3. $b^2 = k/\tau$. In this case, equation (8.2.18) cannot be used at the free boundary but (8.2.10) can be used. We shall try to obtain $q_L(S(t), t)$ occurring in (8.2.10). From the first equation in (8.2.12), $(T_L)_x$ can be written in terms of $(T_L)_t$ and from (8.2.1), $(T_L)_t$ can be expressed in terms of $(q_L)_x$. When this $(q_L)_x$ is substituted in (8.1.3), we get the following equation

$$(q_L)_t + b(q_L)_x = -q_L/\tau, \text{ on } x = S(t), \tag{8.2.21}$$

or

$$\frac{d}{dt}(q_L) + (1/\tau)q_L = 0, \text{ on } x = S(t). \tag{8.2.22}$$

The solution of (8.2.22) is given by

$$q_L = Ae^{-t/\tau}, \ A = \text{arbitrary constant}. \tag{8.2.23}$$

q_L in (8.2.23) does not satisfy the interface condition $\rho l dS/dt = q_L(t)$.

The second law of thermodynamics requires that in any process there should be a positive entropy production. In the case of classical Stefan problem this implies that the condition

$$-T_x(x,t)q(x,t) \geq 0, \tag{8.2.24}$$

must hold at each point of solid and liquid regions. In the classical Stefan problem, the condition (8.2.24) gets satisfied as the flux is defined through the Fourier's law. In the hyperbolic Stefan problem the condition (8.2.24) may not always hold, and an additional condition may have to be imposed to have a physically acceptable solution. If $b^2 < \tau/k$, then the condition (8.2.24) is satisfied but not if $b^2 > \tau/k$.

The model I has been used in [214] to obtain a short-time analytical solution of the problem of melting of a semi-infinite solid $x \geq 0$ (one-dimensional problem). $\partial T_L/\partial x$ is prescribed at $x = 0$ in terms of a 'Dirac delta function'. An analytical solution has been obtained using a suitable Green's function and after making some assumptions about the form of the free boundary, a short-time solution has been obtained.

In a multi-dimensional hyperbolic Stefan problem, when the phase-change is taking place from liquid to solid, the energy balance condition at the interface is given by the equation

$$[H]\vec{V} \cdot \vec{n}_x = (\vec{q}_S - \vec{q}_L) \cdot \vec{n}_x, \tag{8.2.25}$$

Here, $\vec{V}$ is the velocity of the interface and $\vec{n} = (n_t, \vec{n}_x)$ is the unit normal to the interface which points into the liquid (see (1.4.17)) and $[H]$ is the jump in the enthalpy.

8.2.2 Some existence, uniqueness and well-posedness results

A one-dimensional one-phase hyperbolic Stefan problem

Consider the following problem:

$$\tau q_t + KT_x + q = 0, \quad \text{and } C\rho T_t + q_x = 0;\ x_0 < x < S(t),\ S(0) = 0,\ t > 0, \tag{8.2.26}$$

$$T(x,t) = 0, \quad \text{and } \rho l dS/dt = q(x,t);\ \text{on } x = S(t),\ t > 0, \tag{8.2.27}$$

$$T(x,0) = T_0(x), \text{ and } q(x,0) = q_0(x). \tag{8.2.28}$$

The boundary condition could be any of the following type.

$$\text{(i)}\ \ T(x_0,t) = \hat{T}(t),\ t > 0. \tag{8.2.29}$$

$$\text{(ii)}\ \ q(x_0,t) = \hat{q}(t),\ t > 0. \tag{8.2.30}$$

$$\text{(iii)}\ \ q(x_0,t) = h\left[\hat{T}(t) - T(x_0,t)\right],\ t > 0. \tag{8.2.31}$$

By considering the solution of a specific problem, it was shown in § 8.2.1 that the velocity of the free boundary cannot exceed the wave front velocity. It has been argued in [215] that if $|\dot{S}(0)| > (k/\tau)^{1/2}$ then $x = S(t)$ would be a space-like curve for small t. Therefore to have a well-posed problem for (q,T), besides the condition (8.2.27) on $x = S(t)$, one should impose two more conditions on $x = S(t)$ if $\dot{S}(0) > (k/\tau)^{1/2}$ and no condition when $\dot{S}(0) < (k/\tau)^{1/2}$. Otherwise one would lose either uniqueness or existence of the solution.

Depending on the value of x_0, two cases arise.

Case I: $x_0 < 0$. The following result has been proved in [215] which is valid for a short time.

Proposition 8.2.1. Let $T_0,\ q_0 \in C^1(-\infty,0)$. Suppose that at $(x,t) = (0,0)$, the compatibility condition

$$C\rho\frac{\partial}{\partial x}T_0(0)q_0(0) = \rho l\frac{\partial}{\partial x}q_0(0) \tag{8.2.32}$$

is satisfied and also

$$|q_0(0)| < \rho l(k/\tau)^{1/2}. \tag{8.2.33}$$

Then there exists $t_0 > 0$ such that in $[0,t_0]$, the problem (8.2.26)–(8.2.28) with any of the boundary conditions in (8.2.29)–(8.2.31) has a unique solution $(T,q,S) \in C^1\times C^1\times C^2$. If $T_0, q_0 \in C^\infty$ and C^∞ compatibility conditions are satisfied, then the solution $\in C^\infty[0,t_0]$.

The compatibility condition (8.2.32) can be obtained if the time derivative of the isotherm condition is obtained and the derivatives occurring in it are calculated in terms of the known quantities at $x = 0$. Condition (8.2.33) implies that $\dot{S}(t)$ is less than the wave velocity at least initially.

Case II: $x_0 = 0$. In this case no initial condition is required. If $\hat{q}(t)$ and $\hat{T}(t) \in C^1[0,\infty]$ and $q(x_0,t)$ is determined from (8.2.30) or (8.2.31) and satisfies the condition

$$0 < q(0,0) < \rho l (k/\tau)^{1/2}, \tag{8.2.34}$$

then in $[0, t_0]$ a unique solution $(T, q, S) \in C^1 \times C^1 \times C^2$ exists when the boundary condition is either (8.2.30) or (8.2.31). In this case the problem consisting of (8.2.26)–(8.2.28) and (8.2.29) is not well-posed.

Global solution for the one-phase problem

If the following substitutions are made in (8.2.26)–(8.2.28), then the resulting equations become dimensionless and the thermophysical parameters do not appear in the changed equations. Let,

$$\left.\begin{aligned} t = \tau\bar{t},\ x = (K\tau/C\rho)^{1/2}\bar{x},\ \bar{T} = (C/l)T,\ \bar{q} = (C\tau/l^2K\rho)^{1/2}q, \\ \bar{S} = (C\rho/K\tau)^{1/2}S. \end{aligned}\right\} \tag{8.2.35}$$

For further discussion, bar over the changed variables will be dropped. For the analysis of hyperbolic Stefan problems it is more convenient to formulate the problems in terms of *Riemann invariants* A and B which are defined as

$$A = T + q, \text{ and } B = T - q. \tag{8.2.36}$$

In terms of Riemann invariants the system of equations (8.2.26)–(8.2.28) and (8.2.29) can be written as

$$\left.\begin{aligned} A_t + A_x + \tfrac{1}{2}(A - B) &= 0 \\ B_t - B_x + \tfrac{1}{2}(B - A) &= 0 \end{aligned}\right\},\ x_0 < x < S(t), \tag{8.2.37}$$

$$\left.\begin{aligned} (A + B)(x,t) = 0 \\ \dot{S}(t) = A(x,t) \end{aligned}\right\} \text{on } x = S(t), \tag{8.2.38}$$

$$(A + B)(x_0, t) = 2\hat{T}(t),\ t > 0. \tag{8.2.39}$$

Let

$$A(x,0) = A_0(x),\ B(x,0) = B_0(x),\ S(0) = 0. \tag{8.2.40}$$

Proposition 8.2.2. Assume the following : (i) $A_0,\ B_0 \in C^\infty[x_0, 0]$, $x_0 < 0$, and $\hat{T}(t) \in C^\infty[0,\infty)$, (ii) $\hat{T}'(t) \geq 0$, $A_0' < 0$, $B_0' < 0$, $0 < A(0,0) < 1$, and (iii) C^∞-

compatibility conditions are satisfied at $(0,0)$ and at $(x_0, 0)$. Under these assumptions, the system (8.2.36)–(8.2.40) has a unique solution $(A, B, S) \in C^\infty \times C^\infty \times C^\infty$ for all $t > 0$ and the solution satisfies

$$A(S,t) > 0,\; A_x < 0,\; B_x < 0,\;\; 0 < \dot{S}(t) < 1. \tag{8.2.41}$$

From Proposition 8.2.1., a unique solution exists in some interval $[0, t_0]$ and if the condition $0 < \dot{S}(t_0) < 1$ is again satisfied then the time interval in which the solution exists can be extended. It has been proved in [215] that $0 < \dot{S}(t) < 1 - \delta,\ \delta = \delta(t_0) > 0$ for any fixed t_0 and so the Proposition 8.2.2 holds. For the proof of Propositions 8.2.1. and 8.2.2., the technique of integration along characteristics and linear iteration has been adopted.

A Two-phase problem

The following one-dimensional two-phase formulation has been considered in [215] in the regions $-\infty < x < \infty$ and $-x_0 \le x \le x_0$.

$$\left.\begin{aligned} \tau_1(q_1)_t + K_1(T_1)_x + q_1 &= 0 \\ C_1\rho(T_1)_t + (q_1)_x &= 0 \end{aligned}\right\},\quad x < S(t) \tag{8.2.42}$$

$$\left.\begin{aligned} \tau_2(q_2)_t + K_2(T_2)_x + q_2 &= 0 \\ C_2\rho(T_2)_t + (q_2)_x &= 0 \end{aligned}\right\},\quad x > S(t), \tag{8.2.43}$$

$$\left.\begin{aligned} T_1(x,t) &= T_2(x,t) = 0 \\ \rho l\dot{S}(t) &= (q_1 - q_2)(x,t) \end{aligned}\right\},\quad x = S(t),\; S(0) = 0, \tag{8.2.44}$$

$$\left.\begin{aligned} &T_1(x,0) = T_{10}(x),\; q_1(x,0) = q_{10}(x), \quad x < 0, \\ &T_2(x,0) = T_{20}(x),\; q_2(x,0) = q_{20}(x), \quad x > 0. \end{aligned}\right\} \tag{8.2.45}$$

For the boundary conditions at the fixed boundaries, see (8.2.47). For a local-in-time solution of the above problem, the following proposition holds.

Proposition 8.2.3. If T_{10}, q_{10}, T_{20}, and $q_{20} \in C^1$, the compatibility conditions are satisfied at $(0,0)$ and

$$|q_{20}(0) - q_{10}(0)| < \min\left\{\rho l(k_1/\tau_1)^{1/2},\; \rho l(k_2/\tau_2)^{1/2}\right\}, \tag{8.2.46}$$

then a unique solution $(T_1, T_2, q_1, q_2) \in C^1$, $S \in C^2$, of the problem (8.2.42)–(8.2.45) exists in some interval $[0, t_0]$, $t_0 > 0$.

The compatibility conditions at $(0,0)$ can be derived by calculating the total time derivatives of the isotherm conditions and replacing the partial derivatives so obtained by the known quantities. The condition (8.2.46) implies that the velocity of the phase-change interface should be less than the minimum of the two wave velocities in the two phases, atleast initially.

For the global-in-time solution, the case $\tau_1 = \tau_2$ and $k_1 = k_2$ was considered in [215]. The region $[-x_0, x_0]$, $x_0 > 0$ has been considered. To complete the formulation in addition to (8.2.42)–(8.2.45), at the fixed boundaries, temperatures are prescribed as

$$T_1(-x_0, t) = \hat{T}_1(t),\ t > 0;\ T_2(x_0, t) = \hat{T}_2(t),\ t > 0. \tag{8.2.47}$$

Proposition 8.2.4. If A_{J0}, B_{J0}, $\hat{T}_J$, $j = 1, 2$, are all C^∞- functions, C^∞- compatibility conditions are satisfied at $(0, 0)$ and at $(\pm x_0, 0)$, and

$$(-1)^{J+1}\,\hat{T}_J > 0,\ (-1)^{J+1}\,\hat{T}'_J \geq 0,\ J = 1, 2, \tag{8.2.48}$$

$$A'_{J_0} < 0,\ B'_{J_0} < 0,\ J = 1, 2;\ \text{and}\ |S(0)| < (k/\tau)^{1/2}, \tag{8.2.49}$$

then the two-phase problem considered in $[-x_0, x_0]$ has a unique C^∞- solution in $(0, \infty)$. A_{J_0}, B_{J_0} are the values of A_J and B_J at $t = 0$. The Riemann invariants A_J and B_J are defined as

$$q_J = \sqrt{(k_J/\tau_J)}(A_J - B_J),\ J = 1, 2, \tag{8.2.50}$$

$$T_J = \sqrt{1/(C_J\rho)}(A_J + B_J),\ J = 1, 2. \tag{8.2.51}$$

The existence and uniqueness of the solution of a one-phase problem similar to the one described in § 8.2.1 in the region $0 \leq x \leq S(t)$, have been discussed in [216]. Using Riemann functions and hyperbolic-equation theory (cf. [216]) the solution of the problem can be expressed in terms of an integral and it can be shown by taking appropriate limits that if at $x = 0$ any temperature other than the melting temperature is prescribed then the problem is ill-posed. Flux prescribed boundary condition has been considered. The main result of [216] is as follows.

Proposition 8.2.5. If $q = Lip(L_q, R_0)$, $R_0 > 0$ and $r_0 \in (0, 1)$, then the one-phase hyperbolic problem with the initial condition $f(0) = 0$ and $f'(0) = r_0$ has a unique solution on some interval of time $[0, R_1]$, $R_1 > 0$. The solution can be continued to a maximum interval $[0, R_0]$ such that either $R_0 \to +\infty$ or $\min(f'(r),\ 1 - f'(r)) = 0$ as $r \to R_0 - 0$.

In the above proposition $Lip(L_q, R_0)$ is the class of functions satisfying Lipschitz condition with constant L_q on $[0, R_0]$, $f(t/\tau) = S(t)/(k\tau)^{1/2}$ and r is defined by the equation $r - f(r) = t/\tau$, r_0 is some value of r, and q is the prescribed flux at $x = 0$. In addition to other conditions, conditions $f(0) = 0$ and $f'(0) = r_0$ should also be satisfied. The correct choice of r_0 is dictated by extra physical laws [216].

8.3 Model II: Formulation with Temperature Discontinuity at the Interface

8.3.1 The mathematical formulation

In this model also the response of flux to the temperature gradient is delayed and the response of energy to the latent heat is not delayed but unlike model I, temperature could be discontinuous across the phase-change interface [217]. Consider a one-dimensional problem in which a solid material at time $\hat{t} = 0$ occupies the region $\hat{x} \geq 0$. The melting temperature is taken as $\hat{T} = 0$.

The constitutive or governing equations are

$$\rho \frac{\partial \hat{e}}{\partial \hat{t}} + \frac{\partial \hat{q}}{\partial \hat{x}} = 0; \text{ and } \tau \frac{\partial \hat{q}}{\partial \hat{t}} + \hat{q} + K \frac{\partial \hat{T}}{\partial \hat{x}} = 0. \tag{8.3.1}$$

The boundary condition is taken as

$$\hat{T}\left(0, \hat{t}\right) = \hat{T}_0 > 0, \ \hat{t} > 0, \tag{8.3.2}$$

and the initial conditions are taken as

$$\hat{T}(\hat{x}, 0) = \hat{q}(\hat{x}, 0) = 0, \ \hat{x} > 0. \tag{8.3.3}$$

The first equation in (8.3.1) holds in the weak sense as $\hat{e}$ is discontinuous across the phase-change boundary and the second equation in (8.3.1) is now a constitutive equation and not a localized heat flow law which it was in the model I. $\hat{e} = \hat{e}(\hat{T})$ is the specific energy (enthalpy) and is defined by the equations (2.1.22)–(2.1.24). The equations (8.3.1)–(8.3.3) can be normalized by making the following substitutions.

$$\hat{t} = \tau t, \ \hat{x} = (C_V K \tau)^{1/2} / (l \rho^{1/2}) x, \ \hat{e} = l\, e, \tag{8.3.4}$$

$$\hat{q} = (C_V K \rho / \tau)^{1/2} q, \ \hat{T} = \frac{C_V}{l} T. \tag{8.3.5}$$

On using (8.3.4) and (8.3.5), the two equations in (8.3.1) are transformed into

$$\frac{\partial e}{\partial t} + \frac{\partial q}{\partial x} = 0, \text{ and } \frac{\partial q}{\partial t} + q + \frac{\partial T}{\partial x} = 0. \tag{8.3.6}$$

We have

$$e = 1 + T, \text{ if } T > 0, \ e = T, \text{ if } T < 0, \text{ and } e \in (0,1), \text{ if } T = 0. \tag{8.3.7}$$

The initial and boundary conditions become

$$T(x, 0) = q(x, 0) = 0, \ x > 0; \text{ and } T(0, t) = \hat{T}_0 l / C_V = T_0. \tag{8.3.8}$$

The hyperbolic system of equations in (8.3.6) are known to have discontinuous solutions (cf. Problem 1.1.6. and [12]) across shocks. If the shock is denoted by $x = S(t)$, then across the shock, the following Rankine-Hugoniot conditions are satisfied.

$$\dot{S}(e_- - e_+) = q_- - q_+; \text{ and } \dot{S}(q_- - q_+) = T_- - T_+. \tag{8.3.9}$$

For notations used in (8.3.9), see (8.3.10) and (8.3.11). The first condition in (8.3.9) is the familiar Stefan condition which arises from the conservation of energy at $x = S(t)$. If $T_- = T_+$, then we do not have the second condition at $x = S(t)$ and we have to necessarily impose the isotherm condition $T_- = T_+ = T = 0$ on $x = S(t)$.

$$(e_-, T_-, q_-) = \lim_{x \uparrow S(t)} (e, T, q)(x, t) \tag{8.3.10}$$

and

$$(e_+, T_+, q_+) = \lim_{x \downarrow S(t)} (e, T, q)(x, t). \tag{8.3.11}$$

In a pure solid, $e \leq 0$ and in a pure liquid $e \geq 1$. It can be seen from (8.3.9) that the shocks propagate with speeds $\dot{S} = \pm 1$. In the mushy region $0 < e < 1$, $\dot{S} = \pm 0$, $T_- = T_+ = 0$ and $q_- = q_+$. When phase changes from solid to liquid, i.e., when $e_+ \leq 0$ and $e_- \geq 1$ or from liquid to solid ($e_+ \geq 1$ and $e_- \leq 0$), to pick up appropriate shocks, the admissibility conditions given below should be satisfied [218].

Admissibility conditions

A forward shock, $\dot{S} > 0$, with end states $e_- \neq e_+$ is admissible if and only if

$$T(e) - T(e_+) - \frac{T(e_-) - T(e_+)}{e_- - e_+}(e - e_+) \leq 0, \; e_+ < e < e_-, \tag{8.3.12}$$

$$T(e) - T(e_-) - \frac{T(e_-) - T(e_+)}{e_- - e_+}(e - e_-) \geq 0, \; e_- < e < e_+. \tag{8.3.13}$$

A backward shock, $\dot{S} < 0$, with end states $e_- \neq e_+$ is admissible if and only if

$$T(e) - T(e_+) - \frac{T(e_-) - T(e_+)}{e_- - e_+}(e - e_+) \geq 0, \; e_+ < e < e_-, \tag{8.3.14}$$

$$T(e) - T(e_-) - \frac{T(e_-) - T(e_+)}{e_- - e_+}(e - e_-) \leq 0, \; e_- < e < e_+. \tag{8.3.15}$$

The relevance of (8.3.12)–(8.3.15) in the present problem is that when $e_- > 1$ and $\dot{S} > 0$, then necessarily, $e_+ = T_+ = 0$ and we have the following one-phase melting problem.

$$T_t + q_x = 0, \text{ and } q_t + q + T_x = 0; \; 0 < x < S(t), \; t > 0, \tag{8.3.16}$$

$$q(S(t),t) = [T(1+T)]^{1/2}(S(t),t), \text{ and } \frac{dS}{dt} = (\frac{T}{1+T})^{1/2}(S(t),t), \tag{8.3.17}$$

$$T(0,t) = T_0 > 0, \; t > 0, \tag{8.3.18}$$

$$(e,T,q)(x,t) = (0,0,0), \; x > S(t), \; t > 0, \tag{8.3.19}$$

$$e(x,t) = 1 + T(x,t), \; 0 < x < S(t), \; t > 0. \tag{8.3.20}$$

At this point the necessity of considering the present new model can be questioned. Model I seems to be a good model as it is based on the conservation law of energy and incorporates generalized Fourier's law. As $\tau \to 0$ in the telegrapher's equation and in the Stefan condition (8.2.13) ($\tau_S = \tau_L = \tau$), we get back the classical Stefan problem formulation and for $\tau \to 0$ even the solution in (8.2.14)–(8.2.19) is the solution of the corresponding Stefan problem. The neessity of considering different models (we shall be considering some more models) for hyperbolic Stefan problems arose from the fact that we should be able to prove the existence and uniqueness of the solution and the well-posedness of the problem under general initial and boundary data and as $\tau \to 0$, the solution of the hyperbolic Stefan problem should converge to the solution of the classical Stefan problem. The solution given in (8.2.14)–(8.2.19) is the solution of a very special type of problem. There does not exist a general result for the hyperbolic phase-change problems which may tell us that as $\tau \to 0$, the solution of a problem formulated with the help of the model I, or for that matter any other model, will tend to the solution of the classical Stefan problem. Different authors have considered different mathematical models, showed the existence and uniqueness of solutions and tried to establish the convergence as $\tau \to 0$ by considering some specific problems. For model I, some results on the existence, uniqueness, and regularity of the solutions of some hyperbolic Stefan problems have been discussed in § 8.2.1 but no result has been reported in [213, 214, 215, 216] on the convergence of the solution as $\tau \to 0$.

8.3.2 The existence and uniqueness of the solution and its convergence as $\tau \to 0$.

The existence of the weak solution of the problem (8.3.16)–(8.3.20) has been discussed in [217] and for this the functions T, q and e considered in $0 \leq x \leq S(t)$ and satisfying (8.3.16)–(8.3.20) are expressed in terms of Riemann invariants (cf. (8.2.36)). $(T,q,e) = (0,0,0)$ for $x > S(t)$. Temperature is prescribed at $x = 0$. A family of problems parameterized by a time lag τ are considered and the desired solution is obtained as $\tau \to 0$. It has been shown that T, q and e defined by Riemann invariants satisfy the boundary conditions and for every simple closed curve C in $\{0 \leq x \leq S(t), \; t \geq 0\}$ surrounding a domain R(C), we have

$$\int_C \{e(x,t)dx - q(x,t)dt\} = 0, \tag{8.3.21}$$

$$\int_C \{q(x,t)dx - T(x,t)dt\} + \int\int_{R(C)} q(x,t)dxdt = 0. \tag{8.3.22}$$

The integration along C is taken in the clockwise sense. The Rankine-Hugoniot conditions are satisfied in the sense that for every $0 \leq t_1 < t_2 < \infty$

$$\int_{t_1}^{t_2} \left\{ e(S(t),t)\dot{S}(t) - q(S(t),t) \right\} dt = 0, \tag{8.3.23}$$

and

$$\int_{t_1}^{t_2} \left\{ q(S(t),t)\dot{S}(t) - T(S(t),t) \right\} dt = 0. \tag{8.3.24}$$

The short and long term asymptotics of the solution of (8.3.16)–(8.3.20) were also obtained.

The problem considered in [219] is also one-phase with temperature being discontinuous at the interface. The initial and boundary conditions considered are functions of space and time, respectively. In the problem formulation, to start with, energy $e(x,t)$ is taken as

$$e = e_0(T) + Z(T)q^2, \tag{8.3.25}$$

where, $e_0(T)$ is the classical internal energy as considered in (8.3.7) and $Z(T)$ is defined as

$$Z(T) = -\frac{T^2}{2}\frac{d}{dT}\left(\frac{W(T)}{T^2}\right), \; W(T) = \frac{\tau(T)}{K(T)}. \tag{8.3.26}$$

The function $Z(T)$ is a consequence of the second law of thermodynamics combined with the generalized Fourier's law [220]. The coefficients $\tau(T)$ and $K(T)$ depend both on the temperature and the phase-change material under consideration and they can have jump discontinuities at the phase-change temperature. If it is assumed that $\tau(T)/K(T)$ is differentiable at $T = 0$ (zero is the phase-change temperature) and e given in (8.3.25) is substituted in the first equation of (8.3.6), then we have

$$\frac{\partial e_0}{\partial t} + q_x = -\frac{\partial}{\partial t}\left[\left\{\frac{\tau(T)}{TK(T)} - \frac{1}{2}\frac{d}{dT}\left(\frac{\tau(T)}{K(T)}\right)\right\}q^2\right]. \tag{8.3.27}$$

The r.h.s. of (8.3.27) is too complicated and it is difficult to carry out further calculations on retaining it in the present form. If it is assumed that $\tau(T)/K(T) =$ constant for all T, $\tau(T) =$ constant and a small number, and $1/T$ is small, then the r.h.s. in (8.3.27) can be taken to be 0. We have now the same formulation as in (8.3.16) and (8.3.17). However, the initial and boundary conditions considered in [219] are more general.

Dirichlet Problem

$$\left.\begin{aligned} T(x,0) &= \phi(x) > 0, \text{ and } q(x,0)=\psi(x);\ 0 \le x < S(0), \\ T(0,t) &= f(t) > 0,\ t \ge 0. \end{aligned}\right\} \tag{8.3.28}$$

Another type of boundary condition at $x = 0$ could be the 'Neumann boundary condition'.

Neumann Problem

$$\left.\begin{aligned} T(x,0) &= \phi(x) > 0, \text{ and } q(x,0)=\psi(x);\ 0 \le x < S(0), \\ q(0,t) &= f(t),\ t \ge 0. \end{aligned}\right\} \tag{8.3.29}$$

As reported earlier, it is convenient to work with the Riemann invariants A and B which in [219] are defined as

$$A = T + \sqrt{\tau}q; \text{ and } B = T - \sqrt{\tau}q. \tag{8.3.30}$$

Substituting (8.3.30) in (8.3.16) and (8.3.17), we get

$$\sqrt{\tau}A_t + A_x + (A - B)/(2\sqrt{\tau}) = 0,\ 0 < x < S(t),\ t > 0, \tag{8.3.31}$$

$$\sqrt{\tau}B_t - B_x + (B - A)/(2\sqrt{\tau}) = 0,\ \ 0 < x < S(t),\ t > 0, \tag{8.3.32}$$

$$B(S(t),t) = -\frac{A}{1+2A}(S(t),t),\ t > 0, \tag{8.3.33}$$

$$\frac{dS}{dt} = \frac{A}{\sqrt{\tau}(1+A)}(S(t),t),\ t > 0. \tag{8.3.34}$$

For the Dirichlet problem (8.3.28), we have

$$A(x,0) = A_0(x),\ B(x,0) = B_0(x),\ 0 \le x < S(0), \tag{8.3.35}$$

$$A(0,t) + B(0,t) = 2f(t),\ t > 0. \tag{8.3.36}$$

For the Neumann problem (8.3.29), we have

$$A(0,t) + B(0,t) + 2\sqrt{\tau}f(t),\ t > 0, \tag{8.3.37}$$

and the equation (8.3.35).

To discuss the existence, uniqueness, and regularity results, the following assumptions have been made [219].

$$f \in C^2\,[0,\infty]\,,\ \phi \in C^2\,[0,S(0)]\,,\ \psi \in C^2\,[0,S(0)]\,, \tag{8.3.38}$$

$$\phi(x) > -\sqrt{\tau}\psi(x),\ \phi(x) > \sqrt{\tau}\psi(x) - 1/4;\ \text{if } 0 \le x < S(0). \tag{8.3.39}$$

If there exists a C^1 solution of (8.3.31)–(8.3.35) and (8.3.37), then the following compatibility conditions hold good.

$$A_0(0) = B_0(0) + 2\sqrt{\tau} f(0), \tag{8.3.40}$$

$$A_{0,x}(0) + B_{0,x}(0) = -2(\tau f'(0) + f(0)), \tag{8.3.41}$$

and

$$S(0) = S_0,\ S'(0) = S_1,\ S''(0) = S_2, \tag{8.3.42}$$

where $S_0 > 0$ is given and

$$S_1 = \frac{1}{\sqrt{\tau}} \frac{A_0(S_0)}{1 + A_0(S_0)}, \tag{8.3.43}$$

$$S_2 = \frac{1}{2\tau^{3/2}} \frac{\{2\sqrt{\tau}(\sqrt{\tau}S_1 - 1)A_{0,x}(S_0) - (A_0(S_0) - B_0(S_0))\}}{(1 + A_0(S_0))^2}, \tag{8.3.44}$$

$$B_0(S_0) = -\frac{A_0(S_0)}{1 + 2A_0(S_0)}, \tag{8.3.45}$$

$$B_{0,x}(S_0) = \frac{\sqrt{\tau}A_{0,x}(S_0)(1 + 2A_0(S_0)) - 4A_0^2(S_0)(1 + A_0(S_0))^3}{\sqrt{\tau}\,(1 + 2A_0(S_0))^4}. \tag{8.3.46}$$

Some of the results obtained in [219] are given below.

Proposition 8.3.1. If (8.3.38)–(8.3.41), (8.3.45) and (8.3.46) hold, then there exists a solution (A, B, S) of (8.3.31)–(8.3.35) and (8.3.37) with $S \in C^{2,1}$ and $A,\ B$ in $C^{1,1}$ up to the boundary.

Proposition 8.3.2. For any $t_1 > 0$ there exists at most one C^1 solution (A, B, S) of (8.3.31)–(8.3.35) and (8.3.37) for $0 \le t \le t_1$.

The above two results are for the Neumann problem. For the Dirichlet problem, the existence theorem requires that the data satisfy the assumptions

$$\left.\begin{aligned} &f'(t) \ge 0,\ 0 \le t < \infty, \\ -\phi'(x) > \sqrt{\tau}\,&|\psi'(x)|,\ 0 \le x \le S_0,\ \phi(S_0) + \sqrt{\tau}\psi(S_0) > 0. \end{aligned}\right\} \tag{8.3.47}$$

Proposition 8.3.3. If (8.3.38)–(8.3.41) and (8.3.45)–(8.3.47) hold and if one can establish an a priori estimate $A(S(t), t) \ge -1/2 + \varepsilon_\tau$ for some $\varepsilon_\tau > 0,\ 0 \le t \le t_1 < \infty,\ \varepsilon_\tau$

depending on τ, and t_1, then there exists a solution (A, B, S) of the Dirichlet problem with $S \in C^{2,1}$ and $A, B \in C^{1,1}$ up to the boundary.

A uniqueness result similar to that given in Proposition 8.3.2. holds good for the Dirichlet problem also.

Asymptotic behaviour of the one-phase problem as $\tau \to 0$ was also investigated in [219] and the Proposition 8.3.4 given below was proved under some assumptions.

Proposition 8.3.4. Consider the Dirichlet problem stated in (8.3.28). The solution T_τ (T depending on τ) of the Dirichlet problem with $T_\tau = 0$ in $x > S_\tau(t)$, tends to u as $\tau \to 0$ weakly in $(L^\infty\{0 \le x \le M,\ 0 \le t \le t_1\})^*$ for any $M > 0,\ t_1 > 0$, where u is the solution of the one-phase Stefan problem stated in (8.3.48)–(8.3.50).

$$u_t - u_{xx} = 0,\ 0 < x < S(t),\ t > 0, \tag{8.3.48}$$

$$u(0,t) = f(t),\ t > 0;\ u(x,0) = \phi(x),\ 0 < x < S(0), \tag{8.3.49}$$

$$u = 0,\ u_x = -\frac{dS}{dt};\ \text{on } x = S(t). \tag{8.3.50}$$

Here, $f(t)$ and $\phi(x)$ are the same as in (8.3.28).

A result similar to Proposition 8.3.4. could not be proved for the Neumann problem as the solution of Riemann invariants cannot be obtained by integration along characteristics.

A one-phase three-dimensional hyperbolic Stefan problem with discontinuous temperature was studied in [221]. The constitutive equations which hold throughout the region consisting of solid and liquid phases are taken as

$$\frac{\partial e}{\partial t} + \nabla.\ \vec{q} = 0,\ \text{and } \tau\frac{\partial \vec{q}}{\partial t} + \vec{q} + \nabla(KT) = 0;\ \vec{q} = (q_1, q_2, q_3). \tag{8.3.51}$$

The first equation in (8.3.51) holds in the weak sense as the energy $e = e(T)$ is discontinuous across the phase-change boundary whose equation is given by $\phi(x_1, x_2, x_3, t) = x_3 - S(x_1, x_2, t) = x_3 - S(t) = 0$. Across the shock $x_3 = S(t)$, the following Rankine-Hugoniot conditions hold.

$$\left.\begin{aligned} [\tau q_1]\, S_t \ &+\ [KT]\, S_{x_1} = 0, \\ [\tau q_2]\, S_t \ &+\ [KT]\, S_{x_2} = 0, \\ [\tau q_3]\, S_t \ &-\ [KT] = 0, \end{aligned}\right\} \tag{8.3.52}$$

$$[e]S_t + [q_1]\, S_{x_1} + [q_2]\, S_{x_2} - [q_3] = 0. \tag{8.3.53}$$

Here, $[f]$ denotes the jump in f across the interface $\phi = 0$ and $C_S = C_L = C$ and $\rho_S = \rho_L$. It is easy to derive conditions (8.3.52) and (8.3.53) from equations of the type (8.3.9). Since the energy function is non-convex, one cannot expect the two-phase problem to be well-posed even in the one-dimensional case and therefore only one-phase problem has been considered in [221]. The liquid occupies the region $0 \le x_3 < S(t)$ in R^3, $S(x_1, x_2, 0) = S_0(x_1, x_2)$ and the initial conditions for the liquid region are given by

$$\left.\begin{array}{l} T(\bar{x}, 0) = T_0(\bar{x}) > 0, \ \bar{x} = (x_1, x_{2,}x_3) \\ \vec{q}(\bar{x}, 0) = \vec{q}_0(\bar{x}) \end{array}\right\}, \quad 0 \le x_3 \le S_0. \tag{8.3.54}$$

$T = 0$ is the phase-change temperature. The initial conditions for the solid region are given by

$$T(\bar{x}, 0) = 0, \ \vec{q}(\bar{x}, 0) = 0; \ x_3 > S_0. \tag{8.3.55}$$

Let $U = (q_1, q_2, q_3, T)^{T_\gamma}$, where T_γ is the transpose of a matrix. The hyperbolic system consisting of equation (8.3.51) can be written in the form of the following matrix equation

$$\mathcal{L}U = U_t + \sum_{J=1}^{3} A_J U_{x_J} + BU = 0, \tag{8.3.56}$$

where A_J, J=1,2,3 and B are 4×4 matrices which can be easily written. As we saw earlier, for the existence and uniqueness of the solution, some compatibility conditions are to be satisfied at S_0. These are the conditions imposed on the traces of an initial data $(T_0, \vec{q}_0)$ at $x_3 = S(0)$. The compatibility conditions can be obtained by comparing the two values of $U_t(x, 0)$ at $x_3 = S(0)$, one computed from the interior equation (8.3.56) and another from the interface conditions (8.3.52) and (8.3.53). Higher order compatibility conditions can be obtained similarly (cf. [222]).

The compatibility conditions can also be stated in an equivalent way as follows. The initial data in (8.3.52)–(8.3.56), are said to satisfy *k-th order compatibility conditions* if there exist a C^∞- approximate solution $(\bar{U}, \bar{S})$ of (8.3.52)–(8.3.56) such that

$$\bar{U}(\bar{x}, 0) = U_0(\bar{x}), \ \bar{S}(x_1, x_2, 0) = S_0(x_1, x_2), \tag{8.3.57}$$

$$\mathcal{L}\bar{U} = O(t^k), \tag{8.3.58}$$

$$D(\bar{U}, \bar{S}) = O(t^k), \tag{8.3.59}$$

where the system of equations (8.3.52) and (8.3.53) is briefly denoted as $D(U, S) = 0$. $U_0(\bar{x})$ is the value of U at $t = 0$ and the operator $\mathcal{L}$ is the same as in (8.3.56).

The main result of [221] is the following proposition.

Proposition 8.3.5. The hyperbolic Stefan problem (8.3.52)–(8.3.56) has a unique classical solution $(U, S) \in H^k((0, t_0) \times R^3) \times H^{k+1}((0, t_0) \times R^2)$ for some $t_0 > 0$, provided the following conditions are satisfied.

1. $U_0(\bar{x}),\ S_0(x_1, x_2) \in C^\infty$.

2. k-*th* order compatibility conditions (8.3.57)–(8.3.59) are satisfied for $k \geq 4$.

3. $\bar{S}_t(x_1, x_2, 0) < \alpha = \sqrt{K/(\tau C)}$.

In the n-dimensional case, we take $k \geq (n+1)/2 + 2$. There are four main steps in the construction of the proof of Proposition 8.3.5. First, the free boundary is fixed by the transformation $x_3 - S(t) = y_3,\ x_1 = y_1,\ x_2 = y_2$ so that in the new coordinates, equation (8.3.56) becomes

$$\mathcal{L}_1(S)U = 0, \text{ in } y_3 < 0, \tag{8.3.60}$$

where

$$\mathcal{L}_1(S) = \frac{\partial}{\partial t} + \sum_{J=1}^{2} A_J \frac{\partial}{\partial y_J} + (A_3 - S_t - \sum_{J=1}^{2} A_J \frac{\partial S}{\partial y_J}) \frac{\partial}{\partial y_3} + B.$$

The free boundary conditions remain of the same form as stated in (8.3.52) and (8.3.53). The second step is to formulate the given problem in terms of new variables (V, Ψ) which satisfy the homogeneous initial conditions. V and Ψ are defined by the relations

$$U = \bar{U} + V;\ S = \bar{S} + \Psi. \tag{8.3.61}$$

The existence and uniqueness of the solution of the original problem can be established if the same can be proved for the problem formulated in terms of the new variables (V, Ψ).

The third step is to linearize about $(\bar{U}, \bar{S})$ the nonlinear problem formulated in terms of (V, Ψ) and to prove that the linearized problem is uniformly stable. The linear boundary value problem for (V, Ψ) will be uniformly stable if the solution (V, Ψ) satisfies the estimate (cf. [223])

$$\eta\, \|V\|_{0,\eta}^2 + |V|_{0,\eta}^2 + |\Psi|_{1,\eta}^2 \leq \beta(\frac{1}{\eta}\, \|f\|_{0,\eta} + |g|_{0,\eta}), \text{ for } \eta \geq \eta_0, \tag{8.3.62}$$

where $\|V\|_{r,\eta}$ is the *hyperbolic η-weighted interior norm* defined as

$$\|V\|_{r,\eta} = \sum_{|\alpha| \leq r} \int_{R^3 \times R^1_+} \left|\partial^\alpha_{t,x}(e^{-\eta t} V)\right|^2 dx_1 dx_2 dx_3 dt,\ x \in R^3, \tag{8.3.63}$$

and $|V|_{r,\eta}$ is the *boundary norm* defined as

$$|V|_{r,\eta} = \sum_{|\alpha| \leq r} \int_{R^2 \times R^1_+} \left|\partial^\alpha_{t,(x_1,x_2)}(e^{-\eta t} V)\right|^2 dx_1 dx_2 dt. \tag{8.3.64}$$

Here, (f, g) are smooth functions which have zero traces at $t = 0$ upto the order k and are defined as follows:

$$f = \mathcal{L}_1(\bar{S} + \Psi)(\bar{U} + V) - \mathcal{L}_1(\bar{S})(\bar{U}),\ y_3 < 0, \tag{8.3.65}$$

$$g = D(\bar{U} + V,\ \bar{S} + \Psi) - D(\bar{U}, \bar{S}). \tag{8.3.66}$$

It has been proved that under the assumptions about the initial data in Proposition 8.3.5, the smooth solution of the linearized problem satisfies (8.3.62) near $(\bar{y}, t) = (0, 0)$, $\bar{y} = (y_1, y_2, y_3)$.

Proposition 8.3.6. If the linearized problem is uniformly stable, then for any $k \geq 0$ there is a unique solution $(V, \Psi) \in H^k \times H^{k+1}$ satisfying (8.3.62) near $(0, 0)$ with the subscript zero in (8.3.62) replaced by k and the subscript one replaced by $k + 1$; the constant β (depending on k) in (8.3.62) depends on the local H^m norm of the coefficients with $m = \max(k, 4)$.

The fourth step is to use an iterative method to show the existence of a unique classical solution (V, Ψ) in the neighborhood of $(0, 0)$. Only linear iterations can be employed as for the linearized problem the estimate of the boundary function Ψ is one order higher than the estimate of V. For the solution of the nonlinear problem formulated in terms of (V, Ψ), the following result holds good.

Proposition 8.3.7. Consider $k \geq 4$ in the nonlinear problem formulated for (V, Ψ) and assume that the following conditions hold good.

1. $(f, g) \in H^k$ with zero traces at $t = 0$ up to the order $(k - 1)$.

2. At $(V, \Psi) = (0, 0)$, the boundary value problem is uniformly stable, i.e, the estimate in (8.3.62) is satisfied.

Under the above conditions there exists a $t_0 > 0$ such that in $(0, t_0)$, the nonlinear problem for (V, Ψ) has a unique solution $\in H^k \times H^{k+1}$. Moreover,

$$\eta \, \|V\|^2_{k,\eta,t_0} + |V|^2_{k,\eta,t_0} + |\Psi|^2_{k+1,\eta,t_0} \leq \beta_k \left(\frac{1}{\eta} \, \|f\|^2_{k,\eta,t_0} + |g|^2_{k,\eta,t_0} \right), \quad \text{for } \eta \geq \eta_0. \tag{8.3.67}$$

Here, the norms $\|.\|_{m,\eta,t_0}$ and $|.|_{m,\eta,t_0}$ denote, respectively, the restriction in $[0, t_0]$ of the norms $\|.\|_{m,\eta}$ and $|.|_{m,\eta}$.

8.4 Model III : Delay in the Response of Energy to Latent and Sensible Heats

8.4.1 The classical and the weak formulations

In this model, it will be assumed that along the solid-liquid interface, the energy response to the latent heat release (storage) is delayed by the same amount of time as the delay in the response of flux to the temperature gradient. Independently, the specific heat may be delayed or advanced with respect to the total energy by an increment depending on the phase. This seems appropriate in order to match the wave speeds of the hyperbolic equations on either side of the phase-change interface. If it is assumed that the response of energy to the latent heat is instantaneous, following a delay of relaxation time τ, and if the response of energy to the specific heat is not delayed, then the total energy during the phase-change is given by (8.1.6). As discussed earlier this leads to an unrealistic situation.

To have a well-posed problem and render it consistent with the wave speeds in solid and liquid phases and also consistent with the delay in the energy response to the latent heat, the following constitutive relations have been considered in [211].

$$-(\vec{q}\,(t) + \tau\vec{q}\,'(t)) = K\nabla T, \tag{8.4.1}$$

and

$$(1+\tau\frac{d}{dt})e(t) = (1+g(\tau)\frac{d}{dt})\bar{C}(T) + l\,\xi(t),\ \xi \in H_g(T). \tag{8.4.2}$$

Here, $g(\tau) = \tau_S$ if $T < 0$ and $g(\tau) = \tau_L$ if $T > 0$, $\bar{C}(T) = C_S T$ if $T < 0$ and $\bar{C}(T) = C_L T$ if $T > 0$. τ_S and τ_L are relaxation times for the response of energy to the specific heats in the pure solid and the pure liquid phases, respectively. ξ is the fraction of liquid present in any region. $\xi = 0$ for the pure solid, $\xi = 1$ for the pure liquid and $0 < \xi < 1$ for the mushy region whose temperature is $T = 0$. H_g is the Heavyside graph (cf. (4.3.45)). $(\vec{q}\,(t) + \tau\vec{q}\,'(t))$ and $(e(t) + \tau e'(t))$ are assumed to be piecewise smooth in the total region under consideration. The first term on the l.h.s. of (8.4.2) is the first order approximation of $e(t+\tau)$ and the first term on the r.h.s. of (8.4.2) is the first order approximation of $C(T(t+\tau_J))$, J $= S, L$. The essential assumptions in (8.4.2) are: (1) the latent heat affects the energy after a delay of time τ, and (ii) the specific heat affects the energy after a delay of $\tau - \tau_J$, J $= S, L$, in the corresponding phase. When $\tau_S = \tau_L = \tau$, equation (8.4.2) can be integrated and we get

$$e(t) = C(T(t)) + (l/\tau)\int_0^\infty \exp(-p/\tau)\xi(t-p)dp,\ \xi \in H_g(T). \tag{8.4.3}$$

Derivation of energy conservation equation for the two-phase problem

The energy conservation principle and the relation (8.4.2) will now be used to derive energy conservation equation in the phase-change region. We first obtain the classical formulation and consider the case in which only solid and liquid regions exist which are separated by a smooth interface $\Gamma(t)$. The densities of both the phases are taken to be the same . Let $G \subset R^3$, be a fixed open bounded region and $G = G_S(t) \cup G_L(t) \cup \Gamma(t)$, $0 \le t < t_*$. The subscripts S and L stand for solid and liquid regions, respectively. Hyperbolic equations admit discontinuous solutions along shocks. Such shocks will be accommodated in the formulation. The time dependence of G_S, G_L and Γ will not be indicated in the further discussion of this problem. The regions G_S and G_L are disjoint. The external boundary of G is denoted by $\partial G^b = \partial G_S^b \cup \partial G_L^b$ where ∂G_S^b and ∂G_L^b are the external boundaries of solid and liquid regions. Let ∂G_S be the boundary of the solid region when the liquid region is removed from G and let ∂G_L be the boundary of the liquid region when the solid region is removed from G, i.e., $\partial G_S = \partial^b G_S \cup \Gamma$ and $\partial G_L = \partial^b G_L \cup \Gamma$. It will be assumed that ∂G^b and Γ are sufficiently smooth and the outward drawn normal can be defined uniquely on them. The energy conservation principle when applied to G gives

$$\frac{d}{dt}\int_{G} e(t)dx = -\int_{\partial G^b} \vec{q}(t).\ \vec{n}da + \int_{G} F(t)dx. \tag{8.4.4}$$

Here, $\vec{n}$ is the unit outward normal to ∂G^b and $F(t)$ is the intensity of the heat source. By virtue of the constitutive relation (8.4.2), $e(t)$ is differentiable in G and the time derivative on the l.h.s of (8.4.4) can be taken inside the integral. According to (8.4.1), $\vec{q}(t)$ is also differentiable, therefore, on differentiating (8.4.4) again with respect to time, we obtain

$$\frac{d}{dt}\int_{G} e'(t)dx = -\int_{\partial G^b} \vec{q}'(t).\ \vec{n}da + \int_{G} F'(t)dx. \tag{8.4.5}$$

It will be assumed that G, ∂G^b and, $\vec{n}$ on ∂G^b do not change with time. Multiplying (8.4.5) by τ and adding (8.4.4) to it, we obtain

$$\frac{d}{dt}\int_{G}(e(t)+\tau e'(t))dx = -\int_{\partial G^b}(\vec{q}(t)+\tau\vec{q}'(t)).\ \vec{n}da + \int_{G}(F(t)+\tau F'(t))dx. \tag{8.4.6}$$

In G, $e(t)+\tau e'(t)$ is only piecewise smooth but in G_S and G_L, $e(t)+\tau e'(t)$ is smooth. Therefore in (8.4.6) we consider $G_S \cup G_L$. The normal $\vec{n}$ should be defined in such a way that it is consistent with boundaries of both regions G_S and G_L. Therefore $\vec{n}$ will have opposite directions on Γ when Γ is approached from liquid and from solid regions. On Γ, $\vec{n}$ is time dependent and it will be denoted by $\vec{N}$ which points into the liquid. The boundary of G_S includes Γ which is moving with a velocity $\vec{V}(t)$ and we have (see (2.4.12))

$$\frac{d}{dt}\int_{G_S\cup G_L}(e(t)+\tau e'(t))dx = \int_{G_S\cup G_L}\frac{\partial}{\partial t}(e(t)+\tau e'(t))dx + \int_{\Gamma}[e+\tau e'](\vec{V}\cdot\vec{N})\,da \tag{8.4.7}$$

In (8.4.7), we calculate d/dt from the first principle and assume that Γ does not have any surface energy.

Equation (8.4.6) when written for $G_S \cup G_L$ will have the following form

$$\begin{aligned}\int_{G_S\cup G_L}\frac{\partial}{\partial t}(e(t)+\tau e'(t))dx + \int_{\Gamma}[e+\tau e'](\vec{V}\cdot\vec{N})da &= -\int_{\partial G_S^b\cup\partial G_L^b}(\vec{q}(t)+\tau\vec{q}'(t)).\vec{n}da \\ &- \int_{\Gamma}\left[\vec{q}(t)+\tau\vec{q}'(t)\right]\cdot\vec{N}da + \int_{G_S\cup G_L}(F(t)+\tau F'(t))dx.\end{aligned} \tag{8.4.8}$$

Here, $[e+\tau e']$ and $[\vec{q}+\tau\vec{q}']$ denote jumps in $e(t)+\tau e'(t)$ and $\vec{q}(t)+\tau\vec{q}'(t)$ across Γ. The surface integral in (8.4.8) can be converted to a volume integral. The direction of the normal on Γ from the solid side is opposite to its direction from the liquid side. Since (8.4.8) holds for any G and Γ and for any $t > 0$, the following equations should hold.

$$\frac{\partial}{\partial t}(e(t) + \tau e'(t)) + \nabla.\ (\vec{q}(t) + \tau \vec{q}'(t)) = F(t) + \tau F'(t) \text{ in } G\backslash\Gamma,\ 0 < t < t_*. \quad (8.4.9)$$

$$-[e(t) + \tau e'(t)]N_t = [\vec{q}(t) + \tau \vec{q}'(t)].\ \vec{N}_x, \quad \text{on } \Gamma,\ 0 < t < t_*. \quad (8.4.10)$$

Here, $\vec{N} = (N_t, \vec{N}_x)$, and $\vec{N}_x = (N_{x_1}, N_{x_2}, N_{x_3})$, $x = (x_1, x_2, x_3)$. If $\tilde{S}$ denotes the union of Γ and some other surfaces of discontinuities (shocks) of $e(t) + \tau e'(t)$ and $\vec{q}(t) + \tau \vec{q}'(t)$ in G other than Γ, then in (8.4.9) and (8.4.10), Γ should be replaced by $\tilde{S}$.

Next, we assume that G consists of solid, liquid and mushy regions and $G = G_S(t) \cup G_M(t) \cup G_L(t) \cup \Gamma_1(t) \cup \Gamma_2(t)$. Note that there could be surfaces of discontinuities other than Γ_1 and Γ_2 also. In $G_S(t)$, we have $T < 0$, and in $G_L(t)$ we have $T > 0$. $G_M(t)$ is the mushy region in which the temperature is $T = 0$. The liquid fraction $\xi = \xi(x,t)$ is equal to 1 and 0 in liquid and solid regions, respectively, and $0 < \xi(x,t) < 1$ in the mush and is assumed to be a smooth function of t in the mush. $\Gamma_1(t)$ is the solid-mush boundary and $\Gamma_2(t)$ is the liquid-mush boundary. $G_S(t)$, $G_M(t)$ and $G_L(t)$ are mutually disjoint regions. $\Gamma_1(t)$ separates the solid region from the mush and $\Gamma_2(t)$ separates the mush from the liquid region. The external boundary of G will be denoted by ∂G^b and $\partial G^b = \partial G^b_S \cup \partial G^b_M \cup \partial G^b_L$. Here, ∂G^b_S, ∂G^b_M and ∂G^b_L are the exterior boundaries of solid, mush and liquid regions respectively. Let ∂G_S be the boundary of the solid region when liquid and mushy regions are removed from G, i.e., $\partial G_S = \partial G^b_S \cup \Gamma_1$ and similarly let $\partial G_M = \partial G^b_M \cup \Gamma_1(t) \cup \Gamma_2(t)$ and $\partial G_L = \partial G^b_L \cup \Gamma_2(t)$. The outward drawn normal $\vec{n}$ to ∂G_S, ∂G_M and ∂G_L is defined in such a way that it is continuous on the portions common to these boundaries. In order to obtain relations of the form (8.4.10) on $\Gamma_1(t)$ and $\Gamma_2(t)$ and the energy equation (8.4.9) in $G\backslash\Gamma_1 \cup \Gamma_2$, we follow a procedure similar to that used above to obtain (8.4.8). Repeating arguments which lead to (8.4.9) and (8.4.10), we get the energy balance relations on Γ_1 and Γ_2 and the energy equation in $G\backslash\Gamma_1 \cup \Gamma_2$.

Let $\stackrel{\approx}{S} = \Gamma_1 \cup \Gamma_2 \cup$ (shocks other than Γ_1 and Γ_2). The classical formulation of the three region problem (take the densities of the three regions equal to one) is concerned with finding a pair (T, ξ), satisfying the system of equations (8.4.11)–(8.4.16) for $0 < t < t_*$.

$$\frac{\partial}{\partial t}(\tau_S C_S \frac{\partial T}{\partial t} + C_S T) - K_S \nabla^2 T = F + \tau F', \text{ in } G_S - \stackrel{\approx}{S}, \quad (8.4.11)$$

$$l \frac{\partial \xi}{\partial t} = F + \tau F', \text{ in } G_M - \stackrel{\approx}{S}, \quad (8.4.12)$$

$$\frac{\partial}{\partial t}(\tau_L C_L \frac{\partial T}{\partial t} + C_L T + l) - K_L \nabla^2 T = F + \tau F', \text{ in } G_L - \stackrel{\approx}{S}, \quad (8.4.13)$$

$$\left(l(1-\xi) + \tau_L C_L \frac{\partial T}{\partial t} \right) N_t = K_L \nabla T.\ \vec{N}_x, \text{ on } \Gamma_2, \quad (8.4.14)$$

$$\left(-l\,\xi + \tau_S C_S \frac{\partial T}{\partial t}\right) N_t = K_S \nabla T.\ \vec{N}_x,\ \text{on } \Gamma_1, \tag{8.4.15}$$

$$\tau_i C_i \left[\frac{\partial T}{\partial t}\right] N_t = K_i [\nabla T].\ \vec{N}_x,\ \text{on } \overset{\approx}{S} - (\Gamma_1 \cup \Gamma_2). \tag{8.4.16}$$

Here, $i = S$ or L depending on whether T < 0 or $T > 0$. $T = 0$ on Γ_1 and Γ_2 and $\vec{N} = (N_t, \vec{N}_x)$ is the normal on a surface of discontinuity and $\vec{N}$ points into the liquid. $[f]$ denotes the jump in f across a surface of discontinuity. Equations (8.4.11) and (8.4.13) are telegrapher's equations. Equation (8.4.12) regulates the water fraction in the mush. Equations (8.4.14) and (8.4.15) are energy balance conditions at the phase-change boundaries (cf. (8.4.20) for their derivation). The propagation of wave fronts is described by (8.4.16). To complete the formulation, equations (8.4.11)–(8.4.16) should be supplemented with the initial and boundary conditions. For example,

$$T(x,t) = 0,\ x \in \partial G^b,\ 0 < t < t_*, \tag{8.4.17}$$

$$T(x,0) = T_0(x),\ x \in G, \tag{8.4.18}$$

$$\left.\begin{aligned} &(\tau_L C_L \partial T_L/\partial t + C_L T_L + l\,)|_{t=0} = v_1(x),\ \text{where } T_0(x) > 0, \\ &(\tau_S C_S \partial T_S/\partial t + C_S T_S)|_{t=0} = v_2(x),\ \text{where } T_0(x) < 0, \\ &l\,\xi(x, 0+) = v_3(x),\ \text{where } T_0(x) = 0,\ x \in G. \end{aligned}\right\} \tag{8.4.19}$$

Let $u_0(x) = v_1(x)$ for $T_0(x) > 0$, $= v_2(x)$ for $T_0(x) < 0$, $= v_3(x)$ for $T_0(x) = 0$.

To derive energy balance condition (8.4.14) on Γ_2 and (8.4.15) on Γ_1 we consider two one-phase hyperbolic Stefan problems with the assumptions about the delay in the response of flux to the temperature gradient and the delay in the energy response to the sensible heat. At the liquid-mush boundary Γ_2, the energy balance condition without any approximation gives

$$-\vec{q}\,(t+\tau).\bar{N}_x + (l(1-\xi) + C_L T_L(t+\tau_L))\vec{V}.\ \vec{N}_x = 0. \tag{8.4.20}$$

For simplicity we take densities of all the phases equal to unity. The first order approximations of $\vec{q}(t+\tau)$ and $T_L(t+\tau_L)$ when substituted in (8.4.20) give the equation (8.4.14). Similarly by considering the energy balance condition on Γ_1 and considering its approximation, we obtain (8.4.15).

The wave speeds in solid and liquid phases are given by $(k_S/\tau_S)^{1/2}$ and $(k_L/\tau_L)^{1/2}$, respectively. To have a global wave speed $(1/\tau_0)^{1/2}$ in G, $\tau_0 > 0$, independent of the phase, we define

$$\tau_S = \tau_0 k_S;\ \text{and}\ \tau_L = \tau_0 k_L. \tag{8.4.21}$$

To obtain a weak formulation of the classical problem stated in (8.4.11)–(8.4.16), we consider the integral

$$\int_{G_*}\left\{\frac{\partial}{\partial t}\left(\tau_0\frac{\partial K(T)}{\partial t}+\bar{C}(T)+l\xi\right)-\nabla^2(K(T))\right\}\phi\, dxdt. \tag{8.4.22}$$

Here, $G_* = G\times(0,t_*)$, $0<t<t_*<\infty$, $K(T)=K_S T$, if $T<0$ and $K(T)=K_L T$ if $T>0$, $\overline{C}(T)=C_S T$ if $T<0$, $=C_L T$ if $T>0$, and ϕ is the test function, $\phi\in C_0^\infty(G_*)$. By adopting the procedure indicated in (5.2.12)–(5.2.14) it can be shown that

$$\int_{G_*}\left\{\frac{\partial}{\partial t}\left(\tau_0\frac{\partial K(T)}{\partial t}+\bar{C}(T)+l\xi\right)-\nabla^2(K(T))\right\}\phi\, dxdt$$

$$=\int_{G_S}\left\{\frac{\partial}{\partial t}\left(\tau_S C_S\frac{\partial T}{\partial t}+C_S T\right)-K_S\nabla^2 T\right\}\phi\, dxdt$$

$$+\int_{G_M} l\frac{\partial\xi}{\partial t}\phi dxdt+\int_{G_L}\left\{\frac{\partial}{\partial t}\left(\tau_L C_L\frac{\partial T_L}{\partial t}+C_L T\right)-K_L\nabla^2 T\right\}\phi\, dxdt$$

$$+\int_{\approx S}\left[\nabla K(T).\vec{N}_x-(l(1-\xi)+\tau_0\frac{\partial}{\partial t}K(T))N_t\right]\phi\, dxdt. \tag{8.4.23}$$

On using (8.4.11)–(8.4.16) in (8.4.23), we get

$$\frac{\partial}{\partial t}\left(\tau_0\frac{\partial K(T)}{\partial t}+\bar{C}(T)+\ell\xi\right)-\nabla^2(K(T))=F+\tau F',\ \text{a.e. in } G_*. \tag{8.4.24}$$

Equation (8.2.24) is satisfied in G_* in the distributional sense. A pair of functions (T,ξ) such that $T\in W^{1,\infty}(0,t_*;\ L^2(G))\cap L^\infty(0,t_*;\ H_0^1(G))$, $\xi\in L^\infty(G_*)$ is called a generalized solution of the problem (8.4.11)–(8.4.16) and (8.4.17)–(8.4.19) if (8.4.24) is satisfied for *a.a.* $t\in[0,t_*]$. $T(t)\in H_0^1(G)$ and $\xi(t)\in H_g(T(t))$ for *a.a.* $t\in[0,t_*]$; $T(0)=T_0$ and $\left(\tau_0\partial K(T)/\partial t+\bar{C}(T)+\ell\xi\right)_{t=0}=u_0(x)$. If $F\in W^{1,\infty}(0,t_*;\ H^{-1}(G))$ (H^{-1} is the dual space of H^1), then $\tau_0\partial K(T)/\partial t+\bar{C}(T)+\ell\xi(t)\in W^{1,\infty}(0,t_*;\ H^{-1}(G))$ and the initial conditions are meaningful. Under suitable assumptions on F, T_0, $\xi(x,0)$ and u_0, existence of a unique generalized solution has been proved in [211].

A one-phase problem in which the melted ice first forms a mushy region with temperature zero and then becomes water with temperature greater than zero has also been discussed in [211]. The classical, weak and variational inequality formulations have been obtained for this one-phase problem. In terms of the freezing index the existence of the unique weak solution of the variational formulation has been discussed. An example of the non-existence of the generalized solution of a one-phase problem formulated with the above delay assumptions has been constructed in [211].

The one-phase multi-dimensional hyperbolic Stefan problem discussed in [224] consists of finding a hypersurface $S(t) = \{x \in R^n : \phi(x,t) = 0\}$, $0 \le t < \infty$, $n > 1$ and the temperature $T(x,t)$ in $G(t) = \{x \in R^n : \phi(x,t) > 0\}$, $0 \le t < \infty$, such that

$$\tau C T_{tt} + C T_t - K \nabla T = 0, \text{ in } G(t), \tag{8.4.25}$$

$$\left.\begin{aligned} T(x,t) &= 0, \\ (l + \tau C T_t)\phi_t &= K \nabla T \cdot \nabla \phi. \end{aligned}\right\} \text{ on } S(t) \tag{8.4.26}$$

$$\left.\begin{aligned} T(x,0) &= T_0(x), \\ T_t(x,0) &= T_1(x), \\ S(0) = S_0 &= \{x : \phi(x,0) = \phi_0 = 0\}. \end{aligned}\right\} \tag{8.4.27}$$

This model in (8.4.25)–(8.4.27) is based on the delay assumptions of model III.

The existence and uniqueness of H^1- weak solutions were proved in [211]. The existence of the unique classical solution of the problem (8.4.25)–(8.4.27) has been proved in [224]. A solution (T, ϕ) of (8.4.25)–(8.4.27) is called classical if $(T, \phi) \in C^2(R^n \times R^1_+)$. The above problem is called *non-degenerate* if

$$\nabla T_0 \cdot \nabla \phi_0 \neq 0, \text{ on } S_0. \tag{8.4.28}$$

The condition (8.4.28) implies that the interface is really moving. We saw earlier that some compatibility conditions have to be satisfied at S_0 for studying existence and uniqueness of solutions. The zeroth-order compatibility conditions on S_0 are given by

$$\left.\begin{aligned} T_0(x) &= 0, \\ (l + \tau C T_1)T_1 - K |\nabla T_0|^2 &= 0. \end{aligned}\right\} \text{ on } S_0. \tag{8.4.29}$$

The first-order compatibility condition on S_0 can be obtained by comparing the two expressions for T_{tt} on S_0, one obtained from (8.4.25) and another from (8.4.26), and it is given by

$$(l + 2\tau C T_1)(T_{tt} + \nabla T_1 \cdot \vec{\eta}) = 2K \nabla T_0 \cdot \nabla T_1 + K \nabla |T_0|^2 \cdot \vec{\eta}, \tag{8.4.30}$$

where

$$\vec{\eta} = -T_1 \nabla T_0 |\nabla T_0|^{-2}. \tag{8.4.31}$$

The main result of [224] is the following proposition.

Proposition 8.4.1. Assume that the following conditions are satisfied :

(i) The problem (8.4.25)–(8.4.27) is non-degenerate.

(ii) $T_0(x)$, $T_1(x)$ and $\phi_0(x)$ are sufficiently smooth.

(iii) The compatibility conditions are satisfied on S_0 up to the order γ : $\gamma > (n+1)/2$.

(iv) The stability condition

$$|\nabla T_0|^2 > T_1^2 \frac{\tau C}{K}\left(1 + \frac{l}{2\tau C T_1 + l}\right), \tag{8.4.32}$$

is satisfied.

Then there exists a $t_0 > 0$, such that in $(0, t_0)$, there exists a unique classical solution (T, ϕ) of (8.4.25)–(8.4.27) belonging to the space $H^{\gamma+1} \times H^{\gamma+1}(0, t_0;\ R_+^n)$. The method of proof involves fixing the phase-change boundary by the hodograph transformation and then considering a linearized problem. By using the results of the linearized problem, the results for the nonlinear problem have been obtained. For $l = 0$, the stability condition (8.4.32) implies that the velocity of the interface should be less that the velocity of sound. For $l \neq 0$, the condition (8.4.32) is stronger than the usual requirement that the phase interface moves slower that the sound speed. If the condition (8.4.32) is not imposed, then generally, one cannot expect the one-phase problem to be well-posed globally because of the hyperbolic wave property. It has been remarked in [224] that this condition is not satisfied at the time specified in the example considered in [211] to show the non-existence of the solution of the one-phase problem.

Chapter 9

Inverse Stefan Problems

9.1 Introduction

To understand the basic features of an *inverse problem*, which could be or not be a Stefan problem, we first consider a simple heat conduction problem of finding the temperature $T(x,t)$ satisfying the following heat equation, and the initial and the boundary conditions.

$$\frac{\partial T}{\partial t} = \frac{\partial^2 T}{\partial x^2}, \ (x,t) \in \Omega_{t_*} = \Omega \times \{0 < t < t_* < \infty\}, \ \Omega = [0,1], \tag{9.1.1}$$

$$T(x,0) = 0, \ x \in \Omega, \tag{9.1.2}$$

$$\frac{\partial T}{\partial x}(0,t) = 0, \ 0 < t < t_*, \tag{9.1.3}$$

$$T(1,t) = f(t), \ 0 < t < t_*. \tag{9.1.4}$$

For convenience, thermophysical parameters have been taken to be unity in the above formulation. Under fairly general smoothness conditions on $f(t)$, the solution of (9.1.1)–(9.1.4) exists. From the solution $T(x,t)$ of (9.1.1)–(9.1.4), $T(0,t)$ can be obtained. Let

$$T(0,t) = g(t), \quad 0 < t < t_*. \tag{9.1.5}$$

Here, $g(t)$ could be known exactly through an analytical solution or known only approximately, for example, through an experiment or from a numerical solution. We shall call the problem (9.1.1)–(9.1.4), a *direct problem*. Consider another problem consisting of (9.1.1)–(9.1.3) and (9.1.5) and from the solution of this problem (if it exists) obtain $T(1,t) = f(t)$. This is called an *inverse problem* corresponding to the direct problem (9.1.1)–(9.1.4). Several other inverse problems can be formulated corresponding to the direct problem (9.1.1)–(9.1.4). For example, an inverse problem may consist of (9.1.1), (9.1.3), (9.1.4), and the condition

$$T(x, t_0) = T_0,\ x \in \Omega,\ 0 < t_0 < t_*, \tag{9.1.6}$$

and we are required to find $T(x, 0)$.

It is clear from the above discussion that the 'input data' of an inverse problem contains some information about the input data of the direct problem. By interchanging the roles of some known and unknown quantities, a direct problem can be treated as an inverse problem and vice-versa. However, in practical problems, in most cases, there is a quite natural distinction between a direct and an inverse problem. For example, a direct Stefan problem is concerned with finding the temperature field and the free boundary for an exactly known data and an inverse Stefan problem is generally concerned with controlling the free boundary with the help of boundary or initial data. Input data in a direct Stefan problem can be identified with the causes and the temperature and the free boundary are effects. In inverse problems, the objective is to determine causes for a desired or an observed effect. In the context of inverse Stefan problems, a cause could be an unknown thermophysical parameter, an initial condition or a boundary condition and the desired effect could be a specified free boundary or a known temperature distribution at some future time in an evolutionary system.

9.2 Well-posedness of the Solution

The term 'solution' cannot be used loosely in the context of solutions of either direct or inverse problems. The function space to which we expect the solution to belong should be specified. If the thermal conductivity in the heat equation is a discontinuous function of temperature, then we cannot expect $T(x,t)$ to belong to $C^1(\Omega_{t_*})$ ($\Omega_{t_*} = \Omega \times (0, t_*)$). Admissibility of the input data should be specified together with the topology to be used for measuring continuity. We should specify the properties which a solution should possess. For example, we may ask the following questions which are relevant to both direct and inverse problems.

Does a solution exist for all admissible data? (9.2.1)

Is the solution unique for all admissible data? (9.2.2)

Does the solution depend continuously on the data? (9.2.3)

If the answers to the above equations are all positive, then irrespective of the problem being direct or inverse, we say, that the problem is *well-posed* or *correctly-posed or properly-posed* in the sense of Hadamard [225]. If any answer of (9.2.1)–(9.2.3) is negative, then the problem is said to be *ill-posed* or *incorrectly-posed* or *improperly-posed.* Direct problems may or may not be ill-posed but inverse problems are generally ill-posed.

In everyday life also, finding the cause which has given a desired effect may have a non-unique answer. We shall now discuss (9.2.1)–(9.2.3) in the context of the solution of an ill-posed problem.

Non-existence of the solution

For an exact data, the existence of the solution is an important requirement and it can often be achieved by relaxing the requirement of the solution to belong to a desired function space. When the data is only approximately known, such as, through experiments then the problem has to be 'regularized' (cf. § 9.3) and hence changed anyway. The regularized problem is well-posed (cf. § 9.3).

Non-uniqueness of the solution

The non-uniqueness of the solution is considered to be much more serious than non-existence of the solution [226]. If a problem has several solutions, then the solution of interest can be picked up by requiring the solution to satisfy some additional conditions of quantitative or qualitative nature. For example, the solution should have the smallest norm. The qualitative information could be about the smoothness of the solution.

Continuous dependence of the solution on the input data

The rigorous definition of *continuous dependence* of the solution on the data will be given later after formulating the direct and inverse problems as operator equations. We first discuss the significance of the continuous dependence of the solution on the data.

Continuous dependence of a solution on the data is also called 'stability' of the solution and non-stability may create serious numerical difficulties. Real life problems are highly nonlinear which are generally studied numerically. If one wants to study the solution of a problem by traditional numerical methods without 'regularization' and the solution does not depend continuously on the data, then the numerical method becomes unstable as a small error at any step in the numerical procedure goes on magnifying at subsequent steps. As a result of this, either the numerical solution cannot be obtained or the numerical solution obtained is erroneous. A partial remedy for this is the use of 'regularization methods', although one should keep in mind, and this is important, that no mathematical trick can make an inherently unstable problem stable. All that a regularization method can do is to recover partial information about the solution as stably as possible.

Let us now consider the inverse problem of obtaining $f(1,t)$ in (9.1.4) from the solution of (9.1.1)–(9.1.3) and (9.1.5) and show that the inverse problem is unstable. If $T(x,t) \in L^2(\Omega)$ for a. a. t, then the solution of (9.1.1) can be obtained by using 'exponential Fourier transform' which is defined as

$$\hat{T}(x,\alpha) = \frac{1}{\sqrt{2\pi}} \int_{-\infty}^{\infty} e^{-i\alpha t} T(x,t)dt, \ \alpha \in R, \ i = \sqrt{-1}. \tag{9.2.4}$$

On taking the exponential Fourier transform of (9.1.1), we get

$$\frac{\partial^2 \hat{T}}{\partial x^2} = i\alpha\hat{T}, \ x \in \Omega, \ \alpha \in R. \tag{9.2.5}$$

In view of the boundary condition (9.1.3), the solution of (9.2.5) can be taken as

$$\hat{T}(x,\alpha) = A\cosh(x\sqrt{i\alpha}), \tag{9.2.6}$$

and on satisfying (9.1.5), we get

$$\hat{T}(x,\alpha) = \cosh(x\sqrt{i\alpha})\hat{g}(\alpha). \tag{9.2.7}$$

From (9.1.4), it is easy to obtain the equation

$$\hat{f}(x,\alpha) = \cosh(x\sqrt{i\alpha})\hat{g}(\alpha), \tag{9.2.8}$$

and the inverse transform of $\hat{f}(\alpha)$ is given by

$$f(t) = \frac{1}{\sqrt{2\pi}} \int_{-\infty}^{+\infty} e^{i\alpha t}\cosh(\sqrt{i\alpha})\hat{g}(\alpha)d\alpha, \ 0 < t < t_*. \tag{9.2.9}$$

The integral in (9.2.9) will exist only if $\hat{g}(\alpha)$ decays very rapidly as $|\alpha| \to \infty$. This is because

$$\left|\cosh(\sqrt{i\alpha})\right| = \left(\sinh^2\sqrt{\frac{\alpha}{2}} + \cos^2\sqrt{\frac{\alpha}{2}}\right)^{\frac{1}{2}}, \tag{9.2.10}$$

goes to infinity exponentially as $|\alpha| \to \infty$ and the integral in (9.2.9) will exist only if g(t) is very smooth so that it decays very fast as $|\alpha| \to \infty$. Even then arbitrarily small errors in the data $g(t)$ can lead to arbitrarily large errors in the calculation of f which is confirmed by the following example.

If the data is obtained through experiments or numerical methods are employed for the solution, errors are bound to develop. Suppose the data is perturbed as follows [226]:

$$\left.\begin{aligned} \hat{g}^\delta(\alpha) &= \hat{g}(\alpha), \ \alpha \notin [\alpha_0, \alpha_0 + 1], \ \alpha \in R \text{ is arbitrary}, \\ &= \hat{g}(\alpha) + \delta, \ \alpha \in [\alpha_0, \alpha_0 + 1]. \end{aligned}\right\} \tag{9.2.11}$$

Then $\| \hat{g}^\delta(\alpha) - \hat{g}(\alpha)\|_{L^2(R)} = \| g^\delta - g\|_{L^2(R)} = \delta$. The corresponding error in the solution of $f(t)$ can be calculated using (9.2.9) and (9.2.11).

$$\begin{aligned} \| f - f^\delta \|^2_{L^2(R)} = \| \hat{f} - \hat{f}^\delta \|^2_{L^2(R)} &= \delta^2 \int_{\alpha_0}^{\alpha_0+1} \left|\cosh\sqrt{i\alpha}\right|^2 d\alpha, \\ &\geq \delta^2 \int_{\alpha_0}^{\alpha_0+1} \sinh^2\sqrt{\frac{\alpha}{2}}d\alpha \geq \delta^2\sinh^2\sqrt{\frac{\alpha_0}{2}}. \end{aligned} \tag{9.2.12}$$

If $\alpha_0 > 0$, we have

$$\| f - f^{\delta} \| \geq \frac{\delta}{2} \exp\sqrt{\frac{\alpha_0}{2}}. \tag{9.2.13}$$

Since $-\infty < \alpha < \infty$, $\alpha_0 > 0$ could be large and even if δ is small the error in the output function f could be very large. It is interesting to note that some 'a-priori' information about f can stabilize this ill-posed inverse problem. It has been shown in [227] that under suitable assumptions, 'Hölder stability' (cf. § 9.5) of the solution $f(t)$ can be obtained. The inherent reason for the ill-posedness of the above inverse problem lies in the strong smoothing properties of the operator describing the direct problem, i.e., the operator mapping f onto g.

The equation (9.2.9) suggests that in general we should consider an operator equation of the form

$$Az = u; \;\; u \in U, \;\; z \in Z. \tag{9.2.14}$$

Here, A is a linear bounded operator, $A : Z \to U$ and Z and U are suitable metric spaces or Hilbert spaces. u is called an 'input data' or an admissible data and z is called a 'solution' for the input data u, if there exists a $z \in Z$ such that $Az = u$. z need not be unique. If $N(A) = 0$ (if only zero of Z is mapped onto the zero of U under the transformation A), then z is unique and A is invertible, i.e., A^{-1} exists. An equation similar to (9.2.9) can be obtained expressing $g(t)$ in terms of $f(t)$ if $g(t)$ is unknown and $f(t)$ is known. In this case also the operator equation will be of the same form as in (9.2.14). Therefore, in an abstract setting, whether we are dealing with a direct problem or an inverse problem, the operator equation can be taken to be of the form (9.2.14). In principle, A could be a nonlinear operator but unless mentioned otherwise, A will be taken as a linear operator. Z and U will be taken as Hilbert spaces.

In some situations, we may like to identify an inverse problem distinctly from its direct problem. Suppose the direct problem can be formulated as an operator equation of the form (9.2.14). If A is invertible, then the inverse problem corresponding to the direct problem can be stated in the form

$$Gu = A^{-1}u = z; \quad u \in U, \quad z \in Z. \tag{9.2.15}$$

In (9.2.15), z in now an input data and u is the solution. When there is no confusion, for both direct and inverse problems, we shall consider the formulation in the form of an operator equation (9.2.14).

Definition 9.2.1. *Stability of the solution.*

Consider the operator equation (9.2.14). Suppose the concept of solution is defined and to every element $u \in U$ there is a corresponding unique element $z \in Z$ given by the relation $z = A^{-1}u$. Let u_1 and u_2 be any two elements of U and $z_1 = A^{-1}u_1$ and $z_2 = A^{-1}u_2$ with z_1 and z_2 in Z. If for every given positive real number $\varepsilon > 0$, there exists a positive real number $\delta(\varepsilon)$ such that

$$\|u_1 - u_2\| \leq \delta(\varepsilon) \Rightarrow \|z_1 - z_2\| \leq \varepsilon, \tag{9.2.16}$$

then we say that the solution of (9.2.14) depends continuously on the data or the problem formulated in (9.2.14) is stable on the spaces (Z, U). If A^{-1} is defined for all of U, then the well-posedness of the problem is equivalent to the continuity of A^{-1}. For a long time, it was an accepted point of view in mathematical literature that every mathematical problem to be studied has to be well-posed [228]. With the development of regularization methods it is possible to obtain an approximate solution of an ill-posed problem by approximating an ill-posed problem by a well-posed problem. The most important of regularization methods is the 'Tikhonov regularization' (cf. [229, 230]) but before discussing this regularization method, let us explore the possibility of obtaining an approximate solution of the operator equation (9.2.14) by some other methods.

We first consider the case when the input data is exactly known. The operator A is exactly known, is continuous, bounded and has an inverse A^{-1} which is not in general continuous. Let M be a compact subset of Z. Suppose that the input data in (9.2.14) is $u_0 \in U$ and an exact solution $z_0 \in Z$ exists but it is difficult to calculate it. Can z_0 be calculated approximately? Usually, we take for M a set of elements depending on finite number of parameters varying within finite limits in such a way that M is a closed set contained in a finite-dimensional space. Note that if Z is infinite-dimensional, then A^{-1} need not be defined on all of U, i.e., $AZ \neq U$, and secondly A^{-1} defined on $AZ \subset U$ need not be continuous. If the input data belongs to $\hat{N} = AM \subset U$, then there exists a $z_0 \in M$ and

$$\|Az_0 - u_0\|_U = \inf_{z \in M} \|Az - u_0\|_U = 0. \tag{9.2.17}$$

9.2.1 Approximate solutions

To construct an approximate solution $z_0 \in M$ for a given u_0 (as discussed above), we construct a sequence $\{z_n\}$ of elements of M such that the sequence of numbers $\|Az_n - u_0\|_U \to 0$ as $n \to \infty$. If $\{Az_n\} \to u_0$ as $n \to \infty$ in the norm of U and if $\{z_n\} \to z_0$ in the norm of M, then some z_n for sufficiently large n can be taken as a good approximation to z_0. If the conditions of Proposition 9.2.1. (given below) are satisfied, then the sequence $\{z_n\}$ converges and an approximate solution can be obtained. This method of finding an approximate solution is stable and the calculation of A^{-1} which is often difficult can be avoided.

Proposition 9.2.1. Suppose that a compact (in itself) subset M of a Hilbert space Z is mapped onto a subset $\hat{N} = AM$ of a Hilbert space U. If the mapping $M \to \hat{N}$ is continuous and one-to-one, the inverse mapping $\hat{N} \to M$ is also continuous.

For the proof of the continuity of A^{-1}, see [231]. It may be noted if the conditions of Proposition 9.2.1. are satisfied, then the problem stated in (9.2.14) is well-posed on the spaces $(M, \hat{N})$. To have a well-posed problem on the spaces (Z, U), A^{-1} should be defined on all of U and should be continuous.

Generally, the input data is available with some error. Suppose u^δ is the approximate input data which is available in the place of an exact data u_0 and we know that

$$\|u_0 - u^\delta\|_U \leq \delta,\ \delta > 0;\ z_0 = A^{-1}u_0, \text{ and } z^\delta = A^{-1}u^\delta. \tag{9.2.18}$$

Let u_0, $u_\delta \in \hat{N} = AM \subset U$ and the conditions in Proposition 9.2.1. are satisfied. Then there exists a sequence $\{z_n^\delta\} \to z_\delta$ such that $\| Az_n^\delta - u_\delta \| \to 0$ and some z_n^δ for large n can be taken as an approximate solution corresponding to u_δ . z_n^δ will also serve as an approximate solution of (9.2.14) even if u_0 is not exactly known. As $\delta \to 0$, $z^\delta \to z_0$.

Definition 9.2.2. Quasi-solution.

Most of the time, we do not have an effective criterion to determine whether the element u_δ in (9.2.18) belongs to the set $\hat{N} = AM$ or not but we know that $u_0 \in \hat{N}$. In such a case, we cannot write $z_\delta = A^{-1}u_\delta$ and take z_δ as the solution of (9.2.14) for a given u_δ. However, if we know that M is a compact set and A is a *completely continuous operator,* then there does exist an element $\hat{z} \in M$ such that

$$\|A\hat{z} - u\|_U = \inf \|Az - u\|_U\,, \quad u \in U. \tag{9.2.19}$$

$\hat{z}$ is called a *quasi-solution.* The 'Euler equation' for determining the minimum in (9.2.19) or the equation to obtain a quasi-solution has the form

$$A^*Az = A^*u,\ A^* : U \to Z,\ \langle Az, u\rangle = \langle z, A^*u\rangle, \tag{9.2.20}$$

where, A^* is the *conjugate operator* of A. Equation (9.2.20) is called the *normal equation* of the operator equation (9.2.14). If M is compact and $u \in \hat{N}$, then $\hat{z}$ is an exact solution. Quasi-solution may not be unique. It is possible to state the conditions under which a quasi-solution is unique and stable.

Let Q be a subset of a Hilbert space U and u an element of U. An element q of the set Q is called a 'projection' of the element u on Q, if

$$\|u - q\|_U = \inf_{p \in Q} \|u - p\|_U\,. \tag{9.2.21}$$

Proposition 9.2.2. Let M be a compact set of Z and let A be completely continuous on M. A quasi-solution of $Az = u$ exists on M for $u \in U$. If the projection of each element u of U onto the set $\hat{N} = AM$ is unique, then the quasi-solution of the equation $Az = u$ is unique and depends continuously on $u \in U$.

For some theoretical results and approximate determination of quasi-solutions, see [231]. Fundamental to the theory of approximate solutions of ill-posed problems is the notion of regularizing algorithms which are based on a regularizing family of operators which will be discussed now.

9.3 Regularization

9.3.1 The regularizing operator and generalized discrepancy principle

In § 9.2.1 some stable methods were described to obtain approximate solutions of (9.2.14) for a given exact input data as well as for a given approximation of the input data. It was assumed that the class of possible solutions of (9.2.14) is a compact subset M of Z and A^{-1} is continuous on $\hat{N} = AM$ or A is completely continuous on Z. In a number of problems, the set M (the set Z can also be considered) is not compact and the approximation u_δ of the input data u_0 in (9.2.14) may take u_δ ($\|u_0 - u_\delta\| \leq \delta$) outside the set $\hat{N}$ (the set U can also be considered). Such problems are *genuinely ill-posed* and a new approach for the solutions of such problems was developed in [229, 230].

As mentioned earlier, there is no trick by which a genuinely ill-posed problem can be made well-posed. But we can approximate an ill-posed problem by a family of neighbouring well-posed problems. This is done by constructing a family of *regularization operators* $\{R_\alpha\}$, where each R_α is a continuous operator, dependent on a parameter α, and $R_\alpha : U \to Z$.

To motivate the definition of a regularizing operator, we make the following assumptions.

(1) The operator $A : Z \to U$ is only approximately known and its approximation $A_\eta : Z \to U$ is also a linear bounded operator just as A is. Further,

$$\|A_\eta - A\| \leq \eta, \; A_0 = A. \tag{9.3.1}$$

(2) A^{-1} is not continuous on the set AZ.

(3) The set of all possible solutions $M \subset Z$ is not compact.

(4) The input data $u_0 \in U$ is only approximately known and its approximation $u_\delta \in U$ is such that

$$\|u^\delta - u_0\| \leq \delta > 0. \tag{9.3.2}$$

u^δ is called *noisy data*, and δ is the error, also called *noise level.*

(5) A solution $z_0 \in M$ of (9.2.14) for the exact data $u_0 \in U$ exists such that $z_0 = A^{-1}u_0$.

The initial information consists of $\{u_\delta, A_\eta, \delta, \eta\}$. Note that even if a $z_\delta \in Z$ exists such that $Az_\delta = u_\delta$, z_δ may not be a stable solution as we have not assumed the continuity of A^{-1}. From the initial information we are required to obtain an element $z_\beta \in Z$, $\beta = (\delta, \eta)$, such that as $\beta \to 0$, $z_\beta \to z_0 = A^{-1}u_0$. This is possible provided a regularizing operator for A^{-1} exists. In general terms, regularization is the approximation of an ill-posed problem by a family of neighbouring well-posed problems. The precise definition of a regularizing operator is given below, firstly, for the case when A is exactly known.

Definition 9.3.1. Regularization operator.

Let $A : Z \to U$ be a linear bounded operator between Hilbert spaces Z and U, and let $\alpha_0 \in (0, +\infty]$. The initial information consists of $\{A, u^\delta, \delta\}$, and for every $\alpha \in (0, \alpha_0)$

$$R_\alpha \, : \, U \to Z, \tag{9.3.3}$$

is a continuous (not necessary linear) operator. The family $\{R_\alpha\}$ is called a *regularization* or a *regularization (regularizing) operator* for A^{-1} or A^{-1} is said to be *regularizable*, if, for all $u \in D(A^{-1})$ (domain of A^{-1}), there exists a 'parameter choice rule' $\alpha = \alpha(\delta, u^\delta)$ such that

$$\lim_{\delta \to 0} \sup \left\{ \| R_\alpha u^\delta - A^{-1} u \|; \; u^\delta \in U, \; \| u - u^\delta \| \leq \delta \right\} = 0, \tag{9.3.4}$$

holds. Here

$$\alpha : R^+ \times U \to (0, \alpha_0), \tag{9.3.5}$$

is such that

$$\lim_{\delta \to 0} \sup \left\{ \alpha(\delta, u^\delta); \; u^\delta \in U, \; \| u - u^\delta \| \leq \delta \right\} = 0. \tag{9.3.6}$$

For a specific $u \in D(A^{-1})$, a pair (R_α, α) is called a (convergent) regularization method for solving (9.2.14) if (9.3.4) and (9.3.6) hold (cf. [226]).

If the operator A is only approximately known with an approximation A_η satisfying (9.3.1), then in Definition 9.3.1., replace A by A_η and the mapping in (9.3.5) should be defined as

$$\alpha = \{(\beta, u^\delta); \; \beta = (\delta, \eta), \; \delta > 0, \eta > 0, \; \| u^\delta - u \| \leq \delta\} \to (0, \alpha_0). \tag{9.3.7}$$

In view of (9.3.7), we shall have $\beta = (\delta, \eta) \to 0$ in (9.3.4) and (9.3.6).

A *regularization method* consists of constructing a family of regularization operators $\{R_\alpha\}$ and a parameter-choice rule which is convergent in the sense that if the *regularization parameter* α is chosen according to that rule, then the regularized solutions converge in the norm of Z as the error δ in the input data tends to zero (when the operator is exactly known), or as $\beta = (\delta, \eta) \to 0$ when the operator is only approximately known. This convergence is assumed for any collection of noisy input data compatible with the noise level δ and any $u \in U$.

If a regularization method exists for the problem defined by (9.2.14), then there exists $z_\alpha^\delta \in Z$, $\alpha = (\delta, u^\delta)$ such that

$$z_\alpha^\delta = R_\alpha u^\delta, \tag{9.3.8}$$

and as $\delta \to 0$, we have $z_\alpha^\delta \to z_0 \in Z,\ Az_0 = u_0$. If instead of A only A_η is known and a regularization method exists, then there exists $z_\alpha^\beta \in Z,\ \alpha = \alpha(\beta, u^\delta)$, such that

$$z_\alpha^\beta = R_\alpha u^\delta, \tag{9.3.9}$$

and as $(\delta, \eta) \to 0$, we have $z_\alpha^\beta \to z_0 \in Z,\ Az_0 = u_0$.

If the operators $\{R_\alpha\}$ are linear (linearity was not imposed earlier on R_α), then the family $\{R_\alpha\}$ is called a family of linear regularization operators and the corresponding method of obtaining a regularized solution is called a *linear regularization method.* For nonlinear problems, the operator A could be nonlinear. The theory of linear ill-posed problems is very well developed and for both linear and nonlinear regularization methods, we refer the reader to [226].

The parameter-choice rule depends not only on u^δ but also on the exact input data u_0. Since u_0 is generally not known, to take into account the dependence of α on u_0, we should have some qualitative information about the input data such as smoothness properties. The parameter α depends on the operator A also. It may be noted that the regularizing operator $\{R_\alpha, \alpha\}$ is not unique.

Is it possible to construct a stable approximate solution of an ill-posed problem if δ is unknown but u_δ is known and it is known that $\|u_\delta - u_0\| \to 0$ as $\delta \to 0$? The answer is 'negative' for an ill-posed problem but 'positive' for a well-posed problem.

Definition 9.3.2. An a-priori parameter-choice rule.

If the parameter α depends only on δ and not on u^δ, then α is called an *a-priori parameter-choice rule* and we write $\alpha = \alpha(\delta)$. If α is not an a-priori parameter-choice rule, then it is called an *a-posteriori parameter-choice rule.*

An a-priori parameter-choice rule does not depend on the actual computation and can be devised before the actual computations start.

Definition 9.3.3. Least-squares solution of $Az = u$.

Let $A : Z \to U$ be a bounded linear operator and Z and U be Hilbert spaces. An element $\overset{0}{z} \in Z$ is called a *least-squares solution* of $Az = u$ for a given $u \in U$, if

$$\| A \overset{0}{z} - u \| = \inf_{z \in Z} \{ \|Az - u\| \} . \tag{9.3.10}$$

If A is absolutely continuous, then the infimum exists which can be obtained by solving the normal equation (9.2.20).

Definition 9.3.4. Best-approximate solution of $Az = u$.

An element $\hat{z} \in Z$ is called a *best-approximate solution* of $Az = u$ if $\hat{z}$ is a least-squares solution and

$$\|\hat{z}\| = \inf\{ \|\overset{0}{z}\|;\ \overset{0}{z} \text{ is least-squares solution of } Az = u \}. \tag{9.3.11}$$

9.3.2 The generalized inverse

In [226], the operator-theoretic approach for constructing regularizing operators is based on the notion of the *Moore-Penrose generalized inverse* (MP-generalized inverse) which we shall denote by $\hat{A}$. The MP-generalized inverse $\hat{A}$ of $A \in \mathcal{L}(Z, U)$ (the set of linear bounded operators) is defined by restricting the domain and range of A in such a way that the resulting restricted operator is invertible; its inverse is then extended to its maximal domain. In Definition 9.3.1., the generalized inverse $\hat{A}$ can be used in the place of A^{-1}.

In simple terms (for the rigorous definition, see [226]), an operator $\hat{A}$ is the *MP-generalized inverse* of $A \in \mathcal{L}(Z, U)$ if and only if it has the following properties.

$$\begin{aligned} &(i)\ A\hat{A}A = A. \\ &(ii)\ \hat{A}A\hat{A} = \hat{A}. \\ &(iii)\ A\hat{A} = (A\hat{A})^*. \\ &(iv)\ \hat{A}A = (\hat{A}A)^*. \end{aligned} \tag{9.3.12}$$

The MP-generalized inverse always exists, is linear and unique. If A is non-singular, then A^{-1} and $\hat{A}$ are the same. We mention below some of the results associated with the MP-generalized inverse and the operator A (cf. [226]).

Proposition 9.3.1. MP-generalized inverse $\hat{A}$ has a closed graph $g_r(\hat{A}) = \{(u, \hat{A}u);\ u \in D(\hat{A})\}$. Furthermore, $\hat{A}$ is bounded (i.e., continuous) if and only if $R(A)$ (range of A) is closed.

Proposition 9.3.2. Let $u \in D(\hat{A})$. Then $Az = u$ has a unique best-approximate solution $\hat{z}$, which is given by

$$\hat{z} = \hat{A}u. \tag{9.3.13}$$

Note that $\hat{z}$ is not a well-posed solution. The set of all least-squares solution is $\hat{z} + N(A)$.

Proposition 9.3.3. Let $u \in D(\hat{A})$. Then, $z \in Z$ is the least-squares solution of $Az = u$ if and only if the normal equation (9.2.20) holds.

Let $\hat{z} = \hat{A}u$ be a minimum norm solution (best-approximate solution) of $Az = u$ then $A^*A\hat{z} = A^*u$, (from Proposition 9.3.3.). So we have $\hat{A} = (A^*A)^\wedge A^*$.

Proposition 9.3.4. For all $\alpha > 0$, let R_α be a continuous (possibly nonlinear) operator. Then the family $\{R_\alpha\}$ is a regularization for $\hat{A}$ if,

$$R_\alpha \to \hat{A}, \text{ pointwise on } D(\hat{A}) \text{ as } \alpha \to 0. \tag{9.3.14}$$

In this case, there exists, for every $u \in D(\hat{A})$, an 'a-priori' parameter-choice rule α such that (R_α, α) is a convergent regularization method for solving $Az = u$.

The converse of Proposition 9.3.4. holds in the following sense: if $\{R_\alpha, \alpha\}$ is a convergent regularization method, then

$$\lim_{\delta \to 0} R_{\alpha(\delta, u^\delta)} u = \hat{A}u \tag{9.3.15}$$

holds for all $u \in D(\hat{A})$.

The regularizations are pointwise approximations of MP-generalized inverse of A.

9.3.3 Regularization methods

The following questions can be asked about a regularization operator.

(i) Under what conditions can an ill-posed problem be regularized?

(ii) How can a regularization operator be constructed?

(iii) How can one obtain an approximate solution with the help of a regularizing operator?

(iv) How does one perform the steps in the regularizing algorithm in an optimal way?

We shall discuss here briefly the questions (i) and (ii) raised above in the same order. For detailed discussion, see [226, 230, 231, 232]. The third question has been discussed in §§ 9.6 and 9.7 by constructing some approximate solutions. The fourth question is more relevant to the actual numerical computations.

(i) Just as there is no unique regularizing operator, there is no unique method to construct it. Associated with $\{R_\alpha\}$ is also a parameter-choice rule. The conditions under which the regularization of an ill-posed problem is possible depend on the regularization method used and the properties of the operator A. However, a general statement can be made that if the operator A in (9.2.14) is linear, continuous and bijective (one-one and onto), the resulting problem is regularizable.

(ii) Depending on the problem under consideration, several regularization methods have been developed in the literature. We describe here some of them briefly. The applications of these methods will be discussed in connection with the regularization of some specific inverse Stefan problems considered in § 9.7.

We describe below some regularization methods.

(a) A variational method: Tikhonov-regularization

We first introduce *Tikhonov-regularization method* and the motivation behind it will be discussed later. Let $A : Z \to U$, where Z and U are Hilbert spaces, and let $D \subset Z$ be a closed convex set of constraints of the problem ($D = Z$ if there are no constraints), and $0 \in D$. Consider the problem of minimization of a functional $P^\alpha(z)$, called a *smoothing functional*, defined as

$$P^\alpha(z) = \| A_\eta z - u^\delta \|_U^2 + \alpha \|z\|_Z^2 \,. \tag{9.3.16}$$

Here, $\alpha > 0$ is a regularization parameter and A_η and u^δ are defined as in (9.3.1) and (9.3.2), respectively. The variational problem is to find

$$\inf_{z \in D} P^\alpha(z). \tag{9.3.17}$$

Proposition 9.3.5. If A and A_η are linear bounded operators and $\alpha > 0$, then the variational problem given by (9.3.17) is solvable and has a unique solution $z_\beta^{\alpha(\beta)} \in D$, $\beta = (\delta, \eta)$, and

$$\| z_\beta^\alpha \| \leq \| u^\delta \| / (\sqrt{\alpha}). \tag{9.3.18}$$

For $\alpha > 0$, $P^\alpha(z)$ is strictly convex as $((P^\alpha(z))'' z,\, z) = (2A_\eta^* A_\eta\, z + 2\alpha z,\, z) \geq 2\alpha \, \|z\|^2$ for $z \in Z$ and coercive as $\lim\limits_{\|z\| \to \infty} P^\alpha(z) = +\infty$. Therefore there exists a unique element $z_\beta^{\alpha(\beta)} \in Z$ which minimizes $P^\alpha(z)$ (see Theorem 7.2.1.). If z_β^α is an interior point of D, then

$$\left[P^\alpha(z_\beta^\alpha)\right]' z = 0, \ \forall\, z \in Z. \tag{9.3.19}$$

Here, dash denotes Fréchet derivative (see Appendix D). It can be seen that

$$\left[P^\alpha\left(z_\beta^\alpha\right)\right]' z = 2(A_\eta^* A_\eta z_\beta^\alpha - A_\eta^* u^\delta + \alpha z_\beta^\alpha,\ z), \ \forall\, z \in Z. \tag{9.3.20}$$

On using (9.3.20) in (9.3.19), we get

$$A_\eta^* A_\eta z_\beta^\alpha + \alpha z_\beta^\alpha = A_\eta^* u^\delta. \tag{9.3.21}$$

Note that z_β^α may not be a 'regularized solution' of $Az = u$ unless some further conditions are imposed on A and A_η. We have also not given any parameter-choice rule for α. Since $0 \in D$, and z_β^α gives the infimum, we have

$$P^\alpha\left(z_\beta^\alpha\right) \leq P^\alpha(0)\,, \tag{9.3.22}$$

and so (9.3.18) holds. The Proposition 9.3.6. (given after (9.3.33)) suggests that if the operators A and A_η satisfy certain conditions and a parameter-choice rule for α is defined, then a regularized solution can be obtained.

We shall now discuss the motivation for considering the minimization of the functional given in (9.3.16) which comes from the notion of a 'stabilizing functional'. Let $F(z)$ be a continuous, non-negative functional $F : Z \to U$ (take $U = R$) defined on a subset Z_1 of Z which is everywhere dense in Z. We assume that the following conditions hold.

1. z_0 belongs to the domain of Z_1 and $Az_0 = u_0$, $u_0 \in U$ where u_0 is the exact data.

2. For every positive real number d, the set of elements z of Z_1 for which $F(z) \le d$ is a compact subset of Z_1.

A functional $F(z)$ satisfying the above conditions is called a *stabilizing functional.*

Let

$$Z_{1,\delta} = Q_\delta \cap Z_1;\ Q_\delta \equiv \left\{ z : \|Az - u^\delta \| \le \delta \right\}. \tag{9.3.23}$$

Here, we want to consider only those elements of Q_δ on which $F(z)$ is defined. It can be proved (cf. [231]) that there exists a $z^\delta \in Z_{1,\delta}$ such that $z^\delta = \inf_{z \in Z_{1,\delta}} F(z)$ and $z^\delta = \overline{R}_\delta(u^\delta)$, where, $\overline{R}_\delta$ is a regularizing operator and as $\delta \to 0$, $z^\delta \to z_0$ and $Az_0 = u_0$.

We shall now obtain the infimum of $F(z)$ over Z_1. Let

$$z_F^\delta = \inf_{z \in Z_1} F(z). \tag{9.3.24}$$

For simplicity, we take z_F^δ to be the unique minimum but in general it be a set M_0. If z_F^δ belongs to $Z_{1,\delta}$, then we take z_F^δ to be the stable solution on $Z_{1,\delta}$ and $z_F^\delta = z^\delta$. If z_F^δ does not belong to $Z_{1,\delta}$, then $\|Az_F^\delta - u^\delta \| > \delta$. In this case z_F^δ can be obtained by solving a constrained minimization problem which is discussed below.

The stabilizing functional $F(z)$ (defined above), is called *quasimonotonic* if, for every element $z_M \in Z_1$ that does not belong to the set M_0, every neighbourhood of it includes an element z_1 of Z_1 such that $F(z_1) < F(z_M)$. It can be proved (cf. [231]) that if $F(z)$ is quasimonotonic on the set $Z_{1,\delta}$ and $M_0 \cap Z_{1,\delta}$ is empty then the infimum of $F(z)$ is attained at an element z_M^δ for which $\|Az_M^\delta - u^\delta \| = \delta$. We can use this result to find the minimum in (9.3.24) on the set Z_1 under the constraint that the minimizer $z \in Z$ which we are seeking satisfies the condition

$$\|Az - u^\delta \| = \delta. \tag{9.3.25}$$

The method of *Lagrange multipliers* (cf. [82]) can be used to study this minimization problem with a constraint and a functional of the form $P^\alpha(z)$ as in (9.3.16) can be considered in which we take $\alpha\, F(z)$ instead of $\alpha \|z\|^2$ and the parameter α is determined from the condition

$$\|Az_\alpha^\delta - u^\delta \| = \delta. \tag{9.3.26}$$

Here, z_α^δ is the minimizer of $P^\alpha(z)$ and A is exactly known. The Proposition 9.3.6. suggests there is no necessity of considering a constrained minimization problem and z_α^δ can be regarded as the result of applying a regularization R_α such that $z_\alpha^\delta = R_\alpha u^\delta$. The parameter α is determined by a parameter-choice rule.

If instead of the operator A only its approximation A_η is known, then once again, we consider a constrained minimization problem for the functional $P^\alpha(z, u_\delta, A_\eta)$, where

$$P^\alpha(z, u_\delta, A_\eta) = \|A_\eta z - u_\delta\|^2 + (\alpha - \eta^2)F(z). \tag{9.3.27}$$

The parameter α is to be determined from the condition

$$\|A_\eta z_\alpha^\delta - u^\delta\|^2 = \delta^2 + \eta^2 F(z), \tag{9.3.28}$$

where z_α^δ minimizes the functional given in (9.3.27).

The choice of a stabilizing functional $F(z)$ is often prompted by the nature of the problem and in a number of cases, more than one choice is possible. The stabilizing functional considered in (9.3.16) is $\|z\|_Z^2$. If the Hilbert space $W^{p,2}(\Omega)$, $\Omega = \{x;\ a \le x \le b\}$ is considered as the solution space, then we can consider the stabilizing functional $F(z)$ as

$$F(z) = \int_a^b \sum_{r=0}^{p} q_r(x) \left(\frac{d^r(z)}{dx^r}\right)^2 dx,\ \ z = z(x),\ \ z \in W^{p,2}(\Omega)\,. \tag{9.3.29}$$

Here, $q_0(x)$, $q_1(x), \ldots,$ $q_{p-1}(x)$ are given non-negative continuous functions and $q_p(x)$ is a given positive continuous function. Stabilizers of the form where $q_r(x) \ge 0$, for $r = 0, 1, \ldots, p-1$, and $q_p(x) > 0$ are called *stabilizers of p-th order*. If all the functions $q_r(x)$ are constant, then we have a stabilizer of *p-th* order with constant coefficients or a *Tikhonov stabilizer*.

To determine the parameter α in the Tikhonov regularization, a 'generalized discrepancy principle' was first suggested in [233, 234] and later on modified in [235]. We give below few definitions which are related to the definition of the generalized discrepancy principle.

Definition 9.3.5. Incompatibility measure.

Let the set D, the operators A, A_η, and u_0, u^δ be the same as defined in Proposition 9.3.5. Then an *incompatibility measure* $\mu_\beta(A_\eta, u^\delta)$ of (9.2.14) is defined as

$$\mu_\beta(A_\eta, u^\delta) = \inf_{z\in D} \|A_\eta z - u^\delta\|,\ \beta = (\eta, \delta). \tag{9.3.30}$$

If $u^\delta \in \overline{A_\eta D}$ (bar stands for the closure in the space U), then $\mu_\beta(A_\eta, u^\delta) = 0$.

If $\|u^\delta - u_0\| \le \delta$, $Az_0 = u_0$, $z_0 \in D$, $\|A_\eta - A\| \le \eta$, then $\mu_\beta(u^\delta, A_\eta) \to 0$ as $\beta \to 0$. This can be proved by using the following result.

$$\begin{aligned}\mu_\beta\left(u^\delta, A_\eta\right) &= \inf_{z\in D} \|A_\eta z - u^\delta\| \le \|A_\eta z_0 - u^\delta\| \le \|A_\eta z_0 - Az_0 + Az_0 - u^\delta\| \\ &\le \delta + \eta\|z_0\|\,. \end{aligned} \tag{9.3.31}$$

Definition 9.3.6. Discrepancy.

The norm

$$\|A_\eta z_\beta^\alpha - u^\delta\|, \tag{9.3.32}$$

is called *discrepancy*. Here, $z_\beta^\alpha = R_\alpha\left(u^\delta\right)$, R_α is a regularizing operator, $\alpha = (\beta, u^\delta)$, and $\beta = (\eta, \delta)$.

Definition 9.3.7. Generalized discrepancy.

The *generalized discrepancy* denoted here by $\rho_\beta(\alpha)$ is defined as

$$\rho_\beta(\alpha) = \|A_\eta z_\beta^\alpha - u^\delta\|^2 - \left(\delta + \eta \| z_\beta^\alpha \|\right)^2. \tag{9.3.33}$$

Here, z_β^α is the solution of (9.3.21). In earlier works [233, 234], the definition of generalized discrepancy included a term of incompatibility measure also (which can also be computed with an error) but later on in [235] it was pointed out that in the definition of generalized discrepancy the term of incompatibility measure can be taken to be zero even if $u^\delta \notin \overline{A_\eta D}$.

Definition 9.3.8. Generalized discrepancy principle.

The *generalized discrepancy principle* consists of the following rules:

(1) If the condition $\|u^\delta\| > \delta$ is not fulfilled, i.e., $\|u^\delta\| \leq \delta$ then we take $z_\beta^\alpha = 0$ (z_β^α is the solution of (9.3.21)) as an approximate solution of $Az = u_0$. If $z_\beta^\alpha = 0$, then the relation $\|A_\eta z_\beta^\alpha - u^\delta\| \leq \delta$ implies $\|u^\delta\| \leq \delta$.

(2) If the condition $\|u^\delta\| > \delta$ is satisfied, then we proceed as follows:

(a) If some $\alpha^* > 0$ exists such that $\rho_\beta(\alpha^*) = 0$, then we take $z_\beta^{\alpha^*}$ as the approxim -ate solution of $Az = u_0$.

(b) If $\rho_\beta(\alpha) > 0$ for all $\alpha > 0$, then we take $z_\beta = \lim_{\alpha \to 0} z_\beta^\alpha$ as the approximate solut -ion of $Az = u_0$.

Proposition 9.3.6. Let A be a *bijective operator*, (one-one and onto), $A : Z \to U$, and A_η be a bounded linear operator, $A_\eta : Z \to U$, such that $\|A - A_\eta\| \leq \eta$. Further, $\| u_0 - u^\delta \| \leq \delta$, $z_0 = Au_0$, $z_0 \in D \subset Z$, D is a closed convex set and $0 \in D$, $\beta = (\eta, \delta)$ and $\alpha = (\beta, \delta)$. Then $\lim_{\beta \to 0} z_\beta^{\alpha^{**}} = z_0$, where $z_\beta^{\alpha^{**}}$ is the solution of (9.3.21) and it is chosen according to the generalized discrepancy principle stated above. The solutions obtained as above with the help of the generalized discrepancy principle are regularized solutions of the equation $Az = u_0$.

If A is not bijective, then the approximate regularized solution z_β (see Definition 9.3.8.) converges to $\bar{\bar{z}}_0$ on D, where, $\bar{\bar{z}}_0$ is the solution of the normal equation (9.2.21). Tikhonov-regularization method is one of the many regularizing methods used for ill-posed problems. For self-adjoint operators, a class of linear regularization methods can be constructed using *spectral theory* [236]. Tikhonov-regularization method is a particular case of the continuous regularization methods constructed with the help of the spectral theory approach (cf. [226]).

In his original papers [229, 230], Tikhonov considered a regularization method by considering a more general functional of the form

$$\|Az - u\|^2 + \alpha \|Mz\|^2, \ z \in D(M). \tag{9.3.34}$$

Here, M is an operator which could be a differential operator such as a second derivative operator and $D(M)$ is the domain of the operator. In this case, the least-squares solution $\hat{z}_M$ of $Az = u$ minimizes a different (semi) norm, namely,

$$\| M\hat{z}_M \| = \inf\{\|M \overset{0}{z}\|, \ \overset{0}{z} \text{ is the least-squares solution of } Az = u\}. \tag{9.3.35}$$

This leads to the notion of a weighted generalized inverse of A (cf. [226]).

(b) Maximum entropy regularization

Suppose the input data are defined by a random variable X which could be discrete or continuous and $x_1, x_2.., x_n$ be its values (in the discrete case) with probabilities given by

$$P^*(\{x_i\}) = p_i^*, \ \sum p_i^* = 1. \tag{9.3.36}$$

P^* is called a 'prior distribution'. Due to some additional information available, a new probability distribution is assigned to the random values which is given by

$$P\big(\{x_i\}\big) = p_i, \ \sum p_i = 1. \tag{9.3.37}$$

Let $I(p,p^*) = I(p_1, p_2, ..., p_n; p_1^*, p_2^* ..., p_n^*)$ denote the information of P relative to P^*. $I(p,p^*)$ must satisfy some conditions such as continuity, invariance under changes of labelling of the possible values (see [237] and [226] for further conditions). An appropriate form of $I(p,p^*)$ satisfying all the axioms can be taken as.

$$I(p_1, p_2 ..., p_n; p_1^*, p_2^*, .., p_n^*) = \gamma \sum_{i=1}^{n} p_i \log(p_i/p_i^*), \tag{9.3.38}$$

where γ is a constant. The maximum entropy method consists of maximizing the entropy

$$E(p,p^*) = -\gamma \sum p_i \log\left(p_i/p_i^*\right), \tag{9.3.39}$$

subject to the constraint that the sum of all the probabilities equals one. In the continuous case, instead of (9.3.37), we shall have

$$\int_a^b x(t)dt = 1, \tag{9.3.40}$$

where $x(t)$ is the probability density function of X. The data y is given by an (possibly nonlinear) operator equation

$$F(x) = y, \tag{9.3.41}$$

defined on a suitable function space, e.g., on $L^2[a,b]$. The entropy functional to be maximized in the continuous case is defined as

$$E(x, x^*) = -\gamma \int_a^b x(t) \log\left(x(t)/x^*(t)\right) dt, \tag{9.3.42}$$

subject to the constraint (9.3.40). Using Lagrange multipliers, we are led to a problem of minimizing the functional

$$\|F(x) - y\|^2 + \alpha \int_a^b x(t) \log\left(x(t)/x^*(t)\right) dt. \tag{9.3.43}$$

(c) Regularizing methods for equations of convolution integral type

The solutions of many physical problems are obtained by solving integral equations of the first kind and among them one often encounters an equation of the convolution type which is written as $v(t) * z(t) = u(t)$ and which stands for

$$\int_{-\infty}^{\infty} v(t-\tau) z(\tau) d\tau = u(t). \tag{9.3.44}$$

Regularizing operators for operator equations of *convolution integral* type can be constructed by using integral transforms such as Laplace, Fourier and Mellin. Some of these regularizing operators will be discussed in § 9.6 in the context of some inverse Stefan problems.

(d) Regularization by projection

The numerical solutions of ill-posed problems can be obtained by regularization methods using numerical schemes such as finite-difference or finite element, collocation, Galerkin or Ritz approximation. The numerical calculations can be realized only in finite-dimensional spaces. One approach in this direction is the *regularization by projection* (cf. [238]) in which we try to find approximately the best-approximate solution $\hat{z}$ of $Az = u$, in a finite-dimensional subspace of the space Z.

Let there be a sequence $\{Z_n\}_{n=1}^{\infty}$ of finite-dimensional subspaces of Z such that

$$Z_1 \subset Z_2 \subset Z_3 \subset \dots, \tag{9.3.45}$$

whose union is dense in Z. If z_n is the least-squares minimum norm solution of $Az = u$ in Z_n, then

$$z_n = \hat{A}_n u, \tag{9.3.46}$$

where, $A_n = AO_n$, O_n is the orthogonal projector onto Z_n (A is a bounded linear operator and the range of O_n is closed, so the range of A_n is closed). Since the range of A_n is closed, $\hat{A}_n$, the generalized inverse of A_n is bounded (see Proposition 9.3.1.). $\hat{A}_n$ is linear and bounded, hence, continuous. Therefore z_n is a stable approximation of $\hat{z}$. However, as the following proposition suggests, $z_n \to \hat{z}$ only under certain conditions.

Proposition 9.3.7. Let $u \in D(\hat{A})$ ($\hat{A}$ is the generalized inverse of A) and let z_n be as above. Then

(i) $z_n \to \hat{z}$ if and only if $\{\|z_n\|\}$ is bounded. Here, $\hat{z} = \hat{A}u$,

(ii) $z_n \to \hat{z}$ if and only if $\lim\limits_{n\to\infty} \sup \|z_n\| \leq \|\hat{z}\|$.

Proposition 9.3.8. Let $u \in D(\hat{A})$ and z_n be as above. If the condition

$$\lim_{n\to\infty} \sup \|(\hat{A}_n)^* z_n\| = \lim_{n\to\infty} \sup \|(\hat{A}_n^*) z_n\| < \infty, \tag{9.3.47}$$

holds, then $z_n \to \hat{z}$.

In the *dual least-squares method* described below, the convergence $z_n \to \hat{z}$ is always guaranteed. Consider a sequence $\{U_n\}_{n=1}^{\infty}$ of finite-dimensional subspaces of $\overline{R(A)} = N(A^*)^{\perp} \subset U$, whose union is dense in $N(A^*)^{\perp}$ (cf. [238]). Let z_n be the best-approximate solution of the equation

$$A_n z = u_n, \; A_n := Q_n A, \; u_n = Q_n u, \tag{9.3.48}$$

where Q_n is the orthogonal projector onto U_n.

Proposition 9.3.9. Let $u \in D(\hat{A})$ and z_n be as above. Then $z_n = O_n \hat{z}$, where O_n is the orthogonal projector onto $Z_n := A^* U_n$, $\hat{A}_n u = O_n \hat{z}$. Moreover, $z_n \to \hat{z}$ as $n \to \infty$.

If can be shown that $\{\hat{A}_n\}$ with A_n defined in (9.3.48) is a regularizing operator for $\hat{A}$. z_n is a stable approximation to $\hat{z}$ where $\hat{z}$ is the best-approximate solution of $Az = u$.

In the noise free case, $\hat{A}_n u$ is the best-approximation in the space Z_n and $\hat{A}_n u = O_n \hat{z}$ and no further regularization is required. For the noisy data it can be shown (see [226]) that

$$\|z_n^{\delta} - \hat{z}\| \leq \|z_n - \hat{z}\| + \delta/\mu_n, \tag{9.3.49}$$

where μ_n is the smallest singular value of A_n. Since the singular value of A_n decreases rapidly as n increases, the projection method should be used in conjunction with some regularization method, e. g., Tikhonov regularization. In this way spaces with larger dimensions can be used.

In obtaining the numerical solution of an ill-posed problem by Tikhonov regularization, we have to work in a finite-dimensional space Z_n as described in (9.3.46). The minimization of $P^\alpha(z)$ (see (9.3.16)) over the space Z_n gives an approximation of $\hat{z}$. This problem is equivalent to minimizing $P_n^\alpha(z)$ over Z, where

$$P_n^\alpha(z) = \| A_n z - u^\delta \|^2 + \alpha \| z \|^2 . \tag{9.3.50}$$

Here, $A_n = AO_n$ and O_n is the orthogonal projector onto the subspace Z_n. If $Z_{\alpha,n}^\delta$ is the minimizer of $P_n^\alpha(z)$, then

$$z_{\alpha,n}^\delta = (A_n^* A_n + \alpha I)^{-1} A_n^* u^\delta . \tag{9.3.51}$$

It α is chosen according to the parameter-choice rule, then $z_{\alpha,n}^\delta$ is a regularized solution.

9.3.4 Rate of convergence of a regularization method

In obtaining numerical solutions by regularization methods, the rate at which a regularization method converges plays an important role or in other words, an *optimal regularization method* should be used for faster convergence. The rate of convergence of the regularization method (R_α, α) for which $z_\alpha = R_\alpha u$ holds is defined (u is exactly known) as the rate with which

$$\| z_\alpha - \hat{z} \| \to 0 \quad \text{as } \alpha \to 0, \tag{9.3.52}$$

or the rate (if u is approximately known) with which

$$\| z_\alpha^\delta - \hat{z} \| \to 0 \quad \text{as } \delta \to 0. \tag{9.3.53}$$

Here,

$$z_\alpha^\delta = R_\alpha u^\delta, \quad \alpha = (\delta, u^\delta), \quad \text{and } \hat{z} = \hat{A} u. \tag{9.3.54}$$

In both the cases, for simplicity, it has been assumed that A is exactly known.

Let $\mathcal{M} \subset Z$, $A : Z \to U$, $Az = u$, $z \in Z$ and $u \in U$, $\| u - u^\delta \| < \delta$, $\delta > 0$, $\hat{z} = \hat{A} u$. $\hat{A}$ is the generalized inverse of A. We make an a-priori assumption that

$$\hat{z} \in \mathcal{M}. \tag{9.3.55}$$

Under the assumption (9.3.55), the 'worst-case error' for a regularization method $\hat{R}$ for $\hat{A}$ under the information $\| u - u^\delta \| \leq \delta$, is given by

$$\triangle \left(\delta, \mathcal{M}, \hat{R} \right) = \sup \left\{ \| \hat{R} u^\delta - z \| ; z \in \mathcal{M},\ u^\delta \in U,\ \| Az - u^\delta \| \leq \delta \right\} . \tag{9.3.56}$$

An 'optimal method' R_0 in a class of methods $\mathcal{R}$ would be one for which

$$\triangle (\delta, \mathcal{M}, R_0) = \inf \left\{ \triangle \left(\delta, \mathcal{M}, \hat{R} \right) : \hat{R} \in \mathcal{R} \right\} . \tag{9.3.57}$$

The optimality of a method is to be understood with respect to an a priori information (9.3.55) and the class of methods considered.

For some $\mu > 0$, let

$$Z_{\mu,\hat{\rho}} = \{z \in Z : z = (A^*A)^{\mu}\,\omega,\ \|\omega\| \leq \hat{\rho}\}, \tag{9.3.58}$$

and

$$Z_{\mu} = \bigcup_{\hat{\rho}>0} Z_{\mu,\hat{\rho}} = R\left((A^*A)^{\mu}\right), \text{ R stands for the range.} \tag{9.3.59}$$

Definition 9.3.8. Let $R(A)$ (range of A) be non-closed, and (R_α, α) be a regularization operator for $\hat{A}$. For $\mu,\ \hat{\rho} > 0$ and $u \in AZ_{\mu,\hat{\rho}}$, let α be a parameter-choice rule for solving $Az = u$. We call (R_α, α) optimal in $Z_{\mu,\hat{\rho}}$, if

$$\triangle\left(\delta, Z_{\mu,\hat{\rho}}, R_\alpha\right) = \delta^{\frac{2\mu}{2\mu+1}}\ \hat{\rho}^{\frac{1}{2\mu+1}}, \tag{9.3.60}$$

holds for all $\delta > 0$. We call (R_α, α) of optimal order in $Z_{u,\hat{\rho}}$ if there exists a constant $p \geq 1$ such that

$$\triangle\left(\delta, Z_{\mu,\hat{\rho}}, R_\alpha\right) \leq p\,\delta^{\frac{2\mu}{2\mu+1}}\ \hat{\rho}^{\frac{1}{1+2\mu}}, \tag{9.3.61}$$

for all $\delta > 0$

The Tikhonov regularization with an 'a-priori' parameter-choice rule given by

$$\alpha \sim (\delta/\hat{\rho})^{\frac{2}{1+2\mu}}, \tag{9.3.62}$$

is of optimal order in $Z_{\mu,\hat{\rho}}$.

The best possible convergence rate is obtained for $\mu = 1$ and

$$\|z_\alpha^\delta - \hat{z}\| = \mathrm{O}\left(\delta^{2/3}\right), \tag{9.3.63}$$

as soon as $\hat{z} \in Z_{1,\hat{\rho}}$. This is the maximum convergence rate possible in Tikhonov regularization. For further results on convergence, see [226].

9.4 Determination of Unknown Parameters in Inverse Stefan Problems

If some of the thermophysical parameters are unknown in problems of heat conduction with phase-change or without phase-change, then some additional information is required for their determination. This additional information is generally in the form of some over specified boundary conditions and such problems are generally ill-posed. For example, in heat conduction problems without phase-change, if the data is over specified, then we shall be dealing with a *non-characteristic Cauchy problem.* The determination of unknown parameters in parabolic heat transfer problems by the *method of over specified boundary conditions* has been the subject matter of several studies and many such references can be found in [239, 240].

9.4.1 Unknown parameters in the one-phase Stefan problems

The following problem of determining thermal conductivity and some other parameters by prescribing an over specified boundary condition has been considered in [241].

$$\rho C T_t = \nabla\cdot(K(T)\nabla T),\ 0 < x < S(t),\ S(0) = 0,\ t > 0, \tag{9.4.1}$$

$$T(0,t) = T_0 < T_m,\ t > 0,\ \text{and}\ T(x,0) = T_m;\ 0 < x < \infty, \tag{9.4.2}$$

$$T(S(t),t) = T_m,\ t > 0,\ \text{and}\ K(T_m)T_x(S(t),t) = \rho l\dot{S}(t);\ t > 0, \tag{9.4.3}$$

$$K(T_0)T_x(0,t) = q_0/t^{1/2},\ t > 0,\ q_0 > 0. \tag{9.4.4}$$

Here, $K(T) = K_0\{1 + \beta(T - T_0)/\ (T_m - T_0)\}$; $\beta > 0,\ T_0 > 0$ and $T_m > 0$ are constant. The over specified boundary condition is given by (9.4.4) in which $q_0 > 0$ is known. Note that the flux in (9.4.4) is infinite at $t = 0$ which it should be if $S(t)$ is proportional to $t^{1/2}$ (cf. [242]). We make 'a-priori' assumptions that $T(x,t)$ and $S(t)$ can be expressed in the form

$$T(x,t) = T_0 + (T_m - T_0)\Phi_\delta(\eta)/\Phi_\delta(\lambda),\ \eta = x/\left(2a\sqrt{t}\right),\ \delta > -1. \tag{9.4.5}$$

$$S(t) = 2\sigma\sqrt{t} = 2\lambda a\sqrt{t},\ a = \sqrt{K_0/(\rho C)}. \tag{9.4.6}$$

Several combinations of unknown parameters have been considered in (9.4.1)–(9.4.4) but the coefficient β is taken unknown in all the cases and in addition to it two parameters from $K_0,\ \sigma,\ \rho,\ C,\ l$ have been taken as unknown. $\Phi_\delta(x), 0 \leq x < \infty$, is the *modified error function* which is the unique solution of a boundary value problem consisting of (9.4.7) and (9.4.8).

$$[(1 + \delta\, y(x))\, y'(x)]' + 2xy'(x) = 0,\ \delta > -1, \tag{9.4.7}$$

$$y(0+) = 0,\ y(+\infty) = 1. \tag{9.4.8}$$

Here, dash denotes differentiation with respect to x. If will be assumed that $\delta > -1$ is a given real number and $\delta \neq 0$. For $\delta = 0$, $\Phi_0(x) = \text{erf}(x)$. It can be seen that

$$\Phi_\delta(0) = 0,\ \Phi_\delta(\infty) = 1,\ \Phi_\delta'(x) > 0\ \text{and}\ \Phi_\delta''(x) \leq 0,\ 0 \leq x < \infty. \tag{9.4.9}$$

$T(x,t)$ and $S(t)$ given in (9.4.5)–(9.4.6) should satisfy (9.4.1) together with the first condition in both (9.4.2) and (9.4.3). The pair $(T(x,t), S(t))$ will be a solution of the system (9.4.1)–(9.4.4) if the following conditions are satisfied.

$$\beta = \delta\Phi_\delta(\lambda), \tag{9.4.10}$$

$$(1+\delta\Phi(\lambda))\,\Phi_\delta'(\lambda)/\,(\lambda\Phi_\delta(\lambda)) = 2l/(C(T_m - T_0)), \tag{9.4.11}$$

$$\sqrt{K_0\rho C}\;\Phi_\delta'(0)/\Phi_\delta(\lambda)) = 2q_0/(T_m - T_0). \tag{9.4.12}$$

The unknown coefficients are to be determined from (9.4.10)–(9.4.12). The unknowns, for example, can be taken as: (1) β, λ, K_0, or (2) β, λ, l, or (3) β, λ, C. Ten such cases have been investigated in [241]. To illustrate the method used in [241], we consider the following two cases.

Case I. The parameters $q_0 > 0$, $\delta > -1\,(\delta \neq 0)$, $T_0 > 0$, $T_m > T_0$, $\rho > 0$, $C > 0$ and $l > 0$ are taken to be known and parameters $\beta > 0$, $\lambda > 0$ and $K_0 > 0$ are taken as unknown. $\beta > 0$ and $K_0 > 0$ can be determined from the following two equations

$$\beta = \delta\;\Phi_\delta(\lambda), \tag{9.4.13}$$

and

$$K_0 = 4q_0^2\,\Phi_\delta^2(\lambda)/\left\{\rho C(T_m - T_0)^2\,(\Phi_\delta'(0))^2\right\}, \tag{9.4.14}$$

provided λ can be obtained from the equation

$$C(T_m - T_0)F_1(x) = 2lF_2(x), \tag{9.4.15}$$

$$F_1(x) = 1 + \delta\Phi_\delta(x), \text{ and } F_2(x) = x\;\Phi_\delta(x)/\Phi_\delta'(x). \tag{9.4.16}$$

The functions $F_1(x)$ and $F_2(x)$ possess the following properties.

$$F_1(0+) = 1,\; F_1(+\infty) = 1+\delta;\; F_1' > 0 \text{ for } \delta > 0;\; F_1' < 0 \text{ for } -1 < \delta < 0, \tag{9.4.17}$$

$$F_2(0+) = 1,\; F_2(+\infty) = +\infty;\; F_2' > 0 \text{ for } \delta > -1. \tag{9.4.18}$$

In view of (9.4.17) and (9.4.18), equation (9.4.15) has a unique solution $\lambda > 0$.

Case II. The parameters $\beta > 0$, $\lambda > 0$ and $l > 0$ are unknown and all other parameters are known. In this case β is given by (9.4.13) and λ is the solution of the equation

$$\Phi_\delta(x) = (T_m - T_0)\,\sqrt{\rho C K_0}\Phi_\delta'(0)/\!\left(2q_0\right). \tag{9.4.19}$$

In view of (9.4.9), if

$$(T_m - T_0)\sqrt{\rho C K_0}\Phi_\delta'(0) < 2q_0, \tag{9.4.20}$$

then (9.4.19) has a unique solution $\lambda > 0$. The parameter l is given by the equation

$$l = C\,(T_m - T_0)\,\Phi_\delta'(\lambda)\,[1+\delta\Phi_\delta(\lambda)]\,/\,(2\lambda\Phi_\delta(\lambda)). \tag{9.4.21}$$

The condition (9.4.20) is a necessary and sufficient condition for the existence of the solution in case II. The main consideration in the success of the above method is an a-priori assumption that it is possible to obtain both $T(x,t)$ and $S(t)$ in the form of equations (9.4.5) and (9.4.6) which constitute a similarity solution. As mentioned earlier, Neumann-type exact analytical solutions (cf. § 2.3) for the phase-change problems are extremely few. If some other combinations of unknown parameters are considered in the above problem (ten such cases are possible), then for determining λ, we get equations which will involve functions different from F_1 and F_2 given in (9.4.16). The necessary and sufficient conditions for the existence of the unique value of $\lambda > 0$ in these cases can be derived from the equation obtained for determining λ in any particular case.

Another important criterion in the success of this method for the problem (9.4.1)–(9.4.4), and some other related problems discussed below, is that it yields equations of the type (9.4.19) or (9.4.24) (given below) which contain only one unknown (σ/a). Note that both σ and a may be unknown but we consider σ/a as a single unknown. For illustration, consider the derivation of (9.4.24). When (9.4.23) is substituted in (9.4.4), we get (9.4.24). The condition obtained on satisfying (9.4.3) ($T_m = 0$ in this case) with the help of (9.4.22) and (9.4.23) has been split into two equations as in (9.4.25) so as to get an equation of the form (9.4.24).

Determination of two unknown parameters when $S(t)$ is known

In [239], the method described above has been used for the simultaneous determination of two unknown parameters (K is a constant now). Consider the formulation given in (9.4.1)–(9.4.4) with $T_m = 0$ and $K(T) = K=$ constant, and $S(t)$ known. If $S(t)$ is known, then only one boundary condition is required at $x = S(t)$ and thus we have two extra conditions, namely, (9.4.4) and one of the conditions in (9.4.3). There could be six pairs of unknowns: (i) (K,ρ), (ii) (K,C), (iii) (K,l), (iv) (l,C), (v) (l,ρ), or (vi) (C,ρ). We present here the solution for only one pair (l,ρ) but the method of solution for other pairs remains the same. Let

$$S(t) = 2\sigma t^{1/2}, \ \sigma > 0, \ \text{and } \sigma \text{ is known}, \tag{9.4.22}$$

$$T(x,t) = T_0 - \frac{T_0}{f(\sigma/a)} f\left(x/2at^{1/2}\right), \ a^2 = K/\rho C, \ f(y) = \operatorname{erf}(y). \tag{9.4.23}$$

If $\xi = \sigma/a$, then ξ is the solution of

$$f(\xi) = \xi K T_0 / \left(q_0 \sigma \pi^{1/2}\right), \ \xi > 0, \tag{9.4.24}$$

and

$$\rho = K\xi^2 / \left(C\sigma^2\right), \ l = q_0 C \sigma \exp\left(-\xi^2\right) / \left(K\xi^2\right). \tag{9.4.25}$$

A unique solution of (9.4.24) exists if $KT_0/(2q_0\sigma) < 1$. It is not necessary to take q_0 to be known. If q_0 is unknown together with any one of the remaining parameters, even then a solution can be obtained.

9.4.2 Determination of unknown parameters in the two-phase Stefan problems

We consider the two-phase Neumann problem formulated in (1.3.1)–(1.3.7) and take $\rho_S = \rho_L$ and $T_m = 0$ and to match this problem with the problem studied in [243], take the region $0 \leq x < S(t)$ to be liquid and the region $x > S(t)$ to be solid. The overspecified boundary condition is given by

$$K_L \frac{\partial T_L}{\partial x}(0,t) = -q_0/t^{1/2}, \; q_0 > 0. \tag{9.4.26}$$

An exact analytical solution of the problem (1.3.1)–(1.3.7) has been given in (1.3.11) –(1.3.17). Along with $S(t)$ one more thermophysical parameter can be taken to be unknown as an extra condition (9.4.26) is available. The method of solution remains the same as explained in § 9.4.1 but now the necessary and sufficient conditions for the existence and uniqueness of the solution become fairly lengthy. We mention here only some observations about the nature of solutions. If $S(t)$ is unknown and any one of the six parameters ρ, l, C_S, C_L, K_S, or K_L is unknown, then the following results have been proved in [243].

(i) If $S(t)$ and ρ are unknown, then a unique solution of the Neumann-type exists for the two-phase problem.

(ii) If $S(t)$ is unknown and one of the remaining five parameters is unknown (ρ is excluded), then a unique solution of the Neumann-type is possible for the two-phase problem provided in each case a complementary condition (cf. [243]) is satisfied.

If $S(t)$ is known and q_0 in the overspecified condition (9.4.26) is known, then any two of the six thermophysical parameters can be taken as unknown. There will be fifteen such pairs. If ρ and K_L are unknown, then a unique solution of the Neumann-type exists. If (l, K_S), or (l, C_S), or (K_S, C_S) are unknown, then the free boundary problem has infinitely many solutions whenever some complementary conditions are satisfied. In the remaining eleven cases, unique solutions of the Neumann-type can be obtained provided some complementary conditions are satisfied. For complementary conditions, see [243].

If it is not possible to obtain similarity solutions of the type (1.3.11)–(1.3.13), the above method of finding unknown parameters will not work. Short-time analytical solutions based on series expansions of temperatures and the free boundary have been obtained when similarity solutions are not possible (cf. [244, 245]). It is difficult to prove the existence and uniqueness of short-time solutions of Stefan problems but these analytical solutions have been compared in some cases with the numerical solution [246]. Short-time analytical solutions can also be used to obtain approximate analytical solutions of inverse Stefan problems.

9.5 Regularization of Inverse Heat conduction Problems by Imposing Suitable Restrictions on the Solution

In many inverse problems of mathematical physics, an a-priori information about the smoothness of the solution stabilizes the problem. The inverse Stefan problems are generally studied as control problems in which for obtaining the regularized solutions, procedures different from those discussed in this section are adopted. These procedures will be discussed in §§ 9.6 and 9.7. To give some idea of the type of smoothness conditions to be imposed on the solution which may stabilize the problem, some heat conduction problems with and without phase-change are being discussed here. We first consider the following one-dimensional non-characteristic Cauchy problem in heat conduction which has been studied in [247].

$$T_t - a(x)T_{xx} - b(x)T_x - e(x)T = q(x,t),\ x \in (0,d),\ t \in I, \tag{9.5.1}$$

$$T(0,t) = \phi(t),\ t \in I, \tag{9.5.2}$$

$$T_x(0,t) = \psi(t),\ t \in I, \tag{9.5.3}$$

where, $I = R$ or $I = R^+$, and in the latter case, an initial condition

$$T(x,0) = g(x),\ x \in [0,d], \tag{9.5.4}$$

should be prescribed. It may be noted that a boundary condition is required at $x = d$ but instead of that an over specified boundary condition (9.5.3) is prescribed at $x = 0$. The inverse problem consists of obtaining $T(d,t)$. Instead of (9.5.3) and (9.5.4), we can consider

$$T_x(0,t) = 0, \text{ and } T(x,0) = 0, \tag{9.5.5}$$

and take $q(x,t) = 0$ in (9.5.1). This is possible by considering a suitable well-posed problem in the region $0 \le x < \infty$ with (9.5.3) and (9.5.4) and $q(x,t) \neq 0$ in (9.5.1) and subtracting its solution from the solution of the problem (9.5.1)–(9.5.4) which is considered in $0 \le x \le d$.

For further discussion, we shall consider the problem (9.5.1)–(9.5.4) in which we take

$$q(x,t) = 0,\ \psi(t) = 0, \text{ and } g(x) = 0. \tag{9.5.6}$$

The functions, a, b, c and ϕ are given and the continuous dependence of $T(x,t)$ on $\phi(t)$ is to be shown under suitable restrictions. The following smoothness conditions and other restrictions will be assumed.

$$a(x) \in W^{2,\infty}[0,d],\ b(x) \in W^{1,\infty}[0,d],\ e(x) \in L^{\infty}[0,d], \tag{9.5.7}$$

$$e(x) \le 0,\ \lambda \le a(x) \le \gamma,\ \gamma > 0;\ x \in [0,d]\,, \tag{9.5.8}$$

$$\phi \in L^2(R);\ f(t) = T(d,t) \in L^2(R). \tag{9.5.9}$$

In order to obtain stability estimates for $T(x,t)$, the problem is first formulated in terms of the Fourier transform of $T(x,t)$ with respect to time, denoted by $\hat{T}$, and the stability estimate for $\hat{T}$ are obtained (cf. [247]). The stability estimate for $T(x,t)$ which shows the exact Hölder type dependence of $\|T(x,t)\|_{L^2}$ on $\|\phi\|_{L^2}$ can then be obtained which is given by

$$\|T(x,t)\| \le M\,\|\phi\|^{1-A(x)/p}\left(\|\phi\|^{A(x)/p} + \|T(d,t)\|^{A(x)/p}\right), \tag{9.5.10}$$

$$A(x) = \int_0^x a(y)^{-1/2}dy,\ p = A(d). \tag{9.5.11}$$

The $L^2(R)$-norm has been considered in (9.5.10) and M is a suitable constant depending on $\lambda,\ \gamma,\ d$ and the norms of other coefficients.

In the problem considered in [248], $b = e = q = 0$ in (9.5.1), $1 \le a(x) \le \nu,\ \nu > 0,\ a \in L^\infty[0,d],\ 0 \le t \le t_1 < \infty,\ \phi \in C^0[0,t_1],\ \|\phi\|_{L^\infty(0,t_1)} \le \varepsilon$. Further, $T(x,t)$ satisfies an a-priori bound

$$\|T(x,t)\|_{L^2(R)} \le \sqrt{dt_1}E,\ \varepsilon \le E. \tag{9.5.12}$$

Under these assumptions, by considering a weak formulation of the problem, the stability estimates for $T(x,t)$ have been obtained which are of Hölder type in the interior and of logarithmic type at the boundary which are given below. The continuous dependence on the Cauchy data and the coefficient $a(x)$ has also been considered.

$$|T(x,t)| + d\,|T_x(x,t)| \le \beta(x,t)\varepsilon^{\gamma(x,t)}E^{1-\gamma(x,t)},\ 0 \le x < d,\ 0 \le t < t_1, \tag{9.5.13}$$

$$\gamma(x,t) = \exp\left(\frac{-pt_1/t}{1-x/d}\,\frac{x}{d}\right), \tag{9.5.14}$$

$$\beta(x,t) = \left(\frac{pt_1/t}{(1-x/d)}\right)^{p((t_1/t)/(1-x/d))}. \tag{9.5.15}$$

Here, p is a computable constant that depends only on ν and t_1/d^2. For the stability estimate at the boundary, we further require that

$$\max_{t\in[0,t_*]}\left(d\int_0^d T_x^2(x,t)\,dx\right) \le E^2. \tag{9.5.16}$$

For $t \in (0, t_1)$,

$$|T(d,t)| = E \cdot \frac{t_1}{t} o\left(\left(\log\left(\log\frac{E+\varepsilon}{\varepsilon}\right)\right)^{-1/2}\right), \text{ as } \varepsilon/E \to 0. \tag{9.5.17}$$

The following problem of determining an unknown source control $q = q(t)$ has been considered in [249].

$$T_t = a\,(x,t,T,T_x)_x + q(t)T + F(x,t,T,T_x,q(t)), \text{ in } Q_{t_*};$$

$$Q_{t_*} = \{(x,t) : 0 < x < 1,\ 0 < t < t_*\}, \tag{9.5.18}$$

$$T(x,0) = \phi\,(x) \geq 0,\ 0 < x < 1, \tag{9.5.19}$$

$$T(0,t) = f(t) \geq 0, \text{ and } T\,(1,t) = g(t) \geq 0;\ 0 < t < t_*, \tag{9.5.20}$$

$$\int_0^{S(t)} \Phi(x,t)T(x,t)dx = G(t) > 0,\ 0 < t < t_*,\ 0 < S\,(t) \leq 1. \tag{9.5.21}$$

The functions a, $F\,(\geq 0)$, ϕ, f, g, S, $\Phi\,(> 0)$ and G are known. The functions a and F are smooth functions of their arguments. (T,q) is called a solution if there exists an α, $0 < \alpha < 1$, such that $T \in C^{1+\alpha}\left(\bar{Q}_{t_*}\right) \cap C^{2+\alpha}\left(Q_{t_*}\right)$, $q(t) \in C^{\alpha/2}\,[0,t_*]$ and the pair (T,q) satisfies (9.5.18)–(9.5.21). The existence, uniqueness, and continuous dependence of the solution on the data has been shown with the help of some a-priori estimates, compactness arguments, and the strong maximum principle. For the conditions which the data has to satisfy, see [249]. Some problems of recovering a source term or a nonlinear coefficient in the inverse problems of parabolic type have been discussed in [250].

One typical structural restriction for the one-phase melting Stefan Problems is the non-negativity of the solution [251] but the most commonly investigated a priori information concerns norm bounds. In the latter case if the problem is linear, then the stability estimates can be derived by estimating the size of solutions for the data fulfilling such norm bounds. There are two major techniques to obtain stability estimates: (1) complex variable methods [252], and (2) the Fourier transform technique with its own limitations such as in this technique the domain should be cylindrical and the time interval should be infinite.

By using an extension of the complex variable technique, the stability analysis of a one-dimensional one-phase inverse Stefan problem has been done in [253]. The region considered is $D_{t_*} = \{(x,t); 0 < x < S(t),\ 0 < t \leq t_*\}$. The free boundary $x = S(t)$ is assumed to be known and is Lipschitz continuous. The non-characteristic Cauchy problem considered is as follows:

$$\left.\begin{aligned} &T_{xx} - T_t = q(x,t) \text{ in } D_{t_*};\ T(x,0) = g(x), \\ &T(S(t),t) = f_1(t);\ T_x(S(t),t) = f_2(t),\ S(0) = b. \end{aligned}\right\} \tag{9.5.22}$$

The inverse problem consists of determining $T_x(0,t)$ which is assumed to be bounded. The interior estimates of non-uniform Hölder type as well as uniform estimates of logarithmic type have been obtained for the temperature and its gradient under suitable assumptions on the data.

9.6 Regularization of Inverse Stefan Problems Formulated as Equations in the form of Convolution Integrals

The regularization of a one-dimensional one-phase inverse problem concerning melting of ice has been considered in [254]. The formulation of the Stefan problem is as follows:

$$T_{xx} - T_t = 0, \text{ in } D_{t_*} = \{(x,t) : 0 < x \leq S(t),\ 0 < t < t_*\}, \tag{9.6.1}$$

$$T(S(t),t) = 0, \text{ and } T_x(S(t),t) = -\mu \dot{S}(t);\ 0 < t < t_*, \tag{9.6.2}$$

$$T(0,t) = v(t) \geq 0,\ 0 < t < t_*;\ T(x,0) = T_0(x) \geq 0,\ 0 \leq x \leq b, \tag{9.6.3}$$

$$S(0) = b,\ b > 0. \tag{9.6.4}$$

The region $S(t) < x < \infty$ is occupied by ice cold water at the melting temperature zero. For simplicity, some of the thermophysical parameters have been taken to be unity. In (9.6.1)–(9.6.4), $S(t)$ is a given non-decreasing C^1-function and $T_0(x)$ is a given C^1-function with a bounded derivative. The problem is to obtain a regularized solution for $v(t)$ satisfying (9.6.1)–(9.6.4). One of the conditions in (9.6.2) is an overspecified condition as $S(t)$ is known.

Using standard methods (see [9]), an integral equation can be obtained to determine $v(t)$ (see (9.6.9)) which can be studied by Tikhonov regularization but it is difficult to obtain error estimates in this way. Therefore, an equation in the form of *convolution integral* is obtained as follows. Consider the identity

$$\frac{\partial}{\partial \xi}\left(G\frac{\partial T}{\partial \xi} - T\frac{\partial G}{\partial \xi}\right) - \frac{\partial}{\partial \tau}(TG) \equiv 0, \tag{9.6.5}$$

where

$$G(x,t;\xi,\tau) = Q(x,t;\xi,\tau) - Q(-x,t;\xi,\tau), \tag{9.6.6}$$

and

$$Q(x,t;\xi,\tau) = \frac{1}{2\sqrt{\pi(t-\tau)}} \exp\left(-\frac{(x-\xi)^2}{4(t-\tau)}\right). \tag{9.6.7}$$

On integrating (9.6.5) over the domain $0 < \varepsilon < \tau < t - \varepsilon,\ 0 < \xi < S(\tau)$ and letting $\varepsilon \to 0$, after some rearrangement, we get

$$T(x,t) = \int_0^b T_0(\xi) G(x,t;\xi,0) d\xi + \int_0^t v(\tau) \frac{\partial G}{\partial \xi}(x,t;0,\tau) d\tau$$
$$+ \int_0^t \frac{\partial T}{\partial \xi}(S(\tau),\tau)\, G(x,t;S(\tau),\tau) d\tau. \tag{9.6.8}$$

From the condition $T(S(t),t) = 0$, we get

$$\frac{1}{2\sqrt{\pi}} \int_0^t \frac{S(t)}{(t-\tau)^{\frac{3}{2}}} \exp\left(-\frac{S^2(t)}{4(t-\tau)}\right) v(\tau) d\tau = -\int_0^b G(S(t),t;\xi,0) T_0(\xi) d\xi$$
$$+\mu \int_0^t G(S(t),t;S(\tau),\tau) \dot{S}(\tau) d\tau. \tag{9.6.9}$$

Equation (9.6.9) is a linear Volterra integral equation of the first kind which can be transformed into a convolution integral. Using a lemma given in ([9], pp.217), (9.6.9) can be converted into an equivalent equation of the form

$$\frac{m}{2\sqrt{\pi}} \int_0^t (t-\tau)^{\frac{3}{2}} \exp\left(-\frac{m^2}{4(t-\tau)}\right) v(\tau) d\tau = \int_0^t Q_\xi(m,t;S(\tau),\tau) U_0(\tau) d\tau$$
$$- \int_0^t Q(m,t;S(\tau),\tau) \left\{ U_0(\tau) \dot{S}(\tau) + U_1(\tau) \right\} d\tau, \tag{9.6.10}$$

where, m is some real number $> S(t),\ \forall\, t > 0$,

$$U_0(t) = \lim_{x \to S(t)-0} g(x,t), \text{ and } U_1(t) = \lim_{x \to S(t)-0} g_x(x,t), \tag{9.6.11}$$

$$g(x,t) = -\int_0^b G(x,t;\xi,0) T_0(\xi) d\xi + \mu \int_0^t G(x,t;S(\tau),\tau) \dot{S}(t) d\tau + T(x,t),$$
$$0 < x < S(t),\ t > 0. \tag{9.6.12}$$

The equation (9.6.10) can be written as

$$(z * v)(t) = (2\pi)^{-1/2} \int_{-\infty}^{\infty} z(t-\tau) v(\tau) d\tau = F(t),\ \forall\, t \text{ in } R, \tag{9.6.13}$$

where

$$z(t) = \begin{cases} \sqrt{\pi} t^{-3/2} \exp(-m^2/4t), & t > 0, \\ 0, & t \leq 0, \end{cases} \tag{9.6.14}$$

and $F(t)$ is the r.h.s. of (9.6.10) multiplied by $(2/m)(2\pi)^{-1/2}$. A family of regularized solutions $\{v_\varepsilon\}$, $0 < \varepsilon < 1$, can be constructed in which v_ε is stable with respect to variations in the function $F(t)$.

The following proposition describes the regularized solution and gives an estimate of the error involved in it.

Proposition 9.6.1. Suppose the exact solution v_0 of (9.6.13) corresponding to F_0 (given) is in $H^1(R) \cap L^1(R)$ and $\|F - F_0\| < \varepsilon$. There exists a regularized solution v_ε of (9.6.13) which is given by

$$v_\varepsilon(t) = (2\pi)^{-1/2} \int_{-\infty}^{\infty} \Psi(x) e^{itx} dx, \tag{9.6.15}$$

where

$$\Psi(x) = \overline{\hat{z}(x)}.\ \hat{F}(x) / \left(\varepsilon + \|\hat{z}(x)\|^2\right), \tag{9.6.16}$$

and, $\hat{z}(x)$ and $\hat{F}(x)$ stand for the Fourier transforms (cf. (9.6.15)) of z and F, respectively. The error in v is given by

$$\|v_\varepsilon - v_0\| \leq \alpha \left(\ln(1/\varepsilon)\right)^{-1}, \tag{9.6.17}$$

where α is any constant $\geq m\sqrt{\pi}(3 + 2m).\max\left(\left(\|\hat{v}_0\|^2 + 1\right)^{1/2}, \|v_0'\| + M\right)$, with $M = (8/m^3)(2/\pi)^{1/2} e^{-2}$. Here, $\|\cdot\|$ stands for the $L^2(R)$-norm. Furthermore, if $\hat{v}_0 / |\hat{z}| \in L^2(R)$ then we have

$$\|v_\varepsilon - v_0\| \leq \beta\sqrt{\varepsilon}, \tag{9.6.18}$$

where β is any constant $\geq 1 + \|\hat{v}_0/\hat{z}\|$ and $\hat{z}(t)\hat{v}_0(t) = \hat{F}_0(t),\ t \in R$.

The proof of the Proposition 9.6.1. depends on obtaining suitable estimates of

$$\|\hat{v}_\varepsilon - \hat{v}_0\|_{L^2}^2,\ \|v_\varepsilon - v_0\|_{L^2}^2 \text{ and } \varepsilon \|t(\hat{v}_\varepsilon - \hat{v}_0)\|_{L^2}^2 + \|t\hat{z}(\hat{v}_\varepsilon - \hat{v}_0)\|_{L^2}^2. \tag{9.6.19}$$

A numerical example has been considered in which $F(t) = F_0(t) + \varepsilon e^{-t/2}$ and $S(t)$ and $T_0(x)$ are known. Regularized numerical solutions using finite-difference discretization have been obtained for different values of ε and the numerical results indicate convergence as $\varepsilon \to 0$.

A two-phase one-dimensional inverse Stefan problem has been considered in [255]. If some minor changes are incorporated in the Neumann problem (1.3.1)–(1.3.7), then the problem considered in [255] can be obtained. The changes are as follows:

(i) The region $0 \leq x \leq S(0) = b$ is initially occupied by warm liquid and the region $b \leq x \leq d$ is initially occupied by ice and the Stefan problem is concerned with the melting of ice. (ii) $T_L(x, 0) = T_1(x)$, $0 \leq x \leq b$, and $T_S(x, 0) = T_2(x)$, $b \leq x \leq d$. (iii)

$T_L(0,t) = v(t)$, and $\partial T_S/\partial x = 0$, at $x = d$. (iv) $T_m = 0$. (v) Initial and boundary temperatures satisfy compatibility conditions at $x = b$ and d. For a given $S(t)$, the problem is to determine $v(t)$ satisfying all other equations of the problem. With the help of suitable Green's functions, the problem is first reduced to a system of integral equations by obtaining temperatures $T_L(x,t)$ and $T_S(x,t)$ in the form of equations similar to (9.6.8). The Green's functions $G(x,t;\xi,\tau)$ for the liquid and $W(x,t;\xi,\tau)$ for the solid are given below.

$$G(x,t;\xi,\tau) = E(x-\xi, k_L^2(t-\tau)) - E(x+\xi, k_L^2(t-\tau)), \tag{9.6.20}$$

$$W(x,t;\xi,\tau) = E(x-\xi, k_S^2(t-\tau)) - E(x+\xi-2a, k_S^2(t-\tau)), \tag{9.6.21}$$

$$\left.\begin{aligned} E &= \tfrac{1}{2\sqrt{\pi t}} \exp\left(-x^2/4t\right), \; t>0, \; x \in R, \\ &= 0 \qquad\qquad\qquad\quad , \; t \le 0, \; x \in R. \end{aligned}\right\} \tag{9.6.22}$$

The three boundary conditions at $x = S(t)$, give rise to three integral equations involving $v(t)$, $(\partial T_L/\partial x)(S(t),t)$ and $(\partial T_S/\partial x)(S(t),t)$. A crucial step in the regularizing method for this problem is to convert the integral equation for $(\partial T_S/\partial x)(S(t),t)$ into a linear Volterra integral equation of the second kind, and the integral equation for $v(t)$ into an equation in the form of a convolution integral. The integral equation for $v(t)$ contains $(\partial T_L/\partial x)(S(t),t)$. If the solution of the Volterra integral equation of the second kind is known or in other words $(\partial T_S/\partial x)(S(t),t)$ is assumed to be known (note that $S(t)$ is known so we shall be solving a heat conduction problem in the solid region without phase-change), then with the help of the Stefan condition, $(\partial T_L/\partial x)(S(t),t)$ can be determined which is then substituted in the integral equation for $v(t)$. For the details of obtaining convolution integral, see [255]. In (9.6.10), $S(t)$ was replaced by some constant m, $m > S(t)$. In the present two-phase problem, since $S(t) < d$, $\forall\, t > 0$, we can replace $S(t)$ in the convolution integral by the constant d in the convolution integral. A family of regularized solutions $\{v_\varepsilon\}_{\varepsilon>0}$ has been obtained for the present two-phase problem and a proposition similar to Proposition 9.6.1., with some changes, has been proved by obtaining estimates of various expressions. A numerical example has been considered and the regularized numerical solutions have been obtained for different value of ε by the finite-difference discretization of integrals.

A family of regularized solutions has been obtained in [256] for the following one-phase two-dimensional inverse Stefan problem.

$$T_{xx} + T_{yy} - T_t = 0, \; x \in R, \; 0 < y < S(x,t), \; t > 0, \tag{9.6.23}$$

$$T(x, S(x,t), t) = 0, \; x \in R, \; t > 0, \tag{9.6.24}$$

$$T(x,0,t) = v(x,t) \ge 0, \; x \in R, \; t > 0, \tag{9.6.25}$$

$$\frac{\partial T}{\partial n}(x,S(x,t),t)=-\frac{\partial T}{\partial x}(x,S(x,t),t)S_x+\frac{\partial T}{\partial y}(x,S(x,t),t)=\frac{\partial S}{\partial t}, \tag{9.6.26}$$

$$S(x,0)=b\,(x)>0,\ x\in R, \tag{9.6.27}$$

$$T(x,y,0)=T_0(x,y)\geq 0,\ x\in R,\ 0<y<b\,(x)\,. \tag{9.6.28}$$

Here, $y = S(x,t)$, $x \in R$, is the equation of the phase-change boundary which is a known smooth function, and $\vec{n}$ stands for the unit outward normal on the interface. The region $y > S(x,t)$ is at the melting temperature zero. $T_0(x,y)$ is also a known smooth function. The problem is to determine $v(x,t)$. The method of finding a family of regularized solutions depends in obtaining an equation in the form of a convolution integral. By integrating the identity

$$\operatorname{div}\left(T\nabla G-G\nabla T\right)=-\frac{\partial}{\partial\tau}\left(TG\right), \tag{9.6.29}$$

where

$$G(x,y,t;\xi,\eta,\tau)=W(x,y,t;\xi,\eta,\tau)-W(x,-y,t;\xi,\eta,\tau), \tag{9.6.30}$$

and

$$W(x,y,t;\xi,\eta,\tau)=\frac{1}{4\pi\,(t-\tau)}\exp\left(-\frac{(x-\xi)^2+(y-\eta)^2}{4\,(t-\tau)}\right), \tag{9.6.31}$$

over the domain $-\beta<\xi<\beta,\ 0<\eta<S(\xi,\tau)\,,\ 1/\beta<\tau<t-1/\beta$ and taking the limit as $\beta\to\infty$, $T(x,y,t)$ can be obtained in the form of an equation similar to (9.6.8).

$$\begin{aligned}T(x,y,t)=&\int_0^t\int_{-\infty}^{\infty}\frac{y}{(t-\tau)}W(x,y,t;\xi,0,\tau)v(\xi,\tau)d\xi d\tau\\&+\int_{-\infty}^{\infty}\int_0^{b(\xi)}T_0(\xi,\eta)G(x,y,t;\xi,\eta,0)d\eta d\xi\\&+\int_{-\infty}^{\infty}\int_0^{t}G(x,y,t;\xi,S(\xi,\tau),\tau)\frac{\partial S}{\partial\tau}(\xi,\tau)\,d\tau d\xi,\ x\in R,\ t>0,\ 0<y<S(x,t).\end{aligned} \tag{9.6.32}$$

On taking the limit $y\to S(x,t)-0$ in (9.6.32), we get an integral equation to determine $v(x,t)$. For the details of obtaining an equation in the form of a convolution integral, the reader is referred to [256]. A proposition similar to Proposition 9.6.1. defining the regularized solutions and giving error estimate can be proved for the two-dimensional problem also provided some changes are made in the assumptions made in the Proposition 9.6.1. For example, the norm to be considered is $L^2(R^2)$-norm, and the Fourier transform in the place (9.6.15) will now be a two-dimensional Fourier transform defined as:

$$\hat{v}(x,t)=\frac{1}{2\pi}\int_{-\infty}^{\infty}\int_{-\infty}^{\infty}\Psi(\omega,\eta)e^{i(x\omega+t\eta)}d\omega d\eta, \tag{9.6.33}$$

$$\Psi(\omega,\eta) = \overline{\hat{z}(\omega,\eta)}\ \hat{F}(\omega,\eta)/\left(\varepsilon + \|\hat{z}(\omega,\eta)\|^2\right), \tag{9.6.34}$$

$$z(x,t) = \frac{1}{t^2} exp\left(-\frac{x^2+m^2}{4t}\right), \quad m > S(x,t),\ m \in R^+,\ t > 0. \tag{9.6.35}$$

For some other changes in Proposition 9.6.1. which are required for proving the results in the present problem, see [256].

A numerical example has been considered in which $F(x,t) = F_0(x,t)+\varepsilon e^{-|x|-t}$ ($F(x,t)$ is the r.h.s. of the convolution integral in this problem), $S(x,t) = \arctan(x^2+t+1)$, $b(x) = \arctan(x^2+1)$, $T_0(x,y) = |\cos 2x|.(b(x)-y)^2$ and $m = 2$. The numerical method consists of finite-difference discretization of integrals. Regularized solutions for different values of ε have been obtained. The numerical solutions for different ε indicate convergence as $\varepsilon \to 0$.

The regularization of a two-dimensional two-phase inverse Stefan problem in the region $-\infty < x < \infty$, $0 \le y \le a$ has been considered in [257]. The region $0 \le y < S(x,t)$, is occupied by warm water and the region $S(t) < y \le a$ consists of ice. The initial temperatures in the two regions are functions of x and y. The phase-change interface is given by $y = S(x,t)$ and $S(x,0) = b(x)$, $x \in R$. Initially the warm water occupies the region $0 \le y < b(x)$, $x \in R$. The method of obtaining a family of regularized solutions is similar to that considered in [254, 255, 256]. A numerical example has also been considered and regularized solutions have been obtained for different values of ε. The convergence is indicated as $\varepsilon \to 0$.

9.7 Inverse Stefan Problems Formulated as Defect Minimization Problems

A stable and regular propagation of the free boundary in the classical Stefan problem requires application of controls. An application of controls could be through thermo-physical parameters, geometry of the mold, initial temperature or through boundary conditions. In the control of the free boundary, which is our interest in this section, there exist mainly two approaches. The first is only to search for a control which generates a solution to the corresponding direct Stefan problem with a free boundary that is close to the prescribed one. It is assumed in this approach that the solution of the direct problem exists. This approach may yield an approximation but it will not be a good solution as it is well known that the inverse Stefan problems are ill-posed. In the second approach the aim is to regularize the problem in some way and then obtain an approximate solution. If the direct Stefan problem is well-posed, then as a consequence of the stability of the regularized solution, this second approach includes the first one. Both the approaches will be discussed here briefly with reference to some specific Stefan problems.

We consider the following one-dimensional one-phase melting problem which will be studied with the help of the two approaches described above.

$$T_{xx}(x,t) - T_t(x,t) = q(x,t), \text{ in } D_{t_*}(S) = \{(x,t) : 0 < x < S(t), 0 < t < t_*\}, \quad (9.7.1)$$

$$T(x,0) = T_0(x), \ 0 \le x \le b = S(0), \quad (9.7.2)$$

$$\left.\begin{aligned} &(A_i T)(t) = \phi(t), \ 0 < t < t_*, \ i = 1,2, \\ &(A_1 T)(t) = T(0,t), \text{ and } (A_2 T)(t) = \eta T(0,t) - T_x(0,t), \end{aligned}\right\} \quad (9.7.3)$$

$$T(S(t),t) = r_1(S(t),t), \ 0 < t < t_*, \quad (9.7.4)$$

$$T_x(S(t),t) = r_2(S(t),\dot{S}(t),t), \ 0 < t < t_*. \quad (9.7.5)$$

If $T(x,t)$ and $S(t)$ are unknowns to be determined, then we have a direct Stefan problem to be solved and it is well known that problem (9.7.1)–(9.7.5) is well-posed under sign restrictions and smoothness of the data (cf. [59]).

In the above inverse problem, $q(x,t)$, $T_0(x)$, b, r_1, r_2 and $S(t)$ are known and $\phi(t)$ is the control which is to be determined in such a way that for $\phi = \phi^*$, the exact solution of (9.7.1)–(9.7.5) gives exactly prescribed $S(t) = S^*(t)$. Let Q be a free boundary operator defined as

$$Q : \{\phi \in C[0,t_*] : \phi \ge 0\} \to C[0,t_*], \ \phi : \to S; \ Q\phi^* = S^*. \quad (9.7.6)$$

For a given $\phi(t)$, $S(t)$ and $T(x,t)$ can be obtained and the direct problem (9.7.1)–(9.7.5) is solved. Even if ϕ^* exists, it is generally difficult to obtain it exactly. An approximation $\hat{\phi}$ of ϕ^* can be possibly obtained such that $Q\hat{\phi}$ is 'close to' S^*. This closeness to S^* is generally expressed in terms of a norm which involves a function ϕ belonging to some suitable space of functions and which minimizes $\|Q\phi - S^*\|$ or in words we choose a ϕ which minimizes the defect. The questions related to the existence, uniqueness and stability have to be addressed. This approach can be called a 'direct approach' which has been used in [258, 259, 260]. The operator Q is in general a nonlinear operator. The defect minimization problem corresponding to (9.7.1)–(9.7.5) with $r_1 = 0$ and r_2 as given in (9.7.8) has been studied in [261] with the help of a linear auxiliary problem. This auxiliary problem can be stated in terms of a linear operator F (defined in (9.7.14)) such that

$$Q(\phi^*) = S^* \Leftrightarrow F\phi^* = r(S^*), \quad (9.7.7)$$

$$r = r_2 = -\lambda\dot{S}(t) + \mu(t), \ \mu \ge 0. \quad (9.7.8)$$

To obtain the solidification problem studied in [258], we take $q(x,t) = 0$, $r_1 = 0$, $r_2 = -\dot{S}(t)$, $T_x = \phi(t)$ at $x = 0$ and $\phi \le 0$, in (9.7.1)–(9.7.5). For this inverse problem, existence of a ϕ which gives S^* has not been proved in the literature. Therefore, in [258] the inverse problem has been formulated as a nonlinear approximation problem which does not require the existence of the solution to be known and which can be solved by a highly stable iterative Newton-like procedure developed in [262]. The nonlinear approximation problem can be described as follows:

Find $\hat{\phi} \in W$ such that

$$\|Q\hat{\phi} - S^*\| = \inf\{\|Q\phi - S^*\|;\ \phi \in W,\ \phi \leq 0\}. \tag{9.7.9}$$

Here, W represents the set of admissible controls and $\|\cdot\|$ is some norm in $C[0, t_*]$. The operator Q is the same as in (9.7.6). It is difficult to prove the existence of an optimal control if W is an infinite-dimensional space. Therefore, a finite-dimensional subspace V of $C[0, t_*]$ has been considered and let $W = \{\phi \in V :\ \phi \leq 0\}$. The set W is the maximal subset of V for which the existence of a solution of the present direct Stefan problem is known under the conditions that

$$T_0(x) \in C^1[0, b],\ T_0(x) \geq 0,\ T_0(b) = 0. \tag{9.7.10}$$

By virtue of the fact that a unique solution (T, S) of the direct Stefan problem exists under the above conditions, the solution operator $Q : W \to C[0, t_*],\ \phi :\to S(t)$ is Lipschitz continuous, and $S(t)$ depends monotonically on ϕ and is a monotonically nondecreasing fu-
nction (cf. [258]). By considering the uniform norm $\|S\| = \sup_{0 \leq t \leq t_*} |S(t)|$ in W, the existe-
nce of an optimal solution $\hat{\phi} \in W$ has been proved.

Let V_n be an ascending sequence of subspaces, i.e., $V_n \subset V_{n+1},\ n = 1, 2, ..,$ and the union of V_n be dense in $C[0, t_*]$. Let W_n be the cone $\{\phi \in V_n : \phi \leq 0\}$ and

$$\rho_n(S^*) = \inf_{\phi \in W_n} \|Q(\phi) - S^*\|. \tag{9.7.11}$$

Because of the continuity of Q, ρ_n converges to zero as $n \to \infty$ but it is difficult to prove that the sequence $\{\hat{\phi}_n\}$ of optimal controls ($\hat{\phi}_n \in W_n$) converges to the solution ϕ^* of the inverse problem.

If B is the Banach space $C[0, t_*]$ with the uniform norm

$$\|\phi\|_\infty := \sup\{|\phi(t)| : 0 \leq t < t_*\}, \tag{9.7.12}$$

and A is the cone $A : \{\phi \in C[0, t_*] : \phi < 0\}$, then it has been proved in [259] that the operator $Q : A \subset B \to B$ is Fréchet differentiable. This property was found useful in obtaining the numerical solution of a one-dimensional inverse problem in [259] by the generalized Gauss-Newton method. Numerical solution of a two-dimensional inverse Stefan problem has been considered in [260].

The problem considered in [261] is the inverse problem (9.7.1)–(9.7.5) with $r_1 = 0$ and r_2 given in (9.7.8). A melting problem has been considered. $S(t)$ is known and the control $\phi \geq 0$ is to be determined. For the defect minimization, instead of the operator Q, a linear operator F (cf. (9.7.14)) has been considered. Let $S^* \in C^1[0, t_*]$, $S^* > 0$, $b = S^*(0)$, $T_0 \geq 0$, $T_0(b) = 0$, $q \leq 0$, and $\mu \geq 0$ (the last condition can be relaxed). For other regularity conditions on the data, see [261]. As mentioned earlier, for numerical computations, one has to deal with finite-dimensional spaces. Therefore the defect is minimized in the space $X_n \cap A_n$ where X_n is an n-dimensional subspace of $C[0, t_*]$ and $A_n \subset \{\phi : \phi \geq 0\}$, $\phi \in C[0, t_*]$. The minimization problem can be stated as follows:

$$\text{Minimize } \|F\phi - r\|_{L^p}, \text{ for } \phi \in X_n \cap A_n, \ 1 < p < \infty, \tag{9.7.13}$$

$$F : C[0, t_*] \to C[0, t_*], \ \phi \to T_x(S^*(t), t), \tag{9.7.14}$$

where

$$T_x(S^*(t), t) = \left.\begin{matrix} \lim\limits_{x \to S(t)-0} T_x(x, t) & , & t > 0, \\ \lim\limits_{t \to 0+} T_x(S^*(t), t), & & t = 0. \end{matrix}\right\} \tag{9.7.15}$$

$T(x, t)$ solves the problem (9.7.1)–(9.7.5) in the following sense:

$$T \in C(\overline{D_{t_*}(S^*)}); \ T_{xx}, \ T_t \in C(D_{t_*}(S^*)),$$

$T_x(x, t)$ is continuous in (x, t) for $t \in [0, t_*]$, $x \in (0, S^*(t))$ for $i = 1$ and $x \in [0, S^*(t))$ for $i = 2$ (see equation (9.7.3) for $i = 1, 2$) and (9.7.1)–(9.7.4) are satisfied pointwise. Note that the isotherm condition (9.7.4) ($r_1 = 0$) is satisfied exactly in this problem and (9.7.5) is to be satisfied in the sense of (9.7.13).

Some of the results established in [261] are stated below.

1. F in (9.7.14) is well-defined and continuous.

2. If $X_n \cap A_n$ is closed in X_n and not void, then the minimization problem has a solution ϕ_n which is unique if $1 < p < \infty$.

3. For $t \in [0, t_*]$, we have

$$|(Q\phi)(t) - S^*(t)| \leq P(t) \int_0^t |(F\phi)(\tau) - r(\tau)|\, d\tau. \tag{9.7.16}$$

In (9.7.16), $P(t)$ depends on max $q(x, t)$, $\max\left\{(\mu(t))_+, (\mu(t) + (F\phi)(t) - r(t))_+\right\}$, λ and b. This result justifies the choice of the linear auxiliary problem.

4. If $\phi_n \geq 0$, then for $t \in [0, t_*]$, we have

$$|(Q\phi_n)(t) - S^*(t)| \leq P(t) t^{1-1/p} I_n, \tag{9.7.17}$$

where I_n is the minimum of $\|F\phi_n - r\|_{L^p}$; $\phi_n \in X_n \cap A_n$, $1 < p < \infty$. It has been shown that $I_n \to 0$ as $n \to \infty$. To obtain a regularized solution, a numerical method using polynomial splines has been applied and its order of convergence has been discussed. A numerical example has been considered.

The inverse Stefan problem considered in [263] can be obtained if the following changes are made in the Stefan problem (9.7.1)–(9.7.5).

(1) The condition (9.7.3) is replaced by the condition

$$T_x(0,t) \leq 0, \; 0 < t < t_*, \tag{9.7.18}$$

and $q(x,t) = 0$, in (9.7.1).

(2) $S(t),\ T_0(x),\ r_1$ and r_2 are specified and $T(x,t)$ is the control. (9.7.19)

The above control problem with (9.7.18) and (9.7.19) can be formulated as an operator equation as follows:

Let X be the space defined as:

$$X = \left\{T(x,t) \in C^{2,1}(D_{t_*}) \cap C^{1,0}(\overline{D}_{t_*}) \,\middle|\, \mathcal{L}T = T_{xx} - T_t = 0 \text{ in } D_{t_*} = D_{t_*}(S)\right\}. \tag{9.7.20}$$

Let the space X be equipped with the norm

$$\|T\|_X = \max\left\{\|T\|_{\overline{D}_{t_*}}, \|T_x\|_{\overline{D}_{t_*}}\right\}, \tag{9.7.21}$$

where, $\|\cdot\|_{\overline{D}_{t_*}}$ is the supremum norm on $\overline{D}_{t_*}$. It can be proved that X is complete in this norm and X is a Banach space. For $T \in X$, we define an operator $\hat{Q}$ as:

$$\hat{Q}T = \{T(S(t),t),\ T_x(S(t),t),\ T(x,0)\},\ T \in X. \tag{9.7.22}$$

The initial and boundary values are given by the trace operator

$$\hat{Q} : X \to C[0,t_*] \times C[0,t_*] \times C[0,b], \tag{9.7.23}$$

and $C[0,t_*] \times C[0,t_*] \times C[0,b]$ is equipped with the product norm

$$\|(\rho_1, \rho_2, T_0)\| = \max\left\{\|\rho_1\|_{[0,t_*]}, \|\rho_2\|_{[0,t_*]}, \|T_0\|_{[0,b]}\right\}. \tag{9.7.24}$$

Let A and P be defined as:

$$A = \{T(x,t) \in X : T_x \leq 0,\ 0 < t < t_*\}, \tag{9.7.25}$$

$$P = \left\{r_1(S(t),t),\ r_2(\dot{S}(t), S(t), t),\ T_0(x)\right\}. \tag{9.7.26}$$

It can be shown that the set A is closed, convex and nonempty. The inverse problem can be stated as:

$$\text{Find } T(x,t) \in A \text{ such that } \hat{Q}T = P. \tag{9.7.27}$$

As discussed earlier, one of the methods to obtain an approximate numerical solution of the problem stated in (9.7.27) is to minimize the defect over a finite-dimensional space, i.e.,

$$\text{minimize } \|\hat{Q}T - P\| \text{ on } X_n \cap A, \; n \in N. \tag{9.7.28}$$

Here, $X_n \subset X$ and $X_n = \text{span}\{v_1, v_2, \ldots\ldots, v_n\}$; $\{v_n\} \subset X$ is a complete family of functions in X, i.e., $\bigcup_{n \in N} X_n$ is dense in X. The problem (9.7.28) has always a solution as it is a finite-dimensional linear approximation problem of Chebyshev-type [264]. The following proposition has been proved in [263].

Proposition 9.7.1. Let $T^* \in A$ be a solution of $\hat{Q}T = P$. Then T^* is unique and

$$I_n^* \leq \inf\{\|T - T^*\|_X : \; T \in X_n \cap A\}, \tag{9.7.29}$$

where

$$I_n^* = \inf\left\{\|\hat{Q}T - P\| : \; T \in X_n \cap A\right\}, \; n \in N. \tag{9.7.30}$$

$I_n^* \to 0$, as $n \to \infty$.

The temperature giving the infimum is not a regularized solution. The problem stated in (9.7.28) has been regularized in [263] by forcing the finite-dimensional solution to lie in a compact set. For achieving this, we make the following assumptions.

(A1) There exists a unique solution $T^* \in A$ of $\hat{Q}T = P$ such that $T_x^*(0,t) \in C^1[0,t_*]$. Let M be a known constant such that $\|T_{xt}^*(0,t)\|_{[0,t_*]} < M$.

(A2) $\bigcup_{n \in N} X_n$ is dense in X, and $Y = \{T \in X : T_x(0,t) \in C^1[0,t_*]\}$ be a subset of X with the norm

$$\|T\|_Y = \max\left\{\|T\|_X, \; \|T_{xt}(0,t)\|_{[0,t_*]}\right\}. \tag{9.7.31}$$

Let

$$A_M = \left\{T \in Y \cap A : \|T_{xt}(0,t)\|_{[0,t_*]} \leq M, \; \left|T_x(0,0) - T_0'(0)\right| \leq M\right\}. \tag{9.7.32}$$

If $b = 0$, then no initial temperature is to be prescribed and the second condition in (9.7.32) will not be there. Under the above assumptions, A_M is convex and closed and has a non-empty interior in Y. A family of regularized solutions can be obtained by considering the problem:

$$\text{Minimize } \|\hat{Q}T - P\| \text{ on } X_n \cap A_M, \; n \in N. \tag{9.7.33}$$

Let $\hat{T}_n$ be any solution of (9.7.33) which exists as (9.7.33) is a finite-dimensional linear approximation problem of Chebyshev-type with linear constraints and let $\hat{I}_n$ be the minimum in (9.7.33). It has been proved that under assumptions (A1) and (A2) given above,

$\hat{I}_n \to 0$ as $n \to \infty$ and as the following proposition suggests, $\hat{T}_n$ converges to T^* (cf. [263]).

Proposition 9.7.2. Let $S \in C^2[0, t_*]$; for $b = 0$, assume that there exist a constant λ such that $\dot{S}(t) \geq \lambda t$, $\lambda > 0$, $0 \leq t \leq t_*$, and the assumptions (A1) and (A2) hold. Let $\hat{T}_n$ be any solution of (9.7.33). Then $\hat{T}_n$ converges to T^* in $C(\overline{D}_{t_*})$ and $\hat{T}_{nx}(0,t)$ converges to $T_x^*(0,t)$ in $C\,[0, t_*]$ for $n \to \infty$.

The heat polynomials

$$v_i(x,t) = i! \sum_{n=0}^{[i/2]} \frac{x^{i-2n}\, t^n}{(i-2n)!\; n!}\,, \quad (x,t) \in R^2, \; i = 0, 1, 2 \ldots, \tag{9.7.34}$$

have been taken as the complete set of functions in the numerical example considered in [263]. The minimization problem (9.7.33) has been solved by transforming it into an equivalent semi-infinite linear programming problem similar to the usual Chebyshev approximation problem [264]. After finite-difference discretization, we get a finite-dimensional linear programming problem (with constraints) and this problem can be solved by any linear programming package. The controls $\hat{T}_{nx}(0,t)$ $\big($or $\hat{T}_x(0,t)\big)$ are now taken as an input data for the corresponding direct Stefan problem. Numerical solutions of three inverse problems have been presented.

A one-phase two-dimensional inverse Stefan problem in a bounded domain has been considered in [265] with the temperature on a portion of the fixed boundary serving as a control. The two boundary conditions at the free boundary which are prescribed in the form of temperature and its normal derivative are known, and the initial temperature is also known. By imposing some a priori constraints and suitably defining the term 'solution', it has been established that the solution depends continuously on the data. The defect is defined in terms of the maximum of the three norms of the difference of the three prescribed quantities and the respective computed quantities and the maximum norm is to be minimized. In the numerical example considered, the two-dimensional heat polynomials have been taken as the complete set of functions. The basic approach in proving various results in [265] is similar to that used in [263] but for the numerical solution, instead of a semi-infinite linear programming problem considered in [263], a finite linear programming approach has been considered. If the numerical procedure adopted in [263] is followed for a two-dimensional problem it would require solving a problem with thousands of constraints.

The problem considered in [266] is an extended formulation of the one-phase inverse Stefan problem considered in (9.7.1)–(9.7.5). It consists of finding an unknown flux prescribed at $x = 0$ for a given $S(t)$. The formulation of the problem is as follows:

$$d(x,t)T_t = \left(a(x,t)T_x\right)_x + b(x,t)T_x + e(x,t)T + q(x,t), \;\; (x,t) \in D_{t_*},$$

$$D_{t_*} = \{(x,t) : 0 < x < S(t), \;\; 0 < t < t_*\}\,, \tag{9.7.35}$$

$$a(0,t)T_x(0,t) = f(t) \in C^{\alpha}[0,t_*],\ 0 \le \alpha \le 1, \tag{9.7.36}$$

$$T(x,0) = \phi(x),\ 0 \le x \le S(0),\ a(0,0)\phi'(0) = f(0), \tag{9.7.37}$$

$$T(S(t),t) = \mu\,(S(t),t)\,,\ 0 < t < t_*, \tag{9.7.38}$$

$$a\,(S(t),t)\ T_x(S\,(t),t) = -\lambda(S(t),t)\dot{S}(t) + \nu(S(t),t),\ 0 < t < t_*. \tag{9.7.39}$$

Here, $S(t) \in C^1[0,t_*]$ and is known and the inverse Stefan problem consists of determining $f(t)$ and $T(x,t)$. Under suitable assumptions on the data (cf. [266]), a solution of the inverse problem (9.7.35)–(9.7.39) exists and if it exists, then it is stable. The method described below determines the exact solution if it exists, otherwise, a quasi-solution can be obtained. Let the problem (9.7.35)–(9.7.39) be called Problem (P^0) and let its temperature solution be denoted by $T^0(x,t)$. An auxiliary Problem (P_1) consists of equations (9.7.35)–(9.7.38), and let its temperature solution be denoted by $T_1(x,t)$. An auxiliary Problem (P_2) consists of (9.7.35)–(9.7.37) and (9.7.39) and its temperature solution will be denoted by $T_2(x,t)$. We introduce the following notations:

$$J(f) = \iint_{D_{t_*}} (T_1 - T_2)^2 dxdt = \|T_1 - T_2\|^2_{L^2}\,,\ L^2 = L^2(D_{t_*}), \tag{9.7.40}$$

$$G = \left\{f(t) \in W_2^1[0,t_*] : a(0,0)\phi'(0) = f(0)\right\}, \tag{9.7.41}$$

$$E = \left\{f(t) \in W_2^1[0,t_*] : \|f(t)\|_{W_2^1} \le \beta\right\}, \tag{9.7.42}$$

$$G_\beta = G \cap E = \left\{f(t) \in W_2^1[0,t_*] : a(0,0)\phi'(0) = f(0),\ \|f(t)\|_{W_2^1} \le \beta\right\}, \tag{9.7.43}$$

$$J^*(\beta) = \inf_{f \in G_\beta} J(f). \tag{9.7.44}$$

Under some smoothness assumptions on the data, unique solutions of Problem (P_1) and Problem (P_2) exist and these solutions are sufficiently smooth (cf. [266]).

Proposition 9.7.3. The necessary and sufficient conditions for the existence of a solution of Problem (P^0) is that there exists $f^*(t) \in G_\beta$ such that $J^*(\beta) = J(\beta^*) = J(f^*(t)) = 0$.

It has been proved that for the minimization of $J(f)$ on G_β, gradient methods can be used and the convergence is fast (see [267] for gradient methods). Let T_1^β and T_2^β be the solutions of Problems (P_1) and (P_2), respectively, for some β. As $\beta \to \beta^*$, the results in (9.7.45) hold.

$$\|T_1{}^\beta - T^0\|_{W_2^{2,1}} \to 0,\ \text{ and }\ \|T_2{}^\beta - T^0\|_{W_2^{2,1}} \to 0. \tag{9.7.45}$$

If the solution of Problem (P^0) does not exist on G_β, then a quasi-solution can be obtained as follows:

Let $T_\alpha^\beta = \alpha T_1^\beta + (1-\alpha) T_2^\beta$, $0 \le \alpha \le 1$. Then for any fixed α, $0 \le \alpha \le 1$,

$$\|T_\alpha^\beta - T^0\|_{W_2^{2,1}} \to 0 \quad \text{as } \beta \to \beta^*. \tag{9.7.46}$$

We call $T_{\alpha_\beta}^\beta$ a quasi-solution of Problem (P^0) on G_β for $\beta < \beta^*$ if $T_{\alpha_\beta}^\beta$ minimizes the residual $\Phi_\beta(\alpha)$ in $L_2[0,t_*]$, with respect to the parameter α, where,

$$\Phi_\beta(\alpha) = \left\|(1-\alpha)\left\{T_2^\beta(S(t),t) - \mu(S(t),t)\right\}\right\|_{L_2}^2$$
$$+ \left\|\alpha\left\{\left(a(x,t)\frac{\partial T_1^\beta}{\partial x}\right)\Big|_{x=S(t)} + \lambda(S(t),t)\dot{S} - \nu(S(t),t)\right\}\right\|_{L_2}^2. \tag{9.7.47}$$

Proposition 9.7.4. As $\beta \to \beta^*$, $T_{\alpha_\beta}^\beta(x,t)$ converges in the norm of $W_2^{2,1}$ to the solution $T^0(x,t)$ of Problem (P^0).

The inverse Stefan problem considered in [268] is a particular case of the problem considered in (9.7.1)–(9.7.5) but the control is different from that considered in other problems. Let $D_{t_*} = \{(x,t) : a < S(t) < x < d,\ 0 < t < t_*\}$. We take $q(x,t) = 0$ in (9.7.1), $r_1 = 0$ in (9.7.4), $r_2 = \lambda\dot{S}(t)$ in (9.7.5) and instead of (9.7.3), consider

$$T_x(d,t) + \alpha T(d,t) = -v(t). \tag{9.7.48}$$

Here, λ and α are constants. The free boundary $x = S(t)$ is a known monotonically decreasing C^1-function with $S(0) = b > a$. The temperature at $x = d$ is manipulated by a heating (cooling) system according to (9.7.48). In (9.7.48), $v(t)$ depends on an unknown function $u(t)$ and this dependence can be expressed as

$$v'(t) + \gamma v(t) = u(t),\ a.a.\ t \in [0,t_*],\ v(0) = 0,\ u \in U,\ \gamma \text{ (constant)}, \tag{9.7.49}$$

$$U = \{u \in L^\infty(0,t_*) : 0 \le u(t) \le M,\ a.a.\ t \in [0,t_*]\},\ M \text{ (constant)}. \tag{9.7.50}$$

The inverse problem is to find $u \in U$ such that

$$T_u(S(t),t) = 0, \text{ for every } t \in [0,t_*]. \tag{9.7.51}$$

It is understood here that for a given u, the temperature T_u satisfies all other equations exactly except (9.7.51). The optimal control problem is to find u which minimizes the integral

$$\int_0^{t_*} (T_u(S(t),t))^2 dt;\ u \in U. \tag{9.7.52}$$

After obtaining necessary conditions for optimality, a descent algorithm for obtaining a solution for this control problem has been presented. Particular attention has been devoted in [268] to find a starting control by a local variations method described in [269].

Analysis and control of Stefan problems by considering weak enthalpy formulations have been studied in [167]. The report [270] also contains several references on control and identification of free boundary problems of parabolic, hyperbolic and elliptic types (see also the cross references in [167] and [270]). Our main concern here is the classical Stefan problem.

In most of the references in [254]–[268], numerical methods employed to obtain numerical solutions have also been justified and attempts have been made to obtain regularized solutions. In view of the ill-posedness of the inverse problems, justification of the numerical methods becomes necessary so that it is certain that the solution we have obtained could be an approximate solution but is not a bad solution (unstable). For reasons of scope and emphasis in this volume, discussion of numerical solutions is limited.

The Tikhonov regularization method can be used for inverse heat transfer problems with or without phase-change. A given problem is to be formulated first as a regularizing functional. In [254, 255, 256, 257], it is possible to obtain regularizing functionals for the operator equations to which the problems are reduced and then Tikhonov regularization could have been used but for the purpose of calculating error estimates, a different type of regularization was done. The determination of an optimal value of the Tikhonov regularization parameter α (see equation (9.3.16)) requires lot of computational effort. The numerical solution of a one-dimensional inverse heat transfer problem (without phase-change) by Tikhonov regularization has been presented in [271]. The original problem is reformulated in terms of obtaining the solution of a Volterra integral equation of the first kind and a finite-difference discretization has been employed to obtain a stable solution with the help of a Tikhonov regularizing functional. In [272], a one-dimensional heat conduction problem without phase-change has been considered in which both the ends of a plate are considered as free boundaries. On using the transformation given in (9.7.53), a problem on the fixed domain $0 \leq \xi \leq 1$ can be formulated where ξ is given by

$$\xi = \frac{x - S_1(t)}{S_2(t)-S_1(t)}, \quad t > 0. \tag{9.7.53}$$

Here, S_1 and S_2 ($S_2 < S_1$) are the free boundaries but no phase-change is taking place. The fixed domain formulation of the problem on discretization by finite-difference method gets reduced to a system of nonlinear equations with a tri-diagonal matrix. A regularizing functional in which a stabilizing functional is of the form (9.3.29) has been considered. Terms upto second order temperature derivatives have been included in the stabilizing functional. The choice of the regularization parameter has also been discussed.

A one-dimensional two-phase solidification problem similar to the Neumann problem (1.3.1)–(1.3.7) but formulated in the region $0 \leq x \leq b$, has been considered in [273]. The initial temperature T_0 of the melt, $S(t)$, T_m, b, $(\partial T_S/\partial x)(S(t),t) = q_{ys}(t)$, $(\partial T_L/\partial x)(S(t),t) = q_{yl}(t)$ are given and the problem is to find $(\partial T_S/\partial x)(0,t) =$

$q_{0S}(t)$ and $(\partial T_L/\partial x)(b,t) = q_{0l}(t)$. The temperature gradients $q_{yS}(t)$, $q_{yl}(t)$ and $S(t)$ satisfy the Stefan condition

$$K_S q_{yS} - K_L q_{yl} = \rho l \dot{S}(t), \; x = S(t). \tag{9.7.54}$$

In essence, we have two independent inverse Stefan problems to solve. The aim in this problem is to calculate the boundary fluxes at the fixed boundaries that will give the desired freezing front velocity on which depends the liquid feeding to the mold and hence the desired cast structure.

For the numerical computations, a boundary element method with constant elements has been used in conjunction with the *sensitivity analysis* discussed in [274]. By using transformations of the type (9.7.53) (after appropriate modifications), fixed domain formulations can be obtained for both solid and liquid phases. The temperatures in the solid and liquid regions can be expressed in terms of integrals using appropriate Green's functions (see equation (9.6.6)). Note that since $S(t)$ is known, we are solving only parabolic heat equations. These integral representations are required in the numerical method which uses boundary elements. It may be noted that the integral representations of temperatures in the present case are different from (9.6.8) as in the fixed domain formulations, the heat equations will get transformed. Temperatures at the fixed boundaries of the transformed regions can be obtained from their integral representations and on discretization of integrals the matrix equations which contain unknown nodal values of temperatures can be obtained. For the description of sensitivity analysis, we consider here only the solid phase. Let q_{0S}^m be the unknown flux during the time interval (t_{m-1}, t_m), i.e., during the time step m and all $q_{0S}^{m_1}$, for $m_1 < m$ are known. To stabilize the solution of the inverse problem, it is assumed that q_{0S} is constant at the future $(r-1)$ time steps. This assumption is used temporarily until q_{0S}^m is calculated. The *sensitivity coefficients* for this problem are defined as

$$\left\{{T_S}^{m+i-1};\; q_{0S}^m\right\} = \frac{\partial\left\{{T_S}^{m+i-1}\right\}}{\partial q_{0S}^m}, \; i = 1 \text{ to } r. \tag{9.7.55}$$

Here, the notation $\{T_S\}$ stands for a matrix. The error in the prescribed temperature T_m at $x = S(t)$ and its calculated value is to be minimized with respect to q_{0S} and this gives an equation to determine q_{0S}^m, $m > 1$ at time steps other than the initial time step. Similarly q_{0l}^m can be obtained. Starting solutions have been obtained with the help of some approximate analytical solutions developed in [273].

A problem of estimating unknown free boundary in a two-dimensional heat conduction problem with the help of some temperature measurements along a portion of the fixed boundary of the region has been considered in [275]. The formulation of the direct problem is as follows:

$$k\left(\frac{\partial^2 T}{\partial x^2} + \frac{\partial^2 T}{\partial y^2}\right) = \frac{\partial T}{\partial t}, \; 0 < x < b, \; 0 < y < S(x,t), \; 0 < t < t_*, \tag{9.7.56}$$

$$\frac{\partial T}{\partial x} = 0, \text{ at } x = 0, \; 0 < t < t_*, \tag{9.7.57}$$

$$\frac{\partial T}{\partial x} = 0, \text{ at } x = b, \ 0 < t < t_*, \tag{9.7.58}$$

$$\frac{\partial T}{\partial y} = q_0/K, \text{ at } y = 0, \ 0 < t < t_*, \tag{9.7.59}$$

$$T = T_1, \text{ at } y = S(x,t), \ 0 < t < t_*, \tag{9.7.60}$$

$$T = T_0, \text{ at } t = 0. \tag{9.7.61}$$

Here, $y = S(x,t)$ is the equation of the free boundary. In the direct problem considered in [275], $S(x,t)$ is taken as known and the temperature $T(x,y,t)$ is to be determined. In the inverse problem corresponding to this direct problem, $S(x,t)$ is unknown, and in this case we should have two conditions at $y = S(x,t)$. Instead of imposing one more condition at $y = S(x,t)$, we take some temperature readings denoted by $\hat{T}_m(t)$ along $y = 0$. Let

$$T(x_m, 0, t) = \hat{T}_m(t), \ m = 1 \text{ to } M, \ 0 < t < t_*. \tag{9.7.62}$$

The inverse problem consists of determining $T(x,t)$ and $S(x,t)$ satisfying equations (9.7.56)–(9.7.62). For the solution of the inverse problem, conjugate gradient method has been used in conjunction with a boundary element method. For the boundary integral formulation of the present problem and the discretization of equations, see ([275, 276]). The method of solution by the conjugate gradient method for the present inverse problem requires the solution of three problems, namely, the direct, sensitivity and adjoint. The solution of the inverse problem has been obtained in such a way that the following functional is minimized.

$$J\{S(x,t)\} = \int_{t=0}^{t_*} \sum_{m=1}^{M} \left\{T_m(t) - \hat{T}_m(t)\right\}^2 dt. \tag{9.7.63}$$

Here, $T_m(t)$ are the temperatures at $(x_m, 0, t)$ which are obtained by solving a direct problem in which an approximate $\hat{S}(x,t)$ is used in the place of exact $S(x,t)$. Note that in the inverse problem we start with an approximate value of $S(x,t)$. The data $\hat{T}_m(t)$ can have measurement errors. The numerical results indicate that the solution of the inverse problem obtained by the above method remains stable as measurement errors are increased and the number of observed temperature locations are decreased.

A sequential algorithm for the identification of the position of the moving boundary in the one and two-dimensional Stefan problems from discrete measurements of temperatures and fluxes at the fixed boundaries has been presented in [277]. In most cases, the direct measurements of the position of the phase-change boundary is impracticable. Identification of the interface position is, therefore, to be incorporated in the feedback control algorithm. In a two-phase Stefan problem, the physical phenomenon occurring in the liquid phase is difficult to model accurately because of some factors such as surface tension, thermal and solutal convection, and external forces. This drawback can be eliminated by having all the measurements made in the solid phase and considering a one-phase problem consisting of only the solid phase.

The two-dimensional one-phase problem considered in [277] is similar to the problem described in (9.7.56)–(9.7.61) except that in the condition (9.7.60), T_1 is now the fusion (melting) temperature. In the direct Stefan problem described in [277], $S(x,t)$ is known, and either temperature or flux is known at $y = 0$. The temperature is to be determined everywhere. If the temperature is prescribed at $y = 0$, then after calculating the temperature in the direct problem (since $S(x,t)$ is known we are solving a heat conduction problem without phase-change), the flux can be determined at $y = 0$ so that both T and $\partial T/\partial y$ are known at $y = 0$. In the inverse problem both $S(x,t)$ and $T(x,t)$ are to be determined. Therefore, an overspecified boundary condition is imposed at $y = 0$ in terms of either the calculated temperature (if the flux is prescribed) or the calculated flux (if the temperature is prescribed). As mentioned in the earlier problems, for the solution of an inverse problem, a direct problem with some known value of $S(t)$ is first solved to determine the approximate temperature values everywhere. If at $y = 0$, the temperature is taken as prescribed in the direct problem, then this temperature is a model input and the model output will be the flux at $y = 0$. In this way we have a pair $(T, \partial T/\partial y)$. The numerical solution of $\partial T/\partial y$ should be compared for correctness with the prescribed $\partial T/\partial y$ which is an overspecified boundary condition. Similarly we may have another pair $(\partial T/\partial y, T)$. For the given input data, sensitivity coefficients for the output data with respect to $S(t)$ have been calculated. It may be noted that if the flux is prescribed at $y = 0$, then $T(x,0,t)$ is the output and $\partial T/\partial y$ is the input. For a given $S(t)$, output/input sensitivity coefficients have also been calculated. In the one-dimensional case, it was found that the best approach consists of taking prescribed temperature as the model input [277].

Let the value of $S(t)$ at time $t_{m+1} = (m+1)\,\triangle t$ ($\triangle t$ is the time step), $m = 0, 1, 2..$ be denoted by S_{m+1}. The value of S_{m+1} has been obtained in [277] through the minimization of a penalized least-squares criterion evaluated in the time interval $[t_{m+1},\ t_{m+r}]$. The length of the observation horizon is $\tau = r\triangle t$. The functional $J_\alpha(S_{m+1})$ to be minimized is taken as

$$J_\alpha(S_{m+1}) = \frac{1}{r}\sum_{i=1}^{r} e_{m+i}^{T_r} e_{m+i} + \alpha G(S_{m+1}), \tag{9.7.64}$$

where

$$G(S_{m+1}) = [S_{m+1} - S_m]^{T_r}\,[S_{m+1} - S_m], \tag{9.7.65}$$

$$e_{m+i} = \hat{Z}_{m+i} - Z_{m+i}\left(S_{m+i};\ \hat{U}_{m+1}, ..., \hat{U}_{m+i}\right). \tag{9.7.66}$$

Here, T_r stands for the transpose of a matrix, $\hat{Z}$ and $\hat{U}$ are the estimated quantities, α is the regularization coefficient, Z stands for the model output, and U stands for the model input. The matrices Z and U are defined below.

$$Z_{m+i} = [z\{0, (m+i)\triangle t\},\ z\{\triangle x, (m+i)\,\triangle t\}, ..., \ z\{(N-1)\,\triangle x, (m+i)\,\triangle t\}]^{T_r}, \tag{9.7.67}$$

$$U_{m+i} = [u\{0, (m+i)\,\triangle t\}\,,\; u\{\triangle x, (m+i)\,\triangle t\}\,, ...,\; u\{(N-1)\,\triangle x, (m+i)\,\triangle t\}]^{T_r} \tag{9.7.68}$$

In (9.7.67) and (9.7.68), N stands for the number of nodal points in the x-direction in the one-dimensional case. These matrices in the two-dimensional case can be similarly written. Noisy data can also be considered. For optimization, Gauss-Newton algorithm has been considered.

Till now we have discussed ill-posedness of the inverse problems and not of any direct problem. An example showing the ill-posedness of a one-dimensional one-phase oblation problem has been given in [278]. The formulation of this problem is similar to that given in (10.1.34)–(10.1.38) (see Chapter 10) except that f, λ and μ in [278] are not constant but functions of time. The heat source, initial temperature, flux at $x = 0$, flux at $x = S(t)$ and the latent heat are known functions of time and a parameter α, $\alpha > 0$. The prescribed quantities (cf. [278]) are taken in such a way that it is possible to obtain an exact analytical solution of the problem in terms of α and variables x and t, and for any $\alpha > 0$ the prescribed quantities and their derivatives of any order are less than unity. Furthermore, the difference in the derivatives of any order of these data are uniformly close. An exact analytical solution of the free boundary can be obtained in the form $S(t;\alpha) = \alpha t$. The difference in the two values of $S(t;\alpha)$ increases with α and t and the solution becomes unstable. This example shows that the well-posedness of many problems is conditional and if the prescribed data take the solution beyond the limits of well-possedness, the solution becomes unstable. The regularization of this problem has been achieved by defining a suitable solution space.

Chapter 10

Analysis of the Classical Solutions of Stefan Problems

Some results of the analysis (existence, uniqueness, stability and regularity results) of classical solutions of Stefan problems will be presented in this chapter. The analysis aspect of one-dimensional classical Stefan problems has been thoroughly investigated and it may not be possible to add any significant result to the existing literature in the future. This can also be felt from the results reported in §§ 10.1 and 10.2. For analysis, we shall mainly consider the formulation (1.4.3)–(1.4.9) for multi-dimensional problems and formulations such as those given in (3.1.1)–(3.1.5) and (3.2.1)–(3.2.10) for one-dimensional problems. In essence, it will be assumed in the classical formulation that the solid and the liquid phases are separated by a smooth free boundary, the temperatures in the two regions satisfy heat equations and appropriate sign constraints (see § 1.4.6). The analysis of the problem of solidification of a supercooled liquid and some related problems has already been presented in Chapter 4.

In the multi-dimensional Stefan problems, one can easily think of geometries in which the free boundary is initially regular but becomes discontinuous after some time. The results of analysis of classical solutions of multi-dimensional Stefan problems are available mostly for short-time (local-in-time) solutions. However, in principle, it may be possible to study these solutions till the time the free boundary becomes discontinuous. Weak (generalized) solutions of multi-dimensional Stefan problems have been extensively studied under fairly general assumptions about the input data and the coefficients occurring in the formulation. Due to reasons of emphasis, weak solutions of only some selected Stefan problems in which the differentiability or the Lipschitz continuity of the free boundary has been established, will be discussed in Chapter 11.

10.1 One-dimensional One-phase Stefan Problems

The analysis of one-dimensional one-phase problems has a long history and a brief account of it can be found in [21]. There are several factors that affect the results of analysis of Stefan problems, such as, formulation of the problem, the definition of the solution, assumptions about the input data and coefficients, and the mathematical tools used in proving the results. The proof of existence, uniqueness, etc. requires some estimates of temperature derivatives, which involve lengthy calculations. The main result can be proved only after proving several lemmas and propositions. It is not possible to give complete details of the analysis here. As a suitable compromise on length, only some basic approaches used in the proofs will be discussed together with some of the main results. Taking a contemporary perspective, we can consider many of the problems studied earlier as particular cases of the problems studied later. It does not seem necessary to discuss all the particular cases. The reader is requested to look into cross references mentioned in the bibliography.

10.1.1 Analysis using integral equation formulations

We present some results since 1959. Several references of works prior to 1959 with comments on the methods used in proving the results are reported in ([21, 279, 280, 281]). The one-phase melting problem considered in [279] is to find the temperature $T(x,t)$ and the free boundary $S(t)$ satisfying the following system of equations.

$$(p(x,t,T,T_x)T_x)_x = f(x,t,T)T_t, \text{ in } D_{t_*} = \{(x,t) : 0 < x < S(t),\ 0 < t < t_* < \infty\}, \tag{10.1.1}$$

$$p_0(t)T_x(0,t) = -g(t),\ 0 < t < t_*;\ \ p_0(t) = p|_{x=0}, \tag{10.1.2}$$

$$T(S(t),t) = 0,\ 0 < t < t_*, \tag{10.1.3}$$

$$\dot{S}(t) = b - (pT_x)|_{x=S(t)},\ 0 < t < t_*;\ S(0) = 0. \tag{10.1.4}$$

At time $t = 0$, the region $x > 0$ is occupied by ice at the melting temperature which is taken as zero. No initial condition is required as $S(0) = 0$. The coefficients in (10.1.1)–(10.1.4) satisfy the following conditions.

(1) b is a non-negative constant. The functions $g(t), f, p$ have continuous second order derivatives.

(2) $0 < a_0 \leq g(t)/p_0(t) \equiv b(t) < a^0$.

(3) $0 < f_0 \leq f(x,t,T)$.

(4) $0 < p_1 \leq p(x,t,T,T_x) = p_0(t)$ + terms which vanish at $x = 0$.

(5) $p_x \geq 0,\ p_{_T} \leq 0,\ p_z = p_{_{T_x}} \leq 0.$

(6) $S(t)$ is monotonic and it is possible to express $x = S(t)$ as $t = S^{-1}(x)$.

The fifth assumption implies that (10.1.1) can be written as a linear differential equation with non-negative coefficients, i.e.,

$$(p + p_z T_x)T_{xx} + (p_x + p_{_T} T_x)T_x = fT_t\ . \tag{10.1.5}$$

It can be proved that $T_x \leq 0$. Introducing the notations

$$F(x,t,T) = \int_0^T f(x,t,q)dq, \quad \text{and} \quad G(x,t,T) = -\int_0^T f_t(x,t,q)dq, \tag{10.1.6}$$

and integrating (10.1.1) over the region $0 \leq x \leq S(t'),\ 0 \leq t \leq t'$, and writing t in the place of t', the following integral equation is obtained.

$$S(t) = \int_0^t (b + g(q))dq - \int_0^{S(t)} F(x,t,T)dx - \int_0^t \int_0^{S(q)} G(x,q,T(x,q))dxdq. \tag{10.1.7}$$

By using fixed point argument, the existence and uniqueness of the classical solution (T,S) of (10.1.1)–(10.1.4) has been proved in [279]. Let $T(x,t;S(t))$ be the solution of the following 'reduced problem'.

$$(pT_x)_x = fT_t, \text{ in } D_{t_*} \tag{10.1.8}$$

$$p_0(t)T_x(0,t) = -g(t),\ 0 < t < t_*, \tag{10.1.9}$$

$$T(S(t),t) = 0,\ 0 < t < t_*. \tag{10.1.10}$$

Here, $S(t)$ is a given continuous monotonic function which vanishes at $t = 0$. The dependence of the temperature on the solution of the reduced problem for a given $S(t)$ will be expressed in the form $T(x,t;S)$. It may be noted that the given $S(t)$ may not be the solution of the original Stefan problem as (10.1.4) is not satisfied. The following propositions which help in applying fixed point argument have been proved in [279].

Proposition 10.1.1. If $T(x,t;S)$ is the solution of the reduced problem, then there exist numbers $a_1 > 0$ and $a_2 > 0$, independent of $S(t)$, such that

(a) $T(x,t;S) \leq a_1(S(t) - x),\ 0 \leq x \leq S(t),\ 0 < t < t_*,$

(b) $-a_2 \leq p(x,t,T,T_x)T_x \leq 0.$

Proposition 10.1.2. There exists a unique solution of the reduced problem.

By integrating (10.1.4) also an integral equation for $S(t)$ can be obtained but it will involve integral of T_x over a boundary and this is not suitable for using fixed point argument. If $T(x,t)$ and $S(t)$ represent the solution of (10.1.1)–(10.1.4), then (10.1.7) is satisfied exactly. For approximate values of T and S, it is appropriate to write (10.1.7) as

$$z(t) = \int_0^t (b + g(q)dq - \int_0^{S(t)} F(x,t,T)dx - \int_0^t \int_0^{S(q)} G(x,q,T(x,q))dxdq. \qquad (10.1.11)$$

Equation (10.1.11) defines a mapping

$$Z = W(S). \qquad (10.1.12)$$

Here, W is defined on a set of real valued, differentiable boundary curves which vanish at $t = 0$. These conditions on S are enough because

$$\frac{dz}{dt} = b - p(S(t), t, 0, T_x)T_x(S(t), t; S(t)), \qquad (10.1.13)$$

and

$$b \leq \frac{dz}{dt} \leq b + a_2. \qquad (10.1.14)$$

The function $z(t)$ is a differentiable and monotone function which vanishes at the origin, and it can therefore serve as a boundary curve for the reduced problem.

Proposition 10.1.3. Let V be a set of continuously differentiable functions, defined on some finite time interval, which vanish at $t = 0$ and whose derivatives satisfy (10.1.14). If W is the transformation defined by (10.1.12), then W is defined on V and maps V into itself.

Proposition 10.1.4. Under the uniform norm (see (9.7.12)), V is a subset of a Banach space of continuous functions defined on $[0, t_*]$. The set V is convex and equicontinuous. The closure of V denoted by $\overline{V}$ is also convex. Every infinite subset of $\overline{V}$ has an accumulation point in $\overline{V}$, i.e., $\overline{V}$ is compact.

Proposition 10.1.5. W is continuous on the closure $\overline{V}$ of V. W maps $\overline{V}$ into itself.

Since $\overline{V}$ is a compact and convex subset of a Banach space, and W is a continuous mapping of $\overline{V}$ into itself, by *Schauder's fixed point theorem* [282], there exists at least one element of $\overline{V}$ which is left invariant under W, i.e., $S^0(t) = W(S^0(t))$. The function $S^0(t)$ is the solution of (10.1.11) and $T(x,t;S^0)$ is the solution of reduced problem (10.1.8)—(10.1.10).

It can be proved that if $S(t)$ is in $\overline{V}$ and $S(t) = W(S(t))$, then $S(t)$ is differentiable, i.e., $S(t)$ is in V. Further

$$\frac{dS}{dt} = b - (pT_x)|_{x=S(t)},$$

and therefore $(S(t), T(x,t;S(t)))$ is a solution of the problem (10.1.1)–(10.1.4). Uniqueness of the problem (10.1.1)–(10.1.4) has also been proved in [279] but it is not based on *contraction mapping* argument [282].

The problem considered in [280] can be obtained by making some changes in (10.1.1)–(10.1.4), such as, take $p = 1$, $f = 1$, $b = 0$, $S(0) = A$ and $T(x,0) = a(x)$, where $0 \leq a(x) \leq d(A - x)$, $0 \leq x \leq A$, and d is some positive constant. The functions $a(x)$ and $g(t)$ are continuous, $g(t) \leq d$. An integral equation of the type (10.1.7) has been obtained in this case also and now a reduced problem of the type (10.1.8)–(10.1.10) will have a prescribed initial temperature also as $S(0) = A > 0$. The existence of the solution on some finite time interval $[0, t_*]$ has been proved using fixed point theorem as in [279]. However, the uniqueness of the solution has been proved by showing that the iterations done in the numerical solution for calculating the free boundary are converging, i.e., if $S_0 = A$ and $S_{n+1} = F(S_n)$, then F is a contraction.

For the numerical solution of the problem considered in [280], the time interval $[0, t_*]$ is divided into n small time intervals, each of length Δt and in each one of them iterations are done to get better values of $S(t)$. For this purpose an integral equation of the form (10.1.11) is used. $T(x,t;S(t))$ is obtained from the solution of the 'reduced problem' formulated for this problem with appropriate changes. The first iterative process will converge to the solution if the time interval is small (existence holds and uniqueness is proved by contraction argument). Then another iterative process is carried out in the time interval $(0, 2\Delta t]$. The solution of the previous time step is used to obtain $S(t)$ and $T(x,t;S(t))$ in $(0, 2\Delta t]$ by using a suitable iterative process (cf. [280]). The initial temperatures in the 'reduced problems' will go on changing. This procedure is repeated in other time intervals also till the solution is obtained in the time interval $[0, t_*]$. It has been shown that $\lim_{n\to\infty} S_{n+1} = S(t)$ and $S(t)$ is invariant under the transformation of the form (10.1.12) for the present problem also . The subscript n refers here to the n-th iterative process.

The problem considered in [281] can be obtained from (10.1.1)–(10.1.4) if we take $p = 1$ and $f = 1$ in (10.1.1), $b = 0$ in (10.1.4), $S(0) = d$ and replace (10.1.2) by a temperature prescribed boundary condition, e. g.,

$$T(0,t) = v(t) \geq 0, \; 0 < t < t_* < \infty. \tag{10.1.15}$$

Since $S(0) = d$, initial temperature is to be prescribed and let

$$T(x,0) = T_0(x) \geq 0, \; 0 \leq x \leq d. \tag{10.1.16}$$

We shall refer to this problem in [281] as Problem (F). The main result of [281] is the following proposition.

Proposition 10.1.6. Assume that $v(t)$ $(0 \leq t < \infty)$ and $T_0(x)$ $(0 \leq x \leq d)$ are continuously differentiable functions. Then there exists one and only one solution $(T(x,t), S(t))$ of Problem (F) for all $t < \infty$. Furthermore, the function $x = S(t)$ is monotone nondecreasing in t.

By solution we mean here the classical solution discussed in § 1.4.6.

If $S(t)$ exists in any time interval $0 \leq t < \sigma$, then on using maximum principles for parabolic operators [9], it can be proved that $S(t)$ is nondecreasing in this interval. The Problem (F) can be reduced to a problem of solving a nonlinear integral equation. Following the procedure indicated in (9.6.5)–(9.6.8), the temperature can be expressed as in (9.6.8) for the present problem also (substitute d in place of b). We use the notations used in (9.6.8). If both sides of (9.6.8) are differentiated with respect to x and the limit $x \to S(t) - 0$ is taken, then we get the following integral equation.

$$y(t) = 2\left[T_0(0) - v(0)\right] N(S(t), t; 0, 0) + 2\int_0^d \frac{\partial T_0}{\partial \xi} N(S(t), t; \xi, 0) d\xi$$

$$-2\int_0^t \dot{v}(\tau) N(S(t), t; 0, \tau) d\tau + 2\int_0^t y(\tau) G_x(S(t), t; S(\tau), \tau) d\tau, \tag{10.1.17}$$

$$y(\tau) = T_\xi(S(\tau), \tau), \tag{10.1.18}$$

and

$$N(x, t; \xi, \tau) = Q(x, t; \xi, \tau) + Q(-x, t; \xi, \tau). \tag{10.1.19}$$

In (10.1.17), G is given by (9.6.6) and in (10.1.19), Q is given by (9.6.7). In obtaining (10.1.17), the following result has been used.

Proposition 10.1.7. Let $\rho(t)$, $0 \leq t \leq \sigma$, be a continuous function and let $S(t)$, $0 \leq t \leq \sigma$, satisfy a Lipschitz condition. Then for every $0 < t \leq \sigma$

$$\lim_{x \to S(t)-0} \frac{\partial}{\partial x} \int_0^t \rho(\tau) Q(x, t; S(\tau), \tau) d\tau = \frac{1}{2}\rho(\tau) + \int_0^t \rho(\tau) \left[\frac{\partial Q}{\partial x}(x, t; S(\tau), \tau)\right]_{x=S(t)} d\tau. \tag{10.1.20}$$

On integrating the Stefan condition, we get

$$S(t) = d - \int_0^t y(\tau) d\tau. \tag{10.1.21}$$

It can be proved that Problem (F) is equivalent to the problem of finding a continuous function $y(t)$ which is the solution of (10.1.17) where $S(t)$ is defined by (10.1.21) and $S(t)$ is positive. For the set of continuous functions $y(t)$ defined for $0 \leq t \leq \sigma$, σ sufficiently small, the equation (10.1.17) defines a mapping which can be expressed as

$$w(t) = P(y(t)). \tag{10.1.22}$$

The function $w(t)$ is the l.h.s. of (10.1.17) for some y(t) which need not be the value of T_x at the exact solution $x = S(t)$. By using fixed point argument, the existence of

$y(t)$, i.e., the solution of (10.1.17) has been proved in [281] and by showing that P is a contraction, uniqueness of $y(t)$ has been proved.

The validity of the solution obtained above for a short time can be extended to longer times. It has been proved in [281] that there exists an $\varepsilon > 0$ such that if the continuous solution $y(t)$ of the integral equation (10.1.17) exists and is unique for $0 \leq t \leq t_0$, then it exists and is unique for $0 \leq t < t_0 + \varepsilon$. Note that the continuity of $y(t)$ implies the continuity of $\dot{S}(t)$. The existence of the solution in the time interval $[0, t_0 + \varepsilon]$ requires that $T_x(x,t)$ is bounded by a constant which is independent of x and t in the interval $0 \leq x \leq S(t),\ t_0 - \delta \leq t < t_0$ for some $\delta,\ 0 < \delta < t_0$. The solution can then be started with some very small $\eta > 0$ and we get a classical solution for $t_0 - \eta \leq t < t_0 + \varepsilon$. This solution coincides with the classical solution for $t_0 - \eta < t < t_0$ as the solution is unique for $0 \leq t < t_0$.

The one-phase Stefan problem considered in ([21, 283]) has stronger nonlinearity than in [279] and no sign restrictions have been imposed on the boundary and initial data. The problem studied is to find a classical solution (see § 1.4.6 and [21]) $(T(x,t), S(t))$ of problem (10.1.23)–(10.1.27).

$$a^2 \frac{\partial^2 T}{\partial x^2} - \frac{\partial T}{\partial t} + F(x, t, T, T_x, S(t), \dot{S}(t)) = 0,\ (x,t) \in D_{t_*},$$

$$D_{t_*} = \{(x,t) : 0 < x < S(\tau),\ 0 < \tau \leq t < t_*\},\ a > 0, \tag{10.1.23}$$

$$\frac{\partial T}{\partial x} = f(t, T),\ x = 0;\ t > 0, \tag{10.1.24}$$

$$T = \phi(x),\ t = 0;\ 0 \leq x \leq S(0) = d, \tag{10.1.25}$$

$$T = \psi(x),\ x = S(t);\ t > 0, \tag{10.1.26}$$

$$\dot{S} = Z(t, T, T_x, x),\ x = S(t);\ t > 0. \tag{10.1.27}$$

Here, $F,\ f,\ \phi,\ \psi$ and Z are known functions which are defined for $0 \leq x \leq 1,\ 0 \leq t \leq 1$ and $0 \leq S(t) \leq 1$. The functions $F,\ f,$ and Z are differentiable, ϕ is thrice differentiable and ψ is twice differentiable with respect to their arguments everywhere in the domain of definition. The known functions and some of their derivatives are bounded by suitable constants (see [21], pp. 95–140). The compatibility conditions are satisfied, i.e.,

$$f(0, \phi(0)) = \phi_x(0);\ \phi(d) = \psi(d). \tag{10.1.28}$$

We introduce the following notations.

$$T|_{x=0} = w\,;\ T_x = q,\ 0 < x < S(t);\ T_x(S(t), t) = v;\ \dot{S}(t) = p. \tag{10.1.29}$$

On using a method similar to that used in obtaining the temperature in (9.6.8), the temperature given below which is the solution of problem (10.1.23)–(10.1.27) can also be obtained.

$$T(x,t) = -a^2 \int_0^t f(\tau, x) G_1(x, 0, t-\tau) d\tau + \int_0^t \phi(\xi) G_1(x, \xi, t) d\xi$$

$$+ \int_0^t d\tau \int_0^{S(t)} F(\xi, \tau, ..., p) G_1(x, \xi, t-\tau) d\xi - a^2 \int_0^t \psi(S(\tau)) \frac{\partial}{\partial \xi} G_1(x, S(\tau), t-\tau) d\tau.$$

$$+ \int_0^t \left\{ a^2 v(\tau) + \psi(S(\tau)) p(\tau) \right\} G_1(x, S(\tau), t-\tau) d\tau. \qquad (10.1.30)$$

Here, the Green's function $G_1(x, \xi, t) = Q(x, t; \xi, \tau) + Q(-x, t; \xi, \tau)$ and Q is given by (9.6.7). Using (10.1.30), the functions w, q and v can be expressed in terms of appropriate integrals. On integrating $\dot{S}(t) = p$, we get

$$S(t) = d + \int_0^t p(\tau) d\tau. \qquad (10.1.31)$$

The results given below in Propositions 10.1.8. and 10.1.9. have been proved in [21] under suitable assumptions.

Proposition 10.1.8. (Existence and uniqueness). The unique solution $(T(x,t), S(t))$ of the system (10.1.23)–(10.1.28) can be constructed by *Picard iteration method* which can be started with any set of functions $\{T_0, w_0, q_0, v_0, S_0, p_0\}$ having bounded partial derivatives with respect to each of their arguments. If the starting solution satisfies the conditions

$$\phi_x(0) = f(0, w_0(0)); \; \phi_x(d) = q_0(d, 0) = v_0(0); \; \dot{S}_0(t) = p_0, \qquad (10.1.32)$$

then each of the functions $T, w...., p$ obtained as solutions of Picard iteration satisfy a Lipschitz condition of the form

$$|f(t) - f(\tau)| < A\left(\sqrt{t} + \sqrt{\tau}\right)^{-1} |t - \tau|, \text{ A constant}, \qquad (10.1.33)$$

with respect to each of their arguments.

Here, T_0, w_0, q_0, v_0, S_0, p_0 are some initial approximations of the functions T, w, .., p, respectively.

Proposition 10.1.9. The solution (T, w, q, v, S, p) obtained by Picard iteration is stable relative to small variations of all the 'data' of the problem.

The existence of a local-in-time solution has been proved by establishing the convergence of iterations and uniqueness has been proved by showing that the sequences

$\{T_n\}, \{w_n\}...., \{p_n\}$ obtained by iterations are not only uniformly bounded and equicontinuous but are also uniformly convergent. This will imply convergence of the entire iteration process to a solution of the system of integral equations of the problem.

The local-in-time (short-time) solution cannot be extended to any arbitrarity preassigned time interval without assuming the monotonicity of the free boundary and imposing additional conditions of boundedness on the data. The additional assumptions (cf. [21]) imply that if the solution is valid in the time interval $[0, t_0]$, then it can be extended to $[t_0, t_1]$, where $\Delta t = t_1 - t_0$ is dependent only on the magnitude of $T(x,t)$ and its first three derivatives with respect to x at $t = t_0$ and not on the time t, for which the solution has already been constructed.

The approach to prove existence, uniqueness and stability of the solution of the free boundary problem considered in [59], consists of proving the convergence of a sequence of approximating solutions. The formulation of the problem considered in [59] is as follows:

$$\mathcal{L}T = T_{xx} - T_t = q(x,t), \text{ in } D_{t_*} = \{(x,t): \ 0 < x < S(t), \ 0 < t < t_*\}, \quad (10.1.34)$$

$$T(x,0) = \phi(x), \ 0 \le x \le S(0) = d > 0, \quad (10.1.35)$$

$$T(0,t) = u(t), \ 0 < t < t_*, \quad (10.1.36)$$

$$T(S(t),t) = f(S(t),t), \ 0 < t < t_*, \quad (10.1.37)$$

$$T_x(S(t),t) = \lambda(S(t),t)\dot{S}(t) + \mu(S(t),t), \ 0 < t > t_*. \quad (10.1.38)$$

The problem (10.1.34)–(10.1.38) will be called Problem (P1). If instead of (10.1.36), we have

$$T_x(0,t) = g(T(0,t),t), \ 0 < t < t_*, \quad (10.1.39)$$

and all other equations remain the same as in Problem (P1), then we shall call it Problem (P2). The boundary condition (10.1.37) is of 'Cauchy-type'. Several particular cases of Problems (P1) and (P2) have been studied in the literature which have been briefly reviewed in [59]. Let Ω be the quarter-plane $\{(x,t): \ 0 < x < \infty, \ 0 < t < \infty\}$. The input data and the coefficients in (10.1.34)–(10.1.39) satisfy the following assumptions.

(A) $q(x,t)$ is locally Hölder continuous in $\bar{\Omega}$ with respect to x (or t), and

$$|q(x,t)| \le Q, \ (x,t) \in \overline{\Omega}.$$

(B) $f(x,t)$ is continuous and bounded in $\overline{\Omega}$ together with f_x, and the difference $f_{xx} - f_t$ is bounded and locally Hölder continuous in $\bar{\Omega}$ with respect to x (or t).

(C_1) $u(t)$ is (piecewise) continuous for $t \ge 0$ and

$$|u(t)| \le \Phi, \ t \ge 0.$$

(C_2) $g(y,t)$ is Lipschitz continuous with respect to y, i.e.,

$$|g(y_1,t) - g(y_2,t)| \leq m_g |y_1 - y_2|,$$

uniformly with respect to $t \geq 0$ and it satisfies one of the conditions (α_1), (α_2) and one of the conditions (β_1), (β_2) listed below.

(α_1) There exists a constant $Y_1 > \max\{Md, \sup_{\Omega} f(x,t)\}$ such that

$$g(Y_1,t) \geq 0, \; t \geq 0.$$

For M, see (F) given below.

(α_2) There exist two constants Y' and G' such that

$$g(y,t) \geq G', \; \text{for } y \geq Y', \; t \geq 0.$$

(β_1) There exists a constant $Y_2 < \min\{-Md, \; \inf_{\Omega} f(x,t)\}$ such that

$$g(Y_2,t) \leq 0, \; t \geq 0.$$

(β_2) There exist two constants Y'' and G'' such that

$$g(y,t) \leq G'', \; \text{for } y \leq Y'', \; t \geq 0.$$

(D) $\lambda(x,t)$ is continuous in $\overline{\Omega}$ together with its first derivatives and

$$|\lambda(x,t)| \geq \lambda_0 > 0, \; (x,t) \in \overline{\Omega}.$$

(E) $\mu(x,t)$ is continuous in $\overline{\Omega}$, uniformly Lipschitz continuous w.r.t. x in bounded sets, and there exists a constant P such that

$$|\mu(x,t)| \leq P, \; (x,t) \in \overline{\Omega}.$$

(F) $\phi(x)$ is (piecewise) continuous in $[0,d]$ and a positive constant M exists such that

$$|\phi(x) - f(d,0)| \leq M(d-x), \; x \in [0,d],$$

with (for Problem (P1))

$$Md \geq \Phi + |f(d,0)|.$$

By redefining the temperature as $T(x,t) - f(x,t)$ in (10.1.34)–(10.1.39), (10.1.37) can be written as $T(S(t),t) = 0$. The new temperature will also be denoted by $T(x,t)$. The assumptions on the various functions have been made in such a way that they still remain valid after this transformation.

If $\lambda \equiv 0$, then we have an implicit free boundary condition and either this condition can be converted to an explicit condition (cf. § 3.3.1) which can then be studied by the method suggested in [59] or $\lambda = 0$ may yield non-existence, non-uniqueness, or instability of the solution on the input data. The main existence result of [59] is that either the solution exists globally in time, i.e., $t_* = +\infty$ or one of the following cases must occur for some $\hat{t} < \infty$:

$$\text{(i)} \lim_{t \to \hat{t}-0} S(t) = 0, \text{ (ii)} \lim_{t \to \hat{t}-0} \left|\dot{S}(t)\right| = +\infty. \tag{10.1.40}$$

The free boundary condition (10.1.38) can be reformulated as

$$T_x(S(t), t) = (d/dt)\left\{V(S(t), t) - W(t; S)\right\}, \tag{10.1.41}$$

where

$$V(x,t) = \int_0^x \lambda(\xi, t)d\xi, \; (x,t) \in \overline{\Omega}, \tag{10.1.42}$$

and for any $S(t) \in C^1(0, t_*) \cap C\,[0, t_*]$,

$$W(t; S) = \int_0^t \left\{V_\tau(S(\tau), \tau) - \mu(S(\tau), \tau)\right\} d\tau, \; t \in [0, t_*]. \tag{10.1.43}$$

If in (4.4.20), we take P as the heat operator, $P^* = \partial^2/\partial x^2 + \partial/\partial \tau$ and $v(x,t) = V(x,t)$, then on integrating (4.4.20), we get the following equation for Problem (P1).

$$\begin{aligned} \frac{1}{2}V^2(S(t),t) - \frac{1}{2}V^2(d,0) &= \iint_{D_{t_*}} \{V(x,\tau)q(x,\tau) - T(x,\tau)\left[\lambda_x(x,\tau) + V_\tau(x,\tau)\right]\} \, dx d\tau \\ &+ \int_0^{S(t)} V(x,t)T(x,t)dx - \int_0^d V(x,0)\phi(x)dx \\ + \int_0^t V(S(\tau),\tau)\left\{V_\tau(S(\tau),\tau) - \mu(S(\tau),\tau)\right\} d\tau &- \int_0^t \lambda(0,\tau)u(\tau)d\tau, \; t \in [0,t_*]. \end{aligned} \tag{10.1.44}$$

For Problem (P2), take $v = 1$ in (4.4.20) and P and P^* as above. On integrating, we get

$$\begin{aligned} V(S(t),t) - V(d,0) &= \iint_{D_{t_*}} q(x,\tau)dx d\tau + \int_0^{S(t)} T(x,t)dx - \int_0^d \phi(x)dx \\ &+ \int_0^t g(T(0,\tau),\tau)d\tau + \int_0^t \left\{V_\tau(S(\tau),\tau) - \mu(S(\tau),\tau)\right\} d\tau, \; 0 \le t \le t_*. \end{aligned} \tag{10.1.45}$$

Each solution of Problem (P1) satisfies the integral equation (10.1.44). Similarly each solution of Problem (P2) satisfies the integral equation (10.1.45). If $S(t)$ is Lipschitz continuous and $T_x(x,t)$ is continuous upto $x = S(t)$ for $t > 0$, then (10.1.38) is also satisfied by the solution (T, S) of (10.1.34)–(10.1.37) and (10.1.44) ((10.1.45)).

Approximate solutions to Problem (P1) (and similarly for Problem (P2)) can be obtained by induction starting with $S_1(t) = d$ and the temperature determined by the solution of the following system of equations:

$$\mathcal{L}T_k = q(x,t), \text{ in } D_{t_*^{(k)}}^{(k)} = \left\{(x,t) : \; 0 < x < S_k(t), \; 0 < t < t_*^{(k)}\right\}, \tag{10.1.46}$$

$$T_k(x,0) = \phi(x), \; 0 \le x \le S_k(0) = d, \tag{10.1.47}$$

$$T_k(0,t) = u(t), \; 0 < t < t_*^{(k)}, \tag{10.1.48}$$

$$T_k(S_k(t),t) = 0, \; 0 < t < t_*^{(k)}, \tag{10.1.49}$$

$$\lambda(S_k(t),t)\dot{S}_{k+1}(t) = T_{k,x}(S_k(t),t) - \mu(S_k(t),t), \; 0 < t < t_*^{(k+1)} \le t_*^{(k)}, \tag{10.1.50}$$

$$S_{k+1}(0) = d, \; k = 1, 2, \ldots.. \tag{10.1.51}$$

The time $t_*^{(k)}$ is the supremum of the values of τ for which $S_k(\tau) > 0, \; S_k(\tau) \in C^1(0,\tau)$. Note that, for any k such that $t_*^{(k)} > 0$, the problem (10.1.46)–(10.1.51) has a unique solution (cf. [284]) and $T_{k,x}$ is continuous upto the curve $x = S_k(t)$.

The next step in the existence proof is to show the convergence of the sequence of approximate solutions. It can be shown that there exist two constants, t_0 and A_0 such that

$$t_*^{(k)} \ge t_0, \; (t_0 < d/A_0) \text{ and } \left|\dot{S}_k\right| \le A_0, \; 0 < t \le t_0; \; k = 1, 2, \ldots.., \tag{10.1.52}$$

and the sequence $\{S_k(t)\}$ converges uniformly in $[0, t_0]$ to a positive function $S(t)$ with $S(0) = d$ such that

$$|S(t_1) - S(t_2)| \le A_0 \left|t_1 - t_2\right|, \text{ for } t_1, t_2 \in [0, t_0]. \tag{10.1.53}$$

Moreover, the sequence $\{T_k(x,t)\}$ converges uniformly in $R_{t_0} = [0, d + A_0 t_0] \times [0, t_0]$ to a function $T(x,t)$ satisfying (10.1.34)–(10.1.37) with the above $S(t)$. If this pair $(T(x,t), S(t))$ satisfies (10.1.38) or (10.1.44), then the existence of the solution is established for Problem (P1). This has been done in [59] by substituting (T_k, S_k) in (10.1.44) and taking the limit as $k \to \infty$. For Problem (P2) also the same procedure can be followed.

It can be shown that for $0 \le t \le t_0$, we have

$$|T(x,t_0)| \le \alpha_1(S(t_0) - x), \; \alpha_1 S(t_0) \ge \Phi, \; S(t_0) \ge d - A_0 t_0 > 0. \tag{10.1.54}$$

Here, α_1 is a constant independent of the index k. In view of (10.1.55) (given below), the solution can be extended to a time interval $t_0 \le t \le t_1$. For this, we take $S(t_0) = d_1 > 0$

and a new initial temperature in $0 \le x \le d_1$. This process can be extended to other larger time intervals.

Let F_0 be the class of functions which are Lipschitz continuous to which $S(t)$ also belongs. It has been shown that if T and S belong to the class F_0, then the solution depends continuously on the data and the coefficients. The uniqueness of the solution in the class F_0 is an immediate consequence of the stability of the solution.

The local-in-time solution $(t_*, S(t), T(x,t))$ can be extended to larger time intervals provided some further assumptions are made such as: (1) $S(t)$ is nondecreasing in $(0, t_*)$, and (2) the data satisfies some sign constraints such as $\lambda < 0,\ \phi \ge 0,\ u \ge 0,\ \mu \ge 0$ and $q \le 0$ which hold in some time interval $(0, t_0)$, $t_0 < +\infty$. If these assumptions together with the assumptions (A)–(F) mentioned above hold, then the solution of Problem (P1) exists for $t > t_0$ and

$$0 \le \dot{S}(t) \le \hat{A}_0,\ 0 \le t \le t_0 < \infty. \tag{10.1.55}$$

Here, $\hat{A}_0$ can be determined in terms of the data.

An ablation problem concerning melting of a solid in which the melt is instantaneously removed has been considered in ([285, 286]). In order to obtain the formulation of this ablation problem, we take $q(x,t) = 0$ in (10.1.34), $g = g(t)$ in (10.1.39), $f(S(t),t) = 0$ in (10.1.37), retain (10.1.35) as it is, denote μ by $q(t)$ and $\lambda(x,t) = \lambda < 0$ in (10.1.38). The existence, uniqueness and stability results have been proved in [285] and [286] using the approaches followed in [279, 280, 281]. Such results have been obtained for a more general problem in [59]. Some physical aspects of the ablation problem such as the time required for complete melting of the solid have been discussed in [286]. The conditions required for the uniqueness of the solution in a one-dimensional ablation problem with the heat input dependent on time and on melted depth have been studied in [287]. The formulation of this problem is similar to problem (3.1.13)–(3.1.16) except that the problem in [287] is one-dimensional. The heat flux $Q = Q(x,t)$, and the thermal conductivity and the specific heat are functions of temperature. Some examples have been constructed in which the solution is non-unique. The one-phase problem considered in [288] is concerned with a solidification problem in which the region $0 \le x \le d$ is initially ice with temperature $\phi(x) \le 0$ and a liquid at the melting temperature occupies the region $x > d$. The problem formulation is as follows:

$$\left.\begin{aligned} &T_{xx} = T_x,\ 0 < x < S(t),\ t > 0;\ T(x,0) = \phi(x),\ 0 \le x \le d, \\ &-\lambda \dot{S}(t) + T_x(S(t),t) = 0,\ 0 < t \le t_*,\ \lambda > 0;\ T(S(t),t) = 0,\ S(0) = d. \end{aligned}\right\} \tag{10.1.56}$$

At $x = 0$, either

$$T(0,t) = -f(t),\ 0 < t \le t_*,\ f(t) > 0, \tag{10.1.57}$$

is prescribed, or we have

$$T_x(0,t) = f(t),\ 0 < t \le t_*. \tag{10.1.58}$$

To indicate the dependence of the solution on λ the solution will be written as $(T(x,t;\lambda)$, $S(t;\lambda))$.

The main interest in [288] is to investigate the behaviour of the solution $(T(x,t;\lambda)$, $S(t;\lambda))$ as $\lambda \to 0$ (latent heat tends to zero). By using appropriate Green's functions, the solution for the temperature derivative or the temperature can be written as in (10.1.17) or as in (10.1.30) depending on the boundary condition (10.1.57) or (10.1.58) (in this case we take (10.1.30)). $S(t,\lambda)$ can be obtained from the integral equation given below (cf. [71]).

$$\lambda(S(\lambda,t)^2 - d^2) = 2\int_0^t f(\tau)d\tau - 2\int_0^d x\phi(x)dx + 2\int_0^{S(t,\lambda)} xT(x,t,\lambda)dx. \tag{10.1.59}$$

It has been shown that for a fixed $t > 0$, $S(t,\lambda)$ obtained as a solution of (10.1.56)–(10.1.57) or of (10.1.56) and (10.1.58), behaves as

$$S(t,\lambda) \sim [4t\log(1/\lambda)]^{1/2}, \ \lambda \to 0. \tag{10.1.60}$$

The above result has been proved using monotone dependence theorem and some other results obtained in [289]. For a fixed $t > 0$, estimates of $S(\lambda,t)$ have also been obtained.

The problem considered in [290] can be described in terms of (10.1.34)–(10.1.38) (Problem (P1)) or (10.1.34)–(10.1.37) and (10.1.39) (Problem (P2)) provided we take $d = 0$, i.e., $S(0) = 0$. The condition $S(0) = d > 0$ and the hypothesis of Lipschitz continuity of $\phi(x)$ at $x = b$ played a major role in [59] in proving the well-posedness of Problems (P1) and (P2). Both these assumptions essentially resulted in ensuring that $S(t)$ is Lipschitz continuous in $[0,t_*]$. If $d = 0$, then a different approach is needed. Some additional regularity conditions on q, f, λ, μ and u or g and sign constraints on the data and coefficients are needed so that the free boundary actually starts from $x = 0$ at $t = 0$. For the proof of existence of the solution, in addition to the assumptions (A), (B), (C_1), (D), and (E) mentioned earlier in this section, the following assumptions have been made.

Suppose a real number θ exists such that for $t \in [0,\theta]$ the following conditions are fulfilled:

(i) $-\lambda^{(\theta)} \le \lambda(x,t) \le -\lambda_0 < 0, \ x \ge 0, \ 0 \le t \le \theta,$ (10.1.61)

for some positive constants $\lambda^{(\theta)}$ and λ_0.

(ii) $\mu(x,t) \ge 0$, and $q(x,t) \le 0, \ x \ge 0, \ 0 < t \le \theta.$ (10.1.62)

(iii) $u(t) \ge 0, \ u(t) \ne 0$ in each neighbourhood of $t = 0$. (10.1.63)

Conditions (10.1.61)–(10.1.63) are sufficient to ensure that $T(x,t) \ge 0$ in $D_\theta = \{(x,t) : 0 < x < S(t), 0 < t < \theta\}$ and $\dot{S}(t)$ is monotonically increasing.

Let (S^d, T^d) be the solution of (10.1.34)–(10.1.38) in Problem (P1) or in Problem (P2) with $\phi(x) = 0$ in $0 \le x \le d$, for each $d \in (0,1)$. It has been shown in [290] that as $d \to 0$, the solution (S^d, T^d) converges to a solution of the system consisting of (10.1.34) and (10.1.36)–(10.1.38) with $d = 0$. For the existence proof an approach similar to the one used in [59] has been used in [290] also. The case $d > 0$ considered in (10.1.34)–(10.1.38) was reconsidered in [290] and it was observed that it is possible to relax the condition (F) (given earlier in this section) and replace it by the condition (F') given below.

(F') $\phi(x)$ is piecewise continuous in $(0, d)$ and two positive constants M and α $(0 \leq \alpha \leq 1)$ exist such that

$$|\phi(x) - f(d, 0)| \leq M(d - x)^{\alpha}, \ 0 \leq x \leq d, \tag{10.1.64}$$

in Problem (P1). The constant M is such that $Md^{\alpha} \geq \Phi + |f(d, 0)|$.

The assumption (F') can be further relaxed and replaced by (F'') which is more general.

(F'') $\phi(x)$ is piecewise continuous and bounded in $[0, d]$.

Assume that in addition to (10.1.61)–(10.1.63), we have

$$\phi(x) \geq 0, \ x \in [0, d]. \tag{10.1.65}$$

These restrictions ensure the monotonicity of $S(t)$.

It was observed that for proving the stability of the solution when $d = 0$, Lipschitz continuity of $S(t)$ (proved in the case $d > 0$) is not enough. When $d = 0$, another estimate for $\dot{S}(t)$ has been obtained as follows. There exist two constants δ and β ($\beta < 1/2$) such that

$$|\dot{S}(t)| < \delta t^{-\beta}. \tag{10.1.66}$$

The one-phase problem (3.1.1)–(3.1.5) has been studied in [55] by reformulating it as two different problems. Let the problem discussed in (3.1.1)–(3.1.5) be called Problem (FM). If (3.1.1) is replaced by the equation

$$T_t - a^2(x, t, T)T_{xx} = q(x, t, T, T_x), \ (x, t) \ \in \Omega(t_*), \tag{10.1.67}$$

$\psi(x, t) = 0$ in (3.1.4), and conditions (3.1.2), (3.1.3) and (3.1.5) are retained, then the problem so obtained will be called Problem (FMP).

If only (3.1.5) is replaced by

$$\dot{S}(t) = P_t(S, T), \tag{10.1.68}$$

and (3.1.3)–(3.1.4) are retained, then we get a generalization of Problem (FM) and this more general problem with (10.1.68) will be called Problem (GM) in which P_t is a functional. The equation (10.1.68) is more general than (3.1.5). Except the coefficient a, which has to be greater than zero, no other sign restriction has been imposed on the data for proving the existence, uniqueness and stability of the local-in-time solution. For bounds on the various functions, compatibility conditions and spaces to which various functions belong, the reader is referred to [55] as they will occupy considerable space.

The existence proof of the local-in-time solution of Problem (FMP) is based on a method of successive approximations whose convergence has been shown by an argument of contractive type. Under suitable assumptions (cf. [55]), there exists a solution $(t_*, S(t), T(x,t))$ of Problem (FMP) in which $S(t) \in H_{1+\alpha/2}(0, t_*)$ for any $\alpha \in (0,1)$. For $\nu \in (0,1)$, the space $H_\nu[b,d]$ is the space of all functions which are Hölder continuous with exponent ν in $[b,d]$. If for $\nu \in (0,1)$, $G(x) \in H_\nu[b,d]$, then for some constant A and all $\xi_1, \xi_2 \in [b,d]$

$$|G(\xi_1) - G(\xi_2)| \leq A\,|\xi_1 - \xi_2|^\nu . \tag{10.1.69}$$

The norm of $G(x)$ in $H_\nu[b,d]$, $\nu \in (0,1)$ is defined as:

$$\|G(x)\|_{H_\nu} = \sup_{x\in[b,d]} |G(x)| + \sup_{x_1,x_2\in[b,d]} |G(x_1) - G(x_1)| \,/\, |x_1 - x_2|^\nu . \tag{10.1.70}$$

For the definition of a suitable norm in $H_{N+\nu}[b,d]$, $N > 0$, see Appendix B.

The construction of approximating solutions in [55] is similar to that described in (10.1.46)–(10.1.51). The problem corresponding to an approximating solution for any index k, can be easily formulated as a fixed domain problem, for example, if we put $y = x/S(t)$, then for $0 \leq x \leq S(t)$, we have $0 \leq y \leq 1$. By obtaining estimates for S_k, $\dot{S}_k$, v_k, $v_{k,y}$ (the subscript k stands for the *k-th* approximating solution and v_k is the temperature in the new coordinate y), uniform interior Schauder estimates for v_k can be derived and it can be proved that the limit function v of v_k is the solution of the problem. Stability and uniqueness of the solution have also been established. Under suitable assumptions, the methods employed for proving results for Problem (FMP) can be extended to Problem (GM). By using the method discussed in [59] and obtaining uniform estimates, the local-in-time solution of Problem (FMP) can be extended to a solution valid in larger time intervals.

10.1.2 Infinite differentiability and analyticity of the free boundary

The differentiability of the free boundary has been discussed in [281] under assumptions of continuous differentiability of the initial and boundary data. Infinite differentiability of the free boundary in the one-dimensional Stefan problems has also been discussed in ([291, 292, 293]). A simple proof of the infinite differentiability of $S(t)$ which is widely referred has been given in [294] and we give here the main steps in this proof.

The problem formulation is as follows. Let $S(t)$ be a continuous function in $0 < t \leq t_*$ with $S(0) = d$ and let

$$\left.\begin{aligned} &T_t = T_{xx} \text{ in } D_S\,; \quad T(x,0) = \phi(x),\ 0 \leq x \leq d; \quad T(0,t) = f(t), \\ &T(S(t),t) = 0\,; \qquad T_x(S(t),t) = -\lambda \dot{S}(t), \qquad \lambda > 0, \end{aligned}\right\} \tag{10.1.71}$$

$$D_S = \{(x,t) :\ 0 < x < S(t),\ 0 < t < t_{*,}\}. \tag{10.1.72}$$

It is not necessary to impose any specific conditions on ϕ and f as we are concerned with the solution in a neighbourhood of $x = S(t)$. The main result of [294] is the following proposition.

Proposition 10.1.10. If the pair (S, T) satisfies (10.1.71), then $(S, T) \in C^\infty(\varepsilon, t_*)$ for any $\varepsilon > 0$.

By using the transformation

$$\xi = x/S(t),\ \tau = t, \tag{10.1.73}$$

a formulation of the Stefan problem (10.1.71) on the fixed domain $Q = (0,1) \times (0, t_*)$ can be obtained in which $x = S(t)$ becomes $\xi = 1$. This transformation is C^∞ with respect to x and C^1 with respect to t. If $v(\xi, t) = T(x, t)$, then

$$v_\tau = (1/(\hat{\sigma})^2)v_{\xi\xi} + (\sigma/\hat{\sigma})(\xi v_\xi),\ \text{in } Q, \tag{10.1.74}$$

where

$$\sigma = \dot{S},\ \text{and } \hat{\sigma} = S(t). \tag{10.1.75}$$

The initial and boundary conditions for v can be easily written. At $x = S(t)$, we have $v(1, t) = 0$. The Stefan condition is transformed into the condition

$$\sigma(t) = -(1/\hat{\sigma}(t))(v_\xi(1,t))/\lambda,\ 0 < t < t_*. \tag{10.1.76}$$

The following proposition has been proved in [295] and with its help, the proof of Proposition 10.1.10. has been completed in [294].

Proposition 10.1.11. If $T(x,t)$ is a bounded solution of the equation $T_t = aT_{xx} + bT_x + eT$, $(a > 0)$ in the region Q such that $T(1,t) = 0$ for $0 \le t \le t_*$ and if the coefficients a, b and e belong to $H_\alpha(Q)$ $(\alpha > 0)$, then for any $\varepsilon > 0$, we have $T(x,t) \in H_{\alpha+2}(Q_\varepsilon)$, where $Q_\varepsilon = (\varepsilon, 1) \times (\varepsilon, T)$. Here H_α stands for an anisotropic Hölder space (cf. [295]).

Anisotropic Hölder spaces enter in the a priori estimates of Schauder type for the parabolic equations. If for some $\alpha > 0$, $\sigma \in H_\alpha(Q_\delta)$, $\delta \ge 0$, then from (10.1.76) $\hat{\sigma} \in H_{\alpha+1}(Q_\delta)$ (σ can be taken as a function on Q_δ although it is not a function of x). From Proposition 10.1.11., we conclude that $v \in H_{\alpha+2}(Q_{\delta+\epsilon})$ as coefficients in (10.1.74) belong to $H_\alpha(Q_\delta)$. Further, $v_\xi \in H_{\alpha+1}(Q_{\delta+\epsilon})$. From (10.1.76), $\sigma \in H_{\alpha+1}(Q_{\delta+\varepsilon})$. If $\sigma \in H_{\alpha+1}(Q_{\delta+\varepsilon})$, then from (10.1.74) and Proposition 10.1.11. $v_\xi \in H_{\alpha+2}(Q_{\delta+\varepsilon'})$ and if $v_\xi \in H_{\alpha+2}(Q_{\delta+\varepsilon'})$, then $\sigma \in H_{\alpha+2}(Q_{\delta+\varepsilon'})$ from (10.1.76). Proceeding inductively in this way it can be proved that $\sigma \in C^\infty(\varepsilon, t_*)$, for any $\varepsilon < t_*$. To start the induction process one needs an initial estimate for the Hölder continuity (see equation (10.1.69)) of v_ξ which has been obtained in [295]. The infinite differentiability of the temperature follows from the Stefan condition. It has been mentioned in [55] that if a, q and ϕ in (3.1.1)–(3.1.5) are infinitely differentiable, then $S(t)$ is infinitely differentiable.

One of the main results in [296] is that if $f(t)$ in (10.1.71) is an analytic function in $0 \leq t \leq t_*$, then $S(t)$ is also analytic in $0 \leq t \leq t_*$. Assume that in (10.1.71), $f \geq 0$, $f(0) = \phi(0)$, $\phi \geq 0$, $\phi(b) = 0$, $\lambda = 1$, and $f(t)$ is analytic in $0 < t < t_*$. The sign constraints ensure that a unique solution exists in some time interval. If f and ϕ are continuously differentiable, then it can be proved that $\dot{S}(t)$ is continuous. Some of the steps in the proof of analyticity in [296] are: (1) Converting the free boundary problem into a fixed domain problem in which $S(t)$ is fixed at $y = 1$ (see (10.1.77)) and then an application of Proposition 10.1.11. This will ensure that the temperature $v(y, \tau)$ is a C^∞-function in $0 \leq y \leq 1,\ 0 \leq \tau \leq \tau_0$ and $S(\tau)$ is $C^\infty[0, \tau_0]$. (2) Obtaining appropriate estimates for the derivatives of v and S of all orders.

We state below the transformations used in [296] which converts the Stefan problem into an appropriate parabolic problem on a fixed domain for which Proposition 10.1.11 is applicable. Let

$$y = x/S(t), \ \tau = \int_0^t \frac{d\alpha}{S^2(\alpha)}, \ \tau_0 = \int_0^{t_*} \frac{d\alpha}{S^2(\alpha)}, \tag{10.1.77}$$

$$v(y, \tau) = T(x, t) - (1 - x/S(t))f(t). \tag{10.1.78}$$

It can be seen that

$$\frac{dt}{d\tau} = S^2(t), \text{ and } S(t)\dot{S}(t) = f(t) - v_y(1, \tau), \tag{10.1.79}$$

$$\dot{S}(t) = -T_x(S(t), t) = [-v_y(1, \tau) + f(t)]/S(t), \tag{10.1.80}$$

$$v_{yy} - v_\tau = y[v_y(1, \tau) - f(t)][v_y - f(t)] + (1 - y)f'(t)S^2(t), \ 0 < y < 1, \ 0 < \tau < \tau_0, \tag{10.1.81}$$

$$v(0, \tau) = v(1, \tau) = 0, \quad 0 < \tau < \tau_0. \tag{10.1.82}$$

All the coefficients in (10.1.81) are not known but we know their behaviour. By the application of Proposition 10.1.11., it can be concluded that $v(y, \tau)$ and $S(\tau)$ are C^∞-functions for $0 \leq y \leq 1,\ 0 \leq t \leq t_*$. The proof of analyticity of $S(t)$ requires estimates of the derivatives of all orders of several quantities such as $v(y, \tau)$, $v_y(y, \tau)$, $v_{yy}(y, \tau)$ and $S(t(\tau))$. It has been proved that if $f(t)$ is analytic in $0 \leq t \leq t_*$, then $S(t)$ is analytic in $0 < t \leq t_*$ (cf. [296]). Firstly, Schauder type interior-boundary estimates are obtained for the heat equation $T_t - T_{xx} = 0$ in the rectangle $-1 < x < 1,\ 0 < t < t_*$. The way in which negative powers of t (as $t \to 0$) enter into the estimates is crucial. These estimates are then used in proving further results concerning analyticity of $S(t)$. The estimates for $S(t(\tau))$ imply that $S(t(\tau))$ is analytic in τ and so is $S^2(t)$. The first of (10.1.79) implies that $t = t(\tau)$ is analytic in τ and $\tau = \tau(t)$ is analytic in t. Writing $S(t) = S(t(\tau(t)))$, we conclude that $S(t)$ is analytic in t.

The analyticity of $S(t)$ at $t=0$ has been studied in [297]. At $t=0$, the region $x>0$ is at the melting temperature zero and the region $x<0$ is occupied by a warm liquid. We consider the following problem formulation.

$$\left.\begin{aligned} &T_{xx}-T_t=0,\ -\infty<x<S(t),\ t>0;\ T(x,0)=f(x),\ -\infty<x<0,\\ &T(S(t),t)=0,\ t>0;\ \dot{S}(t)=T_x(S(t),t)-\phi(t),\ t>0;\ S(0)=0. \end{aligned}\right\} \tag{10.1.83}$$

In [21], the problem formulated in (10.1.83) has been named as *Cauchy-Stefan problem* and the problem formulated in (10.1.71) has been named *Dirichlet-Stefan problem.* It has been proved in [297] that if $\phi(t)$ in (10.1.83) is analytic in $(0,t_0)$, $t_0>0$, and if the solution of (10.1.83) exists in $(0,t_1)$, $t_1>0$, then $S(t)$ is analytic in $(0,\hat{t}\,)$, $\hat{t}=\min(t_0,t_1)$. For many practical purposes, we require polynomial approximations of $S(t)$ in terms of the initial data. The following results about the analyticity of $S(t)$ have been established in [297].

Proposition 10.1.12. Assume that $\phi(t)$ is analytic with respect to $\sqrt{t}$ in $[0,t_0]$, for some $t_0>0$ and f is an entire function of x such that

$$f(0)=0,\ \text{and}\ |f'(x)|\le \mathrm{M}\exp(\alpha\, x^2), \tag{10.1.84}$$

for some positive constants M and α. Then there exists $t_1\in(0,t_0]$ such that (10.1.83) has a unique solution and the free boundary $x=S(t)$ has the following series representation

$$S(t)=\sum_{n=2}^{\infty} Y_n t^{n/2}/n!,\ t\in[0,t_1). \tag{10.1.85}$$

In (10.1.85), Y_n is defined as follows:

$$y=\sqrt{t},\ Y(y)=S(y^2),\ \text{and}\ Y(y)=\sum_{n=1}^{\infty}\frac{Y_n}{n!}y^n. \tag{10.1.86}$$

To prove Proposition 10.1.12., integral representations of $v(t)=T_x(S(t),t)$ and $S(t)$ (see equations (10.1.17) and (10.1.21)) have been used in conjunction with complex variable techniques. The transformation $y=\rho e^{i\phi}$ introduces a complex variable. Let $C_R=\{y:|y|<\rho_0\}$. The integral representations of $v(y)$ and $Y(y)$ define a mapping $\mathcal{P}$ on a set $M(\rho_0,N)$ of functions $w(y)$, $|w(y)|\le N$, $y\in C_R$ which are analytic in C_R and continuous in $\overline{C}_R$. It has been proved that $\mathcal{P}$ maps $M(\rho_0,N)$ into itself. This mapping is a contraction with respect to the distance metric. If $\overline{V}$ is the unique fixed point of the mapping $\mathcal{P}$, then the restriction of $\overline{V}$ to the real axis is an analytic function in $[0,\rho_0]$. Recursive relations have been developed to determine Y_n. It may be noted that for proving the analyticity, estimates of Y_n are not required. The uniqueness of the one-phase Stefan problem is well-known. The domain of analyticity can be extended under certain assumptions.

If $f(0)\ne 0$, then the following result holds good.

Proposition 10.1.13. Assume that $\sqrt{t}\,\phi(t)$ is analytic with respect to $\sqrt{t}$ in $[0,\,t_0]$, that $\left|\phi(t)\sqrt{t}\right| \leq \phi_0$, and $\lim_{x \to 0-} f(x) = f_0$. Then, if the other assumptions of Proposition 10.1.12. are satisfied, two positive constants ϕ_0^*, f_0^* can be found such that for any $\phi_0 \leq \phi_0^*$, $|f_0| \leq f_0^*$, the problem described in (10.1.83) has a unique solution in the class of solutions whose free boundary is analytic in $[0, t_1)$, for some $t_1 \in (0, t_0]$, with respect to $\sqrt{t}$. If $t^{1/2}\phi(t) \to 0$ as $t \to 0+$ and if $f_0 \geq 1$, then there is no solution of (10.1.83) such that $S(t)/t^{1/2}$ has a bounded limit as $t \to 0+$.

In [297], the problem (10.1.71) has been considered in the region $-d \leq x \leq 0$ ($\lambda = -1$) also. For an unbounded region it was assumed earlier that $f(0) = 0$ but for the problem in a bounded domain this assumption has been relaxed and the analyticity of the free boundary has been proved in $(0, t_0)$ under the assumption that $|f(0)| < f_0^*$, $f_0^* > 0$ and $f(t)$ is analytic for $t > 0$. The proof of this result is on the same lines as the proof of Proposition 10.1.12.

The analyticity of the free boundary in the one-phase Stefan problem, with strong nonlinearity formulated in (10.1.23)–(10.1.27), has been discussed in [298]. The main result of [298] is as follows:

Proposition 10.1.14. The free boundary $S(t)$ is a holomorphic function in some neighbourhood $\cup(t_*) \subset D\left(\hat{t}\right)$ of the interval $0 < t < t_*$, $\forall\ t_* < \hat{t}$, where $D(t_*) = \{t = \rho e^{i\alpha} \in \mathbb{C};\ 0 < \rho < t_*,\ |\alpha| < \pi/32\}$, and $\hat{t}$ is the supremum of all $t_* > 0$ such that there exists a solution to the problem on $[0, t_*]$.

The proof of this proposition is based on the application of Banach contraction mapping theorem to the system of integral equations obtained in the method of Picard iteration (see equations (10.1.29)–(10.1.31)) and extended into the complex plane. The possibility of applying this principle follows from a priori estimates of the heat potentials and their variations in the complex plane. These estimates which are crucial for the proofs developed in [298] also provide generalization of the results obtained in [299]. In order to prove the analyticity of the free boundary in the nonlinear problem (10.1.23)–(10.1.27), the analyticity of Poisson's integral and of the volume heat potentials as well as their variations is needed along the free boundary. If in the complex (x, t) plane, x and t are independent, then the analyticity of Poisson's integral and of other quantities, generally does not hold up to the free boundary. However, the substitution $x = \lambda S(t)$, $\lambda \in [0, 1]$, helps in proving that the integrals are holomorphic in the sector $D(t_*)$, ($t_* > 0$ and small enough) when the integrals are considered as functions of t.

By representing temperature in an infinite series of integrals of error functions, and the free boundary in an infinite series in positive integral powers of $\sqrt{t}$, series solutions of temperatures and the free boundary in a one-dimensional Stefan problem have been obtained in [300]. Proof of the convergence of series expansions considered have been developed in [300] and also in several other works of the same author mentioned in [300]. It has been remarked in ([297, 298]) that the analyticity proofs are unconvincing. A simple proof has been developed in [301] which shows that there are infinitely many temperature solutions satisfying all the equations of the problem considered in [300]. All these temperatures give the same free boundary which has been obtained in [300]. In view of the non-uniqueness of the solution of the temperature in [300], any proof about the convergence of the series does not seem to be important.

10.1.3 Unilateral boundary conditions on the fixed boundary: Analysis using finite-difference schemes

The one-phase problem considered in [302] is concerned with the melting of a solid and its formulation can be obtained if some changes are made in the formulation given in (10.1.56). Take $\lambda = -1$ and prescribe the following unilateral boundary condition at $x = 0$.

$$T_x(0,t) \in \gamma(T(0,t)), \; t > 0. \tag{10.1.87}$$

Here, γ is a maximal monotone graph (see Appendix B) in R^2 with $\gamma(\alpha) \ni 0$ for some non-negative constant α. We give below an example of a unilateral boundary condition.

Unilateral boundary condition at x=0

Let $T(0,t) \geq 2$; $T_x(0,t) \leq -3$, $(T = 2)$; $T_x(0,t) = -3$, $(2 < T < 3)$; $T_x(0,t) = T - 6$, $(T \geq 3)$. Define $\gamma(T)$ as:

$$\left.\begin{aligned} \gamma(T) &= \emptyset, \; T < 2, \\ &= (-\infty, -3], \; T = 2, \\ &= -3, \; 2 < T < 3, \\ &= T - 6, \; T \geq 3. \end{aligned}\right\} \tag{10.1.88}$$

Signorini-type boundary condition at x=0

$$T(0,t) \geq 2, \; T_x(0,t) \leq 0, \text{ and } T_x(0,t)\,(T(0,t) - 2) = 0. \tag{10.1.89}$$

The unilateral boundary conditions model several physical situations in controlling the temperature of a body during heating or cooling at the fixed boundary. The problem described in (10.1.56) with $\lambda = -1$ together with the boundary condition (10.1.87) at $x = 0$ will be called Problem (Y). The existence and uniqueness of the global-in-time solution of Problem (Y) have been investigated in [302]. For the analysis it will be assumed that

$$\phi(x) \geq 0, \text{ is bounded and continuous a.e. for } x \in [0,d]. \tag{10.1.90}$$

We introduce the following notations:

$$\left.\begin{aligned} &D = \{(x,t) : 0 < x < S(t), 0 < t \leq t_*\}, \; \overline{D} = \text{the closure of } D\,, \\ &D^S = \{(x,t) : 0 < x < S(t), \; 0 < t \leq t_*\}, \\ &Z = \{x \in [0,d] : \; x \text{ is a point of the discontinuity of } \phi\} \times \{0\}\,. \end{aligned}\right\} \tag{10.1.91}$$

Definition 10.1.1. The pair (T, S) is a solution of Problem (Y) if the following conditions are satisfied.

(1) $S(0) = d,\ S(t) > 0$ for $t > 0,\ S \in C[0, t_*] \cap C^\infty(0, t_*)$.

(2) T is bounded on $\overline{D},\ T \in C^\infty(D^S) \cap C(\overline{D} - Z)$,

$$\int_\tau^{t_*}\int_0^{S(t)} T_{xx}(x,t)^2 dx dt < \infty, \text{ for each } \tau \in (0, t_*).$$

(3) The pair (T, S) satisfies the heat equation, initial condition and the free boundary conditions in (10.1.56) with appropriate changes as described above.

(4) For a.a. $t \in [0, t_*)$, (10.1.87) is satisfied.

The term 'solution' has been used in the above sense in [302]. The main result of [302] is the following proposition.

Proposition 10.1.15. If $d > 0$, and ϕ satisfies (10.1.90), then there exists a unique solution (S, T) of Problem (Y) satisfying

$$\int_0^{t_*}\int_0^{S(t)} tT_{xx}^2\, dx dt + t_* \int_0^{S(t_*)} T_x^2 dx < +\infty. \tag{10.1.92}$$

Let $d > 0$. The existence proof in this case consist of the following steps.

(1) First an 'implicit' finite-difference discretization of equations in Problem (Y) is done in which the mesh size is of uniform width Δx and time steps $\{\Delta t_n\}, n = 1, 2...$ are of variable size such that the free boundary after time $t_n = \cup_{k=1}^{n} \Delta t_k,\ n = 1, 2, ...,$ is at $x_n = n\Delta x,\ n = 1, 2...,$ i.e., $S(t_n) = x_n$. Thus the position of the free boundary is always known. The variable time step can be obtained approximately from the Stefan condition $\dot{S} = -T_x(S(t), t)$.

(2) The unilateral boundary condition is handled as follows. The discretization of the unilateral boundary condition will have the form

$$\frac{T_1^n - T_0^n}{\Delta x} \in \gamma(T_1^n). \tag{10.1.93}$$

Here, the superscript n stands for the temperature at time t_n and the subscript 0 and 1, stand for temperatures at $x = 0$ and $x = \Delta x$, respectively. Since γ is a maximal monotone graph, $(I + \Delta x\gamma)^{-1},\ \Delta x > 0$, is a contraction mapping from R to itself with $D((I + \Delta x\gamma)^{-1}) = R$ (D stands for the domain and I for the identity map) and $(I + \gamma)(u) = \{u + f, f \in \gamma(u)\}$ (cf. [303]). It can be shown that (10.1.93) is equivalent to $T_0^n = (I + \Delta x\gamma)^{-1}(T_1^n)$. If instead of (10.1.93), we take

$$T_0^n = \xi \in R, \tag{10.1.94}$$

and study the discretized problem with (10.1.94), then this problem has a unique solution (cf. [304]). We define a mapping P as:

$$P : R \ni \xi \mapsto (I + \Delta x \gamma)^{-1}(T_1^{\ n}(\xi)) \in R. \tag{10.1.95}$$

It can be shown that P is a contraction. Therefore, for each J, $1 \leq J \leq m$ (m is the total number of mesh points in space), T_J^n is the unique solution of the discretized Problem (Y) with condition (10.1.93).

(3) The next step is to obtain suitable estimates of T_J^n and $S_n = S(t_n)$. Further conditions are imposed on $\phi(x)$, and $\phi(x)$ satisfies one of the two conditions given below.

(A1) Let $\phi(x)$ satisfy the condition (10.1.90). Further, for some positive constant β, let $\phi(x) \leq \beta(d - x)$, for $x \in [0, d]$, $d > 0$. Note that (A1) $\Rightarrow$ (10.1.90).

(A2) Let $\phi(x)$ satisfy the condition (A1). Further, $-\beta x + \delta^1 \leq \phi(x) \leq \beta x + \delta^2, \delta^1, \delta^2 \in D(\gamma)$. $(A2) \Rightarrow (A1) \Rightarrow$ (10.1.90)

(4) Extend the discretization in space suitably to the region $\overline{G}$, where $G = (0, \infty) \times (0, t_*]$. By using *Ascoli-Arzela theorem* [58], it has been proved that $T_J(x, t)$ (see [302] for the construction of $T_J(x, t)$ on G) converges uniformly on compact subsets in G to a function $T(x, t) \in C(\overline{G})$. The numerical solution should converge to the solution of Problem (Y). The convergence of the numerical solution to $T(S(t), t)$ and $T(x, 0)$ (for those x at which $\phi(x)$ is continuous) has been proved by using estimates of the absolute values of T_J^n and S_n and imposing a further condition on $\phi(x)$. In addition to (A1), $\phi(x)$ is such that

$$\phi \in C^3 \left[d', d\right] \text{ for some } d', \ 0 \leq d' < d. \tag{10.1.96}$$

Existence of the solution under the condition (10.1.90) has also been discussed. The proof of the convergence of the numerical solution to the solution of Problem (Y) satisfying the unilateral boundary condition and the Stefan condition requires L^2-estimates of the finite difference solution. It has been shown that $S(t) \in C^1(0, t_*)$ and $\dot{S}(t) = -T_x(S(t), t)$. As discussed in [294], under these conditions $S(t) \in C^\infty(0, t_*]$ and $T \in C^\infty(D^S)$.

To prove the uniqueness of Problem (Y), first the existence of the unique solution of temperature in an auxiliary problem which we call Problem (M) is proved. In Problem (M), $S(t)$ is assumed to be known and $S(t)$ is a non-decreasing function, positive for $t > 0$, and $S(t) \in C\left[0, t_*\right] \cap C^{0,1}(0, t_*]$. $C^{0,1}(0, t_*]$ denotes the space of Lipschitz continuous functions on $(0, t_*]$. In Problem (M), all the equations of Problem (Y) are present except the Stefan condition ($S(t)$ is known). By using an integral equation formulation as in (10.1.59), it can be proved that if $T(x, t)$ is the solution of Problem (M), and $S(t) \in C\left[0, t_*\right] \cap C^{0,1}(0, t_*]$, then $T_x(S(t), t) = -\dot{S}(t)$, $t \in (0, t_*]$.

The following comparison results have been proved.

Proposition 10.1.16. If $\phi_1 \leq \phi_2$ and $0 \leq d_1 \leq d_2$, then $S_1 \leq S_2$ where (S_i, T_i), $i = 1, 2$ are the solutions of Problem (Y) corresponding to the data $\{d_i, \phi_i\}$.

The uniqueness of $S(t)$ can be easily proved by using the argument that $\phi_1 = \phi_2 \Rightarrow \phi_1 \geq \phi_2$ and it also implies $\phi_1 \leq \phi_2$ and applying the Proposition 10.1.16.

Proposition 10.1.17. Let T and $\hat{T}$ be two solutions of Problem (M) corresponding, respectively, to the data $\{S(t),\ T(x,0)\}$ and $\{\hat{S}(t),\ \hat{T}(x,0)\}$. Let $d > 0$ and ϕ satisfies (10.1.90). Further, let $S(t) \leq \hat{S}(t)$, $T(x,0) \leq \hat{T}(x,0)$ and $\hat{T}(S(t),t) \geq T(S(t),t) = 0$. Then we have $0 \leq \hat{T} - T \leq \max\left\{||\hat{T}(\cdot,0) - T(\cdot,0)||_{L^\infty},\ ||\hat{T}(S(t),t)||_{L^\infty(0,t_*)}\right\}$ in $\overline{D} = \{(x,t) : 0 \leq x \leq S(t),\ 0 \leq t \leq t_*\}$, where $L^\infty = L^\infty(0,d)$, $S(0) = d$. When $d = 0$, we take $\|\hat{T}(\cdot,0) - T(\cdot,0)\|_{L^\infty} = 0$

In the problem considered in [302] (in which $d > 0$) take $d = 0$ and this case has been considered in [305]. The existence of a unique solution has been proved. The asymptotic behaviour of the solution as $t \to \infty$ has also been investigated. The following results have been established.

Proposition 10.1.18. Let $d = 0$ and γ satisfies the assumptions (a) and (b) given below.

(a) $\gamma^{-1}(0) \cap [0,\infty)$ is not an empty set.

(b) $\gamma(0) \subset (-\infty,0)$.

Then there exists a unique solution (S,T) of Problem (Y) with $d = 0$ and the following results hold.

(i) $S \in C[0,\infty) \cap C^\infty(0,\infty)$, and $S(t)$ is non-decreasing in t,

(ii) $0 < T(x,t) \leq \alpha$, in $\overline{D}$,

(iii) $|T(x',t) - T(x,t)| \leq C_\sigma\,|x' - x|$, on $\overline{D} \cap \{t \geq \sigma\}$.

Here, $\alpha = \mathrm{Proj}_{\gamma^{-1}(0)}(0)$, i.e., $\alpha \geq 0$ is an element of $\gamma^{-1}(0)$ which has minimum absolute value. The assumption (b) implies that $\alpha > 0$.

Proposition 10.1.19. Let $\alpha > 0$ and $d \geq 0$ and $\phi(x)$ satisfies assumptions in (10.1.90). Then we have

(iv) $\lim_{t\to\infty} T(x,t) = 0$ uniformly on any compact subset of $[0,\infty)$,

(v) $\lim_{t\to\infty} S(t)/\sqrt{t} = \beta$,

where β is the unique solution of

$$\sum_{n=1}^{\infty}(n!/2n!)\beta^{2n} = \alpha. \tag{10.1.97}$$

When $\alpha = 0$ and $d > 0$, $\lim_{t\to\infty} T(x,t) = 0$ uniformly on $(0,t_*)$ and $\lim_{t\to\infty} S(t) = S^*$ exists such that $d \leq S^* \leq d + \int_0^d \phi(x)dx$.

10.1.4 Cauchy-type free boundary problems

In the Cauchy-type free boundary problems, the temperature and its normal derivative at the free boundary are prescribed as functions of t and $S(t)$. Such problems have been discussed earlier in § 3.3.1. We shall discuss now some more general Cauchy-type problems considered in [306].

Problem (C) Find a triple (t_*, S, T), where (S, T) is a classical solution of the following problem.

$$a(x,t,T,T_x,S)T_{xx} - T_t = q(x,t,T,T_x,S), \text{ in } D_{t_*} = \{(x,t) : 0 < x < S(t),\ 0 < t < t_*\}, \tag{10.1.98}$$

$$T(x,0) = \phi(x),\ 0 \le x \le b = S(0);\ T(0,t) = u(t),\ 0 < t < t_*, \tag{10.1.99}$$

$$T(S(t),t) = f(S(t),t),\ 0 < t < t_*, \tag{10.1.100}$$

$$T_x(S(t),t) = g(S(t),t),\ 0 < t < t_*. \tag{10.1.101}$$

Instead of the temperature, flux can also be prescribed at $x = 0$. Our concern here is to obtain a free boundary condition in which $\dot{S}(t)$ is explicitly appearing.

Problem (C1) Let

$$f_x(x,t) - g(x,t) \ne 0. \tag{10.1.102}$$

On differentiating (10.1.100) with respect to time, we get

$$\frac{dT}{dt} = (T_x\dot{S} + T_t) = g\dot{S} + T_t = f_x\dot{S} + f_t, \text{ at } x = S(t), \tag{10.1.103}$$

or

$$((f_x - g)\dot{S} + T_t - f_t)_{x=S(t)} = 0. \tag{10.1.104}$$

If (10.1.102) holds, then from (10.1.104), $\dot{S}$ can be expressed as

$$\dot{S} = \{(aT_{xx} - q - f_t)/(g - f_x)\}_{x=S(t)}\,. \tag{10.1.105}$$

The Problem (C1) consists of (10.1.98), (10.1.99), (10.1.101) and (10.1.105). From (10.1.103) and (10.1.104), we get

$$\frac{dT}{dt} = T_t + T_x\dot{S} = T_t + g\dot{S} = f_x\dot{S} + f_t = \frac{df}{dt}, \text{ at } x = S(t), \tag{10.1.106}$$

which is the condition (10.1.100).

Problem (C2) Let

$$f_x(x,t) = g(x,t). \tag{10.1.107}$$

If (10.1.107) holds, then from (10.1.104), we have

$$T_t = f_t, \text{ at } x = S(t). \tag{10.1.108}$$

From (10.1.100), (10.1.101) and (10.1.107), we have

$$\frac{d}{dt}(T_x) = T_{xx}\dot{S} + T_{xt} = g_x\dot{S} + g_t = f_{xx}\dot{S} + f_{xt}, \text{ at } x = S(t). \tag{10.1.109}$$

On using (10.1.98) and (10.1.108) in (10.1.109), we get

$$a^{-1}(q + f_t)\dot{S} + T_{xt} - f_{xx}\dot{S} - f_{xt} = 0, \text{ at } x = S(t) \tag{10.1.110}$$

If

$$af_{xx} - f_t - q \neq 0, \tag{10.1.111}$$

then from (10.1.110), we get

$$\dot{S} = a(af_{xx} - f_t - q)^{-1}(T_{xt} - f_{xt}), \text{ at } x = S(t). \tag{10.1.112}$$

The r.h.s. of (10.1.112) involves T_{xxx}. The Problem (C2) consists of (10.1.98)–(10.1.100) and (10.1.112). Several differentiation operations have been done in (10.1.102)–(10.1.112) which will be valid under the following assumptions. These assumptions are also required to prove some results concerning analysis of these problems.

(A1) ϕ, u, a and q are continuous functions of their arguments, $f(x,t)$ is continuous for $x > 0,\ t \geq 0$, and $g(x,t)$ is continuous for $x > 0,\ t > 0$. Further,

$$\phi(0) = u(0), \ \phi(b) = f(b,0). \tag{10.1.113}$$

(A2) The function f is continuously differentiable for $t > 0$.

(A3) g is continuously differentiable for $t > 0$ and

$$\phi'(b) = g(b,0). \tag{10.1.114}$$

If assumptions (A1) and (A2) are satisfied and (10.1.102) holds, then any solution of Problem (C1) solves the Problem (C) (only (10.1.100) is to be satisfied which has been done in (10.1.106)). Under assumptions (A1), (A2), (A3), and the conditions (10.1.107) and (10.1.111), the solution of Problem (C2) will be the solution of Problem (C) if it can be proved that (10.1.101) is satisfied. It can be proved that the solution of Problem (C2) satisfies

$$T_x(S(t),t) = f_x(S(t),t). \tag{10.1.115}$$

From (10.1.103), we have

$$(T_x - f_x)\dot{S}(t) + (T_t - f_t) = 0, \text{ at } x = S(t). \tag{10.1.116}$$

On differentiating (10.1.101), we get

$$\frac{d}{dt} T_x|_{x=S(t)} = T_{xx}\,\dot{S} + T_{xt}. \tag{10.1.117}$$

On substituting T_{xt} from (10.1.110) and T_{xx} from (10.1.98) in (10.1.117), we get

$$\frac{d}{dt} T_x(S(t), t) = \left[a^{-1}(T_t - f_t)\right]_{x=S(t)} \dot{S}(t) + \frac{d}{dt} f_x(S(t), t). \tag{10.1.118}$$

If $X(t) = (T_x - f_x)_{x=S(t)}$, then from (10.1.116) and (10.1.118), we have

$$\dot{X}(t) = -\left[a^{-1}\right]_{x=S(t)} \dot{S}^2 X(t),\ X(0) = 0,\ a > 0. \tag{10.1.119}$$

The solution of (10.1.119) is $X(t) = 0$ which implies (10.1.115). The well-posedness of Problem (C1) has been proved in [306] under the assumption that $T_x(S(t), t) = 0$ and that of Problem (C2) under the assumption that $T_t(S(t), t) = 0$. Since these problems are strongly nonlinear, several assumptions are required to prove the results and the reader is referred to [306] for other assumptions.

Under suitable assumptions about the data and some appropriate compatibility conditions, Problem (C1) has a solution $(\hat{t}, T, S)$ which is unique in $(0, \hat{t}\,)$. Moreover, $S(t) \in H_{1+\nu}[0, \hat{t}\,]$, $\nu \in (0, \alpha]$, $\alpha \in (0, 1)$, and $T_x \in C_{1+\nu}(D_{\hat{t}})$ (see Appendix B for the definition of $C_{1+\nu}$). Continuous dependence of the solution on the data has also been proved. Under slightly different assumptions, similar results have been proved for Problem (C2).

An implicit free boundary problem has been considered in [307] and to obtain this problem formulation we take $a = 1$ and $q = 0$ in (10.1.98), $f(S(t), t) = f(t)$ in (10.1.100), and $g(S(t), t) = g(t)$ in (10.1.101). Such problems are usually reduced to problems in which $\dot{S}(t)$ occurs explicitly and then for analysis either Schauder fixed point theorem or contraction mapping theorem is applied. In [307], the implicit free boundary condition is retained and the existence and uniqueness of the solution has been proved by showing the convergence of a numerical solution obtained by finite-difference discretization of the parabolic heat equation in time but not in space. The discretized equation at $n\Delta t$, $n = 1, 2, \ldots.$, can be written as a second order differential equation in space variable whose explicit solution has been constructed. This solution can be used to determine S_n. In this way we obtain a sequence of interrelated free boundary problems. In each time interval, an appropriate free boundary problem is to be solved in the region $0 \le x \le S_n$, $n = 1, 2, \ldots;$ $S_n = S(n\Delta t)$, where, Δt is the time step. In this way, we construct a two-dimensional function $\hat{T}(x, t)$ and a polygonal path $\hat{S}(t)$ which approximate the solution (T, S) of the original Stefan problem. The convergence of the approximate solution as $\Delta t \to 0$ ($n\Delta t$ remains finite) has been shown under appropriate assumptions.

In the problem studied in [308], instead of an implicit condition at the free boundary, the Stefan condition is prescribed and the problem has been studied with the help of its numerical solution. The numerical solution involves only finite-difference discretization in time as in [307]. In each time interval of length Δt, the free boundary is assumed to be known approximately and the temperature is determined by solving a parabolic heat conduction problem. The free boundary in the next time interval of length Δt is obtained with the help of its earlier value and the temperature derivative already calculated. By solving a sequence of time-independent free boundary problems we obtain a polygonal path $\hat{S}(t)$ which approximates $S(t)$ and also a two-dimensional temperature function. The convergence of the numerical solution has been proved and error estimates have been obtained.

The Problem 3.3.2. described in § 3.3.1 is a particular case of the problem (10.1.98)–(10.1.101). In this case because of the simplicity of the problem the following existence result can be proved (cf. [70]) under some simple assumptions about the data.

Proposition 10.1.20. Suppose in Problem 3.3.2. we have: (i) $0 = a < b$, $t_* = \infty$, $\beta = 0$, and $\alpha = 1$, (ii) $F(x,t) = \hat{h}(x,t) = 0$, (iii) $g(x,t) = g(x) \in C^1$ and there exist constants g_* and g^* such that $0 < g_* \leq -g'(x) \leq g^*$ for $b \leq x < \infty$, (iv) $0 \leq f(t) \in C$ for $0 \leq t < \infty$, (v) $\phi(x) \in C^1$, $\phi(b) = g(b)$, and there exists a constant N_1 such that $0 \leq \phi'(x) \leq N_1(b-x)$ for $0 \leq x \leq b$. If the above data assumptions hold, then there exists a unique classical solution (T,S) of Problem 3.3.2. Furthermore, for each fixed $0 < t_0 < \infty$ there exists a positive constant A depending only on b, g_*, g^*, N_1 and M such that $0 \leq \dot{S}(t) \leq A$ for $0 \leq t \leq t_0$. Here

$$M = \max\left\{ \sup_{0 \leq t \leq t_0} f(t), \quad \sup_{0 \leq x \leq b} \phi(x) \right\}.$$

10.1.5 Existence of self-similar solutions of some Stefan Problems

Some results about the existence and uniqueness of the classical solutions of one-dimensional one-phase Stefan problems have been reported in [309]. These results follow from some theorems proved in the analysis of Stefan problems in $R^n, n \geq 1$. In § 10.4, Meirmanov's [309] method of introducing local coordinates to prove the existence and uniqueness of solutions has been briefly described. Very general results about the analysis of one-dimensional one-phase problems have already been reported in this section. Therefore, instead of the existence and uniqueness results, we present here some results reported in [309] about the asymptotic behaviour of the solution of a Stefan problem whose formulation is as follows:

$$\frac{\partial \Phi(T)}{\partial t} = \frac{\partial^2 T}{\partial x^2}, \quad 0 < x < S(t), \quad S(0) = d, \quad 0 < t < t_*. \tag{10.1.120}$$

Here, the thermal diffusivity $k = (dt/d\Phi)$. Other conditions are:

$$T = 0, \text{ and } \frac{dS}{dt} = -\frac{dT}{dx}, \quad \text{on } x = S(t); \quad T(x,0) = T_0(x), \quad 0 \leq x \leq d, \tag{10.1.121}$$

$$T = f(t) \text{ or } \partial T/\partial x + b(t)T = g(t) \text{ at } x = 0,\ 0 < t < t_*. \tag{10.1.122}$$

Proposition 10.1.21. Let $\Phi \in C^2[0,\infty)$, $\Phi'(T) \geq a_0 > 0$ and non-negative functions $f(t)$ and $T_0(x)$ be such that for some constant M, we have

$$f(t) \leq Md, \text{ and } T_0(x) \leq M(d-x). \tag{10.1.123}$$

Then a unique global classical solution (S,T) exists for all finite $t > 0$ with $\dot{S}(t)$ Hölder continuous for $t > 0$ and $T(x,t) \in H^{r,r/2}(\overline{G_{\delta,t_*}})$, $r > 2$ in each bounded domain $G_{\delta,t_*} = \{(x,t) : x > \delta,\ t > \delta,\ 0 < x < S(t),\ 0 < t < t_*\}$. If $\Phi(T)$ is infinitely differentiable, then $S(t)$ is also infinitely differentiable for $t > 0$. If $\lim\limits_{t\to\infty} f(t) = \beta$, $0 \leq \beta < \infty$ and the above assumptions hold, then $\lim\limits_{t\to\infty} t^{-1/2}\, S(t) = D_*(\beta)$, $\beta > 0$, where $D_*(\beta)$ is given by (10.1.125) (given below).

Proposition 10.1.22. Let $\Phi \in C^2[0,\beta]$, $\Phi'(T) > 0$ for $T \in [0,\beta]$, $f(t) = \beta =$ constant > 0, $d = 0$, and for $x \in (0,\infty)$, $T = 0$ and the specific energy $\mathrm{U} = -1$. Then problem (10.1.120)–(10.1.122) has a unique solution such that

$$S(t) = D_*(\beta)t^{1/2}, \text{ and } T_*(x,t) = T(xt^{-1/2},\beta). \tag{10.1.124}$$

Here, $D_*(\beta)$ depends continuously on β, and $\lim D_*(\beta) = 0$ as $\beta \to 0$. $D_*(\beta)$ is given by the equation

$$\frac{1}{2}D_*^2(\beta) + \int_0^{D_*(\beta)} \xi\Phi(T(\xi,\beta))d\xi = \beta. \tag{10.1.125}$$

10.1.6 The effect of density change

If in the phase-change problems, the density changes sharply, then the effect of the change has to be included in the formulation, although this may make the analysis of such problems more difficult. In the one-phase problem considered in [310], the effect of change in the density on the change of phase has been considered. The formulation given in (10.1.56)–(10.1.57) can be used for the problem considered in [310] provided the Stefan condition is replaced by the following condition.

$$\frac{dS}{dt} - \varepsilon\left(\frac{dS}{dt}\right)^3 = (K_S/(\rho_S l))T_x(S(t),t),\ \varepsilon = (1/2l)\left(1 - \rho_S/\rho_L\right)^2. \tag{10.1.126}$$

Under suitable assumptions about the data (cf. [310]), global existence and stability of the classical solution has been proved using Schauder's fixed point theorem and it has been shown that

$$T(x,t) \in C^{1+1,0+1}(\overline{\Omega}_{t_*}) \cap C^\infty(\Omega_{t_*}),\ S(t) \in C^1[0,t_*];$$

$$\Omega_{t_*} = \{(x,t) : 0 < x < S(t), 0 < t < t_*\}. \tag{10.1.127}$$

As $\varepsilon \to 0$, the convergence of the solution of the present problem to the solution of the classical one-phase Stefan problem has been shown.

10.2 One-dimensional Two-Phase Stefan Problems

10.2.1 Existence, uniqueness and stability results

The classical formulation considered in [57] (see § 3.2.1, equations (3.2.1)–(3.2.10)) is fairly general. In some problems discussed in this section, nonlinear heat equations have been considered but the initial and interface conditions are not as general as in [57]. The results obtained in [57] are generalizations of the results obtained in [311, 312]. If we take $q^{(1)} = q^{(2)} = 0$, $f(x,t) = 0$, $\chi_1^{(1)} = K_1$ and $\chi_2^{(2)} = K_2$ in (3.2.1)–(3.2.10), then the problem considered in [311] and [312] is obtained. General results about the existence of the classical solutions of a two-phase Stefan problem have been obtained in [311] under the main regularity assumptions : (i) $\phi^{(1)}(t)$, $\phi^{(2)}(t) \in C^1[0,t_*]$, (ii) $h^{(1)}(x)$, $h^{(2)}(x)$ are Hölder continuous with exponent $\in (0,1]$, and (iii) compatibility conditions are satisfied at $S(0)$ and at $x = 0, 1$. The proofs in [311] are based upon potential theoretic arguments coupled with Schauder's fixed point theorem. It has been proved that if the data for the Stefan problem belong to a certain class (cf. [311]), then there exists $S(t) \in C^{3/4}[0,t_*] \cap C^1(0,t_*)$ and $T(x,t)$ is continuous on $\overline{\Omega}_{t_*}$ satisfying all the equations of the problem. If also $\phi \in C^{\beta}[0,1], 1/2 \le \beta \le 1$, then $S(t) \in C^{\hat{\beta}}[0,t_*]$ and $\dot{S} \in C_{\varepsilon}^{\hat{\beta}}(0,t_*]$, where $\hat{\beta} = (1+\beta)/2$ and $\varepsilon < \beta/2$. $C_{\varepsilon}^{\nu}(\mathrm{I})$ $(\varepsilon > 0)$ is the subspace of $C_0^{\nu}(\mathrm{I})$ and $C_0^{\nu}(\mathrm{I})$ is the subspace of $\mathrm{C(I)}$. For appropriate norms in these spaces see [311] (the norm of functions in $C_{\varepsilon}^{\nu}(\mathrm{I})$ depends on ε). By using a weak formulation, the existence of a classical solution has been established in [312] which will be discussed in § 11.2.

In [57], the existence of a unique solution has been proved under weaker conditions and differentiability conditions are not imposed on $\phi^{(1)}$ and $\phi^{(2)}$. Only Hölder continuity of $h^{(1)}$ and $h^{(2)}$ is required at $x = b$ as stated in (3.2.11)–(3.2.13). Other assumptions required in [57] for proving well-posedness are (see assumptions (A)–(F) made for problem (10.1.34)–(10.1.38)) as follows:

(i) $q^{(i)}(x,t)$, $i = 1,2$ satisfy condition (A) in which Q will be an upper bound of $\left|q^{(i)}(x,t)\right|$ in $\Omega_{t_*} = \{(x,t) : 0 < x < 1,\ 0 < t < t_* < \infty\}$.

(ii) $f(x,t)$ satisfies assumption (B) and $\mu(x,t)$ satisfies (E).

(iii) The assumption (C_1) is satisfied by both $\phi^{(1)}$ and $\phi^{(2)}$ and $\left|\phi^{(i)}\right| \le \Phi$, $i = 1,2$ and $t \ge 0$ (note that notations in (3.2.11)–(3.2.13) are different from those in (10.1.34)–(10.1.38))

(iv) $\chi^{(i)}$, $i = 1, 2$ and their derivatives satisfy conditions given in (3.2.14).

(v) All the assumptions (A)–(F) hold.

As mentioned in the context of equations (10.1.34)–(10.1.38), with no loss of generality we can set $f \equiv 0$. To develop an existence proof, an integral equation of the form (10.1.7) is required for the present problem also (see (10.2.2) given below). As done in the context of (10.1.34)–(10.1.38), a sequence of approximating solutions can be defined for the present problem also. It has been shown that the limiting solution (S, T) of the sequence of approximating solutions exists (for approximating solutions see the discussion after (10.2.3)). S is Lipschitz continuous, and T_x is continuous upto $x = S(t)$ and satisfies the integral equation for $S(t)$ given in (10.2.2). Hence, the Stefan condition is satisfied. The derivation of an integral equation for $S(t)$ of the form (10.1.7) for the boundary conditions (3.2.3) and (3.2.6) is lengthy so we explain the derivation of this integral equation for the boundary conditions (3.2.9) and (3.2.10). In (4.4.20), a Green's identity has been given. In the present problem, we consider the following Green's identity.

$$\iint_D \{V(T_{xx} - \alpha T_\tau) - T(V_{xx} + \alpha V_\tau)\} dx d\tau$$

$$= \oint_{\partial D} \{(VT_x - TV_x)d\tau + \alpha TV dx\}, \quad \alpha \text{ constant.} \qquad (10.2.1)$$

This identity is valid for sufficiently smooth functions T and V and for sufficiently regular domains. We take $T = T^{(1)}$, $V = \chi^{(1)}$, $\alpha = \delta^{(1)}$, and $D = \{(x, \tau) : \varepsilon < x < S(\tau),\ \varepsilon' < \tau < t\}$. ∂D is the boundary of D. Next we use the identity (10.2.1) by taking $T = T^{(2)}$, $V = \chi^{(2)}$, $\alpha = \delta^{(2)}$ and $D = \{(x, \tau) : S(\tau) < x < 1 - \varepsilon,\ \varepsilon' < \tau < t\}$. Adding the two results and taking the limits as $\varepsilon \to 0$ and $\varepsilon' \to 0$, we get

$$\begin{aligned} S(t) - b = & \int_0^t \int_0^1 (\chi q - \chi^* T) dx d\tau - \int_0^1 \delta\chi(x, 0)\hat{h}(x) dx + \int_0^1 \delta\chi(x, t) T(x, t) dx \\ & + \int_0^t \left\{\chi^{(1)}(0, \tau) g^{(1)}(T^{(1)}(0, \tau), \tau) - T^{(1)}(0, \tau)\chi_x^{(1)}(0, \tau)\right\} d\tau \\ & - \int_0^t \left\{\chi^{(2)}(1, \tau) g^{(2)}(T^{(2)}(1, \tau), \tau) - T^{(2)}(1, \tau)\chi_x^{(2)}(1, \tau)\right\} d\tau \\ & - \int_0^t \mu(S(\tau), \tau) d\tau, \quad 0 < t < t_*. \end{aligned} \qquad (10.2.2)$$

Here, χ, q, δ and $\hat{h}$ take appropriate values in the regions $0 < x < S(t)$ and $S(t) < x < 1$.

$$\left.\begin{aligned} \chi^*(x, t) &= \chi_{xx}^{(1)}(x, t) + \delta^{(1)}\chi_t^{(1)}(x, t), \ 0 \le x \le S(t), \\ &= \chi_{xx}^{(2)}(x, t) + \delta^{(2)}\chi_t^{(1)}(x, t), \ S(t) < x \le 1. \end{aligned}\right\} \qquad (10.2.3)$$

The triple $(S(t), T^{(1)}, T^{(2)})$ satisfies all the equations (3.2.1)–(3.2.10) except (3.2.3) and (3.2.6) as the boundary conditions are taken as (3.2.9) and (3.2.10).

A sequence of approximating solutions $\left(t_*^{(k)}, S_k, T_k^{(1)}, T_k^{(2)}\right)$ can be defined as in (10.1.46) –(10.1.51) provided the approximating problems for $T_k^{(2)}$ are appropriately defined in $S_k(t) < x < 1,\ 0 < t < t_*^{(k)}$ $(S_1(t) = b)$ and $0 < t < t_*^{(k+1)} \leq t_*^{(k)}$. $S_{k+1}(0) = b,\ k = 1, 2, ...$

$$\dot{S}_{k+1}(t) = \chi^{(1)}(S_k(t), t) T_{k,x}^{(1)}(S_k(t), t) - \chi^{(2)}(S_k(t), t) T_{k,x}^{(2)}(S_k(t), t) + \mu(S_k(t), t). \quad (10.2.4)$$

The following uniform estimate for $t > 0$ has been obtained for $\dot{S}(t)$ and

$$|\dot{S}_k(t)| \leq M_1 t^{-(1-\alpha)/2},\ 0 < \alpha < 1,\ M_1 \text{ is some constant; } k = 1, 2, ... \quad (10.2.5)$$

Further, a constant M_2 exists such that

$$|T_k^{(i)}(x, t)| \leq M_2 t^{-(1-\alpha)/2} \,|\, x - S_k(t)|;\ i = 1, 2,\ k = 1, 2. \ . \ . \quad (10.2.6)$$

and

$$|T^{(i)}(x, t)| \leq M_2 t^{-(1-\alpha)/2} \,|\, x - S(t)|,\ i = 1, 2. \quad (10.2.7)$$

First, the existence of the solution (S, T) is established in a small time interval $(0, t_0)$. Then using the estimates obtained in (10.2.5)–(10.2.7), the solution can be extended to larger time intervals. Thus we get a sequence $t_0 < t_1 < t_2 \cdots < t_n$. Since the sequence $\{t_n\}$ is monotonically increasing, we have either $t_n \to \infty$ or $\lim t_n = t_* < \infty$. In the latter case if $\dot{S}(t)$ is finite, then by using the estimates given in (10.2.5)–(10.2.7), the solution can be extended beyond the time t_* and we have a contradiction.

Proposition 10.2.1. Under assumptions (i)–(v) mentioned above in this section, a solution of problem (3.2.1)–(3.2.8) exists in the time interval $[0, t_*]$ and estimates (10.2.5)–(10.2.7) hold in $[0, t_*]$. If $t_* < \infty$, then

$$\lim_{t \to t_*-} \{S(t), 1 - S(t)\} = 0, \quad \text{and/or} \quad \lim_{t \to t_*-} \sup \dot{S}(t) = +\infty. \quad (10.2.8)$$

The following simple two-phase Stefan has been studied by many authors with or without some changes in the initial and boundary conditions.

$$k_1 \frac{\partial^2 T_1}{\partial x^2} = \frac{\partial T_1}{\partial t},\ 0 < x < S(t),\ S(0) = b,\ t > 0, \quad (10.2.9)$$

$$k_2 \frac{\partial^2 T_2}{\partial x^2} = \frac{\partial T_2}{\partial t},\ S(t) < x < d,\ t > 0, \quad (10.2.10)$$

$$T_1(0, t) = f_1(t) \geq 0,\ t > 0, \quad (10.2.11)$$

$$T_2(d, t) = f_2(t) \leq 0,\ t > 0, \quad (10.2.12)$$

$$T_1(x,0) = \phi_1(x) \geq 0,\ 0 \leq x \leq b,\ \phi_1(b) = 0, \tag{10.2.13}$$

$$T_2(x,0) = \phi_2(x) < 0,\ b \leq x \leq d,\ \phi_2(b) = 0, \tag{10.2.14}$$

$$T_1(S(t),t) = T_2(S(t),t) = 0,\ t \geq 0, \tag{10.2.15}$$

$$\lambda \dot{S}(t) = -K_1 \partial T_1/\partial x + K_2 \partial T_2/\partial x, \text{ at } x = S(t). \tag{10.2.16}$$

The method proposed in [283] for the analysis of a one-phase problem described earlier in § 10.1, can be suitably extended to the above two-phase problem (10.2.9)–(10.2.16). The problems described below which can be obtained by making some changes in the above two-phase problem have been investigated in ([21], Part two, Chapter II).

Problem (A1) Temperature boundary conditions are prescribed at $x = 0$ and at $x = 1$ (take $d = 1$), $S(0) = 0$. $T_2(x,0) = \phi(x) > 0$, $0 \leq x \leq 1$, $f_1(t) < 0$, $f_2(t) > 0$ and $\lim_{t\to\infty} |f_i(t)| = \alpha_i > 0$, $i = 1,2$. Note that there is only one initial temperature for the two phases.

Problem (A2) Radiation type boundary condition is prescribed at $x = 0$ and temperature is prescribed at $x = 1$ (see (10.2.18)). Other conditions are the same as in Problem (A1).

Problem (B1) Consider the problem described in (10.2.9)–(10.2.16) with $f_1(t)$ and $f_2(t)$ as in Problem (A1), $S(0) = b, 0 < b < d$. $\phi_1(x) < 0$ in $0 \leq x \leq b$ and $\phi_2(x) > 0$ in $b \leq x \leq d$.

Problem (B2) Radiation type boundary condition is prescribed at $x = 0$ and temperature is prescribed at $x = 1$ as in Problem (A2). Other conditions are the same as in Problem (B1).

Problem (C) (Cauchy-Stefan problem). Consider the region $-\infty < x < \infty$ with $T_1(x,0) = \phi_1(x)$, $-\infty < x < 0$ and $T_2(x,0) = \phi_2(x)$, $0 < x < \infty$. $T_1(x,t)$ and $T_2(x,t)$ satisfy heat equations, isotherm conditions and the Stefan condition.

For the regularity and compatibility conditions to be satisfied by the data, the reader is referred to [21]. By using heat potentials, the temperatures $T_1(x,t)$ and $T_2(x,t)$ and their first derivatives w.r.t. x can be expressed in the form of integrals similar to those obtained in (10.1.17) and (10.1.30). For Problem (B1), which is easier to handle, the existence, uniqueness and stability of the local-in-time solution has been proved under suitable assumptions using Picard iteration which was used earlier for the problem (10.1.23)–(10.1.27). By assuming suitable a-priori uniform estimates of the temperature derivatives and other quantities, it is possible to extend this solution to longer times. The constant b in Problem (B1) cannot be taken to be zero as one of the phases degenerates to a point which destroys the convergence of Picard iteration which is the main step in the proof of the existence of the solution in [21]. By imposing additional conditions (see

[21]) on the initial and boundary conditions, the existence and uniqueness of solutions of Problems (A1) and (A3) have been discussed. Since fairly general results related to the existence and uniqueness have already been reported, we discuss the asymptotic behaviour of the solution. For the asymptotic behaviour of the solution of Problem (B1), the following result holds good.

Proposition 10.2.2. If $\lim_{t\to\infty} f_1(t) = \alpha_1 < 0$, and $\lim_{t\to\infty} f_2(t) = \alpha_2 > 0$, then as $t \to \infty$, the solution of Problem (B1) tends uniformly in x to the following limits.

$$\lim_{t\to\infty} T_i(x,t) = -\alpha_1 + (\alpha_1+\alpha_2)x, \ i = 1,2; \ \lim_{t\to\infty} S(t) = \frac{\alpha_1}{\alpha_1+\alpha_2}. \tag{10.2.17}$$

The solution in (10.2.17) is also the solution of the corresponding stationary problem. In Problems (A2) and (B2), we take

$$\left(\frac{\partial}{\partial x} - g_0\right) T_1 = -g_0 f_1(t), \text{ at } x = 0, \text{ and } T_2|_{x=1} = f_2(t); \ t > 0. \tag{10.2.18}$$

The existence, uniqueness and stability of the solution of Problem (B2) have been proved locally-in-time as well as globally-in-time using Picard iteration method (see [311] for a direct proof of the existence of the classical solution). Concerning asymptotic behaviour of the solution of Problem (B2), the following results have been proved.

$$\lim_{t\to\infty} T_1(x,t) = -g_0\alpha_1(\lambda - x)/(1+\lambda g_0), \ g_0 \text{ is constant}, \tag{10.2.19}$$

$$\lim_{t\to\infty} T_2(x,t) = \alpha_2(x-\lambda)/(1-\lambda), \tag{10.2.20}$$

$$\lim_{t\to\infty} S(t) = \lambda = (g_0\alpha_1 - \alpha_2)/(g_0(\alpha_1+\alpha_2)). \tag{10.2.21}$$

The limiting solution in (10.2.19)–(10.2.21) is also a stationary solution of Problem (B2). In Problem (C), if $\phi_1(0) = \phi_2(0) = 0$; $S(0) = 0$, $\lim_{x\to-\infty} \phi_1(x) = -\alpha_1 < 0$, $\lim_{x\to+\infty} \phi_2(x) = \alpha_2 > 0$, $\lim \dot{\phi}_i(x) = 0$ as $|x| \to \infty$ and $\phi_i(x)$ are thrice differentiable, then existence, uniqueness and stability of the solution have been established for large time. If $\alpha_2 - a\alpha_1 >$ 0, where $a^2 = k_2$ $(k_1 = 1)$, then $\lim_{t\to\infty} S(t) = -\infty$. If $\alpha_2 - a\alpha_1 < 0$, then $\lim_{t\to\infty} S(t) = +\infty$.

The initial velocity of the free boundary in Problems (B1), (B2) and (C) is known if the compatibility condition

$$\phi_i(b) = 0, \ i = 1,2, \tag{10.2.22}$$

holds. In this case

$$\dot{S}(0) = \dot{\phi}_1(b) - \dot{\phi}_2(b). \tag{10.2.23}$$

If $\beta = \dot{S}(0) \neq 0$, then

$$S(t) = b + \beta t(1 + S^*(t)); \ \lim_{t \to 0+} S^*(t) = 0. \tag{10.2.24}$$

But for $\beta = 0$, the order of contact of the curves $x = b$ and $x = S(t)$ at $t = 0$ becomes unknown. In addition, it is unknown when the condition (10.2.22) is omitted, or in the case of Problems (A1) or (A2) when the domain of existence of one of the phases degenerates to a point.

A detailed study of the interface initial velocity, has been done in [21]. We present here just one result. In Problem (A1), assume that as $t \to 0$,

$$\dot{f}_1(t) = -\psi_1(t)t^{m_o}, \ m_o > -1; \ \dot{\phi}(x) = \psi_2(x)x^{n_o}, \ n_o > -1. \tag{10.2.25}$$

Further, for any $\varepsilon > 0$, let

$$\lim_{t \to \infty} \psi_i(t)t^{\varepsilon} = 0, \text{ and } \lim \psi_i(t)t^{-\varepsilon} = \infty, \ i = 1, 2. \tag{10.2.26}$$

If $f_1(0) \neq 0$ and $\sqrt{t}\dot{S}(t)$ is continuous for $t \geq 0$, then

$$\dot{S}(t) = \beta t^{-/2}(1 + S_1^*(t)); \ \lim_{t \to 0} S_1^*(t) = 0, \tag{10.2.27}$$

where β is the root of a transcendental equation (cf. [21]). In most of the practical problems, $S(t) \sim O(t^{1/2})$ or $O(t)$ or $O(t^{3/2})$ (cf. [60]).

The existence and uniqueness of Problem (A1) or Problem (A2) cannot be investigated easily as the convergence of Picard iteration process is destroyed by the degeneracy of one of the phases. Some additional conditions are to be imposed on the data to prove the existence of the solution of Problem (A1) by Picard iteration ([cf. [21]). To construct the existence proof the solution of Problem (B1), $(T_{1,n}, T_{2,n}, S_n)$ is obtained for $t > t_n$ by choosing an arbitrary monotonically decreasing sequence

$$t_0 > t_1 . \ldots > t_n > \ldots; \ \lim_{n \to \infty} t_n = 0; \ t_0 < t^*. \tag{10.2.28}$$

The definition of t^* is complicated as it depends on the time for which the solution of an auxiliary problem exists and simultaneously some other conditions are satisfied and for this information the reader is referred to [21]. We now consider a sequence of Problems of the type B_1 in the regions, $0 < x < S_n(t)$, (S_n, $n = 1, 2, ..$ are known) and $S_n(t) < x < d$, $t > t_n$, in which temperature at t_n is substituted from the solution of an auxiliary problem in which $S(t_n)$ is used. It has been shown that the sequence of solutions $\{T_{1,n}, T_{2,n,}, S_n\}$ of these problems converges to the unique solution of Problem (A1). Problem (A2) can be similarly studied.

By using Picard iteration method, the existence and uniqueness of a one-dimensional two-phase problem in a region with cylindrical symmetry has also been studied in [21]. In this case the Green's function used to construct the solution of the heat equation is given by

$$E_0(r,\xi,a^2(t-\tau)) = \frac{\exp\left(-\frac{r^2+\xi^2}{4a^2(t-\tau)}\right)}{2a^2(t-\tau)} I_0\left(\frac{r\xi}{2a^2(t-\tau)}\right). \tag{10.2.29}$$

Here, r is the radial coordinate in a cylinder of radius a and $I_0(x)$ is the modified Bessel's function of the first kind of zeroth order.

A two-phase Stefan problem in an unbounded domain has been considered in [313]. The heat equations in the regions $D_{t_*}^{(i)}$, $i = 1, 2$, have been considered as follows:

$$\mathcal{L}^{(1)}T^1(x,t) = q^{(1)}(x,t), \text{ in } D_{t_*}^{(1)} = \{(x,t) : -\infty < x < S(t),\ S(0) = d > 0,\ 0 < t < t_*\}, \tag{10.2.30}$$

$$\mathcal{L}^{(2)}T^2(x,t) = q^{(2)}(x,t), \text{ in } D_{t_*}^{(2)} = \{(x,t) : S(t) < x < \infty,\ 0 < t < t_*\}, \tag{10.2.31}$$

$$\mathcal{L}^{(i)}T^{(i)}(x,t) = a^{(i)}(x,t)T_{xx}^{(i)} + b^{(i)}(x,t)T_x^{\ (i)} + e^{(i)}(x,t)T^{\ (i)} - T_t^{\ (i)},\ i = 1,2. \tag{10.2.32}$$

The problem formulation consists of (10.2.30)–(10.2.32), (3.2.2), (3.2.5), (3.2.7) with $f(x,t) = 0$, and the equation

$$a^{(1)}(S(t),t)X^{(1)}(S(t),t)T_x^{\ (1)}(S(t),t) - a^{(2)}(S(t),t)X^{(2)}(S(t),t)T_x^{\ (2)}(S(t),t)$$
$$= \dot{S}(t) + \mu(S(t),t), \quad 0 < t < t_*, \quad S(0) = d. \tag{10.2.33}$$

This problem will be called Problem (YA). It is assumed that: (i) $q^{(i)}(x,t)$ and other coefficients in (10.2.31) and (10.2.32) are Hölder continuous in $\Omega = \{(x,t) : -\infty < x < \infty,\ 0 < t < \infty\}$ with $q^{(1)}(x,t) \geq 0$ and $q^{(2)}(x,t) \leq 0$ in Ω, (ii) $h^{(1)}(x)$ and $h^{(2)}(x)$ are continuous and $h^{(1)}(x) \geq 0$ and $h^{(2)}(x) \leq 0$, (iii) $X^{(i)}(x,t)$, $X_x^{(i)}$ and $X_t^{(i)}$ are continuous and $-N_1 \leq X^{(1)}(x,t) \leq 0$ and $N_2 \geq X^{(2)}(x,t) \geq 0$, (iv) $\mu(x,t)$ is continuous and differentiable in (x,t), $-\mu_0 \leq \mu(x,t) \leq 0$ $(\mu_0 > 0)$, $\mu(d,0) = X^{(i)}(d,0) = 0$ $(i = 1,2)$, (v) $e^{(i)} \leq 0$, $i = 1,2$.

For the complete set of assumptions, the reader is referred to [313]. It has been shown that a unique classical solution $\left\{t_*, S(t), T^{(1)}, T^{(2)}\right\}$ exists, $S(t)$ is differentiable and $0 \leq \dot{S}(t) \leq \beta$, β constant. To prove this result, solutions of a sequence of auxiliary problems are constructed (by retarding the argument) as follows. For an arbitrary θ, $0 < \theta < d$, let,

$$h^{(1)\theta}(x) = \begin{cases} h^{(1)}(x) & \text{for} \quad x \leq d - \theta, \\ 0 & \text{for} \quad x > d - \theta. \end{cases} \tag{10.2.34}$$

Similarly $h^{(2)\theta}(x)$ can be defined for $x \geq d - \theta$ and $x < d - \theta$. We take $S^\theta(t) \equiv d$ on the time interval $[0, \theta]$ and solve Problem (YA) without the condition (10.2.33) (free boundary is now known). Let $T^{(1)\theta}$ and $T^{(2)\theta}$ be the temperature solutions in the two regions $-\infty < x < S^\theta(t)$ and $S^\theta(t) < x < \infty$ for $0 \le t \le \theta$. To construct the solution for $0 \le t \le 2\theta$, $S^\theta(t)$ is determined in the time interval $[\theta, 2\theta]$ from (10.2.33) after integrating it w.r.t. to time and using the value of $S^\theta(t)$ in $[0,\theta]$ in the temperature derivatives at $x = S(t)$. Now this $S^\theta(t)$ for $\theta \le t \le 2\theta$ is used to determine temperatures $T^{(1)\theta}$ and $T^{(2)\theta}$ for $0 \le t \le 2\theta$. It may be noted that temperatures are uniquely determined as they are solutions of parabolic heat equations with boundary conditions prescribed at the known boundaries and with known initial temperatures. By induction we obtain $S^\theta(t), n\theta \le t \le (n+1)\theta$ and continue till we get a solution in $[0, t_*]$ The conditions required for the existence of a unique solution are met by the assumptions already made about the data. Proceeding inductively in this way, we get a sequence of approximate solutions $\left\{S^\theta,\ T^{(1)\theta},\ T^{(2)\theta}\right\}$ in $0 \le t \le n\theta$, $n = 1, 2, \ldots$. It can be proved that the approximate solutions S^θ are such that $0 \le \dot{S}^\theta(t) \le \alpha$ (constant) and the sequence $\{S^\theta\}$ is uniformly bounded and equicontinuous on a compact set. On using Ascoli-Arzela theorem it can be concluded that a subsequence converges to $S(t)$ (the conditions required for the application of Ascoli-Arzela theorem have been established in [313]). With this known $S(t)$, we determine the temperatures $T^{(1)}(x,t)$ and $T^{(2)}(x,t)$ by solving heat equations. To show that this limiting solution is the solution of Problem (YA), it should be proved that the limiting solution satisfies an equation of the form (10.2.2). For the present problem, an integral equation for $S(t)$ can be obtained by using the following Green's formula.

$$\iint_{\Omega_{t_*}} (v\mathcal{L}T - T\mathcal{L}^*v)\,dx\,dt = \int_{\partial\Omega_{t_*}} Tv\,dx + \int_0^{t_*} \Big\{a(x,t)(vT_x - Tv_x) + \left(b(x,t) - \frac{\partial a}{\partial x}\right)Tv\Big\}\,dt. \tag{10.2.35}$$

Here, Ω_{t_*} could be $D^{(1)}_{t_*}$ or $D^{(2)}_{t_*}$ and v is a smooth function in Ω_{t_*}.

For the operator $\mathcal{L}$ in (10.2.32), the operator $\mathcal{L}^*$ (conjugate) is given by

$$\mathcal{L}^*v = \frac{\partial^2}{\partial x^2}(av) - \frac{\partial}{\partial x}(bv) + e(x,t)v + v_t. \tag{10.2.36}$$

For the application of Green's formula and deriving an integral equation for $S(t)$ in Problem (YA), the reader is referred to [313].

In [314], instead of the Stefan condition in (10.2.9)–(10.2.16), the following condition is prescribed at the free boundary.

$$\partial T_1(x)/\partial x = F_t(S, T_2, T_{2,x}), \text{ on } x = S(t). \tag{10.2.37}$$

Here, F_t is a functional acting on the triple $\{S(\tau),\ T_2(x,\tau),\ T_{2,x}(x,\tau)\}$, $x \in [S(\tau), 1]$, $\tau \in [0,t]$. Let $k_1 = k_1(T_1)$, and $k_2 = k_2(T_2)$. Introduce heat source or sink terms $q^{(1)}(T_1)$ and $q^{(2)}(T_2)$ in (10.2.9) and (10.2.10), respectively. In the place of (10.2.16), prescribe (10.2.37) and let other conditions be the same as in (10.2.11)–(10.2.15) but without sign constraints. This problem will be called Problem (FAP). If instead of (10.2.37), we take

$$\frac{\partial T_1}{\partial x}\Big|_{x=S(t)} = g(t), \tag{10.2.38}$$

in which $g(t)$ is known, then we have the formulation of a one-phase Stefan problem in $0 < x < S(t)$, $0 < t < t_*$ with an implicit free boundary condition (see § 10.1.4, equation (10.1.101)). Under suitable assumptions, this one-phase problem is well-posed and the unique $S(t)$ obtained as the solution of this problem can be used to solve a parabolic heat conduction problem to determine $T_2(x,t)$ in the region $S(t) \le x \le 1$, $0 < t < t_*$. The unique solution of $T_2(x,t)$ can also be obtained provided appropriate assumptions are made about the data in the region $S(t) \le x \le 1$. Here, $S(t)$ is known for determining $T_2(x,t)$. Now the relationship between $g(t)$ and the functional F_t in (10.2.37) should be explored. Note that we have taken $g(t)$ to be some known function but it may not satisfy (10.2.37) exactly and in this situation it is appropriate to write

$$G(t) = F_t(S, T_2, T_{2,x}). \tag{10.2.39}$$

We express $G(t)$ as

$$G(t) = Mg. \tag{10.2.40}$$

The operator M is acting on a suitable function space to which g belongs. It has been proved in [314] that a $t_0 > 0$ exists such that M has at least one fixed point on $(0, t_0)$. If $g \in H_{1/2+\varepsilon}$, $\varepsilon \in (0, 1/2)$, then a unique classical solution (S, T_1, T_2) of Problem (FAP) exists in the interval $(0, t_0)$. Moreover, $S(t) \in H_{1+\varepsilon}[0, t_0]$ with $\varepsilon \in (0, 1/2)$.

A two-phase Stefan problem in the region $0 \le x \le 1$ with unilateral boundary conditions at $x = 0$ and $x = 1$ has been considered in [315, 316]. The formulation of the problem considered in these references can be obtained from equations (10.2.9)–(10.2.16) if instead of (10.2.11) and (10.2.12) (take $d = 1$), the following boundary conditions are considered.

$$T_x(0,t) \in \gamma_0(T(0,t)), \; 0 < t < t_*, \tag{10.2.41}$$

and

$$-T_x(1,t) \in \gamma_1(T(1,t)), \; 0 < t < t_*, \tag{10.2.42}$$

and all the thermophysical parameters are taken to be unity. γ_0 and γ_1 are maximal monotone graphs in R^2 such that both $\gamma_0^{-1}(0) \cap [0, \infty)$ and $\gamma_1^{-1}(0) \cap (-\infty, 0]$ are non-empty sets. This implies that there is a kind of heater at $x = 0$ and a kind of cooler at $x = 1$. The time t_*, $0 < t^* < \infty$, is defined to be the first time that the free boundary $x = S(t)$ touches $x = 0$ or $x = 1$.

We shall call this Stefan problem with (10.2.41) and (10.2.42) and other conditions in (10.2.9)–(10.2.16) (except (10.2.11) and (10.2.12)), Problem (YU). The existence and uniqueness of the solution of Problem (YU) has been proved in [315] under the following assumptions.

$$\left.\begin{aligned} &\phi_1(x) \ge 0 \; (0 < x < b), \; \phi_2(x) \le 0 \; (b < x < 1), \\ &\phi_1 \text{ and } \phi_2 \text{ are bounded and continuous for a.e. } x \in [0,1]. \end{aligned}\right\} \tag{10.2.43}$$

Several properties of $S(t)$ and of temperatures in the two-phases have been discussed. For example, it has been proved that:

(1) $S \in C^{0,1/3}[0, t_*] \cap C^{0,2/3}(0, t_*] \cap C^{\infty}(0, t_*)$, (10.2.44)

(2) $0 \leq T_1 \leq \max\left\{\|\phi_1\|_{L^\infty(0,b)}, \alpha_1\right\}$ on $\overline{D}_{t_*}^{(1)} = \{(x,t) : 0 \leq x \leq S(t),\ 0 \leq t \leq t_*\}$,

$$\alpha_1 = \min\left\{\alpha_1 \geq 0;\ \alpha_1 \in \gamma_0^{-1}(0)\right\}, \tag{10.2.45}$$

(3) $\min\left\{-\|\phi_2\|_{L^\infty(b,1)}, \alpha_2\right\} \leq T_2 \leq 0$ on $\overline{D}_{t_*}^{(2)} = \{(x,t) : S(t) \leq x \leq 1,\ 0 \leq t \leq t_*\}$,

$$\alpha_2 = \max\left\{\alpha_2 \leq 0;\ \alpha_2 \in \gamma_1^{-1}(0)\right\}. \tag{10.2.46}$$

The space $C^{0,\alpha}$, $\alpha \in (0,1)$ is the space of Hölder continuous functions.

The existence and uniqueness of Problem (YU) have been proved using the finite-difference discretization of equations. Several assumptions about the data and, estimates of the nodal values in the finite-difference numerical solutions of $S(t)$ and temperatures, are required which cannot be given here and the reader is referred to [315] for them. The convergence of the numerical solution has been established. The free boundary in a two-phase problem cannot be attached to mesh points which could be easily done in the one-phase problem discussed in [302]. For the existence of local-in-time solution no sign constraints on $\phi(x)$ are required.

The Problem (YU) has been considered in [316] also but the emphasis in [316] is on the asymptotic behaviour of its solution. The stationary solution of Problem (YU) is a solution of the following elliptic problem.

$$\left.\begin{aligned} &w_{xx}(x) = 0,\ 0 < x < 1, \\ &w_x(0) \in \gamma_0(w(0)),\ \text{and}\ -w_x(1) \in \gamma_1(w(1)), \\ &w(\mu) = 0\ \text{ for some } \mu \in [0,1]. \end{aligned}\right\} \tag{10.2.47}$$

It has been shown that if a stationary solution does not exist, then $t_* < \infty$, and if (10.2.47) has a solution, then there exist a minimum solution $\underline{w}(x)$ and a maximum solution $\overline{w}(x)$. Let

$$\underline{S} = \begin{cases} 0, & \text{if } \underline{w}(x) \equiv 0 \\ \text{the unique zero of } \underline{w}(x), & \text{if } \underline{w}(x) \not\equiv 0. \end{cases}$$

and

$$\overline{S} = \begin{cases} 1, & \text{if } \overline{w}(x) \equiv 0 \\ \text{the unique zero of } \overline{w}(x), & \text{if } \overline{w}(x) \not\equiv 0 \end{cases}$$

The following proposition describes some properties of the asymptotic solution.

Proposition 10.2.3. Suppose that $t_* = \infty$. Then there exists a real number S^* with $\underline{S} \leq S^* \leq \overline{S}$, and a solution $w^*(x)$ of (10.2.47) such that

$$\text{(i)} \lim_{t\to\infty} S(t) = S^*, \quad \text{(ii)} \lim_{t\to\infty} T(x,t) = w^*(x) \text{ in } C[0,1], \quad \text{(iii)} \; w^*(x) = w_x(0)(x - S^*).$$

In the multi-phase phase-change problems, the density change in the phases gives rise to mass transport. The formulation of problems with both heat and mass transfer is not our concern here. Our interest in the problem discussed in [317] which is described below in (10.2.48)–(10.2.54), arises from the fact that after suitable transformation, the formulation becomes an extended formulation of the type (3.2.1)–(3.2.8). The local-in-time existence of the solution and well-posedness of the freezing problem (10.2.48)–(10.2.54) has been established in [317]. The following equations are to be satisfied.

$$\beta_S(T_S)(T_{S,t} + V_S T_{S,x}) - (K_S(T_S)T_{S,x})_x = 0, \; 0 < x < S(t), \; 0 < t < t_*, \tag{10.2.48}$$

$$T_S(x,0) = \phi_1(x), \; 0 \le x \le S(0) = b; \; T_S(0,t) = f_1(t), \; 0 < t < t_*, \tag{10.2.49}$$

$$\beta_L(T_L)(T_{L,t} + V_L T_{L,x}) - (K_L(T_L)T_{L,x})_x = 0, \; S(t) < x \le d, \; 0 < t < t_*, \tag{10.2.50}$$

$$T_L(x,0) = \phi_2(x), \; b \le x \le d; \; T_L(0,t) = f_2(t), \; 0 < t < t_*, \tag{10.2.51}$$

$$T_S(S(t),t) = T_L(S(t),t) = T_m(P), \; 0 < t < t_*, \tag{10.2.52}$$

$$[\rho_S l + (\rho_L C_L - \rho_S C_S)T_m(P)]\,\dot{S}(t) = (\rho_L C_L V_L - \rho_S C_S V_S)T_m(P)$$

$$+K_S T_{S,x} - K_L T_{L,x}, \text{ at } x = S(t), \; 0 < t < t_*, \; \dot{S}(t) > 0, \tag{10.2.53}$$

$$(\rho_S - \rho_L)\dot{S}(t) = \rho_S V_S - \rho_L V_L, \text{ at } x = S(t), \; 0 < t < t_*. \tag{10.2.54}$$

Here, $P = P(x,t)$ is the pressure and the phase-change temperature $T_m(P)$ in (10.2.52) is a pressure dependent known quantity. Both solid and liquid phases are compressible and V_S and V_L are velocities of the two phases and are assumed to be known from the hydrodynamical and thermoelastic considerations. $\beta_i(T_i) = C_i + a_i\gamma_i + (-1)^i T_i\gamma_i(\rho_i\alpha_i)^{-1}$, $i = S, L$ and equal to $1, 2$ in $(-1)^i$; a_i are some specified positive constants, α_i is the compressibility of the ith phase and γ_i is the volumetric thermal expansion coefficient. The equation (10.2.52) describes the local thermodynamic equilibrium temperature of the two phases, (10.2.53) describes the condition of dynamical compatibility for heat transfer and (10.2.54) is the mass balance condition at the interface. The coefficient of $\dot{S}(t)$ in (10.2.53) is the jump in the enthalpy across $S(t)$ and the first term on the right represents the difference in rates at which heat enters and leaves across $S(t)$ by convection. The second term on the r.h.s is the difference in the fluxes. Flux prescribed boundary conditions can also be considered at the fixed boundaries.

On using the transformation (1.4.29) in formulations of both phases, making use of some thermodynamical relations and adopting suitable notations, equations (10.2.48)–

(10.2.54) get reduced to equations similar to (3.2.1)–(3.2.8) (provided velocity terms are added in (3.2.1)–(3.2.8) and the coefficients are redefined suitably). The small-time existence of the classical solution has been proved in [317] using the method of approximating solutions (see (10.2.46)–(10.2.51)) and making use of some results given in [295]. It has been proved that $S(t) \in H^{1+\alpha/2}(0, t_*)$ for any $\alpha \in (0,1)$. The continuous dependence of the solution on the data and coefficients has been established from which the uniqueness of the solution also follows.

10.2.2 Differentiability and analyticity of the free boundary in the one-dimensional two-phase Stefan problems

We assume that a classical solution of the two-phase problem (10.2.9)–(10.2.16) exists and the Stefan condition (10.2.16) holds for $< t \leq t_*$. For the one-phase problem formulated in (10.1.71), the proof of the infinite differentiability of $S(t)$ for $0 < t \leq t_*$, depends mainly on the application of Proposition 10.1.11. which tells us about the function space to which the temperature belongs. We assert from (10.1.76) that $\dot{S}(t) \in H_\alpha(Q_\delta)$, $\delta \geq 0, 0 < \alpha < 1$. If appropriate assumptions on coefficients in the heat equations after transformations (see (10.1.73)) in both the two-phases are made, then Proposition 10.1.11. can be used to conclude that the second order temperature derivatives in the two-phases are Hölder continuous with exponent α, $0 < \alpha < 1$. Arguments similar to those used in §10.1.2 for proving the infinite differentiability of $S(t)$ and of the temperature in the one-phase problem can be put forward for proving the infinite differentiability of $S(t)$ and of temperatures in both phases in a two-phase problem. The infinite differentiability of the free boundary in a two-phase problem under unilateral boundary conditions has been discussed in [315, 316].

The analyticity of the free boundary in a two-phase ice-water system described by (10.2.9)–(10.2.16) (with all thermophysical parameters taken as unity) has been described in [296]. Let $f_1(t) > 0$, $f_2(t) < 0$, $\phi_1(x) \geq 0$, $\phi_2(x) \leq 0$; f_i and ϕ_i, $i = 1, 2$, be continuously differentiable functions. Further, $f_1(0) = \phi_1(0)$, $f_2(0) = \phi_2(d)$, $\phi_1(b) = \phi_2(b) = 0$. Making use of the transformation $y = x/S(t)$ and using Schauder estimates for parabolic equations satisfied by $V_1(y,t) = T_1(x,t)$ and $V_2(y,t) = T_2(x,t)$, one can deduce (as done in the one-phase problem (see § 10.1.2 and [296]) that if a classical solution of (10.2.9)–(10.2.16) exists, then $S(t)$ is a C^∞-function. Concerning analyticity of $S(t)$, the following proposition has been proved in [296].

Proposition 10.2.1. If $f_1(t)$ and $f_2(t)$ are analytic functions for $0 \leq t \leq t_*$, in (10.2.9)–(10.2.16) then $S(t)$ is analytic for $0 < t \leq t_*$.

To prove the above proposition, in addition to quantities defined in (10.1.77)–(10.1.79), we define the following transformations.

$$Z = \frac{(d-x)}{(d-S(t))}, \quad \text{and } \sigma = \int_0^t \frac{d\lambda}{(d-S(\lambda))^2}, \tag{10.2.55}$$

$$V(y,\tau) = T_1(x,t) - (1 - x/S(t))f_1(t), \tag{10.2.56}$$

$$W(y,\sigma) = T_2(x,t) - (x - S(t))/(d - S(t))f_2(t). \tag{10.2.57}$$

With the help of the above transformations, the heat equation in each phase is reduced to a form suitable for the application of Proposition 10.1.11. and various estimates for $S(t)$ and derivatives of temperatures can also be obtained. The local-in-time unique classical solution can be extended to a global-in-time solution provided the a priori estimates as given below can be obtained.

$$\left|\frac{\partial T_1}{\partial x}(x,t)\right| \le A_t,\ 0 < x < S(t),\ \text{and}\ \left|\frac{\partial T_2}{\partial x}(x,t)\right| \le A_t,\ S(t) < x < d, \tag{10.2.58}$$

where, A_t is a bounded function of t. It has been proved in [318] that A_t is bounded under suitable assumptions.

The analyticity of the free boundary in a strongly nonlinear two-phase Stefan problem has been discussed in [298] by using the method of proof (expressing temperatures in terms of heat potentials and treating the problem in a complex plane by defining a suitable complex variable) discussed earlier in § 10.1.2 concerning the analyticity of a one-phase problem. The heat source terms have been taken in the form $F_i = F_i(t, x, T_1, T_2, T_{1x}, T_{2x}, S, \dot{S})$, $i = S, L$. The one-phase formulation (10.1.23)–(10.1.27) can be easily generalized to obtain the formulation of a two-phase problem defined in $D_{t_*} = \{(x,t) : (0 \le x \le S(t)) \cup (S(t) \le x \le 1), 0 \le t \le t_*\}$ and is not being given here. Under suitable assumptions it has been proved that $S(t)$ is an analytic function of $t^{1/2}$ in $0 < t < \hat{t}$, and $T_i(\lambda^* S(t), t)$ and $T_{i,x}(\lambda^* S(t), t)$ are analytic in $t^{1/2}$ in $0 < t < \hat{t}$, for each $\lambda \in [0,1]$. Here $\lambda^* = \lambda$ if $i = 1$ $(x < S(t))$ and $\lambda^* S(t) = 1 + \lambda(S(t) - 1)$ if $i = 2$ $(x > S(t))$.

10.2.3 One-dimensional n-phase Stefan problems with n > 2

In this subsection, an n-phase problem refers to a problem with n phases and $(n-1)$, $n > 2$ distinct free boundaries. Some n-phase problems have already been discussed in §§ 3.2.3 and 3.2.4. We present below a simple formulation of an n-phase Stefan problem. Find $(n-1)$ free boundaries $S_i(t)$, $i = 1, 2.., (n-1)$ and temperatures $T_r(x,t)$, $r = 1, 2.., n$, such that

$$0 < S_1(t) < ... < S_{n-1}(t) < d;\ S_i(0) = b_i,\ i = 1, 2, .., (n-1), \tag{10.2.59}$$

$$k_r T_{r,xx} - T_{r,t} = 0,\ \text{ if } S_{r-1}(t) < x < S_r(t),\ 0 < t < t_*;\ 1 \le r \le n, \tag{10.2.60}$$

$$S_0(t) = 0,\ \text{and}\ S_n(t) = d, \tag{10.2.61}$$

$$T_1(0,t) = f_1(t),\ \text{and}\ T_n(d,t) = f_2(t);\ 0 < t < t_*, \tag{10.2.62}$$

$$T_r(x,0) = \phi_r(x),\ b_{r-1} \le x \le b_r,\ r = 1,2,..,n;\ b_0 = 0,\ b_n = d, \tag{10.2.63}$$

$$T_r(S_i(t),t) = 0,\ 0 < t < t_*;\ r = 1,2,..,n;\ i = 1,2,..,(n-1),$$
$$(\text{for } i = m,\ r = m \text{ and } (m+1)), \tag{10.2.64}$$

$$-K_r\frac{\partial T_r}{\partial x}(S_i(t)-0) + K_{r+1}\frac{\partial T_{r+1}}{\partial x}(S_i(t)+0) = (-1)^{i-1}\frac{dS_i(t)}{dt},\ 0 < t < t_*;$$

$$r = 1,2,..,(n-1),\ i = 1,2,..,(n-1),\ (\text{for } i = m,\ r = m \text{ and } (m+1)). \tag{10.2.65}$$

The above formulation can be easily generalized to more complicated problems. When the number of phases is more than two, in addition to the study of those aspects of the analysis discussed earlier, the questions related to the disappearance of phases should also be addressed. A three-phase problem in which regions $-\infty < x < S_1(t)$, $S_1(t) < x < S_2(t)$ and $S_2(t) < x < \infty$ are initially occupied by water, ice, and water, respectively, has been studied in [319]. Thermal properties of the two water regions are the same but their initial temperatures are taken be different. Depending on the initial temperatures, the piece of ice can melt away entirely at some finite time t_* or water can freeze at each ice-water interface. Let $T_1(x,t)$, $T_2(x,t)$ and $T_3(x,t)$ be temperatures of water, ice and water, respectively, with $T_1(x,0) = \phi_1(x) \ge 0$, $-\infty < x \le b_1$, $T_2(x,0) = \phi_2(x) \le 0$, $b_1 \le x \le b_2$, and $T_3(x,0) = \phi_3(x) \ge 0$, $b_2 \le x < \infty$. Here the latent heat has been taken as unity. Suppose that after time $t > t_*$, the ice phase disappears and $T(x,t)$ be the temperature of water for $t \ge t_*$ with $T(x,t_*) = T_1(x,t_*)$ for $-\infty < x < S_1(t_*) = S_2(t_*)$, and $T(x,t_*) = T_3(x,t_*)$ for $S_2(t_*) < x < \infty$. The global existence and uniqueness of the solution $(S_1, S_2, T_1, T_2, T_3, T)$ has been proved. It has been shown that S_1 and S_2 depend continuously and monotonically on the data. The main tool used in the proofs is the maximum principle, both in its strong form [320] and in the form of the parabolic version of Hopf's lemma [321]. The constructive element in the approach in [319] is based on the idea of retarding the argument in the Stefan conditions at the free boundary (see the discussion concerning equation (10.2.34)).

It has been shown that under suitable assumptions, free boundaries are continuously differentiable and $|\dot{S}_i(t)| \le$ a constant, for $0 \le t < t_*$, $i = 1,2$. Let $E(t) = H(t) - (S_2(t) - S_1(t))$, where $H(t)$ is the sensible heat of the system at time t. It can be proved that $E(t) = E(0)$. The following proposition holds.

Proposition 10.2.2. Suppose that $\lim_{t\to\infty} H(t) = 0$. If $E(0) > 0$, then t_* is finite. If $E(0) = 0$, then t_* is plus infinity and $\lim_{t\to\infty}(S_2 - S_1) = 0$. If $E(0) < 0$, then $\lim_{t\to\infty}(S_2 - S_1) = -E(0) < 0$ and there does not exist a finite t_*. If $t_* = \infty$, and if the initial temperatures are bounded and have compact support in $-\infty < x < \infty$, then $\lim_{t\to\infty} H(t) = 0$.

A three-phase problem similar to that considered in [319] has been considered in [322] in the bounded region $-d \le x \le d$. A piece of ice with temperature $T_2(x,t)$ occupies the region $S_2(t) \le x \le S_1(t)$ and $T_2(x,0) = \phi_2(x) < 0$. The region $-d \le x \le S_2(t)$ is water having the temperature $T_1(x,t)$ and $T_1(x,0) = \phi_1(x) > 0$. The region $S_1(t) \le x \le d$ is also water having the temperature $T_3(x,t)$ and $T_3(x,0) = \phi_3(x) > 0$. $T_1(-d,t) = f_1(t) > 0$ and $T_3(d,t) = f_3(t) > 0$. The classical formulation of this problem can be easily written. Let this problem be called Problem (AK). The main result of [322] is the following proposition.

Proposition 10.2.3. Let $\{T_1, T_2, T_3, S_1, S_2\}$ be the classical solution of Problem (AK). There exist constants $\alpha \in R^+$, $\beta \in R^+$, $\lambda_1 \in R^-$, $\lambda_2 \in R^+$, $\hat{t} \in R^+$ and $t_* \in R^+$ such that

$$|T_2(x,t)| \le \alpha e^{\beta/(t-t_*)} \text{ for } t \in (0,\hat{t}); \; S_2(t) \le x \le S_1(t), \tag{10.2.66}$$

$$\lim_{t \to t_*-0} \left.\frac{\partial T_2}{\partial x}\right|_{x=S_i(t)} = 0, \; i = 1,2; \; S_1(t_*) = S_2(t_*), \tag{10.2.67}$$

and

$$\lim_{t \to t_*-0} \frac{dS_i}{dt} = \lambda_i, \; i = 1,2; \; -\infty < \lambda_1 < 0, \; 0 < \lambda_2 < \infty. \tag{10.2.68}$$

The temperature in the solid phase decreases to zero as t approaches t_*.

The existence and uniqueness of similarity solutions of a one-dimensional multi-phase Stefan problem has been discussed in [323]. The formulation of the problem is the same as in [65] (see § 3.2.3) except that phases are not in motion. The sufficient conditions for the existence of the similarity solution have been obtained in [323] but are not being discussed here as they can be described only after describing the complete solution of the problem.

Analyticity of $(n-1)$ non-intersecting phase-change boundaries in an n-phase problem has also been discussed in [296]. The formulation of the problem is similar to that given in (10.2.59)–(10.2.65) except that all the thermophysical parameters have been taken equal to unity. It has been assumed that $(-1)^{m-1}\phi_m(x) \ge 0$, $f_1 > 0$, $(-1)^n f_2 < 0$, $\phi_m(b_m+0) = \phi_{m+1}(b_m-0) = 0$; $1 \le m \le n$, $\phi_{n+1} \equiv 0$, $\phi_1(0) = f_1(0)$, $\phi_n(d) = f_2(0)$; $\phi_r(1 \le r \le n)$, f_1 and f_2 are continuously differentiable, and b_i $(1 \le i \le n-1)$ are given constants. Under suitable assumptions, using the arguments given in the case of one-phase and two-phase problems the existence of the unique solution of this n-phase problem in a small time interval can be proved. If a suitable transformation is used in each phase, a fixed domain formulation in each phase can be obtained. On making suitable assumptions and using Proposition 10.1.11., it can be shown that free boundaries are C^∞-curves (cf. [296]). To prove the existence of a global classical solution it suffices to establish a priori bounds (10.2.58) for the temperature $T(x,t)$ defined in $0 \le x \le d$, $x \ne S_i(t)$. For the present problem, A_t (see (10.2.58)) is given by

$$A_t = \max\left\{ \sup_{0<\tau<t} |T_x(0,\tau)|, \sup_{0<\tau<t} |T_x(d,\tau)|, \max_{1\le r\le n} \sup_{b_{r-1}\le x\le b_r} |\phi_r'(x)| \right\}, \quad (10.2.69)$$

for all t for which $S_i(\tau)$, $1 \le i \le n-1$, $\tau < t$ do not intersect each other.

Proposition 10.2.4. If $f_1(t)$, $f_2(t)$ are analytic functions for $0 \le t \le t_*$ in (10.2.59)–(10.2.65), then $S_1(t),..,S_{n-1}(t)$ are analytic functions for $0 < t \le \sigma$, where σ is the first time such that

$$\lim_{t\to\sigma-} \min_i [S_i(t) - S_{i-1}(t)] = 0. \quad (10.2.70)$$

To prove the above proposition, functions V_r have been introduced in [296] which are defined as follows:

$$V_r(y_i, \tau_i) = T(x,t),\ 1 \le r \le n, \quad (10.2.71)$$

where $T(x,t)$ is the temperature in the region $0 \le x \le d$, and y_i and τ_i are related to x, t in the ith phase by the relations

$$\tau_{i+1} = \int_0^t \frac{d\lambda}{[S_{i+1}(\lambda) - S_i(\lambda)]^2},\ y_{i+1} = \frac{x - S_i(t)}{S_{i+1} - S_i};\ 1 \le i \le n-1. \quad (10.2.72)$$

To prove the analyticity, the inductive estimates are then obtained for derivatives of V_r and $S(t)$ with respect to τ.

10.3 Analysis of the Classical Solutions of Multi-Dimensional Stefan Problems

10.3.1 Existence and uniqueness results valid for a short time

The earliest results of general nature available on the existence and uniqueness of the classical solutions of multi-dimensional Stefan problems seem to be those reported in [324]. A two-dimensional two-phase Stefan problem has been considered in a rectangular region Ω, where

$$\Omega = \{(x_1, x_2) : 0 \le x_1 \le d_1, 0 \le x_2 \le d_2\} \subset R^2;\ \text{and } \Omega_{t_*} = \Omega \times (0, t_*). \quad (10.3.1)$$

The classical formulation of this problem will not be presented here as it can be obtained by making appropriate changes in the formulation given in § 3.2.4. In the present problem, it has been assumed that a single free boundary exists which has a parametric representation

$$x_1 = x_1(\lambda, t),\ x_2 = x_2(\lambda, t);\ x_{1,\lambda}^2 + x_{2,\lambda}^2 \ne 0;\ \lambda_0 \le \lambda \le \lambda_1. \quad (10.3.2)$$

The functions $x_{1,t}$, $x_{2,t}$, $x_{1,\lambda}$, and $x_{2,\lambda}$ are continuous and all the interior points of the surface obtained after eliminating λ in (10.3.2) are the interior points of $\overline{\Omega}_{t_*}$.

On eliminating λ from the two equations in (10.3.2), we get an equation of the free boundary S as $\Phi(x_1, x_2, t) = 0$. In view of the parametric representation in (10.3.2), instead of a Stefan condition of the type (1.4.8), the following equations have been considered at the free boundary.

$$\frac{\partial x_1}{\partial t} = K_1 \frac{\partial T_1}{\partial x_1} - K_2 \frac{\partial T_2}{\partial x_1}, \tag{10.3.3}$$

$$\frac{\partial x_2}{\partial t} = K_1 \frac{\partial T_1}{\partial x_2} - K_2 \frac{\partial T_2}{\partial x_2}. \tag{10.3.4}$$

We now consider a classical two-phase Stefan problem in Ω_{t_*} in which the Stefan condition is replaced by the conditions (10.3.3) and (10.3.4). Let the free boundary S be given by $\Phi(x_1, x_2, t) = 0$ and temperatures $T_1(x_1, x_2, t)$ and $T_2(x_1, x_2, t)$ in the two-phases satisfy all the conditions mentioned in the definition of the classical solution given in § 1.4.6. Then the solution (T_1, T_2, S) in which S has the parametric representation and the Stefan condition is replaced by (10.3.3)–(10.3.4) also forms a classical solution of the Stefan problem (substitute (10.3.3) and (10.3.4) in the Stefan condition).

Homogeneous parabolic heat equations have been considered in [324] in the two phases and heat fluxes have been prescribed on the fixed boundaries. The fluxes and their derivatives, initial temperature and the function representing the initial position of the free boundary have been taken to be continuous functions. On using suitable Green's functions, temperatures at the free boundary together with the temperature derivatives, can be expressed in terms of nonlinear Volterra equations of the second kind (see equation (9.6.32)). It has been reported that under suitable assumptions a unique classical solution exists for a short time. However no details of the proof are available. The classical solution is stable in the norm $C(\Omega_{t_*})$ with respect to perturbations of the data in $C(\Omega_*)$. It has been proposed that similar results about the existence, uniqueness and stability will hold good for problems considered in $\Omega \subset R^n$, $n > 2$.

A problem more general than the one considered in [324] has been considered in [325]. There are m-phases in the region $0 \leq x_2 \leq d$, $-\infty < x_1 \leq S^{(m)}$, where $S^{(1)}, S^{(2)}, .., S^{(m-1)}$ are simple continuous, piecewise smooth, non-intersecting free boundaries and $S^{(m)}$ is a given surface. The parabolic equations considered in different phases are nonhomogeneous and temperatures at the free boundaries are unequal and are not constant. Each free boundary has a parametric representation of the form (10.3.2). At each free boundary, instead of the Stefan condition, conditions given in (10.3.3) and (10.3.4) have been considered in which thermal conductivities are functions of (x_1, x_2, t). On obtaining solutions of temperatures with the help of appropriate heat potentials and using contraction mapping argument, the existence, uniqueness and stability of the solution have been investigated for a short time. It has been reported that the method and the results obtained can be extended to problems in R^n, $n > 2$.

The multi-dimensional multi-phase problem considered in [325] has been further generalized in [326]. A general linear parabolic equation of second order has been considered in a non-cylindrical region occupied by each phase. The boundary of the region could be piecewise smooth. Instead of a Stefan condition, conditions of the form (10.3.3) and (10.3.4) have been considered at the free boundaries. On using suitable heat potentials, at the free boundaries, temperatures and temperature derivatives can be obtained in terms of Volterra integral equations of the second kind. Under suitable assumptions (see [326]), contraction mapping theorem can be applied to the system of integral equations to which the problem is reduced and it has been proved that the existence, uniqueness and stability of the solution holds for a short time

The existence and uniqueness proofs have been constructed in [309] using local coordinates and parametrization of the free boundary. This requires 'regularization' of the Stefan condition as after transformation in the local coordinates the compatibility conditions at the free boundary are not satisfied. The classical solutions of both one-phase and two-phase problems have been investigated and the existence of solutions has been studied for both short time and long times. We begin with the main steps in the proof of the existence of short-time solutions of one-phase Stefan problems. Let $\Omega_{t_*} = \Omega \times (0,t_*) \subset R^n_{t_*} = R^n \times (0,t_*)$, $n \geq 1$ and $\Omega(t) \subset R^n$, $t \in (0,t_*)$, $\Omega(0) = G \subset R^n$. Ω_{t_*} is a domain lying between two hypersurfaces $F_{t_*} = F \times (0,t_*)$, $F \subset R^n$, $n \geq 1$ and $\Gamma_{t_*} = \{(x,t); x \in \Gamma(t) \subset R^n, t \in (0,t_*)\}$. F_{t_*} is a known surface and Γ_{t_*} is the unknown free boundary. The classical solution of one-phase Stefan problem consists of finding the temperature $T(x,t)$, $x = (x_1, x_2, .., x_n) \in R^n$, $n \geq 1$, $0 < t < t_*$ and the free boundary $\Gamma(t)$, $0 < t < t_*$, satisfying the following system of equations.

$$a(T)\frac{\partial T}{\partial t} = \sum_{i=1}^{n} \frac{\partial^2 T}{\partial x_i^2} + f(x,t), \text{ in } \Omega_{t_*}, \tag{10.3.5}$$

$$T = T_1(x,t), \text{ or } \sum_{i=1}^{n} b_i(x,t)\frac{\partial T}{\partial x_i} + g(x,t)T = T_2; \text{ on } F_{t_*}, \tag{10.3.6}$$

$$T(x,t) = 0, \text{ on } \Gamma_{t_*}, \tag{10.3.7}$$

$$\vec{V} \cdot \vec{\nu} = -\frac{\partial T}{\partial t} / |\nabla T| = \sum_{i=1}^{n} \nu_i \frac{\partial T}{\partial x_i}, \text{ on } \Gamma(t),\ t \in (0,t_*);\ |\nabla T| \neq 0,\ (x,t) \in \Gamma_{t_*}, \tag{10.3.8}$$

$$\Gamma(0) = S;\ T(x,0) = T_0(x),\ x \in \Omega(0) = G. \tag{10.3.9}$$

Here, $\vec{\nu} = (\nu_1, \nu_2, .., \nu_n)$ is the unit vector normal to $\Gamma(t)$, $\vec{\nu} = \nabla T / |\nabla T|$ and $\vec{V}$ is the velocity of the free boundary. The surface $S\,(= \Gamma(0))$ is a C^2- surface and does not intersect F. The compatibility conditions of a suitable order which follow from (10.3.5)–(10.3.9) are satisfied. The basic step in the proof of the existence of the solution of multi-dimensional Stefan problems consists of establishing some a priori estimates for the solution in a neighbourhood of the free boundary $\Gamma(t)$. Therefore, with the help of a scalar function $R(x_0, t)$, $x_0 \in S$, $\Gamma(t)$ is defined by the equality

$$x = x_0 + R(x_0,t)\vec{\nu}(x_0), \ \vec{\nu}(x_0) = D\,T_0(x_0)/\left|DT_0(x_0)\right|, \ \vec{\nu} = (\nu_1, \nu_2, .., \nu_n). \quad (10.3.10)$$

Here, $D = (\partial/\partial x_1, \partial/\partial x_2, ..., \partial/\partial x_n)$ and $|R(x_0,t)| \leq 2B$, B is a constant which is chosen small enough so that $\Gamma(t)$ defined in (10.3.10) does not intersect F. In view of (10.3.10), the isotherm condition (10.3.7) becomes

$$T(x_0 + R(x_0,t)\vec{\nu}(x_0), t) = 0. \quad (10.3.11)$$

On differentiating (10.3.11) w.r.t. t and using (10.3.8), the Stefan condition becomes

$$\frac{\partial}{\partial t}R(x_0,t) = -\left|DT(x,t)\right|^2 / (DT(x,t)\,\vec{\nu}(x_0)). \quad (10.3.12)$$

The main result of [309] concerning the existence and uniqueness of the solution of problem (10.3.5)–(10.3.9) is the following proposition.

Proposition 10.3.1. Assume the following hypotheses:

(i) Closed surfaces F and S bound the sets $W_F \subset R^n$ and $W_S \subset R^n$ respectively such that $W_F \subset W_S$, $F \cap S = \emptyset$, and F and S belong to the class H^{2r}, $[r] = m+1 = n+5$, $[r]$ stands for the greatest integer in r (for m see the next assumption).

(ii) $a \in C^{m+2}[0,\infty)$, and functions T_1, g, and b_i $(i = 1,2,..n)$ are in the space $H^{2r,r}(F_\infty)$ and

$$a \geq M_0^{-1} = \text{constant} > 0, \ \sum_{i=1}^{n} b_i q_i \geq M_0^{-1}, \quad (10.3.13)$$

where $\vec{q} = (q_1, q_2, .., q_n)$ is the normal to the surface F at the point x.

(iii) $f \in H^{2r,r}(R^n_\infty)$, and $T_0 \in H^{2r}(\overline{G})$. On the surfaces F and S the compatibility conditions up to the order $m+1$ which follow from (10.3.5)–(10.3.7) and (10.3.12) hold. Moreover,

$$T_0(x) > 0, \ x \in G; \ |\log|\,DT_0(x)\,|| \leq N_0, \ x \in S. \quad (10.3.14)$$

(iv) Norms of the functions f, T_0, T_1, T_2 defined in appropriate spaces are bounded by a common constant N_0, and norms of a, g and b_i $(i = 1,2,..n)$ and norms of functions defined in local coordinates of surfaces F and S are bounded by a constant M_0.

Then there exists a sufficiently small $t_* > 0$, dependent only on M_0, N_0 and B, such that problem (10.3.5)–(10.3.9) admits a unique solution $(R(x_0,t), T(x,t))$ with $R(x_0,t) \in C^{2,1}(S \times (0,t_*))$ and $T \in C^{2,1}(\overline{\Omega}_{t_*})$.

It may be noted that if $f(x,t) = 0$, then it follows from the condition $T_0(x) > 0$ and the maximum principle that $T(x,t)$ is strictly positive at least in a sufficiently small time interval $(0, t_*)$. If $f(x,t) \neq 0$, then $T_0(x) > 0$ implies the positivity of the short-time solution outside a small neighborhood of the free boundary Γ_{t_*}. The second inequality in (10.3.14) allows us to use local coordinates. In a simple problem setting in which the surface F is given by $x_n = 1$ and S is given by $x_n = 0$; $f = 0$, $T_1(x) = 1$, $T_0(x)$ is 1-periodic with respect to $x' = (x_1, x_2, .., x_{n-1})$ and

$$|\log|\, D_n T_0(x)\,|| \leq N_0 \text{ for } x \in G,\ D_n = \partial/\partial x_n, \tag{10.3.15}$$

the following local coordinates (von Mises variables) can be used.

$$\tau = t,\ y' = (y_1, y_2, .., y_{n-1}) = x',\ y_n = T(x,t). \tag{10.3.16}$$

The surface F_{t_*} is invariant under the mapping $(x,t) \to (y,\tau)$ and the surface $S_{t_*} = S \times (0, t_*)$ corresponds to the free boundary Γ_{t_*}. The region $G_{t_*} = G \times (0, t_*)$ corresponds to Ω_{t_*}. A new dependent variable $u(y,\tau) = x_n$, 1-periodic with respect to x' is defined and the Stefan problem (10.3.5)–(10.3.9) can be formulated in terms of $u(y,\tau)$. Compatibility conditions satisfied earlier according to (10.3.5)–(10.3.7) and (10.3.12) are lost in the new formulation which can be restored by an appropriate 'regularization' of the Stefan condition transformed in terms of $u(y,\tau)$.

The regularization of the Stefan condition can be explained only with the help of the problem formulated in terms of von Mises variables and therefore the reader is referred to [309] for further details. Let u^ε be the solution of the regularized problem. If has been proved in [309] that on a sufficiently small time interval $(0, t_*^\varepsilon)$, a unique solution of the regularized problem exists and $u \in H^{2r,r}(\overline{G}_{t_*^\varepsilon})$, $[r] = m + 1 = n + 5$.

For any arbitrary surface S, a unique representation of the free boundary $\Gamma(t)$ in the form $x_n = S(x', t)$ is generally not available and the regularization of the compatibility conditions which was possible in a simple problem setting discussed above, is not possible in a general case. However, on a sufficiently small time interval, a regularized problem can be defined in a different way as follows.

We look for a scalar function $R^\varepsilon(x_0, t)$ defined on the surface S_{t_*} which determines the surface $\Gamma_{t_*}^\varepsilon$ by the equation

$$x = x_0 + R^\varepsilon(x_0, t)\vec{\nu}(x_0),\ \text{for } x_0 \in S. \tag{10.3.17}$$

Here, $\vec{\nu}(x_0)$ is the normal as defined in (10.3.10). The temperature $T^\varepsilon(x,t)$ and R^ε satisfy the following system of equations.

$$a(T^\varepsilon)\frac{\partial T^\varepsilon}{\partial t} = \sum_{i=1}^{n} \frac{\partial^2 (T^\varepsilon)}{\partial x_i^2} + f,\ \text{for } (x,t) \in \Omega_{t_*}^\varepsilon, \tag{10.3.18}$$

$$T^\varepsilon = T_1^\varepsilon \text{ or } \sum_{l=1}^{n} b_i \frac{\partial T^\varepsilon}{\partial x_i} + gT^\varepsilon = T_2^\varepsilon,\ \text{for } (x,t) \in F_{t_*}, \tag{10.3.19}$$

$$T^\varepsilon(x,t) = 0,\ \text{for } (x,t) \in \Gamma_{t_*}^\varepsilon, \tag{10.3.20}$$

$$T^\varepsilon(x,t) = T_0^\varepsilon(x), \text{ for } x \in G; \; R^\varepsilon|_{t=0} = 0, \tag{10.3.21}$$

$$\frac{\partial R^\varepsilon}{\partial t} - \varepsilon \nabla_S^2 R^\varepsilon = -\left|DT^\varepsilon\right|^2 / (DT^\varepsilon \vec{\nu}(x_0)), \; (x_0,t) \in S_{t_*}. \tag{10.3.22}$$

Here, $\Omega_{t_*}^\varepsilon$ is the region bounded by the surfaces F_{t_*}, $\Gamma_{t_*}^\varepsilon$ and the planes $t=0$ and $t=t_*$. The derivative on the r.h.s. in (10.3.22) is calculated at (x,t) with x given by (10.3.17) ∇_S^2 is the *Laplace-Beltrami operator* on the surface S (cf. [309]).

At $\varepsilon = 0$, the problem (10.3.18)–(10.3.22) coincides with the original Stefan problem and it is expected that for any $\varepsilon > 0$, the solution of (10.3.18)–(10.3.22) gives an approximate solution to the Stefan problem. The condition (10.3.22) is no more a Stefan condition and $T_0^\varepsilon(x)$ may not satisfy the compatibility conditions. Therefore, the initial temperature for T_ε should be changed to satisfy the compatibility conditions which follow from (10.3.18), (10.3.20) and (10.3.22). It has been proved in [327] that for each $T_0(x) \in H^{2r}(\overline{G})$ satisfying the compatibility conditions up to the order $[r] = m+1$ that follow from (10.3.5), (10.3.7) and (10.3.8) on the surface S, a $T_0^\varepsilon(x) \in H^{2r}(\overline{G})$ exists. For each $\varepsilon > 0, T_0^\varepsilon(x)$ coincides with $T_0(x)$ outside a small neighbourhood of S and satisfies the compatibility conditions up to the order $m+1$ that follow from (10.3.18), (10.3.20) and (10.3.22). Further, $\lim\limits_{\varepsilon \to 0} \|T_0 - T_0^\varepsilon\|_G^{(2r)} = 0$, $R^\varepsilon \in H^{2r,r}(S_{t_*})$ and $T^\varepsilon \in H^{2r,r}(\overline{\Omega}_{t_*}^\varepsilon)$.

Using a variational inequality formulation of the classical one-phase multi-dimensional problem, the regularity of the free boundary has been proved in [328]. If for a fixed time t_0, the point x_0 is a density point for the coincidence set (ice) in a ice-water system, then in a neighbourhood in space and time of (x_0, t_0), the free boundary is a surface of class C^1 in space and in time and all the second derivatives (in space and in time) of the solution are continuous up to the free boundary. The solution is hence classical in that neighbourhood.

All the results on the solvability of the one-phase Stefan problem (10.3.5)–(10.3.9) apply without any change to the problem in a two-phase setting (cf. [309]). The formulation of the two-phase problem can be easily written and it will not be presented here. If the free boundary $\Gamma(t)$ is expressed as in (10.3.10), then the Stefan condition on the surface Γ_{t_*} becomes

$$\frac{\partial}{\partial t} R(x_0,t) = X^-(x_0, DT) - X^+(x_0, DT), \text{ for } (x_0,t) \in S_{t_*}. \tag{10.3.23}$$

In (10.3.23), DT is to be calculated at (x,t) and

$$X^\pm = \lim_{r \to 0} \left|DT(x_r^\pm, t)\right|^2 / (DT(x_r^\pm, t)\vec{\nu}(x_0)), \tag{10.3.24}$$

$$x_r^\pm = x_0 + \vec{\nu}(x_0)(R(x_0,t) \pm r) \in \Omega_{t_*}^\pm, \; \Omega_{t_*} = \Omega_{t_*}^+ \cup \Omega_{t_*}^- \cup \Gamma_{t_*}. \tag{10.3.25}$$

Here, $\Omega_{t_*}^+$ and $\Omega_{t_*}^-$ are regions occupied by the two phases. $T_0(x)$ is strictly positive in G^+ and strictly negative in G^- and the second condition in (10.3.14) should be taken as

$$\lim_{r \to 0} |\log| DT_0(x \pm r\nu(x))|| < \infty, \text{ for } x \in S. \tag{10.3.26}$$

10.3.2 Existence of the classical solution on an arbitrary time interval

How can a short-time solution be extended in time? It has been remarked in [324] that following the arguments proposed in [9] (see also [281]), the solution of the problem considered in [324] can be extended till the time the free boundary reaches $x_2 = d_2$. It may be noted that to construct a regular solution even on a short time interval $(0, t_*)$, the initial temperature $T_0(x)$ in problem (10.3.5)–(10.3.9) has to be in $H^{2r}(\overline{G})$, where $[r] = m + 1 = n + 5$. Therefore to extend the solution $T(x,t)$ beyond $t > t_*$, $T(x, t_*)$ should belong to $H^{2r}(\overline{\Omega}_{t_*})$. Such a regularity of the solution cannot be expected in general. However, the following statement can be made. The existence interval $(0, t_*^{\infty})$ of the classical solution of the one-phase Stefan problem is characterized by the relations

$$\left.\begin{aligned} J_0(t_*) &= |\log|\, D_n T\, \|_{\Omega_{t_*}}^{(0)} < \infty, \\ J_1(t_*) &= |T(x,t)|_{\Omega_{t_*}}^{(2r)} < \infty, \; t_* < t_*^{\infty}, \\ \lim_{t_* \to t_*^{\infty}} &\{J_0(t_*) + J_1(t_*)\} = \infty . \end{aligned}\right\} \tag{10.3.27}$$

It turns out that under some restrictions on the data of the problem which ensure that the first equality in (10.3.27) is satisfied on an infinite time interval, the classical solution of the reference Stefan problem exists for all positive times.

Chapter 11

Regularity of the Weak Solutions of Some Stefan Problems

11.1 Regularity of the Weak Solutions of One-dimensional Stefan Problems

A weak formulation (also called a weak enthalpy formulation) of the one-dimensional classical Stefan problem was first developed in [329] which was further explored in [330]. Since the publication of these two fundamental works, the theory of weak solutions (also called generalized solutions) of Stefan problems and Stefan-like problems had a phenomenonical growth. Since the emphasis in this volume is on the classical solutions, our interest in this chapter in dealing with weak solutions is limited to the extent of exhibiting the regularity of some weak solutions under suitable assumptions. By considering a one-dimensional Stefan problem it has been shown earlier in § 5.2 that a classical solution is a weak solution. Some conditions under which a weak solution becomes a classical solution were also discussed. We shall state in this chapter some properties of the weak solutions and discuss under what conditions a weak solution becomes a classical solution. Some questions related to the behaviour of the mush will also be examined.

The weak formulation of a classical problem was given in § 5.2. For the analysis of weak solutions, the following formulation is more convenient. We consider the two-phase melting problem (10.2.9)–(10.2.16) and for simplicity take densities and specific heats of the two phases equal to unity. Equations (10.2.9)–(10.2.16) will now be considered with these changes. Let enthalpy $H(T)$ and functions $\phi(x)$ and $f(t)$ be defined as follows (cf. [331]).

Set

$$\Omega = \{x : 0 \leq x \leq d\}, \text{ and } \Omega_{t_*} = \Omega \times \{0 < t < t_*\}, \tag{11.1.1}$$

and let $T(x,t)$ be the temperature in Ω_{t_*}, where $T = K_1 T_1$ at points where $T_1 >$

0 and $T = K_2T_2$ at points where $T_2 < 0$. Let,

$$\left.\begin{aligned} H(T) &= T/K_1, \quad \text{if} \quad T > 0, \\ &= T/K_2 - l, \quad \text{if} \quad T \leq 0, \\ &= \beta(x,t), \quad \text{if} \quad T = 0, \quad -l \leq \beta(x,t) \leq 0, \end{aligned}\right\} \tag{11.1.2}$$

where $\beta(x,t)$ is an arbitrary function,

$$\left.\begin{aligned} T(x,0) = \phi(x) &= K_1\phi_1 \quad , \quad \text{for} \quad 0 \leq x \leq b, \\ &= K_2\phi_2 \quad , \quad \text{for} \quad b \leq x \leq d, \end{aligned}\right\} \tag{11.1.3}$$

$$\left.\begin{aligned} T(x,t)\big|_{x=0,\ d} &= K_1 f_1(t) \quad , \quad \text{on} \quad x = 0, \\ &= K_2 f_2(t) \quad , \quad \text{on} \quad x = d. \end{aligned}\right\} \tag{11.1.4}$$

In (11.1.2), the enthalpy of the mushy region belongs to the interval $[-l, 0]$, whereas in § 5.2, the enthalpy of the mushy region belongs to the interval $[0, l]$. This is possible by choosing the reference enthalpy differently in different problems but the jump in the enthalpy across the free boundary should remain l.

Definition 11.1.1. A pair of bounded measurable functions $(T(x,t),\ H(T(x,t)))$, on Ω_{t_*} is called a weak solution (or a generalized solution) of (10.2.9)–(10.2.16) if the equality

$$\iint\limits_{\Omega_{t_*}} [T\psi_{xx} + H(T)\psi_t]\, dxdt = \int\limits_0^{t_*} [\phi\psi_x]_{x=0}^{x=d}\, dt - \int\limits_{\Omega} H(\phi(x))\psi(x,0)dx, \tag{11.1.5}$$

holds for any $\psi(x,t)$ such that ψ_x, ψ_{xx}, ψ_t are continuous in $\overline{\Omega}_{t_*}$ and $\psi = 0$ at $x = 0, d$ and at $t = t_*$.

It may be noted that if no sign restrictions are imposed on the initial temperature, then the uniqueness of the weak solution is guaranteed only if the initial enthalpy is known exactly. This implies that if in the classical formulation the initial temperature is equal to the melting temperature in any region, then it should be specified whether it is a solid region or a liquid region.

The energy equation has the form

$$\frac{\partial H(T)}{\partial t} = \frac{\partial^2 T}{\partial x^2}, \quad (x,t) \in \Omega_{t_*}, \tag{11.1.6}$$

which is satisfied in the distributional sense as at $T = 0$, $H(T)$ is not differentiable. It has been shown in § 5.2 that a classical solution is a weak solution. The converse that a weak solution is a classical solution was discussed in § 5.2 under the assumptions that a smooth interface $x = S(t)$ having the temperature $T = 0$ exists which separates the solid region from the liquid region. Examining the conditions under which the mushy region in a weak solution in R^n, $n \geq 1$ has measure zero, and it coincides with the phase-change interface and the Stefan condition holds on the interface, is the main consideration in this chapter.

The definition of a weak solution in R^n, $n > 1$ and some properties of the weak solution will be discussed in § 11.2. In [331], the existence and the continuity of the free boundary in the weak solution of (10.2.9)–(10.2.16) were established and several other results were also proved. On using some of the results established in [332], the continuous differentiability of the free boundary was proved in [333]. Some of the results which were used in [333] to prove the continuous differentiability of the free boundary are given below in Proposition 11.1.1.

Proposition 11.1.1. Assume that ϕ' belongs to $L^2(0,d)$ and there exists a function $\Psi(x,t)$ in $\overline{\Omega}_{t_*}$ with Ψ_x, Ψ_{xx}, Ψ_t continuous in $\overline{\Omega}_{t_*}$ and $\Psi(0,t) = f_1(t)$, $\Psi(d,t) = f_2(t)$, $\Psi(x,0) = \phi(x)$ for $0 \leq t \leq t_*$ and x close to 0 or 1. Then the following results hold.

(1) There exists a unique weak solution $T(x,t)$.

(2) $T(x,t)$ is continuous on $\overline{\Omega}_{t_*}$, smooth in $\Omega_{t_*} \backslash \{T = 0\}$, and satisfies $T_t = k_1 T_{xx}$ ($T_t = k_2 T_{xx}$) in $\Omega_{t_*} \cap \{T > 0\}$(resp. $\Omega_{t_*} \cap \{T < 0\}$).

(3) There exists a constant $p > 0$, which depends only on the data, such that $T > 0$ ($T < 0$) for $|x| \leq p$ (resp. $|1-x| \leq p$).

(4) There exist a constant q such that

$$\left(\int_0^1 |T_x(x,t)|^2 \, dx \right)^{1/2} \leq q, \; 0 \leq t \leq t_*.$$

(5) For each $t \in [0, t_*]$ and $x_1, x_2 \in [0, d]$, there exists a constant q' such that

$$|T(x_1,t) - T(x_2,t)| \leq q' \, |x_1 - x_2|^{1/2}.$$

(6) For each $t \in [0, t_*]$, there exist a unique $S(t)$ such that $T(S(t),t) = 0$.

(7) $S(t)$ is a continuous function of t for $0 \leq t < t_*$.

(8) The energy balance condition at $x = S(t)$ is satisfied for each $t \in [0, \; t_*]$ in a weak form of (10.2.16) as follows.

$$l(S(t) - S(0)) = -(1/k_1) \int_0^{S(t)} T_1(x,t) dx - (1/k_2) \int_{S(t)}^{d} T_2(x,t) dx + (1/k_1) \int_0^{b} \phi_1(x) dx$$

$$+(1/k_2)\int_b^d \phi_2(x)dx + \int_0^t \frac{\partial T_2}{\partial x}(d,\sigma)d\sigma - \int_0^t \frac{\partial T_1}{\partial x}(0,\sigma)d\sigma. \qquad (11.1.7)$$

The above equation represents the energy balance in the region $0 \le x \le d$ (apply energy conservation law) if the parameters are defined suitably (take $K_S = K_L = 1$ as in [331]). Let $W_\sigma = \{x \in \Omega : T(x,\sigma) = 0\}$ and $W = \cup_{0\le\sigma<t_*} W_\sigma$. W is called a mushy region or a cloud. The main result of [333] is the following proposition.

Proposition 11.1.2. Let $T(x,t)$ and $S(t)$ satisfy the conditions given above in (2)—(7). Assume (without loss of generality) that $S(t)$ is continuous for $0 \le t \le t_*$ for some t_*. Then $S(t)$ is continuously differentiable for $0 < t \le t_*$, $T_x(S(t) \pm 0, t)$ are well defined, bounded and continuous for $0 < t \le t_*$ and

$$l\,\dot{S}(t) = -K_1 T_x(S(t)-0,t) + K_2 T_x(S(t)+0,t),\ 0 < t \le t_*. \qquad (11.1.8)$$

In particular, Propositions 11.1.1. and 11.1.2. guarantee the existence of a unique classical solution without size restrictions on the data which have been imposed in some other studies, see [292, 318]. As discussed earlier, continuous differentiability of the free boundary in the classical solution of (10.2.9)—(10.2.16) implies that $S(t)$ is a C^∞-function. The first step in the proof of Proposition 11.1.2., as discussed in [334], is to prove that if the results of Proposition 11.1.1. hold, then there exists a constant M such that (Hölder continuity with exponent 3/4)

$$|S(t_1) - S(t_2)| \le M\,|t_1 - t_2|^{3/4},\ 0 \le t_1, t_2 \le t_*. \qquad (11.1.9)$$

On using the uniform continuity of $S(t)$ on $[0, t_*]$ and the weak form of the energy equation (11.1.7), an estimate of the maximum oscillation of $S(t)$ over a time interval of variable length can be obtained. The following proposition asserts the behaviour of the derivative of the temperature in the solution of the heat equation considered in the region $0 \le x \le S(t),\ 0 < t \le t_*$. Proposition 11.1.3. when used in (11.1.8) establishes that $\dot{S}(t)$ is continuous.

Proposition 11.1.3. Let $S(t)$ be such that $S(t) \ge \hat{d} > 0\ (0 \le t \le t_*),\ S(0) = b$ and

$$|S(t_1) - S(t_2)| \le M\,|t_1 - t_2|^{\lambda},\ 0 \le t_1, t_2 \le t_*, \qquad (11.1.10)$$

where $1/2 < \lambda \le 1$. Let $V(x,t)$ be the solution of the following problem: (i) $V_t = V_{xx},\ 0 < x < S(t),\ 0 < t \le t_*$, $S(t)$ is known, (ii) $V(x,0) = \phi(x),\ 0 \le x \le b$, (iii) $V(0,t) = f(t),\ 0 \le t \le t_*$, (iv) $V(S(t),t) = 0,\ 0 \le t \le t_*$, (v) $f(t)$ and $\phi(x)$ are continuous with $f(0) = \phi(0)$ and $|\phi(x)| \le \alpha(b-x),\ 0 \le x \le b$, α constant Then $V_x(x,t)$ converges as $x \to S(t)-0$ to a limit $V_x(S(t)-0,t)$ which is a bounded continuous function of t for $0 < t \le t_*$. Moveover, the convergence is uniform on $[\delta, t_*]$ for any $\delta > 0$.

The proof of the above proposition has been developed in [334] by using the results which follow from Proposition 11.1.1., and the integrated (with respect to time over the interval $[t_1,t_2] \subset (0,t_*]$) form of the Stefan condition. As the following proposition suggests, the infinite differentiability of the free boundary can be proved under weaker conditions also (cf. [45]).

Proposition 11.1.4. Assume that $\phi_i(x)$, $f_i(t)$ $(i = 1, 2)$ are bounded piecewise continuous functions. Then there exists a $t_* > 0$ such that problem (10.2.9)–(10.2.16) possesses a unique solution in $(0, t_*)$ and $t_* = \sup(t;\ 0 < S(t) < d,\ -\infty < \dot{S}(t) < +\infty)$. Moreover $S(t) \in C^\infty(0, t_*)$ and the solution exhibits continuous and monotone dependence upon the data.

To prove the above proposition firstly a smooth chain of data (e.g., C^∞-functions) with appropriate compatibility conditions at (0, 0) and (1, 0) is considered. The existence of a classical solution in some small time interval is guaranteed. By showing the convergence of a suitable sequence of weak solutions with the assumed data, the Lipschitz coefficient of the classical temperature solution, its derivative and $S(t)$ can be estimated. It has been shown that

$$|T(x,t)| \leq M\,|x - S(t)|\,,\ \tau < t \leq t_*,\ \tau \in (0, t_*/2),\ 0 \leq x \leq d, \tag{11.1.11}$$

and

$$|S(t') - S(t'')| \leq \nu M\,|t' - t''|\,,\ \tau \leq t',\ t'' \leq t_*, \tag{11.1.12}$$

for some constant M and $\nu = (k_1 + k_2)/l$ which depend only on the L_∞ norm of the data. Next, consider sequences $\{f_i^{(k)}\}$, $\{\phi_i^{(k)}\}$ forming a smooth chain of data which approximate f_i, ϕ_i and preserve L_∞ norms and signs. In this way we can construct sequences of classical solutions $\left\{S^{(k)}(t), T^{(k)}(x,t)\right\}$. It has been shown that these sequences converge to the classical solution of the Stefan problem in which the Stefan condition (11.1.8) is satisfied in the integrated (w.r.t. time) form.

The regularity of the weak solutions in the degenerate Stefan problems has also been considered in some references. For the classical formulation of parabolic-elliptic degenerate Stefan problem, we make some changes in (10.2.9) and (10.2.10) and consider the following equations in their place.

$$\alpha_1(T)T_t = T_{xx} + q(T),\ 0 < x < S(t),\ 0 < t < t_*, \tag{11.1.13}$$

$$\alpha_2(T)T_t = T_{xx} + q(T),\ S(t) < x < d,\ 0 < t < t_*, \tag{11.1.14}$$

$$\left.\begin{aligned} T &= T_1,\ \text{if } T_1 > 0, \\ &= T_2,\ \text{if } T_2 \leq 0. \end{aligned}\right\} \tag{11.1.15}$$

Here, $\alpha_i(T) \geq 0$ and the equality holds if and only if $T = 0$ $(i = 1, 2)$. For the formulation of a parabolic-elliptic Stefan problem, we shall consider equations (11.1.13)–(11.1.15) and equations (10.2.11)–(10.2.16). To define a weak formulation of this degenerate problem, we introduce the following notations.

$$H(T) \in Q(T),\ \text{for a.e. } (x,t) \in \Omega_{t_*}\ (\text{see (11.1.1) for } \Omega_{t_*}), \tag{11.1.16}$$

$$\left.\begin{aligned} Q(T)=\int_0^T \alpha_1(\xi)d\xi &= C_+(T),\ \ T>0,\\ &= [-l,0],\ \ T=0,\\ &= -l+\int_0^T \alpha_2(\xi)d\xi = C_-(T),\ \ T<0. \end{aligned}\right\} \tag{11.1.17}$$

Let

$$\left.\begin{aligned} V_0(x) &= C_+(\phi(x)),\ \text{ if } \ x<S(0)=b,\\ &= C_-(\phi(x)),\ \text{ if } \ x>b. \end{aligned}\right\} \tag{11.1.18}$$

Here, $\phi(x)$ is the initial temperature (see (11.1.3)). Let $\psi(x,t)$ be a test function as defined in (11.1.5). A weak solution of the degenerate problem (11.1.13)—(11.1.15) and (10.2.11)—(10.2.16) is defined as a pair of bounded measurable functions (H,T) such that the following equation

$$\iint_{\Omega_*}(H\psi_t+T\psi_{xx}+\psi q)dt=\int_0^{t_*}(f_2(t)\psi_x(1,t)-f_1(t)\psi_x(0,t))dt-\int_0^{t_*}V_0(x)\psi(x,0)dx, \tag{11.1.19}$$

is satisfied for all test functions $\psi(x,t)$. Note that in the degenerate case, the temperature is not a well-defined function of enthalpy as at $T=0,\ \alpha_i(T)=0,\ i=1,2$.

The uniqueness of the weak solution of the above degenerate problem has been proved in [335]. The Lipschitz continuity of the free boundary in the degenerate problem has been proved in [336] under the following assumptions.

(1) $\alpha_i(T)\in C^\infty(R),\ \alpha_i(T)\geq 0$ and $\alpha_i(T)=0$ if and only if $T=0\ (i=1,2)$.

(2) $f_1(t),\ f_2(t)\in C^2[0,t_*],\ \phi(x)\in C^3[0,1]$.

(3) $f_1(t)$ is strictly positive, while $f_2(t)$ is strictly negative.

(4) $\phi(x)(x-b)\leq 0$ and the equality holds only at $x=b$.

(5) The compatibility conditions hold at $x=0$ and at $x=d$, i.e., $f_1(0)=\phi(0), f_2(0)=\phi(1)$.

(6) $\phi'(b)<0$.

If in addition to assumptions (1)—(6), we have also

(7) $f_1(t),\ f_2(t)$ and $\phi'(x)$ piecewise monotone,

then using the lap number theory [337] it has been proved in [338] that for each $t\in[0,t_*]$, the limits $T_x(S(t)\pm,t)$ exist and $S(t)$ is differentiable.

A weak formulation of an n-phase ($n > 2$) classical Stefan problem (10.2.59)–(10.2.65) with (n–1) free boundaries can be easily obtained if enthalpy and temperature are suitably defined (see (11.1.2)–(11.1.4)) in the union of regions occupied by different phases. It has been shown in [331] that under suitable assumptions, the free boundaries are C^{∞}-functions.

Using a weak formulation, the criterion for the disappearance of a mushy region in finite time or its existence for all times has been studied in many references. In § 5.2, this aspect has been discussed for some one-dimensional problems, and in § 11.2 this aspect will be further discussed for multi-dimensional problems. It may be noted that in the weak formulation, mushy region came into existence because of the formulation of the classical problem in terms of enthalpy defined in (11.1.2). We can say that mushy region in (11.1.2) is a mathematical novelty. But in CEF (classical enthalpy formulation) discussed in § 5.1, the physics tells us that the mush is formed before the liquid phase comes into existence. The mushy region in the weak formulation of the classical problem has been introduced artificially, in the sense that in the classical formulation there was no mushy region. No energy equation is written for the mushy region introduced in (11.1.2). In CEF discussed in § 5.1 (equations (5.1.1)–(5.1.12)), an energy equation is to be solved in the mushy region and energy balance is done at the boundaries of the mushy region, i.e., at the solid-mush and liquid-mush boundaries.

We shall now briefly discuss the characterization of mushy regions which may arise in the weak formulation or in CEF. It may be noted that the solution of CEF (we call this solution CES) is not a solution of the classical problem in which a single phase-change boundary exists but CES could be a weak solution. A weak solution will be CES provided smooth interfaces $S_1(t)$ and $S_2(t)$ exist in the weak solution and energy balance conditions are satisfied on them. The relationship amongst these three solutions was briefly discussed in (5.2.16).

The non-existence of a classical solution, i.e., the existence of a mushy region in the two-phase formulation of the Joule heating problem (see § 5.1) has been discussed in [339]. If a constant heat source is present, then the infinite differentiability of the solid-mush boundary $S_1(t)$ and of the liquid-mush boundary $S_2(t)$ has been established in [146]. Several results on the structure of the weak solution have been presented in [309] and we present some of them here. When a volumetric heat source $f(x,t)$ is present, the differentiability of $S_1(t)$ and $S_2(t)$ has been proved in a small time interval $(0, t_*)$ in ([309], Chapter VI) under the assumption that at $t = 0$, $S_1(0) = S_2(0) = x_0 \in [-1, 1]$ and $f(x_0, 0) > 0$.

The existence of the weak solution of a phase-change problem considered in $\Omega = \{x : |x| \leq 1\}$ in which an initial specific internal energy is arbitrarily prescribed has been proved in [309]. Let $T(\pm 1, t) = T^{\pm}(t)$ for $t \in (0, \hat{t})$. Then for $t > t_*$ (see [309] for the definition of t_* as it involves many quantities) there exists only one phase (solid or liquid) if $T^+T^- > 0$. If $T^+T^- < 0$, then for $t > t_*$ both solid and liquid phases are present and the weak solution coincides with the classical solution with only one free boundary. For some more results about the disappearance of the mushy region see [309] and § 5.1.

The evaluation of the lifetime of a mush has been studied in [340]. If a suitable energy criterion holds, then the mushy zone disappears in finite time. Otherwise, it exists for all times.

11.2 Regularity of the Weak Solutions of Multi-dimensional Stefan Problems

As mentioned earlier, we are in principle interested only in those weak solutions which under suitable assumptions are as good as classical solutions, i.e., the phase-change boundary exists and is differentiable. This requirement is not met by the weak solutions of multi-phase multi-dimensional Stefan problems studied in the references. Therefore multi-dimensional problems are being discussed here for an expository reason. We first present a weak formulation of a simple two-phase Stefan problem. Weak formulation of a one-phase Stefan problem can be easily obtained as a particular case of the formulation discussed below.

11.2.1 Weak solutions of some two-phase Stefan problems in R^n, $n > 1$

The weak solution of a two-phase Stefan problem is being discussed first. For the one-phase problems stronger results are available than for two-phase problems. The geometry and notations of the two-phase Stefan problem formulated in § 1.4.1 will be retained . Let ∂G_1 lie in the interior of ∂G_2. For any t_*, $0 < t \le t_* \le \infty$, set $\Omega_{t_*} = G \times (0, t_*)$. For simplicity, the parabolic heat equations in the two-phases will be taken as

$$\alpha_i \frac{\partial T_i}{\partial t} = \mathcal{L}_i T_i = \nabla^2 T_i + q^{(i)}(x,t) T_i, \ (x,t) \in \Omega_{t_*}^{(i)} = G_i(0) \times (0, t_*); \ i = 1, 2, \quad (11.2.1)$$

where α_i is a positive constant. We introduce the following notations.

$$\left.\begin{aligned} H(T) &= \alpha_1 T, && \text{if } T > 0, \\ &= [-l, 0], && \text{if } T = 0, \\ &= \alpha_2 T - l, && \text{if } T \le 0. \end{aligned}\right\} \quad (11.2.2)$$

$$\left.\begin{aligned} T &= T_1/\alpha_1, && \text{if } T > 0, \\ &= T_2/\alpha_2, && \text{if } T < 0. \end{aligned}\right\} \quad (11.2.3)$$

$$T = g = g_i/\alpha_i, \text{ on } \partial_i G(0) \times (0, t_*), \ i = 1, 2. \quad (11.2.4)$$

$$T = f = f_i/\alpha_i, \text{ on } G_i(0), \ i = 1, 2. \tag{11.2.5}$$

$$q = q^{(i)}, \ (x,t) \in \Omega_{t_*}^{(i)}, \ i = 1, 2. \tag{11.2.6}$$

With the help of (11.2.2)–(11.2.6), the two equations in (11.2.1) can be written as a single equation

$$\frac{\partial H}{\partial t} = \nabla^2 T + qT, \ (x,t) \in \Omega_{t_*}, \tag{11.2.7}$$

which holds in the distributional sense in Ω_{t_*}. Following the procedure indicated in § 5.2 to obtain a weak solution in the one-dimensional case, a weak formulation for the problem (11.2.2)–(11.2.6) can also be obtained. In the present case, integration by parts is to be done using Green's formulas [82]. A pair of bounded measurable functions (H,T) is called a weak or a generalized solution of the equation

$$\iint\limits_{\Omega_{t_*}} \left[T\mathcal{L}^*\phi + H(T)\frac{\partial \phi}{\partial t} \right] dxdt = \int\limits_0^{t_*}\int\limits_{\partial G} g\frac{\partial \phi}{\partial n} ds_x dt - \int\limits_{G(0)} T(f)\phi dx, \tag{11.2.8}$$

if (11.2.8) is satisfied for all test functions $\phi(x,t)$ with $\nabla_x\phi, \nabla_x^2\phi, D_t\phi$ continuous in $\overline{\Omega}_{t_*}$ and $\phi = 0$ on $G(t_*)$ and on $\partial G \times (0, t_*)$. $\vec{n}$ is the unit outward normal on the lateral surface and ds_x is the elementary surface area. $\mathcal{L}^* = \mathcal{L}_1$ at points where $T > 0$ and $\mathcal{L}^* = \mathcal{L}_2$ at points where $T < 0$. The set W defined as

$$W = \{(x,t) \in \Omega_{t_*} : -l < H(T) < 0\}, \tag{11.2.9}$$

is called a mushy region or a cloud. It has been proved in [332] that under suitable assumptions such as appropriate sign restrictions, continuity (in some cases smoothness is also required) of initial-boundary data, smoothness of ∂G, the condition $q = q(x) \le 0$, and holding of compatibility conditions, there exists a unique weak solution of (11.2.8) which belongs to $W^{1,2}(\Omega_{t_*})$ (see [332] for the complete set of assumptions).

Following the procedure indicated in § 5.2 for a one-dimensional Stefan problem, it can be proved that a classical solution of the present multi-dimensional problem is also its weak solution. If a smooth function $\Phi(x,t)$ or a smooth surface $\Gamma(t)$ exists (see § 1.4.1) which satisfies the conditions mentioned in § 1.4.1, then it can be proved that a weak solution is also a classical solution. In § 5.2, it was assumed that the weak solution satisfies initial-boundary conditions and $T_i = 0$, $i = 1, 2$, on $\Gamma(t)$. These assumptions are not necessary. If Φ and $\Gamma(t)$ satisfy appropriate smoothness assumptions together with other assumptions mentioned in § 1.4.1, then by choosing test functions suitably, it can be proved that a weak solution satisfies initial-boundary conditions in addition to the Stefan condition.

The following result plays an important role in the justification of numerical solutions of the classical Stefan problems obtained with the help of weak formulations. Since weak solutions are fixed domain formulations without phase-change boundaries, it is easier to obtain these solutions numerically.

Proposition 11.2.1. Assume that a unique weak solution of (11.2.8) exists in Ω_{t_*}. In any open subset M of Ω_{t_*} where $T \geq 0$ and $H(T) \geq 0$ (resp. $T \leq 0$ and $H(T) \leq -l$), T is a classical solution of $\alpha_1 \partial T_1/\partial t = \mathcal{L}_1 T_1 = \nabla^2 T_1$ (resp. $\alpha_2 \partial T/\partial t = \mathcal{L}_2 T_2 = \nabla^2 T_2$).

In essence, the above proposition tells us that under certain assumptions, away from the mushy region, the weak solution is as good as a classical solution. More general parabolic operators can also be considered in the above proposition. The stability of the weak solution and some properties of the weak solution have also been discussed in [332].

In several references, more general parabolic equations have also been considered. For further discussion, we shall continue with the notations given in (11.2.2)—(11.2.6). In [341], parabolic equations of the following form have been considered in both solid and liquid phases.

$$C(x,t,T)\frac{\partial T}{\partial t} - \nabla(K(x,t,T)\nabla T) = q(x,t,T), \; (x,t) \in \Omega_{t_*}. \tag{11.2.10}$$

On using Kirchhoff's transformation given in (1.4.29) and notations used in (11.2.2)—(11.2.6), an equation of the form (11.2.7) can be obtained in terms of θ (see (1.4.29) for θ). The existence and uniqueness of the weak solution and its continuous and monotone dependence on the initial-boundary data has been proved in [341] under weaker conditions and piecewise smoothness of ∂G is sufficient. It has been proved that

$$\theta \in H^1(\Omega_{t_*}) \cap L^\infty(0, t_*; \; H^1(G)).$$

See Appendix D for the definition of $L^\infty(0, t_*; H^1(G))$.

Instead of a classical formulation of the two-phase Stefan problem, the following singular nonlinear partial differential equation has been considered in [342].

$$\frac{\partial H(T)}{\partial t} - \text{div}\; \vec{a}(x,t,T,\nabla T) + b(x,t,T,\nabla T) \ni 0. \tag{11.2.11}$$

The weak derivatives have to be considered in (11.2.11). By a weak solution of (11.2.11), we mean a function $T(x,t) \in W_2^{1,1}$ defined by $T = H^{-1}(w)$, where w is a function defined in Ω_{t_*} such that $w \subset H(T)$, the inclusion being intended in the sense of graphs and w and T satisfy the equation

$$\int_G w(x,\tau)\phi(x,\tau)\Big|_{t_0}^{t} dx + \int_{t_0}^{t}\int_G \{-w(x,\tau)\frac{\partial \phi}{\partial \tau} + \vec{a}\cdot\nabla_x\phi + b(x,\tau,T,\nabla_x T)\phi\} dx d\tau = 0, \tag{11.2.12}$$

for all test functions $\phi \in W_2^{1,1}(\Omega_{t_*})$, whose trace is zero on $\partial G \times (0, t_*)$ and on all intervals $[t_0, t] \subset (0, t_*]$. Under suitable assumptions (cf. [342]), the continuity of $T(x,t)$ has been established which can be extended upto $\overline{\Omega}_{t_*}$. By considering a singular partial differential equation of the form

$$\frac{\partial H(T)}{\partial t} \ni \nabla^2 T(x,t), \tag{11.2.13}$$

which holds in Ω_{t_*} in the sense of distributions, continuity of the temperature in the weak solution of a two-phase problem has been proved in [343]. The method of proof in [343] relies strongly on the properties of the Laplacian operator and the absence of lower order terms in the energy equation. The approach in [342] is different from [343] and the method of proof consists of a suitable modification of the parabolic version of De Giorgi estimates reported in [295]. For further references on weak solutions, see the bibliography in [342].

The characterization of the mushy region in CEF in R^n, $n > 1$ and in the weak solution of the classical Stefan problem will now be discussed briefly. The first attempt to investigate the behaviour of the mushy region in a multi-dimensional two-phase problem was made in [344]. The non-increase of the mushy region in the homogeneous Stefan problem (no heat sources) with constant Dirichlet data and almost uniformly continuous initial data was proved. The results of [344] have been generalized in [345] by considering a nonhomogeneous Stefan problem with heat sources. Let $(T, H(T))$ be a bounded generalized solution of the following problem.

$$\left.\begin{aligned} &\frac{\partial H}{\partial t} - \nabla^2 T = f(H), \text{ in } \Omega_{t_*}, \\ &T|_{\partial G\times(0,t_*)} = \overline{T}(x,t), \ x \in \partial G, \\ &H|_{t=0} = H_0(x), \ x \in G. \end{aligned}\right\} \tag{11.2.14}$$

Let the mushy region $W(x,t)$ be defined by (11.2.9) and let $W(t_0) = W \cap (t = t_0)$. If the function $f(H)$ in (11.2.14) is uniformly Lipschitz continuous, $f(0) \geq 0$ and $f(-l) \leq 0$, then for every $\overline{T} \in L_\infty(\partial G \times (0, t_*))$, $H_0 \in L_\infty(G)$, the mushy region $W(x,t)$ in the bounded generalized solution of problem (11.2.14) is nonincreasing in time. $W(t_2) \subset W(t_1)$ for every $t_2 > t_1$, in the sense that $| W(t_2) \backslash W(t_1) |= 0$. Furthermore, the mushy region can be described in the following way: There exists a non-negative function $P : G \to R \cup \{+\infty\}$ such that

$$W = \{(x,t) : x \in W(0), \ 0 \leq t < P(x)\}. \tag{11.2.15}$$

For every $x \in W(0)$ on the interval $t \in [0, P(x))$, the function $H(T)$ is a solution of the Cauchy problem

$$H_t = f(H), \ H|_{t=0} = H_0(x). \tag{11.2.16}$$

Criteria for the disappearance of the mushy region after some finite time in the one-phase and two-phase Stefan problems in the absence of heat sources in domains with C^1 boundary have been discussed in [346]. The behaviour of the mushy region in the corner points of a domain from different angles has also been discussed.

A CEF in R^n, $n \geq 1$ has been considered in [157] and under the assumption that both regular and weak solutions exist, the behaviour of the solution has been investigated. A weak formulation in the form of the following singular equation has been considered.

$$\frac{\partial H}{\partial t} - \text{div}\vec{K}(x,t,T,\nabla_x T) = Q(x,t,H), \text{ in } \Omega_{t_*}. \tag{11.2.17}$$

Here, the function $Q \in C(\Omega_{t_*} \times R)$. The main interest in the weak solution in [157] is in the behaviour of the solution near the free boundaries and in the growth of the mushy region. To investigate the behaviour of the solution near solid-mush and liquid-mush boundaries, it has been assumed that the classical enthalpy solution exists. It has been further assumed that the mush keeps expanding into the solid phase, until, eventually, it is invaded by a new liquid phase. To investigate the behaviour of the mush near the points where $H = 0$, $(H \geq 0$ in the liquid region) it is assumed that a weak solution exists with $T \geq 0$ in Ω_{t_*} (temperature of the solid is taken as zero). Let (T, H), where $H = T + w$ with $T \in L^\infty(\Omega_{t_*}) \cap W_2^{1,1}(\Omega_{t_*})$, $w \in H_v(T)$ a.e. in Ω_{t_*} and $w \in L^\infty(\Omega_{t_*})$ be a local solution of (11.2.17) in Ω_{t_*} (see [157] for the definition of a local solution). Here, H_v is Heavyside graph in which $H_v(T) = 0,\ T > 0,\ H_v(T) = -l,\ T < 0$ and $H_v(T) \in [0, -l]$ for $T = 0$. If $\vec{K}$ and Q satisfy some assumptions (cf. [157]), then

$$w(x,t) - w(x,\tau) \geq \int_\tau^t Q(x,p,w(x,p))\chi_{\{T=0\}}(x,p)\, dp, \tag{11.2.18}$$

a.e. in Ω_{t_*}, $\tau < t$. Also for $0 < \tau < t < t_*$, we have

$$\text{meas}\,\{x \in G |\, T(x,t) = 0\} \leq \text{meas}\,\{x \in G |\, T(x,\tau) = 0\}\,. \tag{11.2.19}$$

Similar results can be obtained near the points where $H = -l$ (see [157] for other results). An example has been constructed in which the mushy region is enclosed by two solid phases. The mushy region disappears and reappears immediately after extinction.

11.2.2 Regularity of the weak solutions of one-phase Stefan problems in $\mathbf{R^n}$, $\mathbf{n > 1}$

In view of (11.2.2), we shall call a Stefan problem one-phase, if either $H \leq -l\ (H \geq 0)$ in one of the phases and in another phase $H = 0\ (H = -l)$. We consider the second case and let $G_1(t)$ be the liquid phase at any time $t \in (0, t_*)$ with temperature $T(x,t) \geq 0$ and $H(T) \geq 0$. In the solid phase $H(T) = -l$. $G_1(t)$ is the set of points (x,t) with x outside $\partial_1 G$. Note that ∂G_1 lies inside ∂G_2. Take $\phi_2 = 0,\ f_2 = 0,\ H(\phi_2) = -l$ and $H(\phi_1) > 0$. The existence and uniqueness of the weak solution (under suitable assumptions) of the one-phase Stefan problem has been proved in [332]. The continuity of the temperature in Ω_{t_*} in the weak solution can also be proved under suitable assumptions some of which are : (i) $\partial_1 G \in C^{2+\eta}$, $\eta > 0$, (ii) f_1 and ϕ_1 are continuous functions on $\partial_1 G \times [0, \infty)$ and $\overline{(G_1(0))}$ respectively, coinciding on $\partial_1 G$, (iii) $\phi_1 > 0$ in $G_1(0)$ (see [332] for some other assumptions). The non-occurrence of the mushy region in the one-phase problem has also been proved.

Let $W = \{(x,t) \in \Omega_{t_*}; -l < H(T) < 0\}$ and $W(\sigma) = W \cap \{t = \sigma\}$. Then $W(\sigma) \subset \overline{(G_1(\sigma))} - G_1(\sigma)$. W can be called a *weak free boundary*. It has been proved that: (i) a weak free boundary has no interior points in Ω_{t_*}, and (ii) W is determined only upto a set of measure zero.

The measure of the mushy region in the weak solution of a one-phase multi-dimensional Stefan problem has been estimated in [347] with the help of a refined method of isoperimetric inequalities. There are two disjoint connected components of the boundary. On one part of the boundary, the temperature $T = 0$ is prescribed and on the other part of the boundary, a constant temperature $T > 0$ is prescribed. The results obtained in [347] hold till the liquid phase reaches the second component of the boundary. The weak solution of a spherically symmetric one-phase Stefan problem in R^n has been considered in [348]. The main goal is to obtain conditions on the prescribed boundary temperature and initial enthalpy for the disappearance of the mushy region after a finite time. A sketch of the proof has been given for a one-phase problem considered in a regular bounded open set of R^n.

The behaviour of the mush in a one-phase Stefan problem has been discussed in [349]. A one-phase problem in a region $\Omega \subset R^n$ with a piecewise-smooth boundary has been considered. It has been shown that the lifetime of the transient phase is uniformly bounded for every $\hat{t}$ and $\gamma > 0$ on the Lebesgue set $\left\{G_1(\hat{t}, x) < -\gamma\right\}$ for some specific function G_1 (cf. [349]). If the boundary of Ω is C^2-smooth and a suitable smooth boundary temperature is prescribed, then the necessary and sufficient conditions for the disappearance of the mushy region in some finite time have been obtained. The behaviour of the mush near corner points of a two-dimensional domain has also been investigated.

The equivalence of the weak enthalpy formulation of a one-phase Stefan problem in R^n and its weak variational inequality formulation has been established in [197]. It may be noted that a weak enthalpy formulation of a one-phase Stefan problem can be extended to a two-phase problem but the variational inequality formulation given in (7.4.36) has no natural extension to an inequality formulation of a two-phase problem. Let Ω be a bounded domain in R^n, $n \geq 1$, whose boundary consists of two smooth connected hypersurfaces Γ_1 and Γ_2 with Γ_1 lying inside Γ_2 and bounding a simply connected domain G. Let B be a large ball with center 0 containing Ω and set $D = B - G$. The one-phase Stefan problem consists of finding the free boundary $t = S(x)$ and the temperature $T(x,t)$, $0 < t < t_*$, $x \in D$, such that

$$T_t - \nabla^2 T = 0, \text{ in } \{(x,t) : x \in D,\ t > S(x)\}\,;\ S(x) = 0, \text{ if } x \in \Omega, \tag{11.2.20}$$

$$T = 0, \text{ and } \nabla T \cdot \nabla S = -l;\ t = S(x),\ x \in D - \Omega, \tag{11.2.21}$$

$$T = \phi(x) > 0,\ x \in \Omega,\ t = 0, \text{ and } T = g(x,t) > 0,\ x \in \Gamma_1,\ 0 < t < t_*. \tag{11.2.22}$$

Here, $l > 0$ is a constant, and $g(x,t)$ and $\phi(x)$ are C^2-functions in $\Gamma_1 \times (0, t_*]$ and $\overline{\Omega}$, respectively. To define a weak formulation of the problem (11.2.20)–(11.2.22), we define enthalpy $H(T)$ as:

$$\left.\begin{aligned} H(T) &= T, & T > 0 \\ &= T - l, & T \leq 0 \end{aligned}\right\}, \text{ and } \left\{\begin{aligned} H(T(x,0)) = f &= \phi(x),\ x \in \Omega, \\ &= -l,\ x \in D - \Omega. \end{aligned}\right. \tag{11.2.23}$$

The region $D-\Omega$ is occupied by ice at temperature $T=0$. The weak formulation of the problem (11.2.20)−(11.2.22) in D can be easily obtained and is given by

$$\int_0^{t_*}\int_D \left(T\nabla^2 v + H(T)v_t\right) dx dt = \int_0^{t_*}\int_{\partial D} g\frac{\partial v}{\partial n} ds_x dt - \int_D H(T(x,0))v(x,0)dx. \quad (11.2.24)$$

Here, $v(x,t)$ is a test function in $\overline{D}\times(0,t_*)$ with a definition similar to that given in (11.1.5) (make appropriate changes). The unique solution of (11.2.24) exists. On using the transformation

$$u(x,t)=\int_0^t T(x,\tau)d\tau,\ T\geq 0,\ (x,t)\in D_{t_*}=D\times(0,t_*), \quad (11.2.25)$$

we have $u_t = T(x,t)$. Writing the energy equation in terms of enthalpy and then integrating (11.2.20) with respect to time, we get

$$u_t - \nabla^2 u = \gamma(u_t) + f,\ \text{a.e. in } D_{t_*}. \quad (11.2.26)$$

Here, we have set $H(T)=T-\gamma(T)$, γ is a monotone graph, and $\gamma(T)\geq 0$ (cf. [332]).

For all test functions $v\geq 0$, a.e. in D_{t_*} with appropriate initial-boundary values (make appropriate changes in (7.4.19) to define $v(x,t)$), we have

$$(u_t-\nabla^2 u)(v-u_t)\geq f(v-u_t),\ \text{a.e. in } D_{t_*},\ u_t\geq 0. \quad (11.2.27)$$

Note that $(u_t-\nabla^2 u - f)v=\gamma(T)v\geq 0$ and $(u_t-\nabla^2 u-f)u_t=0$, in D_{t_*}. Conversely, if u satisfies (11.2.27) and $T(x,t)$ is defined through (11.2.25) and satisfies the boundary and initial conditions ($T(x,0)=\phi(x), x\in\Omega$ and $T(x,0)=0,\ x\in D-\Omega$), then enthalpy equation (11.2.7) with $q=0$ can be recovered from (11.2.27) and $(T(x,t),H(T))$ is a solution of (11.2.24). In other words, $T(x,t)$ is a weak solution if and only if $u(x,t)$ in (11.2.25) is a solution of the 'variational inequality' (11.2.27). Note that the variational inequality (7.4.36) is different from the 'variational inequality' (11.2.27) as we have $(v-u_t)$ in (11.2.27) and not $(v-u)$ as in (7.4.36) and u is also defined differently in (7.4.36). The variational inequality corresponding to the problem described above is given by (7.4.36) whose unique solution exists (under suitable assumptions). It can be proved that the unique solution of (7.4.36) satisfies (11.2.27). Therefore $T(x,t)$ is a solution of (11.2.24) if and only if it is a solution of (7.4.36). Using the variational inequality formulation of the one-phase problem, it has been proved in [197] that the free boundary arises as a boundary of a set and so it has no interior points. It has also been shown that under suitable hypotheses, the domain occupied by water in a ice-water system is star shaped and the free boundary is star shaped with the representation $\rho=\rho^*(\theta,t)$, (ρ,θ) are polar coordinates; ρ^* is a continuous function of θ and t and uniformly Lipschitz continuous in the angles θ, and is monotonically increasing in t.

A two-phase two-dimensional Stefan problem in a rectangular region $\Omega_{t_*}=\Omega\times(0,t_*)$, $\Omega=\{(x_1,x_2):\ 0<x_1<1, 0<x_2<b\}$ has been considered in [350]. Linear

homogeneous parabolic equations are considered in the two phases occupying the regions Ω_1 and Ω_2, and $\Omega = \Omega_1 \cup \Omega_2 \cup S(x, 0)$; $x_2 = S(x_1, t)$ is the equation of the free boundary. $\Omega_1 = \{(x_1, x_2, t) : 0 < x_2 < S(x_1, t)\}$ and $\Omega_{t_*} \cap \partial\Omega_1 \times (0, t_*)$ is the free boundary. Let $T(x_1, x_2, t)$ be the temperature in Ω_{t_*} defined as in (11.2.3) with appropriate changes (to formulate this problem see § 3.2.4 and make appropriate changes). Prescribe suitable boundary conditions on the edges so that the free boundary divides Ω_{t_*} into two disjoint regions, each consisting of a single phase. Let,

$$T(x_1, x_2, 0) = \psi(x_1, x_2), \text{ in } \Omega; \text{ and } S(x_1, 0) = y_0(x_1),\ 0 < x_1 < 1. \tag{11.2.28}$$

Here, ψ and y_0 are given functions. It has been proved that if the classical Stefan problem described above is formulated in terms of the temperature $T(x_1, x_2, t)$ in Ω_{t_*}, then $T(x_1, x_2, t) \in W_2^{1,1}(\Omega_{t_*}) \cap H^{\delta,\delta/2}(\overline{\Omega}_{t_*})$, $0 < \delta < 1$, the free boundary is a Lipschitz surface and the Stefan condition written in terms of the temperature $T(x_1, x_2, t)$ is satisfied in almost everywhere sense. The Stefan condition has been considered in the form

$$\left[\alpha(T)\frac{\partial T}{\partial x_i}\right] \cos(\vec{n}, x_i) + l\cos(\vec{n}, t) = 0,\ x_2 = S(x_1, t),\ i = 1, 2, \tag{11.2.29}$$

$\vec{n}$ is the unit outward normal to the free boundary, $\alpha(T) = K_1$ in Ω_1 and $\alpha(T) = K_2$ in Ω_2, the notation $[v]$ stands for the jump in v across the free boundary. The next step is to obtain a weak formulation by multiplying the energy equation formulated in terms of $T(x_1, x_2, t)$ by a suitable test function $\eta(x_1, x_2, t)$ and carry out the integration by parts. We obtain

$$\int_{\Omega_{t_*}} \{\alpha(T)\nabla T \cdot \nabla\eta + T_t\eta + l\chi_t\eta\}\, dxdt + \beta\int_{\Gamma_{1t_*}} T\eta ds = 0. \tag{11.2.30}$$

Here, $\chi(x, t)$ is the characteristic function of the region Ω_1, $\Gamma_{1t_*} = \{(1, x_2);\ 0 < x_2 < b\} \times (0, t_*)$ and $K_i\partial T/\partial n + \beta T = 0$ on $\Gamma_{1t_*} \cap \partial\Omega_i$. Other boundary conditions have been taken to be either temperature prescribed or no flux conditions. The second term in (11.2.30) arises during the integration by parts and using the boundary condition, $\eta = 0$ on the boundary on which temperature is prescribed. When equation (11.2.30) is discretized in time and not in space (cf. [307], [351]), a sequence of elliptic free boundary problems is obtained. The solution of each free boundary problem has been obtained as the minimum of a suitable functional. To obtain existence, uniqueness and regularity results some conditions on the initial temperature $\psi(x_1, x_2)$ are imposed such $\psi(x_1, x_2) \in W_2^1(\Omega) \cap C^\nu(\overline{\Omega})$, $0 < \nu < 1$, $\psi_{x_1} \le 0$, $\psi_{x_2} \ge 0$ (cf. [350] for further details).

A regularity theory for the weak free boundary which is defined as $\partial(T > 0)$ for the parabolic two-phase free boundary problems ($T > 0$ in one of the phases and $T \le 0$ in another phase) has been developed in [352]. The regularity theory has several approaches:

(i) Lipschitz minimal surfaces are smooth.

(ii) 'Flat' minimal surfaces (in some 'Lebesgue' differentiability sense) are smooth.

(iii) Generalized minimal surfaces are smooth except on some small set.

The part (i) of the theory, that is, free boundaries, Lipschitz in space and time, are regular, has been developed in [352] (see [389] also). *Viscosity solutions* have been considered (weak solution is a viscosity solution but every viscosity solution is not a weak solution) whose free boundaries are given (locally) by a Lipschitz graph. In this case they enjoy further regularity and other properties such as the free boundary is a C^1 graph in space and time and temperature is a classical solution. The regularity of the free boundary is possible for those two-phase problems in which the two fluxes from both sides at the free boundary are not vanishing simultaneously. An example has been constructed which shows that a Lipschitz free boundary may remain Lipschitz for an interval of time and may not regularize instantaneously although both the phases may have non-zero temperatures.

Appendix A

Preliminaries

Some functional analytic material used in the chapters but not defined or explained there is being presented in Appendices A–D. This material is intended as an *aide-memoire.* For supplementary reading the reader is referred to references [9, 21, 58, 282, 309, 353, 354, 355, 356]

We start with the definition of a vector space as the concept of a vector space, also called a linear space, is fundamental to the functional analytic results which hold in vector spaces of some special type such as Hilbert spaces, Banach spaces and Sobolev spaces.

1. Abstract space. An abstract space, also called simply a 'space' is a set of (unspecified) elements satisfying certain axioms. By choosing different sets of axioms, a variety of abstract spaces can be defined.

2. Linear space. Let F be a field, generally taken to be R (real line) or $\mathcal{C}$ (complex plane). The elements of F are called 'scalars'. Let W be a nonempty set whose elements are called 'vectors'. If x, y belong to W, define a mapping from $W \times W$ into W as $(x, y) \to x + y$, where $x + y$ is a unique element of W. We call this mapping 'addition' of vectors which is a 'binary operation' on W. Let the addition of vectors satisfy the following axioms.

(i) $x + y = y + x$, for all $x, y \in W$. This axiom is called 'commutative law'.

(ii) $(x + y) + z = x + (y + z)$, for all $x, y, z \in W$. This axiom is called 'associative law'.

(iii) For each $x \in W$, there exists a unique element 0 in W such that $x + 0 = 0 + x = x$. This implies the existence of a 0 (zero) element in W.

(iv) For each $x \in W$, there exists a unique element $-x \in W$ such that $x + (-x) = 0$ $= (-x) + x$. This implies the existence of an additive 'inverse' for each x.

Scalar multiplication axioms. To every scalar $\alpha \in F$ and a vector $x \in W$, there corresponds a unique vector $\alpha x \in W$ such that

(v) $\alpha(\beta x) = (\alpha\beta)x$, for every $\beta \in F$,

(vi) $1x = x$ and $0x = 0$, for all $x \in W$,

(vii) $\alpha(x + y) = \alpha x + \beta y$, (distributive law),

(viii) $(\alpha + \beta)x = \alpha x + \beta x$, (distributive law),

The element '1' belonging to F is called the 'multiplicative identity'.

A set W with the operations of addition of vectors and the scalar multiplication of a scalar and a vector defined on it and satisfying axioms (i)–(vii) is called a *vector space* or a *linear space* over the field F.

The concept of a *metric* or 'distance' in the vector space W can be introduced through the notion of norm which is a real valued function defined on W.

3. Norm. A norm (a length function) on a vector space W is a real valued function defined for all $x \in W$, denoted by $\|x\|$. It is called *norm* of x if it satisfies the following axioms:

(i) $\|x\| \geq 0$; $\|x\| = 0$, if and only if $x = 0$,

(ii) $\|x + y\| \leq \|x\| + \|y\|$, for all $x, y \in W$,

(iii) $\|\alpha x\| = |\alpha| \, \|x\|$, $\alpha \in F$.

If the property, $\|x\| = 0$ implies $x = 0$ is not true, and all other properties described above hold, then we call the function $\|\cdot\|$, a *semi-norm* on W.

A vector space with a norm defined on it is called *a normed space* and is denoted as $(W, \|\cdot\|)$. The notation $\|\cdot\|_W$ means the norm is defined on W. A vector space can be equipped with different norms. The *distance* $d(x, y)$ between $x, y \in W$ or a distance metric on W may be defined as

$$d(x, y) = \|x - y\|.$$

4. Completeness of a vector space. Let $\{x_n\}$ be a sequence of vectors in W. $\{x_n\}$ is said to be a *Cauchy sequence* if for every $\varepsilon > 0$ there exists a positive integer $N = N(\varepsilon)$ such that

$$d(x_n, x_m) < \varepsilon \text{ for every } m, n > N.$$

A linear space W is said to be a *complete linear space* if every Cauchy sequence in W converges to a vector in W or in other words it has a limit which is an element of W. A normed space which is complete in the distance metric defined by the norm is called a *Banach space.*

The concept of the length of a vector has been generalized in the definition of the norm of a function but what is missing is the analogue of the definition of familiar 'dot product' of two vectors in vector calculus. This is taken care by the definition of an inner product.

5. Inner product or Scalar product. An *inner product* on W is a mapping from $W \times W$ into the scalar field F of the linear space W which associates with any two elements $x, y \in W$ a scalar which we denote by (x, y) such that

(i) $(x + y, z) = (x, z) + (y, z)$, for all $x, y, z \in W$,

(ii) $(\alpha x, y) = \alpha(x, y)$, and $(x, \alpha y) = \overline{\alpha}(x, y)$, $\alpha \in F$,

(iii) $(x, y) = \overline{(y, x)}$ (bar denotes the complex conjugate),

(iv) $(x, x) \geq 0$; $(x, x) = 0$ if and only if $x = 0$.

An inner product defines a norm as well as a metric on W which are given by

$$\|x\| = \sqrt{(x,x)}; \text{ and } d(x,y) = \sqrt{(x-y, x-y)} = \|x - y\|.$$

An inner product space which is complete in the norm defined by the inner product is called a *Hilbert space*. All Hilbert spaces are Banach spaces.

If can be easily proved that: (i) $(x, y) \leq \|x\| \|y\|$, (ii) $|\|x\| - \|y\|| \leq \|y - x\|$. The following results hold good for normed spaces.

(1) On a finite-dimensional vector space W, any norm $\|\cdot\|$ is equivalent to any other norm $|\!|\!|\cdot|\!|\!|$. Two norms $\|\cdot\|$ and $|\!|\!|\cdot|\!|\!|$ are said to be equivalent if there exist positive numbers a and b such that for all $x \in W$, we have

$$a\,|\!|\!|x|\!|\!| \leq \|x\| \leq b\,|\!|\!|x|\!|\!|.$$

(2) Every finite-dimensional normed space is complete.

6. Compactness. A subset V in a normed space W is *compact* if every sequence in V has a convergent subsequence with a limit point in V.

The same definition holds for the normed space W. In a finite-dimensional normed space W, any subset V of W is compact if and only if V is closed and bounded. A compact subset V of W is closed and bounded but the converse of this statement is in general false for infinite-dimensional normed spaces. V is said to be *weakly compact* if every sequence of its elements contains a subsequence which converges weakly to an element of V (see **26**. of this appendix for the definition of weak convergence).

7. Compact support. A function $f : W \to F$ (field F) has a *compact support* in W if it is zero outside a compact subset of W. The closure of the set $\{x \in W : F(x) \neq 0\}$ is called the support of F.

8. Denseness. A subspace V of a normed space W is *dense* or 'everywhere dense' in W if its closure with respect to the norm is equal to W or contains W as a subset. W is said to be *separable* if it has a countable subset which is dense in W.

9. Linear operator. In the case of normed spaces, a mapping is called an 'operator'. A *linear operator* $P : D(P) \subset W_1 \to W_2$ is an operator such that: (i) the domain $D(P)$ of P is a normed vector space and the range $R(P)$ lies in a normed vector space over the same field F over which $D(P)$ is a vector space, (ii) for all $x, y \in D(P)$ and scalars $\alpha \in F$,

$$P(x + y) = Px + Py; \; P(\alpha x) = \alpha Px.$$

The *null space* or *kernel* of P is the set of all $x \in D(P)$ such that $Px = 0$. If P is one-to-one, then a mapping P^{-1}, called an *inverse mapping* of P, can be defined as

$P^{-1} : R(P) \to D(P)$. If $Px_0 = y_0$, then $P^{-1}y_0 = x_0$. If P^{-1} exists, then it is linear. P^{-1} exists if and only if the kernel of P consists of only the zero element.

10. Bounded linear operator. Let W_1 and W_2 be two normed spaces and $P : D(P) \subset W_1 \to W_2$ be a linear operator. The operator P is said to be a *bounded linear operator* if there is a real number $\delta > 0$ such that for $x \in D(P)$,

$$\|Px\| \le \delta \|x\| .$$

The norm on the left is on W_2 and the norm on the right is on W_1. A bounded linear operator P maps bounded sets in W_1 onto bounded sets in W_2. If W is a finite-dimensional normed space, then a linear operator on W is bounded.

11. Norm of a bounded linear operator. The norm $\|P\|$ of a bounded linear operator is defined as

$$\|P\| = \sup_{\substack{x \in D(P) \\ \|x\|=1}} \|Px\|$$

By taking $\delta = \|P\|$, we have $\|Px\| \le \|P\| \, \|x\|$.

The vector space $\mathcal{L}(X,Y)$ of all bounded linear operators from a normed space X into a normed space Y is itself a normed space under the norm defined as

$$\|Q\| = \sup_{\substack{x \in X \\ \|x\|=1}} \|Qx\| , \quad Q \in L(X,Y) .$$

If Y is a Banach space, then $\mathcal{L}(X,Y)$ is a Banach space.

12. Continuity of an operator. Let W_1 and W_2 be two normed spaces and $P : D(P) \subset W_1 \to W_2$ be an operator not necessarily linear. The operator P is said to be continuous at a point $x_0 \in D(P)$ if for every $\varepsilon > 0$ there is a $\delta > 0$ such that $\|Px - Px_0\| < \varepsilon$ for all $x \in D(P)$ satisfying $\|x - x_0\| < \delta$. P is continuous if P is continuous at every $x \in D(P)$.

The mapping $x \to \|x\|$ is continuous, i.e., the mapping $(W, \|\cdot\|_W) \to R$ is continuous. If P is linear, then it has some interesting properties.

(i) P is continuous if and only if P is bounded. If P is bounded, then $x_n \to x$ implies $Px_n \to Px$.

(ii) If P is continuous at a single point, then it is continuous.

13. Compact and completely continuous operators. Let W_1, W_2 be normed spaces and P an operator from W_1 into W_2. The operator P is called compact if $P(A)$ is *precompact* in W_2 whenever the set A is bounded in W_1. The term relatively compact is also used for precompact. A is precompact in W_1 if $\overline{A}$ is compact in W_1.

If P is continuous and compact, then it is called a *completely continuous operator.*

Every compact operator is bounded. A linear operator $P : W_1 \to W_2$ has the property that if W_1 is finite-dimensional, P is compact.

Every linear compact operator is continuous.

Let P be a linear operator from W_1 to W_2. Then P is compact if and only if it maps every bounded sequence $\{x_n\}$ in W_1 onto a sequence $\{Px_n\}$ in W_2 which has a convergent subsequence in W_2. If a sequence $\{T_n\}$ of compact linear operators from $W_1 \to W_2$ is uniformly operator convergent, i.e., $||T_n - T|| \to 0$, then the limit operator T is compact provided W_2 is a Banach space.

14. Restriction (extension) of an operator. Let $P : D(P) \subset X \to Y$ and $A \subset D(P)$. Let $P|_A$ be a mapping from $A \to Y$ defined by

$$P|_A : A \to Y, \quad P|_A x = Px \text{ for all } x \in A.$$

The operator $P|_A$ is called the *restriction of an operator P*. Let $M \supset D(P)$. An operator $\hat{P}$ is called an extension of P if $\hat{P} : M \to Y$ is such that $\hat{P}|_{D(P)} = P$, i.e., $\hat{P}x = Px$ for all $x \in D(P)$.

15. Orthogonal complement. In an *inner product space* (a space with a inner product defined on it) Z, two vectors x and y are said to be orthogonal, written as $x \perp y$, if $(x, y) = 0$. The vector x is said to be orthogonal to a set $Y \subset Z$, written as $x \perp Y$, if x is orthogonal to every $y \in Y$. The subsets X and Y of Z are said to be orthogonal, written as $X \perp Y$, if $(x, y) = 0$ for all $x \in X$ and $y \in Y$. The set $Y^\perp = \{z \in Z : z \perp Y\}$ is called the orthogonal complement of Y.

16. Direct sum and projection operator. A vector space W is said to be the *direct sum* of two subspaces X and Y of W, written as $W = X \oplus Y$, if each $w \in W$ has a unique representation

$$w = x + y, \; x \in X, \; y \in Y.$$

If Y is any closed subspace of a Hilbert space H, then $H = Y \oplus Z, \; Z = Y^\perp$ and for every $x \in H$ there exist a $y \in Y$ such that

$$x = y + z, \; z \in Y^\perp \; \text{(direct sum)}.$$

The above equation defines a mapping $P : H \to Y, \; Px = y$. P is called *orthogonal projection* or *projection operator* of H onto Y. It can be proved that P is a bounded linear operator and P is *idempotent*, i.e., $P^2 = P$.

17. Functional. A *functional* is defined to be an operator whose range lies on the real line R or in the complex plane $\mathbb{C}$ or in other words a functional is a real or complex valued function. Some authors use the term functional for a continuous linear real valued operator. In this volume, we have used the term functional as a real valued function and whenever the term functional is used as a continuous linear real valued function, it has been explicitly indicated.

18. Dual space. Let W be a normed space. Then the set of all functionals f (bounded linear real valued functions on W) on W constitute a normed space with the norm defined by

$$\|f\| = \sup_{\substack{x \in W \\ \|x\|=1}} |f(x)|$$

This space of functionals is called the *dual space* of W and is denoted by W'.

The dual space W' of a normed space W is a Banach space (whether or not W is). We can consider $(W')'$, i.e., the dual space of the dual space W' denoted by W''. For each $x \in W$, we define a mapping g_x which is such that if $f \in W'$, then $g_x(f) = f(x)$, $g_x \in W''$ and to each $x \in W$ there exists a $g_x \in W''$. Thus we have defined a mapping $G : W \to W''$, $x \mapsto g_x$. G is called the *canonical mapping* or the canonical embedding of W into W''. G is linear.

19. Isomorphism. Let W_1, W_2 be two given vector spaces over the same field. We would like to know whether W_1 and W_2 are essentially identical, i.e., do they have the same structure in an abstract sense.

An *isomorphism* of a Hilbert space W_1 onto a Hilbert space W_2 over the same field is a *bijective* (one-to-one and onto) linear operator $P : W_1 \to W_2$ such that for $x, y \in W_1$,

$$(Px, Py) = (x, y).$$

W_1 and W_2 are called 'isomorphic Hilbert spaces'. Isomorphisms in normed spaces preserve norms. If W_1 and W_2 are two vector spaces (not necessarily normed spaces), then P should preserve the two algebraic operations of a vector space.

20. Reflexivity. It can be proved that the canonical mapping $G : X \to X''$, $x \mapsto g_x$ is linear and one-to-one. G need not be onto. A normed space X is said to be *reflexive* if the canonical mapping G is onto. The following results hold.

(i) If a normed space X is reflexive, it is complete and hence a Banach space.

(ii) Every finite-dimensional normed space is reflexive.

(iii) Every Hilbert space is reflexive.

21. Riesz representation theorem for functionals. For any bounded linear functional f on a Hilbert space W there exists a unique vector $y \in W$ such that

$$f(x) = (x, y) \text{ for all } x \in W.$$

Here, y depends on f and is uniquely determined by f and has the norm

$$\|y\| = \|f\|.$$

22. Bilinear form. A mapping $a(u, v) : U \times V \to R$, where U and V are vector spaces over the field R of real numbers is called a *bilinear form* if the mapping $a(u, v)$ is linear in both the arguments, i.e.,

$$a(\alpha_1 u_1 + \alpha_2 u_2, \beta_1 v_1 + \beta_2 v_2) = \sum_{i,J=1}^{2} \alpha_i \beta_J a(u_i, v_J), \ \alpha_i, \beta_J \in R, \ u_i \in U, v_J \in V;$$

$$i = 1, 2, \ \mathrm{J} = 1, 2.$$

We can also consider a bilinear form from $U \times U \to R$. If U and V are normed spaces and there exists a real number δ such that

$$|a(u,v)| \leq \delta \, \|u\|_U \, \|v\|_V \,, \ u \in U, \ v \in V,$$

then $a(u,v)$ is said to be bounded. The norm $\|a\|$ of a bounded bilinear form is defined as

$$\|a\| = \sup_{\substack{\|u\|=1 \\ \|v\|=1}} |a(u, \, v)| \, .$$

It is easy to prove that

$$|a(u,v)| \leq \|a\| \, \|u\|_U \, \|v\|_V \, .$$

A bilinear form is said to be continuous if it is bounded or if there exists a $\beta \in R, \ \beta > 0$ such that $|a(u,v)| \leq \beta \, \|u\|_U \, \|v\|_V$.

23. Coercivity. Let $a(u,v) : U \times U \to R$, and U be a normed space. $a(u,v)$ is said to be coercive on U if there exists an $\alpha \in R, \ \alpha > 0$ and $a(u,u) \geq \alpha \, \|u\|^2$, for all $u \in U$.

Let $a(u,v) : U \times U \to R$, where U is a Hilbert space. Define $a(u,v) = (u,v)$. Then $a(u,v)$ is symmetric, i.e., $a(u,v) = a(v,u)$, is continuous (take $\beta = 1$ in **22.**) and has coercive property (take $\alpha = 1$) as $a(u,u) = (u,u) = \|u\|^2 \geq \|u\|^2$. Note that a bilinear form is not symmetric in general.

24. Adjoint operator. Let W_1 and W_2 be two Hilbert spaces and P be a bounded linear operator, $P : W_1 \to W_2$. Then the *Hilbert-adjoint operator* P^* of P is an operator from $W_2 \to W_1$ such that for all $x \in W_1$, and $y \in W_2$,

$$(Px, y) = (x, P^* y).$$

It can be proved that P^* exists, is unique, is a bounded linear operator, and $\|P^*\| = \|P\|$.

25. Strong convergence of a sequence. A sequence $\{x_n\}$ in a normed space X is said to be *strongly convergent* or convergent in the norm if there exists an $x \in X$ such that $\lim\limits_{n \to \infty} \|x_n - x\| = 0$. The strong convergence is indicated as $x_n \to x$ or $\lim\limits_{n \to \infty} x_n = x$.

26. Weak convergence of a sequence. A sequence $\{x_n\}$ in a normed space X is said to be *weakly convergent* if there exists an $x \in X$ such that for every $f \in X'$ (dual space), we have $\lim\limits_{n \to \infty} f(x_n) = f(x)$. Weak convergence is indicated as $x_n \rightharpoonup x$ or $x_n \overset{w}{\to} x$. The element x is unique and is called the weak limit of $\{x_n\}$. Note that in the weak convergence we are dealing with a sequence of numbers.

The following results hold.

(i) If $\{x_n\}$ converges weakly to x, then every subsequence of $\{x_n\}$ converges weakly to x.

(ii) If $x_n \rightharpoonup x$, then the sequence $\{\|x_n\|\}$ is bounded.

(iii) Strong convergence implies weak convergence with the same limit. The converse is generally not true.

(iv) If X is a finite-dimensional normed space, then weak convergence implies strong convergence.

A sequence of bounded linear operators $\{P_n\}$, $P_n \in \mathcal{L}(X, Y)$ (X and Y are normed spaces) can be considered for defining strong and weak operator limits. If an operator $P \in \mathcal{L}(X, Y)$ exists such that if, (i) $\|P_n - P\| \to 0$, then P is called the *uniform operator limit* of $\{P_n\}$, (ii) $\|P_n x - Px\| \to 0$ for all $x \in X$, then P is called the *strong operator limit* of $\{P_n\}$, and (iii) $\|f(P_n x) - f(Px)\| \to 0$ for all $x \in X$ and $f \in Y'$, then P is called the *weak operator limit* of $\{P_n\}$.

Appendix B

Some function spaces and norms

1. Convexity. A set V in a vector space W is said to be a *convex set* if the line segment joining any two points in V is contained in V. A real valued function $f(x)$ defined on a convex set V is said to be a *convex function* if $f(\lambda x + (1-\lambda)y) \leq \lambda f(x) + (1-\lambda) f(y)$, $0 < \lambda < 1$ for all $x, y \in V$. The function is said to be concave if the inequality is reversed. If the equality is excluded, then we have a *strictly convex (concave) function.*

A Hilbert space X is *strictly convex* in the sense that if $x, y \in X$ and $\|x\| = \|y\| = 1$, $x \neq y$, then $\|x + y\| < 2$.

2. Hölder continuity and Lipschitz continuity. A function $f(x)$ defined on a bounded closed set Ω of R^n is said to be *Hölder continuous* in Ω with exponent α, $0 < \alpha < 1$, if there exists a constant δ such that $|f(x) - f(y)| \leq \delta\, |x - y|^\alpha$ for all $x, y \in \Omega$. The smallest value δ_0 of δ for which the inequality holds is called a *Hölder coefficient.*

If $\alpha = 1$, then f is called a *Lipschitz continuous* function.

3. Equicontinuity. Let F be a set of real or complex-valued functions such that $f \in F$ $(f(x) = f(x_1, x_2, .., x_n))$ is defined on a compact subset B of R^n. The functions in F are *uniformly bounded* on B if there exists a constant M with the property that $|f(x)| \leq M$ holds for all $x \in B$ and all $f \in F$.

The collection of functions F is *equicontinuous* on B if for every $\varepsilon > 0$ there exists a

$\delta > 0$ which depends only on ε, such that for $x', x'' \in B$, $\|x' - x''\| < \delta$ implies

$$|f(x') - f(x'')| < \varepsilon \text{ for all } f \in F.$$

Note that equicontinuity of F implies uniform continuity of each member of F, but not vice versa (δ may depend on f). If instead of F, a sequence $\{x_n\}$ of functions is considered, then $\{x_n\}$ is said to be equicontinuous on B if for every $\varepsilon > 0$ there exists a $\delta > 0$, depending only on ε such that for all x_n and $y_1, y_2 \in B$ satisfying $\|y_1 - y_2\| < \delta$, we have

$$|x_n(y_1) - x_n(y_2)| < \varepsilon.$$

4. Lower semicontinuity. Let W be a normed space and $f : W \rightarrow R$ and let $N(x_0)$ be the family of neighbourhoods of a point $x_0 \in W$. f is said to be *lower semicontinuous* (l.s.c.) at $x_0 \in W$ if for all $\varepsilon > 0$, there exists a $V_\varepsilon \in N(x_0)$ such that for all $y \in V_\varepsilon$, $f(y) \geq f(x_0) - \varepsilon$. An *upper semicontinuous function* can be defined in an analogous manner.

5. The space $C^m(\overline{\Omega})$. The space $C^m(\Omega)$, where m is a non-negative integer and $\Omega \subset R^n$, is a vector space of all functions $f(x)$, $x \in \Omega$ which together with all their partial derivatives $D^\beta f$ of orders $0 \leq |\beta| \leq m$, are continuous on Ω. Here, $\beta = (\beta_1, \beta_2, \cdots, \beta_n)$ and $|\beta|$ is defined below. $C^0(\Omega) = C(\Omega)$ and $C^\infty(\Omega) = \bigcap_{m=0}^{\infty} C^m(\Omega)$. The subspaces $C_0(\Omega)$ and $C_0^\infty(\Omega)$ consist of all those functions in $C(\Omega)$ and $C^\infty(\Omega)$, respectively, which have compact support in Ω.

If $f \in C(\Omega)$ is bounded and uniformly continuous on Ω (Ω is bounded), then it possesses a unique, bounded continuous extension to $\overline{\Omega}$. The vector space $C^m(\overline{\Omega})$ consists of all those functions $f \in C^m(\Omega)$ for which $D^\beta f$, $0 \leq |\beta| \leq m$, are bounded and uniformly continuous on Ω. $C^m(\overline{\Omega})$ is a Banach space if the norm of $f \in C^m(\overline{\Omega})$ is defined as

$$\|f\|_m = \sum_{|\beta| \leq m} \sup_{x \in \Omega} \left| D^\beta f(x) \right|,$$

$D^\beta = D_1^{\beta_1} D_2^{\beta_2} ... D_n^{\beta_n}$, $|\beta| = \sum_{J=1}^{n} \beta_J$ and $D_J = \partial / \partial x_J$, $J = 1, 2...n$. Here, all β_j's are non-negative integers.

6. The space $H_\alpha(\overline{\Omega})$ **or** $C^{0,\alpha}(\overline{\Omega})$, $0 < \alpha < 1$. The set of all Hölder continuous functions on $\overline{\Omega}$ with exponent α, $0 < \alpha < 1$ is denoted by $H_\alpha(\overline{\Omega})$ or $C^\alpha(\overline{\Omega})$ or $C^{0,\alpha}(\overline{\Omega})$. If Ω is a bounded open set in R^n, then $f(x)$ is *locally Hölder continuous* on Ω if $f(x)$ is Hölder continuous in every bounded closed set B of Ω. The constant δ may depend on B. If the constant δ (δ as in **2**. of this Appendix) is independent of the set B, then f is said to be *uniformly Hölder continuous* with exponent α. If $f \in H_\alpha(\overline{\Omega})$, then we define its norm as

$$\|f\|_{H_\alpha} = \|f\|_0 + \sup_{\substack{x,y \in \Omega \\ x \neq y}} |f(x) - f(y)| \,/\, |x - y|^\alpha .$$

Here, $\|f\|_0$ is the uniform norm of f defined as

$$\|f\|_0 = \sup_{x \in \Omega} |f(x)| .$$

7. The space $H_{m+\alpha}(\overline{\Omega})$ or $C^{m,\alpha}(\overline{\Omega})$, $m \geq 0$, $0 < \alpha < 1$. For $0 < \alpha < 1$, the space $H_{m+\alpha}(\overline{\Omega})$ or $C^{m,\alpha}(\overline{\Omega})$ or $C^{m+\alpha}(\overline{\Omega})$ is a subspace of $C^m(\overline{\Omega})$, $m \geq 0$ and consists of those functions f for which $D^\beta f$, $0 \leq |\beta| \leq m$, satisfies in Ω a Hölder condition of exponent α, i.e., there exists a constant $\delta > 0$ such that

$$\left|D^\beta f(x) - D^\beta f(y)\right| \leq \delta \left|x - y\right|^\alpha, \ x, y \in \Omega.$$

Here, $\beta = (\beta_1, \beta_2, \cdots, \beta_n)$, $D^\beta = D_1^{\beta_1} D_2^{\beta_2} ..., D_n^{\beta_n}$, $|\beta| = \sum_{J=1}^n \beta_J$ and $D_J = \partial/\partial x_J$, $J = 1, 2, .., n$.

If $f \in C^{m,\alpha}(\overline{\Omega})$ and $\|f\|$ is defined as

$$\|f\|_{m+\alpha} = \|f\|_m + \sup_{\substack{x,y\in\Omega \\ x\neq y}} \frac{\left|D^{|\beta|}f(x) - D^{|\beta|}f(y)\right|}{|x-y|^\alpha} = \|f\|_{m-1} + \|D^m f\|_{H_\alpha}, \ |\beta| = m,$$

is finite, then $C^{m,\alpha}(\overline{\Omega})$ is a Banach space in this norm.

For norms in anisotropic Hölder spaces the reader is referred to [295].

In the one-dimensional parabolic problems, we consider $\Omega_{t_*} = D\times(0, t_*)$, $D \subset R$, $0 < t < t_*$. Let $P = (x, t)$, $P_1 = (x_1, t_1)$ and $P_2 = (x_2, t_2)$, where $P_0, P_1, P_2 \in \Omega_{t_*}$. Define the distance P_1P_2 as

$$P_1P_2 = \left\{(x_1 - x_2)^2 + |\, t_1 - t_2 \,|\right\}^{1/2}.$$

Let $f(x, t)$ be a continuous function in $\overline{\Omega}_{t_*}$. We say that $f \in C_\alpha(\overline{\Omega}_{t_*})$, $0 < \alpha < 1$, if the norm of f defined below is finite.

$$\|f\|_{C_\alpha(\overline{\Omega}_{t_*})} = \|f\|_0 + \sup_{P_1, P_2\in\Omega_{t_*}} |f(P_1) - f(P_2)| / (P_1P_2)^\alpha.$$

Here, $\|f\|_0 = \sup_{P\in\Omega_{t_*}} |f(P)|$. The spaces $C_{1+\alpha}(\overline{\Omega}_{t_*})$ and $C_{2+\alpha}(\overline{\Omega}_{t_*})$ are Banach spaces of functions f provided the norms defined below are finite.

$$\|f\|_{C_{1+\alpha}}(\overline{\Omega}_{t_*}) = \|f\|_{C_\alpha(\overline{\Omega}_{t_*})} + \|f_x\|_{C_\alpha(\overline{\Omega}_{t_*})}, \ \text{(norm in } C_{1+\alpha}(\overline{\Omega}_{t_*})).$$

$$\|f\|_{C_{2+\alpha}}(\overline{\Omega}_{t_*}) = \|f\|_{C_{1+\alpha}(\overline{\Omega}_{t_*})} + \|f_{xx}\|_{C_\alpha(\overline{\Omega}_{t_*})} + \|f_t\|_{C_\alpha(\overline{\Omega}_{t_*})}, \ \text{(norm in } C_{2+\alpha}(\overline{\Omega}_{t_*})).$$

8. Imbedding. A normed space X is said to be *embedded* in the normed space Y and written as $X \hookrightarrow Y$, provided

(i) X is a subspace of Y,

(ii) the identity operator defined on X into Y by $Ix = x$ for all $x \in X$ is continuous.

9. The space $L^P(\Omega)$. Let Ω be a domain in R^n and let p be a positive real number. We denote by $L^p(\Omega)$ the class of all measurable functions f defined on Ω such that

$$\int_\Omega |f(x)|^p \, dx < \infty.$$

Here, the integration is taken in the Lebesgue sense.

Two functions in $L^p(\Omega)$ are equal if they are equal almost everywhere (a.e.) on Ω, i.e., they are equal except on a set of measure zero. If $1 \leq p \leq \infty$, then the norm of a function $f \in L^p(\Omega)$ can be defined as

$$\|f\|_{L^p(\Omega)} = \left\{ \int_\Omega |f(x)|^p \, dx \right\}^{1/p}.$$

The space $L^p(\Omega)$, $1 \leq p \leq \infty$ is a Banach space in the above norm. $L^2(\Omega)$ is a Hilbert space with respect to the inner product defined as

$$(f, g) = \int_\Omega f(x)\overline{g(x)}dx, \ \ f, g \in L^2(\Omega), \text{ bar stands for the complex conjugate.}$$

10. Essentially bounded function. A function f, measurable on Ω is said to be *essentially bounded* on Ω if there exists a constant δ such that $|f(x)| \leq \delta$ a.e. on Ω. The set of all essentially bounded functions on Ω denoted by $L^\infty(\Omega)$ is a vector space. The greatest lower bound of all such constants δ is called the *essential supremum* of $|f|$ on Ω and is denoted by $\operatorname{ess\,sup}_{x\in\Omega} |f(x)|$. If $f \in L^\infty(\Omega)$ and

$$\|f\|_\infty = \operatorname*{ess\,sup}_{x\in\Omega} |f(x)| < \infty,$$

then $L^\infty(\Omega)$ is a Banach space in the norm $\|f\|_\infty$.

11. A locally integrable function. A function f defined a.e. on Ω is said to be *locally integrable* on Ω written as $f \in L^1_{loc}(\Omega)$ provided $f \in L^1(A)$ for every measurable compact subset A of Ω.

12. Locally compact space. A normed space is said to be *locally compact* if each point of the space has a compact neighbourhood.

13. Graph of an operator. The graph of a linear operator $P : H_1 \to H_2$, where H_1 and H_2 are normed spaces is the set of points G_A such that

$$G_A = \{(x, y) : x \in \text{Domain}(A), \ y = Ax\}.$$

14. Maximal monotone graph. Let A be a multivalued operator, i.e., $A : H \to 2^H$ from H to itself. A will be viewed as a subset of $H \times H$ and A will not be distinguished from its graph. A subset $A \subset H \times H$ is called *monotone* if

$$\forall u, v \in H, \ \forall \xi \in A(v), \ \eta \in A(v), \ (\xi - \eta, \ u - v) \geq 0.$$

A monotone subset of $H \times H$ is called *maximal monotone* if it is not properly contained in any other monotone subset of $H \times H$.

15. The boundary $\partial\Omega$ of Ω is $C^{m+\alpha}$. If each point x of $\partial\Omega$ has a neighbourhood B such that the graph of the intersection of B with $\partial\Omega$ belongs to $C^{m+\alpha}$, then $\partial\Omega \in C^{m+\alpha}$.

Appendix C

Fixed Point Theorems and Maximum Principles

Fixed point theorems play a key role in proving the existence and uniqueness theorems in the analysis. They also provide rigorous justification for the convergence of iterations in the numerical computations. In many situations, the behaviour of the temperature and/or its derivatives can be analysed by simple application of maximum principles.

1. Banach fixed point theorem. A *fixed point of a mapping* $P : X \to X$ of a set X into itself is an element $x \in X$ which is mapped onto itself, i.e., $Px = x$. A mapping $P : X \to X$, where X is a normed space is called a *contraction* on X if there exists a positive real number $\alpha < 1$ such that $\|Px - Py\| \leq \alpha \|x - y\|$ for all $x, y \in X$. This definition implies that the mapping P is uniformly continuous on X.

Let X be a Banach space and $P : X \to X$, a contraction. *Banach fixed point theorem* which is also called *Banach contraction mapping theorem* states that there exists a unique element $x^* \in X$ such that $P(x^*) = x^*$. x^* is called the fixed point of P.

The following results hold for contraction mappings.

(i) If X is a Banach space and $P : X \to X$ is such that $P^r = PP....r$ times is a contraction for some integer $r > 1$, then P has a unique fixed point.

(ii) Let Z be a closed subset of a Banach space X and $P : Z \to Z$ be such that $\|Px - Py\| \leq \alpha \|x - y\|$, $0 < \alpha < 1$, for all $x, y \in Z$. Then there exists a unique vector $x^* \in Z$ such that $P(x^*) = x^*$ and x^* may be obtained as the limit of a sequence $\{x_n\}$ where $x_n = P(x_{n-1})$, $x_0 \in Z$.

2. Schauder's fixed point theorem. Let P be a continuous operator on a Banach space X which maps a closed convex set Z of X into itself. Assuming that the image set PZ is relatively compact, P has at least one fixed point in Z.

Several versions of fixed point theorem are available in the literature and for this the reader is referred to the functional analysis books mentioned in the bibliography.

3. Ascoli-Arzela theorem. Let Z be a compact metric space, and $W(Z)$ a Banach space of real or complex valued continuous functions $f(x)$ normed by $\|f\| = \sup_{x \in Z} |f(x)|$. Then a sequence $\{f_n(x)\} \subseteq W(Z)$ is relatively compact (also called *precompact* in $W(Z)$) if the following two conditions are satisfied:

(i) $f_n(x)$ is *equibounded* (in n), i.e., $\sup_{n \geq 1} \sup_{x \in Z} |f_n(x)| < \infty$,

(ii) $f_n(x)$ is equicontinuous (in n), i.e.,

$$\lim_{\delta \to 0} \sup_{n \geq 1,\ \mathrm{dis}(x', x'') \leq \delta} |f_n(x') - f_n(x'')| = 0.$$

Another way of stating Ascoli-Arzela theorem is as follows. Let Ω be a bounded domain in R^n. A subset F of $C(\overline{\Omega})$ is relatively compact in $C(\overline{\Omega})$ provided the following two conditions hold:

(i) There exists a constant α such that for every $f \in F$ and $x \in \Omega$, $|f(x)| \leq \alpha$.

(ii) For every $\varepsilon > 0$, there exists a $\delta > 0$ such that if $f \in F$, $x, y \in \Omega$, and $|x - y| < \delta$, then $|f(x) - f(y)| < \varepsilon$.

4. The strong maximum principle. The strong maximum principle is associated with Nirenberg [320] and is concerned with the parabolic operators. Consider the parabolic operator L defined as

$$LT = \sum_{i,J=1}^{n} a_{iJ}(x,t)\frac{\partial^2 T}{\partial x_i \partial x_J} + \sum_{i=1}^{n} b_i(x,t)\frac{\partial T}{\partial x_i} + d(x,t)T - \frac{\partial T}{\partial t},$$

in an $(n+1)$-dimensional parabolic domain $D_{t_*} = \Omega \times (0, t_*)$, where Ω (open and bounded) $\subset R^n$ and $t_* < \infty$. D_{t_*} is bounded by the planes $t = 0$ and $t = t_* < \infty$. Let $D_0 = \Omega \times \{t = 0\}$. The lateral surface of this parabolic cylinder is given by $D_l = \partial\Omega \times (0, t_*)$. $D_l \cup \overline{D}_0$ is called the parabolic boundary of D_{t_*}

We make the following assumptions:

(a) L is parabolic in D_{t_*} (see (7.3.26) for parabolicity).

(b) The coefficients of L are continuous functions in D_{t_*}.

(c) $T(x,t) \in C^{2,1}(D_{t_*})$ and the coefficient $d(x,t) \leq 0$ in D_{t_*}.

Let $P^0 = (x^0, t^0)$ be any point in D_{t_*} and let $Z(P^0)$ be the set of all points in D_{t_*} which can be connected to P^0 by a 'simple' continuous curve in D_{t_*} along which the $t-$ coordinate is nondecreasing as we move from any point in $Z(P^0)$ to P^0.

The *strong maximum principle* asserts the following:

Let the assumptions (a)–(c) given above hold. If $LT \geq 0$ in D_{t_*} or $LT \leq 0$ in D_{t_*} and the temperature T has a positive maximum (negative minimum) in D_{t_*} which is attained at a point $P^0(x^0, t^0)$, then $T(P) = T(P^0)$ for all $P \in Z(P^0)$.

The strong maximum principle holds even if $P^0 \in \Omega \times \{t_*\}$ provided $T(x,t)$ is continuous in $D_{t_*} \cup \Omega \times \{t_*\}$.

Let $L = a^2\partial^2/\partial x^2 - \partial/\partial t$, $a > 0$, then we say T is subparabolic if $LT \geq 0$ and superparabolic if $LT \leq 0$.

5. The weak maximum principle. Let the assumptions (a)–(c) hold and $T(x,t)$ be a continuous function in $\overline{D}_{t_*}$. Let $LT \geq 0$ ($LT \leq 0$) in D_{t_*}. Then the *weak maximum principle* asserts that the maximum (minimum) of $T(x,t)$ is attained on the parabolic boundary of D_{t_*}. Note that the same maximum (minimum) of $T(x,t)$ can be attained in D_{t_*} also.

For further extensions of these principles see [9].

Appendix D

Sobolev Spaces

One of the several reasons to enlarge the family of spaces discussed in the earlier appendices is the incompleteness of some of the linear spaces in some norms. For example, the space of functions $C[a,b]$ is a normed space in the norm

$$\|f\| = \left(\int_a^b f^2 dx \right)^{1/2}.$$

However, in this norm $C[a,b]$ is not complete. If the norm on $C[a,b]$ is defined as $L^2[a,b]$ norm, then $C[a,b]$ becomes a Banach space. This is called 'completion' of an incomplete normed space. The sequence of functions $\{f_m(x)\}, f_m(x) = \sin mx/\sqrt{m}, x \in R$, converges uniformly to $f = 0$ in the 'distance norm' but $\{f'_m(x)\}$ does not converge in the distance norm. If we want $\{f'_m(x)\}$ to converge, then we cannot consider derivative as the classical derivative. The sequence does converge if a *weak derivative* (defined below), also called a *distributional derivative* or a *generalized derivative*, is considered. The space of those functions whose weak derivatives exist should also contain all those functions whose classical derivatives exist. We shall first define a weak derivative and then define a suitable norm on the space of functions whose weak derivatives exist and discuss the completeness of this normed space.

1. The weak derivative. Let Ω be a bounded domain in R^n. If $f \in L^p(\Omega)$, then the weak derivative $D^\alpha f$ of order $\alpha = (\alpha_1, \alpha_2, .., \alpha_n)$, $|\alpha| = \alpha_1 + \alpha_2 + ...\alpha_n$ and $D^\alpha = D_1^{\alpha_1} D_2^{\alpha_2} .. D_n^{\alpha_n}$, where $D_i = \partial/\partial x_i$, $i = 1, 2, ..n$, is defined as follows.

A function $g \in L^p(\Omega)$ is said to be the α*-th weak derivative* of $f \in L^p(\Omega)$, $1 \leq p < \infty$ in the sense of $L^p(\Omega)$, if

$$\int_\Omega g(x)\phi(x)dx = (-1)^{|\alpha|} \int_\Omega f(x) D^\alpha \phi(x) dx,$$

for all $\phi \in C_0^{|\alpha|}(\Omega)$. $C_0^{|\alpha|}(\Omega)$ is the subspace of $C^{|\alpha|}(\Omega)$ and consists of functions which have compact support in Ω.

The weak derivative if it exists is unique. If for all $|\alpha| \leq m$ the weak derivatives $D^\alpha f$ exist, then we say that f possesses weak derivatives of all orders $|\alpha| \leq m$.

2. The strong derivative in $L^p(\Omega)$. A function $f \in L^p(\Omega)$ is said to be the α*-th strong derivative* of $u \in L^p(\Omega)$, if there exists a sequence $\{u_m\}$ in $C^{|\alpha|}(\overline{\Omega})$ such that $\{u_m\} \to u$ in $L^p(\Omega)$, and $D^\alpha\{u_m\} \to f$ in $L^p(\Omega)$. If the strong derivative exists, then the weak derivative also exists and is unique.

3. The space $V(\Omega)$ of testing functions. The space $C_0^m(\Omega)$ is a subspace of $C^m(\Omega)$ and $C^m(\bar{\Omega})$ is complete in the norm defined earlier in Appendix C. The space

$C_0^m(\Omega)$ is a topological vector space in the induced topology but is not complete. We consider $C_0^\infty(\Omega)$, the space of all C^∞-functions ϕ such that: (i) ϕ has a compact support in Ω, and (ii) ϕ has continuous derivatives of all orders. The convergence of a sequence $\{\phi^{(n)}\}$ in $C_0^\infty(\Omega)$ to a function ϕ in $C_0^\infty(\Omega)$ is defined by demanding that for every non-negative integer r, $D^r\phi^{(n)} \to D^r\phi$ uniformly. Let $V(\Omega)$ be the space of functions in $C_0^\infty(\Omega)$ with the above convergence criterion. $V(\Omega)$ is a complete topological vector space but is not a normed space. $V(\Omega)$ is reflexive. The elements of $V(\Omega)$ are called *testing functions* or *test functions*.

4. Distributions and the space of distributions. The continuity of a linear real valued function v on $V(\Omega)$ is defined by specifying that v is continuous if and only if $v(\phi_n) \to v(\phi)$ whenever $\phi_n \to \phi$ in the sense of $V(\Omega)$. A functional (continuous linear real valued function) on $V(\Omega)$ is a called a *distribution*. The space of distributions on $V(\Omega)$ is the dual space denoted by $V(\Omega)'$. If $v \in V(\Omega)'$, then for any $\phi \in V(\Omega)$, the value of v at ϕ is denoted by $v(\phi)$. There is no natural norm in $V(\Omega)'$ and this space can only be given a weak-star topology as dual of $V(\Omega)$. $V'(\Omega)$ is a locally convex topological vector space (cf. [58]) with this topology.

We shall not dwell on the space $V(\Omega)'$, the space of distributions, as it is too big for our purpose. As mentioned earlier bigger spaces could be an advantage for the study of existence results, but they are disadvantageous for the study of uniqueness and stability results. We need some complete normed subspaces of the space of distributions. These subspaces are called *Sobolev spaces*. Sobolev spaces were introduced in the analysis earlier than the space $V(\Omega)'$ (also called the 'space of Schwartz distributions').

5. Sobolev space $W^{m,p}(\Omega)$. We shall consider only those spaces in which m is a non-negative integer, $1 \le p \le \infty$, and Ω is a bounded domain in R^n. The space $W^{m,p}(\Omega)$ is the space of all functions f in $L^p(\Omega)$ whose weak derivatives $D^\alpha f$ of order $|\alpha| \le m$ belong to $L^p(\Omega)$. $W^{m,p}(\Omega)$ is a Banach space in the norm

$$\|f\|_{m,p} = \|f\|_{p,\Omega}^{(m)} = \left(\sum_{|\alpha|\le m} \int_\Omega |D^\alpha f|^p \, dx \right)^{1/p}, \quad \|f\|_p = \|f\|_{0,p} \,.$$

For $p = \infty$, we define the norm of f as

$$\|f\|_{m,\infty} = \sum_{|\alpha|\le m} \|D^\alpha f\|_{L^\infty(\Omega)} = \sum_{|\alpha|\le m} \operatorname{ess\,sup}_{\Omega} |D^\alpha f| \,.$$

$W_{loc}^{l,p}(\Omega)$ denotes the space of functions which belong to $W^{l,p}(Q)$, where $Q \subset \Omega$ is an arbitrary bounded domain. Note that $C^\infty(\Omega) \cap W^{m,p}(\Omega)$, is dense in $W^{m,p}(\Omega)$.

6. The spaces $H^m(\Omega)$ and $H_0^m(\Omega)$. It is customary to use the notation $H^{m,p}(\Omega)$ for $W^{m,p}(\Omega)$ and $H^m(\Omega) = W^{m,2}(\Omega)$. The closure of $C_0^\infty(\Omega)$ in $W^{m,p}(\Omega)$ is denoted by $W_0^{m,p}(\Omega)$ and $H_0^m(\Omega) = W_0^{m,2}(\Omega)$. The dual space of $H^m(\Omega)$ is denoted by $H^{-m}(\Omega)$.

7. Sobolev space $W_p^{2m,m}(\Omega_{t_*})$. In the parabolic problems, functions are defined on the set $\Omega_{t_*} = \Omega \times (0, t_*) = \{(x,t) : \ x \in \Omega \subset R^n, \ 0 < t < t_*\}$. $W_p^{2m,m}(\Omega_{t_*})$, where m is a non-negative integer and $p \geq 1$, is the closed subspace of $L^p(\Omega_{t_*})$ consisting of functions whose weak derivatives $D_t^r D_x^q$ with r and q satisfying $2r + q \leq 2m$ have finite norms. If $f \in W_p^{2m,m}(\Omega_{t_*})$, then

$$\|f\|_{p,\Omega_{t_*}}^{2m} = \sum_{2r+q=0}^{2m} \|D_t^r D_x^q\|_{p,\Omega_{t_*}} \cdot$$

If $f \in L^p(\Omega_{t_*})$, then

$$\|f\|_{p,\Omega_{t_*}} = \left(\int_{\Omega_{t_*}} |f|^p \, dxdt \right)^{1/p} .$$

The space $W_2^{1,0}(\Omega_{t_*})$ is a Hilbert space with the scalar product

$$(f,g)_{W_2^{1,0}(\Omega_{t_*})} = \int_{\Omega_{t_*}} (fg + \nabla f . \nabla g) dxdt \,,$$

and the space $W_2^{1,1}(\Omega_{t_*})$ is the Hilbert space with the scalar product

$$(f,g)_{W_2^{1,1}(\Omega_{t_*})} = \int_{\Omega_{t_*}} (fg + D_t f D_t g + \nabla f . \nabla g) dxdt.$$

8. The space $L^p(0,t_*;H^{m,q}(\Omega))$. Functions dependent on x and t, where $x \in \Omega$ and $t \in (0,t_*)$ are often treated as functions of t with values in some Banach space for a.a. $t \in (0,t_*)$. For example, the space $L^p(0,t_*;H^{m,q}(\Omega))$ consists of functions $u(t)$ (u is infact a function of x and t) with values in $H^{m,q}(\Omega)$ for a.a. $t \in (0,t_*)$ and is equipped with the norm

$$\|u\| = \left(\int_0^{t_*} \left(\|u(t)\|_{q,\Omega}^{(m)} \right)^p dt \right)^{1/p}, \quad 1 \leq p \leq \infty, \; 1 \leq q < \infty.$$

The space $L^p(0,t_*;W)$, where W is a Banach space, is a Banach space in the norm defined above.

9. The space $L^2(\Omega_{t_*})$. We write $L^2(0,t_*;\, L^2(\Omega)) = L^2(\Omega_{t_*})$.

10. Gâteaux derivative or G-derivative. Let $f : H \to R$ be a linear continuous real valued function, and H a Hilbert space. If there exists an element $f'(u) \in H'$ (dual of H), $u \in H$, such that for all $v \in H$, we have

$$\frac{f(u+\lambda v) - f(u)}{\lambda} \to {}_{H'}\langle f'(u)\,, v\rangle_H \text{ when } \lambda \to 0$$

then $f'(u)$ is called the *Gâteaux derivative* of f at u or *G-derivative* of f at u.

11. Fréchet derivative. The function f as defined in **10.** is said to have a *Fréchet derivative* or a *F-derivative* of f at $u \in H$ if there exists a $\phi \in H'$ such that

$$f\,(u+v) = f\,(u) + \phi\,(v) + o\,(\|v\|)\,\|v\|\,.$$

If the F-derivative exists then the G-derivative also exists and both are equal.

12. Sobolev imbedding theorem. The elements of $W^{m,p}(\Omega)$ are strictly speaking not functions defined everywhere on Ω. Equivalence classes of such functions are defined and the functions are equal upto a set of measure zero. When we say that $W^{m,p}(\Omega) \hookrightarrow C^J(\overline{\Omega})$, it means that each $f \in W^{m,p}(\Omega)$ when considered as a function can be redefined on a set of measure zero in Ω such that the modified function $\hat{f}$, which equals f in $W^{m,p}(\Omega)$, belongs to $C^J(\overline{\Omega})$ and satisfies $\| \hat{f};\ C^J(\overline{\Omega}) \| \leq M \|f\|_{p,\Omega}^{(m)}$, with M independent of f.

We give here just two results pertaining to Sobolev imbedding (also called embedding). For further details, see [22, 58]. Let m, n and r be non-negative integers.

(1) If $m > n/2 + r$, $\Omega \subset R^n$, then $H^m(\Omega) \subset C^r(\overline{\Omega})$ with continuous injection. Hence, if $\Omega \subset R^2$, then $f \in H^2(\Omega) \Rightarrow f \in C(\overline{\Omega})$ and f is almost everywhere equal to a unique function in $C(\overline{\Omega})$.

(2) If $mp > n$, then $W^{J+m,p}(\Omega) \hookrightarrow C^J(\Omega)$. In particular, if $f \in W^{1,p}$, $\Omega \subset R^n$, $p > n$, then f is almost everywhere equal to a unique function in $C(\overline{\Omega})$.

13. Trace operator. For an arbitrary function $f \in L^p(\Omega)$, $\Omega \subset R^n$, $1 \leq p < \infty$, how to define the values of f on $\partial\Omega$. If $f \in L^p(\Omega)$ is continuous up to the boundary $\partial\Omega$ of Ω, then one can say that the value f takes on $\partial\Omega$ is the restriction to $\partial\Omega$ of the function f. In general, however, the elements of $W^{m,p}(\Omega)$ are defined except on a set of measure zero and it is meaningless therefore to speak of their restrictions to $\partial\Omega$ which has an n-dimensional measure zero. As mentioned earlier $H^1(\Omega)$ is the closure of $C^\infty(\overline{\Omega})$ with respect to the norm

$$\|f\|^2 = \int_\Omega \left(|f|^2 + |\nabla f|^2\right) dx.$$

If $f \in C^\infty(\bar{\Omega})$, then it can be proved that there exists a unique continuous linear operator γ_0 from $C^\infty(\bar{\Omega})$ to $L^2(\partial\Omega)$ (provided $\partial\Omega$ is Lipschitz continuous) such that $\gamma_0(f(x)) = f(x)$ for $x \in \partial\Omega$. If $f \in H^1(\Omega)$, we call γ_0, a *trace operator*, denoted here by T_r, such that

$$T_r :\ H^1(\Omega) \to L^2(\partial\Omega);\ T_r f(x) = f(x) \text{ for } x \in \partial\Omega,\ f \in C^\infty(\overline{\Omega}).$$

In particular $H_0^1(\Omega) = \{f \in H^1(\Omega);\ f = 0 \text{ on } \partial\Omega\}$.

Bibliography

[1] Friedman, A., *Variational Principles and Free-Boundary Problems*, John Wiley & Sons, New York, 1982.

[2] Streeter, V.L. and Wylie, E.B., *Fluid Mechanics*, McGraw-Hill, Singapore, 1983.

[3] Ockendon, H. and Tayler, A.B., *Inviscid Fluid Flows*, Springer-Verlag, Berlin, 1983.

[4] Stoker, J.J., *Water Waves*, Interscience, New York, 1957.

[5] Wrobel, L.C. and Brebbia, C.A., (Eds), *Computational Modelling of Free and Moving Boundary Problems*, Vol. 1, Fluid Flow, Proceedings of the First International Conference held 2-4 July, 1991, Southampton, U.K., Computational Mechanics Publications, Southampton, 1991.

[6] Xu, X., A free boundary problem arising from reproductive toxic mass spreading, Nonlinear Anal. Theory. Methods. Appl., 23, 1994, pp. 577—603.

[7] Aronson, D.G., Some properties of the interface for a gas flow in porous media, in: *Free Boundary Problems: Theory and Applications*, Volume I, (Eds: Fasano A. and Primicerio, M.), Research Notes Math., N. 78, Pitman, London, 1983.

[8] Darcy, H., *Les Fontaines Publiques de la ville de Dijon*, Paris, 1856.

[9] Frideman, A., *Partial Differential Equations of Parabolic Type*, Prentice-Hall, Englewood Cliffs, N.J., 1964.

[10] Aronson, D.G., Regularity properties of flows through porous media: the interface, Arch. Rat. Mech. Anal., 37, 1970, pp. 1—10.

[11] Knerr, B.F., The porous media equation in one dimension, Trans. Amer. Math. Soc. 234, 1977, pp. 381—415.

[12] Williams, W.E., *Partial Differential Equations*, Clarendon Press, Oxford, 1980.

[13] Courant, R. and Friedrichs, K.O., *Supersonic Flow and Shock Waves*, Interscience, New York, 1948.

[14] Jeffrey, A. and Taniuti, T., *Nonlinear Wave Propagation with Applications to Physics and Magnetohydrodynamics*, Academic Press, New York, 1964.

[15] Elliot, C.M. and Ockendon, J.R., *Weak and Variational Methods for Moving Boundary Problems*, Pitman, London, 1982.

[16] Barenblatt, G.I. and Ishlinskii, A. Iu., On the impact of a visco-plastic bar on a rigid obstacle, J. Appl. Math. Mech. (PMM), 26, 1962, pp. 497–502.

[17] Šiškin, G.I., On a heat transfer problem with free boundary, Soviet Math. Dokl., 12, 1971, pp. 682–686.

[18] Šiškin, G.I., On a problem of Stefan type with discontinuous moving boundary, Soviet Math. Dokl. 16, 1975, pp. 1409–1412.

[19] Fasano, A., Meyer, G.H. and Primicerio, M., On a problem in the polymer industry: Theoretical and numerical investigation of swelling, SIAM J. Math. Anal., 17, 1986, pp. 945–960.

[20] Andreucci, D., Fasano, A., Primicerio, M. and Ricci, R., Mathematical problems in polymer crystallization, Surv. Math. Ind., 6, 1996, pp. 7–20.

[21] Rubinstein. L.I., *The Stefan Problem*, American Mathematical Society, Providence, 1971.

[22] Baiocchi, C. and Capelo, A., *Variational and Quasivariational Inequalities*, Translated by Jayakar, L., John Wiley and Sons, New York, 1984.

[23] Howison, S.D., Morgan, J.D. and Ockendon, J.R., A class of codimension-two free boundary problems, SIAM Rev., 39, 1997, pp. 221–253.

[24] Carslaw, H.S. and Jaeger, J.C., *Conduction of Heat in Solids*, 2nd ed. Clarendon Press, Oxford, 1959.

[25] Hill, J.M. and Dewynne, J.N., *Heat Conduction*, Blackwell, Oxford, 1987.

[26] Widder, D.V., *The Heat Equation*, Academic Press, New York, 1975.

[27] Neumann, F., Unpublished lecture notes in the 1860's. Riemann-Weber, Die partiellen Differentialgleichungen der mathematischen physik, (edn. 5, 1912), Vol. 2, pp. 121.

[28] Stefan, J., Über einige Probleme der Theorie der Wärmeleitung, S. -B. Wien. Akad. Mat. Natur., 98, 1889, pp. 173–184.

[29] Abuaf, N. and Gutfinger, C., Heat transfer with moving boundary: Application to fluidized-bed coating of thin plates, Int. J. Heat Mass Transfer, 16, 1973, pp. 213–216.

[30] Langer, R.E., A problem in diffusion or in the flow of heat for a solid in contact with fluid, Tôhoku Math. J., Ser. 1, 35, 1932, pp. 260–275.

[31] Aiki, T., Multi-dimensional Stefan problems with dynamic boundary conditions, Applicable Analysis, 56, 1995, pp. 71–94.

[32] Kenmochi, N., Two-phase Stefan problems with nonlinear boundary conditions described by time-dependent subdifferentials, Control and Cybernatics, 16, 1987, pp. 7–31.

[33] Bathelt, A.G. and Viskanta, R., Heat transfer at the solid-liquid interface during melting from a horizontal cylinder, Int. J. Heat Mass Transfer, 23, 1980, pp. 1493–1503.

[34] Kay, J.M. and Nedderman, R.M., *Fluid Mechanics and Transfer Processes*, Cambridge University Press, Cambridge, 1985.

[35] Sears, F.W., and Salinger, G.L., *Thermodynamics, Kinetic Theory and Statistical Thermodynamics*, 3rd ed., Narosa Publishing House, New Delhi, 1988.

[36] Keizer, J., *Statistical Thermodynamics and Nonequilibrium Processes*, Springer-Verlag, New York, 1987.

[37] Tien, C.L. and Lienhard, J.H., *Statistical Thermodynamics*, Hemisphere, Washington, 1979.

[38] Cottrell, A., *An Introduction to Metallurgy*, 2nd ed., Edward Arnold (Publishers) Ltd., London, 1975.

[39] Reed-Hill, R.E., *Physical Metallurgy Principles*, 2nd ed., Affiliated East-West Press Pvt. Ltd., New Delhi, 1968.

[40] Kurz, W. and Fisher, D.J., *Fundamentals of Solidification*, 3rd ed., Trans Tech Publications, Switzerland, 1989.

[41] Horway, G. and Cahn, J.W., Dendritic and spheroidal growth, Acta Metall., 9, 1961, pp. 695–706.

[42] Chalmers, B., *Principles of Solidification*, John Wiley & Sons, New York, 1964.

[43] Jones, D.R.H., The free energies of solid-liquid interfaces, J. Material Science, 9, 1974, pp. 1–17 .

[44] Minkoff, I., *Solidification and Cast Structure*, John Wiley & Sons, New York, 1986.

[45] Fasano, A., Primicerio, M. and Kamin, S., Regularity of weak solutions of one-dimensional two-phase Stefan problems, Annal. Mat. Pura. Appl., 115, 1978, pp. 341–348.

[46] Glicksman, M.E., What we do not know about solidification theory, Material Science and Engineering, 25, 1976, pp. 93–101.

[47] Gurtin, M.E. and Murdoch, A.I., A continuum theory of elastic material surfaces, Arch. Rat. Mech. Anal., 57, 1975, pp. 291–323.

[48] Moeckel, G.P., Thermodynamics of an interface, Arch. Rat. Mech. Anal., 57, 1975, pp. 255–280.

[49] Baldoni, F. and Rajagopal, K.R., A continuum theory for the thermodynamics of solidification, Int. J. Non-linear Mechanics, 32, 1997, pp. 3–20.

[50] Baldoni, F. and Rajagopal, K.R., Conditions of compatibility for the solid-liquid interface, Quart. Applied Mathematics, 55, 1997, pp. 401–420.

[51] Fernandez-Diaz, J. and Williams, W.O., A generalized Stefan condition, J. Applied Mathematics Physics (ZAMP), 30, 1979, pp. 749–755.

[52] Prosperetti, A., Boundary conditions at the liquid-vapor interface, Meccanica, 14, 1979, pp. 34–47.

[53] Shuttleworth, R., The surface tension of solids, Proc. Phys. Soc. A., 63, 1950, pp. 444–457.

[54] Oleinik, O.A., Primicerio, M. and Radkevich, E.V., Stefan-like problems, Meccanica, 28, 1993, pp. 128–143.

[55] Fasano, A. and Primicerio, M., Free boundary problems for nonlinear parabolic equations with nonlinear free boundary conditions, J. Math. Anal. Appl. 72, 1979, pp. 247–273.

[56] Jiang, Li-Shang, Remarks on the Stefan-Signorini problem, in: *Free Boundary Problems: Applications and Theory*, (Eds: Bassavit, A., Damlamian, A. and Fremond, M.), Vol. III, IV, Research Notes Math., N. 120, 121, Pitman, London, 1985.

[57] Fasano, A. and Primicerio, M., General free boundary problems for the heat equation III, J. Math. Anal. Appl., 59, 1977, pp. 1–14.

[58] Adams, R.A., *Sobolev Spaces*, Academic Press, New York, 1975.

[59] Fasano, A. and Primicerio, M., General free-boundary problems for heat equation I, J. Math. Anal. Appl., 57, 1977, pp. 694–723.

[60] Sikarskee, D.L. and Boley, B.A., The solution of a class of two-dimensional melting and solidification problems, Internat. J. Solids Structures 1, 1965, pp. 207–234.

[61] Boley, B.A. and Yogoda, H.P., The three-dimensional starting solution for a melting slab, Proc. Roy. Soc. London, Ser. A, 323, 1971, pp. 89–110.

[62] Gupta, S.C., Temperature and moving boundary in two-phase freezing due to an axisymmetric cold spot, Quart. Applied Mathematics, 45, 1987, pp. 203–222.

[63] Gupta, S.C., Solidification or melting initiation over a limited portion of an edge in a plate, Flow, Turbulence and Combustion, 62, 1999, pp. 1–27.

[64] Friedman, A., Free boundary problems for parabolic equations, Bull. Amer. Math. Soc., 76, 1970, pp. 934–941.

[65] Wilson, D.G., Lagrangian coordinates for moving boundary problems, SIAM J. Appl. Math., 42, 1982, pp. 1195–1201.

[66] Zerroukat, M. and Chatwin, C.R., *Computational Moving Boundary Problems*, Research Studies Press Ltd., Taunton, 1994.

[67] Ozisik, M.N. and Uzzell, J.C., Jr., Exact solution for freezing temperature range, ASME J. Heat Transfer, 101, 1979, pp. 331–334.

[68] Gupta, S.C., Laitinen, E. and Valtteri, T., Moving grid scheme for multiple moving boundaries, Comput. Methods Appl. Mech. Engrg., 167, 1999, pp. 345–353.

[69] Gupta, S.C., A moving grid numerical scheme for multi-dimensional solidification with transition temperature range, Comput. Methods. Appl. Mech. Engrg., 189, 2000, pp. 525–544.

[70] Schatz, A., Free boundary problems of Stefan type with prescribed flux, J. Math. Anal. Appl., 28, 1969, pp. 569–580.

[71] Cannon, J.R. and Hill, C.D., Existence, uniqueness, stability and monotonic dependence in a Stefan problem for the heat equation, J. Math. Mech., 17, 1967, pp. 1–20.

[72] Crank, J. and Gupta, R.S., A moving boundary problem arising from the diffusion of oxygen in absorbing tissues, J. Inst. Maths Appls, 10, 1972, pp. 19–33.

[73] Rogers, J.C.W., A free boundary problem in diffusion with non-linear absorption, J. Inst. Maths Appls, 20, 1977, pp. 261–268.

[74] Baiocchi, C. and Pozzi, G.A., An evolution variational inequality related to a diffusion-absorption problem, Appl. Math. Optim., 2, 1976, pp. 304–314.

[75] Fasano, A. and Primicerio, M., New results on some classical parabolic free-boundary problems, Quart. Applied Mathematics, 38, 1981, pp. 439–460.

[76] Andreucci, D. and Zhengzhong, D., Existence of the solution for the oxygen diffusion consumption problem in a cylindrical domain, Control and Cybernatics, 13, 1984, pp. 39–57.

[77] Ockendon, J.R., The role of the Crank-Gupta model in the theory of free and moving boundary problems, Advances in Computational Mathematics, 6, 1996, pp. 281–293.

[78] Crank, J., *Free and Moving Boundary Problems*, Oxford University Press, Oxford, 1984.

[79] Saffman, P.G. and Taylor, S.G., The penetration of a fluid into a porous medium or Hele-Shaw cell containing a more viscous fluid, Proc. Royal Soc. London, 245A, 1958, pp. 312–329.

[80] Landau, L.D. and Lifshitz, E.M., *Statistical Physics*, Addison-Wesley Publishing, Massachusetts, 1958.

[81] Caginalp, G., An analysis of a phase field model of a free boundary, Arch. Rat. Mech. Anal., 92, 1986, pp. 205–245.

[82] Dettman, J.W., *Mathematical Methods in Physics and Engineering*, McGraw-Hill, New York, 1962.

[83] Hohenberg, P.C. and Halperin, B.I., Theory of dynamic critical phenomenon, Rev. Mod. Phys, 49, 1977, pp. 44–62.

[84] Penrose, O. and Fife, P.C., Thermodynamically consistant models of phase-field type for the kinetics of phase transitions, Physica D, 43, 1990, pp. 44–62.

[85] Langer, J.S., Models of pattern formation in first-order phase transitions, in : *Directions in Condensed Matter Physics*, World Scientific, Singapore, 1986, pp. 165–186.

[86] Caginalp, G., Length scales in phase transition models: Phase field, Cahn-Hilliard and blow-up problems, Fields Institute Communications, 5, 1996, pp. 67–83.

[87] Caginalp, G. and Xie, W., An analysis of phase-field alloys and transition layers, Arch. Rat. Mech. Anal., 142, 1998, pp. 293–329.

[88] Caginalp, G. and Socolovsky, E., Efficient computation of a sharp interface by spreading via phase field methods, Appl. Math. Lett., 2, 1989, pp. 117–120.

[89] Elliot, C.M. and Songmu, Z., Global existence and stability of solutions to the phase field equations, International Series of Numerical Mathematics, Birkhauser-Verlag, Basel, 95, 1990, pp. 46–57.

[90] Caginalp. G. and Chen, X., Convergence of the phase field model to its sharp interface limits, Euro. J. Appl. Maths, 9, 1998, pp. 417–445.

[91] Caginalp, G., Phase field models and some differences in subtle situations, Rocky Mountain J. Math., 21, 1991, pp. 603–615.

[92] Caginalp, G., The role of microscopic anisotropy in the macroscopic behavior of a phase boundary, Annals. Physics, 172, 1986, pp. 136–155.

[93] Caginalp, G. and Jones, J., A derivation of a phase field model with fluid properties, Appl. Math. Lett., 4, 1981, pp. 97–100.

[94] Penrose, O. and Fife, P.C., Thermodynamically consistent models of phase-field type for the kinetics of phase transitions, Physica D, 43, 1990, pp. 44–62.

[95] Penrose, O. and Fife, P.C., On the relation between the standard phase-field model and a thermodynamically consistent phase field model, Physica D, 69, 1993, pp. 107–113.

[96] Alt, H.W. and Pawlow, I., A mathematical model of dynamics of non-isothermal phase separation, Physica D, 59, 1992, pp. 389–416.

[97] Wang, S.L., Sekerka, R.F., Wheeler, A.A., Murray, B.T., Coriell, S.R., Braun, R.J. and McFadden, G.B., Thermodynamically-consistent phase-field models for solidification, Physica D, 69, 1993, pp. 189–200.

[98] Kobayashi, R., Modelling and numerical simulations of dendritic crystal growth, Physica D, 63, 1993, pp. 410–423.

[99] Ruelle, D., *Statistical Mechanics*, Benzamin, New York, 1969.

[100] Colli, P., Gentili, G. and Giorgi, C., Nonlinear systems describing phase transition models compatible with thermodynamics, Math. Models Methods Appl. Sciences, 9, 1999, pp. 1015–1037.

[101] Visintin, A., Stefan problem with phase relaxation, IMA J. Appl. Maths., 34, 1985, pp. 225–245.

[102] Visintin, A., Supercooling and superheating effect in phase transitions, IMA J. Appl. Maths, 35, 1985, pp. 233–256.

[103] Wheeler, A.A. and McFadden, G.B., A ξ-vector formulation of anisotropic phase-field models: 3D asymptotics, Euro. J. Appl. Maths, 7, 1996, pp. 367–381.

[104] Fasano, A., Primicerio, M., Howison, S.D. and Ockendon, J.R., One-phase Stefan problems in one dimension, Quart. Applied Mathematics, 48, 1990, pp. 153–168.

[105] Fasano, A., Primicerio, M., Howison, S.D. and Ockendon, J.R., On the singularities of one-dimensional Stefan problems with supercooling, *Mathematical Models for Phase Change Problems*, (Ed: Rodrigues, J.F.) International Series of Numerical Mathematics, 188, Birkhauser-Verlag, Basel, 1989, pp. 215–226.

[106] Marangunic, P. and Turner, C., The behavior of the solutions of a two-phase Stefan problem and the value of an energy integral, Boll. UMI Suppl. Fis. Mat. 5, 1986, pp. 215–217.

[107] Fasano, A. and Primicerio, M., A critical case for the solvability of Stefan like problems, Math. Meth. Appl. Sci., 5, 1983, pp. 84–96.

[108] Fasano, A. and Primicerio, M., Convexity of the free boundary in some classical parabolic free boundary problems, Riv. Mat. Univ. Parma (4), 5, 1979, pp. 635–645.

[109] Friedman, A. and Jensen, R., Convexity of the free boundary in the Stefan problem and in the dam problem, Arch. Rat. Mech. Anal., 67, 1977, pp. 1–24.

[110] Petrova, A.G., Tarzia, D.A. and Turner, C.V., The one-phase supercooled Stefan problem with temperature boundary condition, Advances Math. Sci. Appl., Gakkotosho, Tokyo, 4, 1994, pp. 35–50.

[111] Tarzia, D.A., and Turner, C.V., The one-phase supercooled Stefan problem with a convective boundary condition, Quart. Applied Mathematics, 55, 1997, pp. 41–50.

[112] Mannucci, P. and Tarzia, D.A., The supercooled one-phase Stefan problem in spherical symmetry, Demostratio Mathematica, 32, 1999, pp. 67–76.

[113] Andreucci, D., Fasano, A. and Primicerio, M., The supercooled Stefan problem in radial symmetry, Ricerche di Matematica 41, 1992-Supplemento, pp. 21–37.

[114] Comparini, E., Ricci, R. and Tarzia. D.A., Remarks on a one-dimensional Stefan problem related to the diffusion-consumption model, Z. Angew. Math. Mech., (ZAMM), 64, 1984, pp. 543–550.

[115] Howison, S.D. and Xie, W., Kinetic undercooling regularization of supercooled Stefan problems, in: *Mathematical Models for Phase Change Problems*, (Ed: Rodrigues, J.F.) International Series of Numerical Mathematics, 88, 1989, pp. 227–237, Birkhauser-Verlag, Basel, PP. 227–237.

[116] Dewynne, J.N., Howison, S.D., Ockendon, J.R. and Xie, W., Asymptotic behaviour of solutions to the Stefan problem with a kinetic condition at the free boundary, J. Austral. Math. Soc., Ser. B, 31, 1989, pp. 81–96.

[117] Primicerio, M. and Xie, W., On a class of Stefan-like problems, Rendiconti di Matematica, Series VII, 3, Roma, 1989, pp. 353–367.

[118] Guan, Z.C., On the global solutions to two-phase nonequilibrium problems, Meccanica, 28, 1993, pp. 323–326.

[119] Guan, Z.C., Existence of solutions to the problem of nonequilibrium phase change, Rend. Inst. Mat. Univ. Trieste, 21, 1989, pp. 39–50.

[120] Guan, Z.C., and Wang, X.J., Existence and blow-up of solutions to two-phase nonequilibrium problems. SIAM J. Math. Anal., 27, 1996, pp. 1038–1048.

[121] Guan, Z.C., and Wang, X.J., Critical exponent in a Stefan problem with kinetic condition, Euro. J. Appl. Maths, 8, 1997, pp. 525–532.

[122] Jun, L., Stefan problems with kinetic conditions at the free boundary, Applied Mathematics, A J. Chinese Universities, 8, 1993, pp. 32–44.

[123] Gotz, I.G. and Zaltzman, B., Two-phase Stefan problem with supercooling, SIAM J. Math. Anal., 28, 1995, pp. 694–714.

[124] Charach, Ch., Zaltzman, B. and Gotz, I.G., Interfacial kinetic effect in planar solidification problems without initial undercooling, Math. Models Meth. Appl. Sci., 4, 1994, pp. 331–354.

[125] Meirmanov, A.M., The Stefan problem with surface tension in the three dimensional case with spherical symmetry: nonexistence of the classical solution, Euro. J. Appl. Maths, 5, 1994, pp. 1–19.

[126] Gotz, I.G. and Primicerio, M., Spherically symmetric Stefan problem with the Gibbs-Thomson law at the moving boundary, Euro. J. Appl. Maths, 7, 1996, pp. 249–275.

[127] Xie. W., The Stefan problem with a kinetic condition at the free boundary, SIAM J. Math. Anal., 21, 1990, pp. 362–373.

[128] Velazquez, J.L., Cusp formation for the undercooled Stefan problem in two and three dimensions, Euro. J. Appl. Maths, 8, 1997, pp. 1–21.

[129] Baiocchi, C., Su un problema di frontiera libera connesso a questioni di idraulica, Annal., Mat. Pura Appl., 92, 1972, pp. 107–127.

[130] Zuhan, L. and Guangwei, Y., The multi-dimension one phase Stefan problem with kinetic condition, J. Math. Research Exposition, 14, 1994, pp. 479–485.

[131] Chen, X. and Reitich, F., Local existence and uniqueness of solutions of the Stefan problem with surface tension and kinetic undercooling, J. Math. Anal. Appls, 164, 1992, pp. 350–362.

[132] Radkevich, E.V., The Gibbs-Thomson correction and conditions for the existence of a classical solution of the modified Stefan problem, Soviet Math. Dokl, 43, 1991, pp. 274-278.

[133] Hanzawa, Ei-Ichi, Classical solution of the Stefan problem, Tôhoku Math. J., 33, 1981, pp. 297–335.

[134] Friedman, A. and Reitich, F., The Stefan problem with small surface tension, Trans. Amer. Math. Soc., 328, 1991, pp. 465–575.

[135] Mullins, W.W. and Sekerka, R.F., Morphological stability of a particle growing by diffusion or heat flow, J. Appl. Phys., 34, 1963, pp. 323–329.

[136] Chadam, J. and Ortoleva, P., The stabilizing effect of surface tension on the development of the free boundary in a planar one-dimensional Cauchy-Stefan problem, IMA J. Appl. Math., 30, 1983, pp. 57–65.

[137] Rubinstein, L, and Zaltzman, B., Morphological instability of similarity solution to the Stefan problem with undercooling and surface tension, Quart. Applied Mathematics, 56, 1998, pp. 341–354.

[138] Gurtin, M., Some remarks on the Stefan problem with surface structure, Quart. Applied Mathematics, 50, 1992, pp. 291–303.

[139] Luckhaus, S., Solutions for the two-phase Stefan problem with the Gibbs-Thomson law for the melting temperature, Euro. J. Appl. Maths, 1990, pp. 101–111.

[140] Plotnikov, P.I. and Starovoytov, V.N., Stefan problem as the limit of the system of the phase field equations, Differential equations, 3, 1992, pp. 354–362.

[141] Radkevich, E.V., On the heat Stefan wave, Dokl. Acad. Nauk USSR., 47, 1993, pp. 150–155.

[142] Visinitin, A., Introduction to the models of phase transitions, Bullettino U.M.I., 8, I-B, 1998, pp. 1–47.

[143] Visinitin, A., Stefan problem with surface tension, in: *'Mathematical Models of Phase Change Problems'* (Ed: Rodrigues, J.F.), Birkhäuser, Basel, 1989, pp. 191–213.

[144] Lacey, A.A. and Shillor, M., The existence and stability of regions with superheating in the classical two-phase one-dimensional Stefan problem with heat sources, IMA J. Appl. Math., 30, 1983, pp. 215–230.

[145] Atthey, D.R., A finite difference scheme for melting problems, J. Inst. Math. Applics, 13, 1974, pp. 353–365.

[146] Primicerio, M., Mushy region in phase change problems, in: *Applied Non-linear Functional Analysis*, (Eds: Gorenflo, R. and Hoffman, K.H.), Lang, Frankfurt, 1982, pp. 251–269.

[147] Gupta. S.C., Moving boundaries due to distributed sources in a slab, Appl. Scientific Research, 54, 1995, pp. 137–160.

[148] Ughi, M., A melting problem with a mushy region: Qualitative properties, IMA J. Appl. Math., 33, 1984, pp. 135–152.

[149] Fasano, A. and Primicerio, M., Phase-change with volumetric heat sources: Stefan's scheme vs enthalpy formulation, Fisica Mathematica, Suppl. B.U.M.I., IV-5-N.1, 1985, pp. 131–149.

[150] Primicerio, M. and Ughi, M., Phase-change problems with mushy regions, in : *Free Boundary Problems: Applications and Theory III*, (Eds: Bossavit, A., Damlamian, A. and Fremond, M.), Res. Notes, Math., 120, Pitman, London, 1985, pp. 61–71.

[151] Fasano, A. and Primicerio, M., Mushy regions with variable temperature in melting processes, Bolletino U.M.I., 6, 4-B, 1985, pp. 601–626.

[152] Bertsch, M., Mottoni, P.De and Peletier, L.A., The Stefan problem with heating: Appearence and disappearance of a mushy region, Trans. Amer. Math. Soci. 293, 1986, pp. 677–691.

[153] Bertsch, M and Klaver, M.H.A., The Stefan problem with mushy regions: Differentiability of the interfaces, Annal. di Matematica Pura Aplicata, 166, 1994, pp. 27–61.

[154] Ding, Z.Z. and Ughi, M., A model problem for the diffusion of oxygen in living tissues, Bollettino U.M.I., 7 (I-B), 1987, pp. 111–127.

[155] Gelder, D. and Guy, A.G., Current problems in the glass industry, in: *Moving Boundary Problems in Heat Flow and Diffusion*, (Eds: Ockendon, J.R. and Hodgkins, W.R.), Clarendon Press, Oxford, 1975.

[156] Ricci, R., A Stefan-type free boundary problem in a thermal switch, Meccanica, 1984, pp. 196–200.

[157] Andreucci, D., Behaviour of mushy regions under the action of a volumetric heat source, Math. Methods Appl. Sci., 16, 1993, pp. 35–47.

[158] Lacey, A.A. and Tayler, A.B., A mushy region in a Stefan problem, IMA J. Appl. Math., 30, 1983, pp. 303–313.

[159] Lacey, A.A. and Herraiz, L.A., Macroscopic models for melting derived from averaging microscopic Stefan problems I: Simple geometries with kinetic undercooling or surface tension, Euro. J. Appl. Math., 11, 2000, pp. 153–169.

[160] Aiki, T. and Imai, H., Behaviour of blow-up solutions to one-phase Stefan problems with Dirichlet boundary conditions, in : *Free Boundary Problems: Theory and Applications*, (Eds: Niezgodka, M. and Strezelecki, P.), Volume I, Longman, 1996, Volume I, pp. 1–14.

[161] Lacey, A.A. and Ockendon, J.R., Ill-posed free boundary problems, Control and Cybernatics, 14, 1985, pp. 275–296.

[162] Garguichevich, G.G. and Tarzia, D.A., A steady-state two-phase Stefan problem with an internal energy and some related problems, Atti Sem. Mat. Fis. Univ. Modena, 39, 1991, pp. 615–634.

[163] Tarzia, D.A., An inequality for the constant heat flux to obtain steady state two-phase Stefan problem, Engineering Analysis, 5, 1988, pp. 177–181.

[164] Kufarev, P.P., On one-parameter families of analytic functions, Mat. Sb., 13, 1943, pp. 87–118, (in Russian).

[165] Obbink, B.K., Theorem on the existence of solutions of quasi-static moving boundary problems, Euro. J. Appl. Maths, 6, 1995, pp. 529–538.

[166] Fasano, A., Primicerio, M. and Radkevich, E.V., Transitional zone problems, Soviet Math. Dokl., 44, 1992, pp. 507–511.

[167] Pawlow, I., Analysis and control of free boundary problems, The Bulletin of the Faculty of Education, Chiba University, 36, 1988, pp. 9–38.

[168] Duijn, C.J. Van and Hulshof, J., An elliptic parabolic problem with a nonlocal boundary condition, Arch. Rat. Mech. Anal., 99, 1987, pp. 61–73.

[169] Cannon, J. and Yin, H.M., On the existence of the weak solution and the regularity of the free boundary to a one-dimensional two-phase degenerate Stefan problem, J. Differential Equations, 73, 1988, pp. 104–118.

[170] Xu, X., Regularity results for a one-dimensional two-phase degenerate Stefan problem, J. Differential Equations, 87, 1990, pp. 96–114.

[171] Richardson, S., Some Hele-Shaw flows with time-dependent free boundaries, J. Fluid Mechanics, 102, 1981, pp. 263–278.

[172] Di-Beneditto, E. and Friedman, A., The ill-posed Hele-Shaw model and the Stefan problem for supercooled water, Trans. Amer. Math. Soc., 282, 1984, pp. 183–204.

[173] Radkevich, E.V., and Melikulov, A.C., *Free Boundary Problems*, "Fan", Tashkent, 1992, (in Russian).

[174] Li, D. Regularity of solutions for a two-phase denerate Stefan problem, J. Differential Equations, 102, 1993, pp. 402–418.

[175] Lacey, A.A., Bounds on the solutions of one-phase Stefan problem, Euro. J. Appl. Maths, 6, 1995, pp. 509–516.

[176] Howison, S.D., Ockendon, J.R. and Lacey, A.A., Singularity development in moving-boundary problems, Quart. J. Mech. Appl. Math., 38, 1985, pp. 343–359.

[177] Howison, S.D., Cusp development in Hele-Shaw flow with a free surface, SIAM J. Appl. Math., 46, 1986, pp. 20–26.

[178] Kinderlehrer, D. and Stampacchia, G., *An Introduction to Variational Inequalities and their Applications*, Academic Press, New York, 1980.

[179] Cryer, C.W., The solution of a quadratic programming problem using systematic overrelaxation, SIAM J. Control, 9, 1971, pp. 385–392.

[180] Rodrigues, J.F., *Obstacle Problems in Mathematical Physics*, Math. Studies, 134, North-Holland, 1987.

[181] Chipot, M., *Variational Inequalities and Flow in Porous Media*, Springer-Verlag, New York, 1984.

[182] Lang, S., *Introduction to Differentiable Manifolds*, Interseience, New York, 1962.

[183] Mosco, U. and Troianiello, G.M., On the smoothness of solutions of unilateral Dirichlet problems, Boll. U. Mat. Ital., (4) 8, 1973, pp. 57–67.

[184] Biot, M.A. and Agrawal, H.C., Variational analysis of ablation, J. Aero-Space, 29, 1962, pp. 227–229.

[185] Duvait, G., Equation aux derivees partielles-rèsolution d'un problème de Stefan (fusion d'un bloc de glace à zèro degré), C.R. Acad. Sci. Paris, 276A, 1973, pp. 1461–1463.

[186] Fremond, M., Variational formulation of the Stefan-problem-coupled Stefan problem-frost propagation in porous media, in: *Computational Mechanics*, (Ed: Oden, J.T.), Lecture Notes in Mathematics, 461, Springer-Verlag, Berlin, 1974.

[187] Lions, J.L., *'Sur Quelques Questions d' Analyse de Mechanique et de Controle Optimal*, Les Presses Univ. Montreal, 1976.

[188] Ichikawa, Y., A numerical analysis of a one-phase Stefan problem by variational inequalities, M.S. Thesis, Univ. of Texas at Austin, 1978.

[189] Ichikawa, Y. and Kikuchi, N., A one-phase multi-dimensional Stefan problem by the method of variational inequalities, Intern. J. Numerical Methods Engng, 14, 1979, pp. 1197–1220.

[190] Friedman, A. and Kinderlehrer, D., A one-phase Stefan problem, Ind. Univ. Math. J., 24, 1975, pp. 1005–1035.

[191] Rodrigues, J.F., The variational inequality approach to the one-phase problem, Acta Applicandae Mathematicae, 8, 1987, pp. 1–35.

[192] Baiocchi, C., Sur un problème à frontiàre libre traduisant lefiltrage de liquides à travers des milieux poreux, C.R. Acad. Sci. Paris, 273-A, 1971, pp. 1215–1217.

[193] Cafarelli, L.A., The regularity of free boundary in higher dimensions, Acta Math., 139, 1977, pp. 155–184.

[194] Cafarelli, L.A., Some aspects of the one-phase Stefan problems, Indiana Univ. Math. J., 27, 1978, pp. 73–77.

[195] Caffarelli, L.A. and Riviere, N.M., Smoothness and analyticity of free boundaries in variational inequalities, Annal. Scoula Norm. Sup. Pisa, 3, 1976, pp. 289–310.

[196] Rodringues, J.F., An evolutionary continuous casting problem of Stefan type, Quart. Applied Mathematics, 44, 1986, pp. 109–131.

[197] Friedman, A. and Nirenberg, L., The smoothness of the free boundary in the one phase Stefan problem, Commun. Pure Appl. Math., 31, 1978, pp. 257–282.

[198] Brezis, H., Problemes Unilateraux, J. Math. Pures et Appl., 51, 1972, pp. 1–168.

[199] Friedman, A., Parabolic variational inequalities in one space dimension and smoothness of the free boundary, J. Funct. Anal., 18, 1975, pp. 151–176.

[200] Bensoussan, A., Goursat, M. and Lions, J.L. Contrôle. impulsionnel et inéquations quasi-variationnelles Stationnaires, C.R. Ac. Sci. Paris, 276, 1973, pp. 1279–1284.

[201] Friedman, A. and Kinderlehrer, D., A class of parabolic quasi-variational inequalities, J. Differential Equations, 21, 1976, pp. 395–416.

[202] Friedman, A., A class of parabolic quasi-variational inequalities, II, J. Differential Equations, 22, 1976, pp. 379–401.

[203] Friedman, A., Free boundary problems for parabolic equations. I, Melting of solids, J. Math. Mech., 8, 1959, pp. 499–505.

[204] Pawlow, I., *Analysis and Control of Evolution Multi-phase Free Boundary Problems*, Ossolineum, Wroclaw, Warszawa, 1987.

[205] Duvaut, G., The solution of a two-phase Stefan problem by a variational inequality, in: *Moving Boundary Problems in Heat Flow and Diffusion,* (Eds: Ockendon J.R. and Hodgkins, W.R.), Clarendon Press, Oxford, 1975.

[206] Cattaneo, C., Sulla conduzione del calore, Atti Sem. Mat. Fis. Univ. Modena, 3, 1948, pp. 3–21.

[207] Boley, B.A., in: *High Temperature Structures and Materials,* (Eds: Freudenthal, A., Boley, B.A. and Liebowitz, H.), MacMilan, New York, 1964, pp. 260–315.

[208] Joseph, D.D. and Preziosi, L., Heat Waves, Reviews of Modern Physics, 61, 1989, pp. 41–73.

[209] Beckurtz, K. and Wirtz, K., *Neutron Physics*, Springer-Verlag, New York, 1964.

[210] Gurtin, M. and Pipkin, A.C., A general theory of heat conduction with finite speeds, Arch. Rat. Mech. Anal., 31; 1968, pp. 113–126.

[211] Showalter, R.E. and Walkington, N.J., A hyperbolic Stefan problem, Quart. Applied Mathematics, 45, 1987, pp. 769–781.

[212] Bogy, D.B. and Naghdi, P.M., On heat conduction and wave propagation in rigid solids, J. Mathematical Physics, 11, 1970, pp. 917–923.

[213] Solomon, A.D., Alexiades, V. and Wilson, D.G., On the formulation of a hyperbolic Stefan problem, Quart. Applied Mathematics, 43, 1985, pp. 295–304.

[214] Socio, L.M.De, and Gualtieri, G., A hyperbolic Stefan problem, Quart. Applied Mathematics, 41, 1983, pp. 253–260.

[215] Li, D., The well-posedness of a hyperbolic Stefan problem, Quart. Applied Mathematics, 47, 1989, pp. 221–231.

[216] Letavin, M.I., Correctness of the formulation of a homogeneous one-dimensional hyperbolic Stefan problem, Differentsial'nye Uravneniya, 27, 1991, pp.1395–1402.

[217] Gereenberg, J.M., A hyperbolic heat transfer problem with phase changes, IMA J. Appl. Maths, 38, 1987, pp. 1–21.

[218] Dafermos, C.M., Polygonal approximations of solutions of the initial value problem for a conservation law, J. Math. Anal. Applics, 38, 1972, pp. 33–41.

[219] Friedman, A. and Hu, B., The Stefan problem for a hyperbolic heat equation, J. Math. Anal. Applics, 138, 1989, pp. 249–279.

[220] Coleman, B.C., Fabrizio, M. and Owen, D.R., Thermodynamics and the constitutive relations for second sounds in crystals, in: *Lecture Notes in Physics,* (Ed. Grioli, G.), Springer-Verlag, Heidelberg, 1983, pp. 20–43.

[221] Li, D., A three-dimensional hyperbolic problem with discontinuous temperature, Quart. Applied Mathematics. 49, 1991, pp. 577–589.

[222] Rauch, J. and Massey, F., Differentiability of solutions to hyperbolic initial-boundary value problems, Trans. Amer. Math. Soc., 189, 1974, pp. 303–318.

[223] Majda, A., The stability of multi-dimensional shock fronts, Mem. Amer. Math., Soc. No. 275, 1983.

[224] Li, D., A one-phase hyperbolic Stefan problem in multi-dimensional space, Trans. Amer. Math. Soc., 318, 1990, pp. 401–415.

[225] Hadamard, J., *Le Problème de Cauchy et les Equations aux Dérivées Partièlles Linéaires Hyperbolique*, Hermann, Paris, 1932.

[226] Engl, H.W., Hanke, M. and Neubauer, A., *Regularization of Inverse Problems*, Kluwer Academic Publishers, Dordrecht, 2000.

[227] Engl, H.W. and Manselli, P., Stability estimates and regularization for an inverse heat conduction problem in semi-infinite and finite time intervals, Numer. Funct. Anal. Optim., 10, 1989, pp. 517–540.

[228] Courant, R. and Hilbert, D., *Methods of Mathematical Physics*, II, *Partial Differential Equations*, Interscience, New York, 1962.

[229] Tikhonov, A.N., Regularization of incorrectly posed problems, Soviet Math. Dokl., 4, 1963, pp. 1624–1627.

[230] Tikhonov, A.N., Solution of incorrectly formulated problems and the regularization method, Soviet Math. Dokl., 4, 1963, pp. 1035–1038.

[231] Tikhonov, A.N. and Arsenin, V.Y., *Solutions of Ill-posed Problems*, John Wiley, New York, 1977.

[232] Neubauer, A., Tikhonov-regularization for non-linear ill-posed problems: optimal convergence and finite-dimensional approximation, Inverse problems, 5, 1989, pp. 541–557.

[233] Goncharsky, A.V., Leonov, A.S. and Yagola, A.G., A generalised discrepancy principle for the case of operators specified with an error, DAN SSSR, 203, 6, 1972, pp. 1238–1239, (in Russian).

[234] Goncharsky, A.V., Leonov, A.S. and Yagola, A.G., A generalized discrepancy principle, Zh. Vychisl. Mat. i Mat. Fiz., 13, 2, 1973, pp. 294–302, (in Russian).

[235] Tanana, V.P., A projection-iterative algorithm for operator equations of the first kind with perturbed operator, DAN SSSR, 224, 5, 1975, pp. 1028–1029, (in Russian).

[236] Helmberg, G., *Introduction to Spectral Theory in Hilbert Spaces*, North-Holland, Amsterdam, 1969.

[237] Hobson, A., *Concepts in Statistical Mechanics*, Gordon and Breach Science Publishers, New York, 1971.

[238] Natterer, F., Regularisierung schlecht gestellter Probleme durch Projecktionsverfahren, Num. Math., 28, 1977, pp. 329–341.

[239] Tarzia, D.A., Simultaneous determination of two unknown thermal coefficients through an inverse one-phase Lame-Clapeyron (Stefan) problem with an overspecified condition on the fixed face, Int. J. Heat Mass Transfer, 26, 1983, pp. 1151–1157.

[240] Cannon J.R. and Duchateau, P., Determination of unknown coefficients in a nonlinear heat conduction problem, SIAM J. Appl. Math., 24, 1973, pp. 298–314.

[241] Tarzia, D.A., The determination of unknown thermal coefficients through phase change process with temperature-dependent thermal conductivity, Int. Comm. Heat Mass Transfer, 25, 1998, pp. 139–147.

[242] Gupta, S.C., Axisymmetric melting of a long cylinder due to an infinite flux, Proc. Indian Acad. Sci. (Math. Sci.), 95, 1986, pp. 1–12.

[243] Stampella, M.B. and Tarzia, D.A., Determination of one or two unknown thermal coefficients of a semi-infinite material through a two-phase Stefan problem, Int. J. Engng. Sci., 27, 1989, pp. 1407–1419.

[244] Boley, B.A., A method of heat conduction analysis of melting and solidification problems, J. Math. Phy., 40, 1960, pp. 300–313.

[245] Gupta, S.C., Axisymmetric solidification in a long cylindrical mold, Appl. Sci. Res., 42, 1985, pp. 229–244.

[246] Gupta, S.C., Analytical and numerical solutions of radially symmetric inward solidification problems in spherical geometry, Int. J. Heat Mass Transfer, 30, 1987, pp. 2611–2616.

[247] Knabner, P. and Vessella, S., The optimal stability estimate for some ill-posed Cauchy problems for a parabolic equation, Math. Methods Appl. Sci., 10, 1988, pp. 575–583.

[248] Knabner, P. and Vessella, S., Stabilization of ill-posed Canchy Problems for parabolic equations, Annal. Mat. Pura Appl., IV, 149, 1987, pp. 393–409.

[249] Lin, Y., An inverse problem for a class of quasilinear parabolic equation, SIAM J. Math. Anal., 22, 1991, pp. 146–156.

[250] Prilepko, A.I., Dmitry, G.O. and Vasin, I.A., *Methods for Solving Inverse Problems in Mathematical Physics*, Marcel Dekkar, Inc., New York, 2000.

[251] Knabner, P., Regularizing the Cauchy problem for the heat equation by sign restriction, Intern. Ser. Numer. Math., 63, 1983, pp. 165–177.

[252] Cannon, J.R. and Douglas, J. Jr., The Cauchy problem for the heat equation, SIAM J. Numer. Anal., 4, 1967, pp. 317–336.

[253] Knabner, P., Regularizing the Cauchy problem for the heat equation by norm bounds, Applied Analysis, 17, 1984, pp. 295–312.

[254] Ang, D.D., Dinh, A.P.N. and Thanh, D.N., An inverse Stefan problem: identification of boundary value, J. Comput. Appl. Math., 66, 1996, pp. 75–84.

[255] Ang, D.D., Dinh, A.P.N. and Thanh, D.N., Regularization of an inverse two-phase Stefan problem, Nonlinear Analysis, 34, 1998, pp. 719–731.

[256] Ang, D.D., Dinh, A.P.N. and Thanh, D.N., A bidimensional inverse Stefan problem : Identification of boundary value, J. Comput. Appl. Math., 80, 1997, pp. 227–240.

[257] Ang, D.D., Dinh, A.P.N. and Thanh, D.N., Regularization of a two-dimensional two-phase inverse Stefan problem, Inverse Problems, 13, 1997, pp. 607–619.

[258] Jochum, P., The inverse Stefan problem as a problem of nonlinear approximation theory, J. Approximation Theory, 30, 1980, pp. 81–98.

[259] Jochum, P., The numerical solution of the inverse Stefan problem, Numer. Math., 34, 1980, pp. 411–429.

[260] Jochum, P., To the numerical solution of an inverse problems in two-space variables, in : *Numerical Treatment of Free Boundary Value Problems*, (Eds: Albrecht, J., Collatz, L., and Holfmann, K.H.), ISNMN. 58, Birkhäuser, Basel, 1982, pp. 127–136.

[261] Knabner, P., Control of Stefan problems by means of linear-quadratic defect minimization, Numer. Math., 46, 1985, pp. 429–442.

[262] Osborne, M.R. and Watson, G.A., An algorithm for minimax approximation in nonlinear case, Comput. J., 12, 1969, pp. 63–68.

[263] Reemtsen, R. and Kirsch, A., A method for the numerical solution of the one-dimensional inverse Stefan problem, Numer. Math. 45, 1984, pp. 253–273.

[264] Collatz, L. and Wetterling, W., *Optimierungsaufgaben*, Springer-Verlag, Heidelberg, 1971.

[265] Colton, D. and Reemtsen, R., The numerical solution of the inverse Stefan problem in two space variables, SIAM J. Appl. Math., 44, 1984, pp. 996–1013.

[266] Budak, B.M. and Vasiléva, V.N., On the solution of the inverse Stefan problem, Soviet Math. Dokl., 13, 1972, pp. 811–815.

[267] Hestenes, M.R. *Conjugate Direction Methods in Optimization*, Springer-Verlag, New York, 1980.

[268] Arnăutu, V., On approximation of the inverse one-phase Stefan problem, International Series of Numerical Mathematics, 99, Birkhäuser, Basel, 1991.

[269] Sibony, M., Sur l'approximation d'équations et inéquations aux dérivées partielles non linéaires de type monotone, J. Math. Anal. Appl., 34, 1971, pp. 502–564.

[270] Hoffman, K.H. and Sprekels, J., Control of evolutionary free boundary problems (Rapporteur's report), in : *Free Boundary Problems: Theory and Applications*, Volume II, Longman Scientific and Technical, New York, 1990, pp. 439–450.

[271] Alifanov, O.M., Regularization of solutions of inverse problems of heat conduction, Heat Transfer-Soviet Research, 5, 1973, pp. 163–169.

[272] Alifanov, O.M., Artyukhin, Ye, A. and Pankratov, B.M., Solutions of nonlinear inverse problems of heat conduction, Heat Transfer-Soviet Research, 10, 1978, pp. 63–71.

[273] Zabaras, N., Mukherjee, S. and Richmond, O., An analysis of inverse heat transfer problems with phase changes using an integral method, ASME J. Heat Transfer, 110, 1988, pp. 554–561.

[274] Beck, J.V., Litkouhi, B. and St. Clair, C.R., Jr., *Inverse Heat Conduction, Ill-Posed Problems*, Wiley-Interscience, New York, 1985.

[275] Huang, C.C., and Tsai, C.C., A transient inverse two-dimensional geometry problem in estimating time-dependent irregular boundary configurations, Int. J. Heat Mass Transfer, 41, 1998, pp. 1707–1718.

[276] Brebbia, C.A. and Dominguez, J., *Boundary Elements, An Introductory Course*, McGraw-Hill, New York, 1989.

[277] Bénard, C. and Afshari, A., Inverse Stefan problem: Tracking of the interface position from measurements on the solid phase, Intern. J. Numer. Methods Engineering, 35, 1992, pp. 835–851.

[278] Iskenderov, A.D., On conditional well-posedness of problems with an unknown boundary of the domain, Soviet Math. Dokl. 42, 1991, pp. 588–592.

[279] Kyner, W.T., An existence and uniqueness theorem for a nonlinear Stefan problem, J. Math. Mechanics, 8, 1959, pp. 483–498.

[280] Kyner, W.T., On a free boundary value problem for the heat equation, Quart. Applied Mathematics, 17, 1959, pp. 305–310.

[281] Friedman, A., Free boundary problems for parabolic equations I. Melting of solids, J. Math. Mechanics, 8, 1959, pp. 499–516.

[282] Larsen, R., *Functional Analysis*, Marcel Dekker, INC, New York, 1973.

[283] Rubinšteĭn, L.I., On a case of Stefan's problem, Soviet Physics Dokl., 7, 1962, pp. 21–22.

[284] Sherman, B., General one-phase Stefan problems and free boundary problems for the heat equation with Cauchy data prescribed on the free boundary, SIAM J. Appl. Math., 20, 1971, pp. 555–570.

[285] Sherman, B., A free boundary problem for the heat equation with prescribed flux at both fixed face and melting interface, Quart. Applied Mathematics, 25, 1967, pp. 53–63.

[286] Sherman, B., Continuous dependence and differentiability properties of the solution of a free boundary problem for the heat equation, Quart. Applied Mathematics, 27, 1970, pp. 427–439.

[287] Boley, B.A., Uniqueness in a melting slab with space-and time dependent heating, Quart. Applied Mathematics, 27, 1970, pp. 481–487.

[288] Sherman, B., Limiting behaviour in some Stefan problems as the latent heat goes to zero, SIAM J. Appl. Math., 20, 1971, pp. 319–327.

[289] Sherman, B., Free boundary problems for the heat equation in which the moving interface coincides initially with the fixed face, J. Math. Anal. Appl. 33, 1971, pp. 449–466.

[290] Fasano, A. and Primicerio, M., General free-boundary problems for the heat equation, II, J. Math. Anal. Appl., 58, 1977, pp. 202–231.

[291] Cannon, J.R. and Hill, C.D., On the infinite differentiability of the free boundary in a Stefan problem, J. Math. Anal. Appl., 22, 1968, pp. 385–397.

[292] Cannon, J.R. and Primicerio, M., A two phase Stefan problem: Regularity of the free boundary, Annal. Mat. Pura Appl., 88, 1971, pp. 217–228.

[293] Jiang, Li-Shang., Existence and differentiability of the solution of a two-phase Stefan problem for quasi-linear parabolic equations, Chinese Math.–Acta., 7, 1965, pp. 481–496.

[294] Schaeffer, D.G., A new proof of the infinite differentiability of the free boundary in the Stefan problem, J. Differential Equations, 20, 1976, pp. 266–269.

[295] Ladyzhenskaja, O.A., Solonnikov, V.A. and Uralceva, N.N., *Linear and Quasi-linear Equations of Parabolic type,* American Mathematical Soc., Providence, R.I., 1968.

[296] Friedman, A., Analyticity of the free boundary for the Stefan problem, Arch. Rat. Mech. Anal., 61, 1976, pp. 97–125.

[297] Rubinštein, L., Fasano, A. and Primicirio, M., Remarks on the analyticity of the free boundary for the one-dimensional Stefan problem, Annal. Math. Pura Appl. (IV), 125, 1980, pp. 295–311.

[298] Rubinštein, L. and Shillor, M., Analyticity of the free boundary for the one-phase Stefan problem with strong nonlinearity, Fisica Matematica, Suppl. B.U.M.I., 1, 1981, pp. 47–68.

[299] Gevrey, M., Sur les équations aux derivées partielles du type parabolique, J. Math., Ser. 6, 9, 1913, pp. 305–471.

[300] Tao, L.N., The Stefan problem with arbitarary initial and boundary conditions, Quart. Applied Mathematics, 36, 1978, pp. 223–233.

[301] Gupta, S.C., *Analytical and Numerical Solutions of Some Moving Boundary Problems, Vol. I,* Doctor of Science Thesis, Indian Institute of Science, Bangalore, 1997.

[302] Yotsutani, S., Stefan problems with the unilateral boundary condition on the fixed boundary I, Osaka J. Math., 19, 1982, pp. 365–403.

[303] Brezis, H., *Opérateurs Maximaux Monotones et Semigroupes de Contractions dans les Espaces de Hilbert,* Math. Studies, 5, North-Holland, 1973.

[304] Petrovskii, I.G., *Partial Differential Equations,* Iliffe Books Ltd, London, 1967.

[305] Yotsutani, S., Stefan problems with the unilateral boundary condition on the fixed boundary III, Osaka J. Math., 20, 1983, pp. 845–862.

[306] Fasano, A. and Primicerio, M., Cauchy type free boundary problems for nonlinear parabolic equations, Riv. Mat. Univ. Parma (4), 5, 1979, pp. 615–634.

[307] Sackett, G.G., An implicit free boundary problem for the heat equation, SIAM J. Numer. Anal. 8, 1971, pp. 80–96.

[308] Borgioli, G., Di-Benedetto, E. and Ughi, M., Stefan problems with nonlinear boundary conditions: The polygonal method, ZAMM, 58, 1978, pp. 539–546.

[309] Meirmanov, A.M, *The Stefan Problem,* Walter De Gruyter, Berlin, 1992.

[310] Yuan, Xu, Stefan problem with change of density upon change of phase, J. Partial Differential Equations, 7, 1994, pp. 330–338.

[311] Cannon, J.R., Henry, D.B. and Kotlow, D.B., Classical solutions of the one-dimensional two-phase Stefan problem, Annal. Mat. Pura Appl. 107, 1975, pp. 311–341.

[312] Fasano A., Primicerio, M., and Kamin, S., Annal. Mat. Pura Appl., IV, 115, 1977, pp. 341–348.

[313] Radyno, Ya.V. and Bong, T.F., A two-phase problem with a free boundary for a parabolic equation in an infinite domain, Translated from Differentsial'nye Uravneniya, 27, 1991, pp. 1941–1947.

[314] Fasano, A. and Primicerio, M., Classical solutions of general two-phase parabolic free boundary problems, in: *Free Boundary Problems : Theory and Applications,* S(Eds: Fasano, A. and Primicerio, M.) Vol. II, Research Notes Math., N. 79, Pitman, London, 1983, pp. 644–657.

[315] Yotsutani, S., Stefan problems with the unilateral boundary condition on the fixed boundary II, Osaka J. Math., 20, 1983, pp. 803–844.

[316] Yotsutani, S., Stefan problems with the unilateral boundary condition on the fixed boundary IV, Osaka J. Math., 21, 1984, pp. 149–167.

[317] Mohamed, F.A., Two-phase nonlinear free boundary problems for parabolic equations, Nonlinear Analysis Theory Methods Appl., 12, 1988, pp. 389–407.

[318] Jiang, Li-Shang, The two-phase Stefan problem II, Chinese Math.—Acta, 5, 1964, pp. 35–53.

[319] Cannon, J.R., Douglas, J. Jr. and Hill, C.D., A multi-boundary Stefan problem and the disappearance of phases, J. Math. Mech., 17, 1967, pp. 21–33.

[320] Nirenberg, L., A strong maximum principle for parabolic equations, Comm. Pure Appl. Math., 6, 1953, pp. 167–177.

[321] Friedman, A., Remarks on the maximum principle for parabolic equations and its applications, Pacific J. Math., 8, 1958, pp. 201–211.

[322] Aronson, D.G. and Kamin, S., Disappearance of phase in the Stefan problem: One space dimension, Euro. J. Appl. Math., 1, 1990, pp. 301–309.

[323] Wilson, D.G., Existence and uniqueness for similarity solutions of one-dimensional multi-phase Stefan problems, SIAM J. Appl. Math., 35, 1978, pp. 135–147.

[324] Budak, B.M. and Moskal, M.Z., On the classical solution of the Stefan multidimensional problem, Soviet Math. Dokl., 10, 1969, pp. 219–222.

[325] Budak, B.M. and Moskal, M.Z., Classical solution of the multidimensional multi-front Stefan problem, Soviet Math. Dokl., 10, 1969, pp. 1043–1046.

[326] Budak, B.M. and Moskal, M.Z., On a classical solution of a multidimensional multiphase Stefan type problem in a domain with piecewise smooth boundary, Soviet Math. Dokl., 11, 1970, pp. 427–431.

[327] Meirmanov, A.M., On the classical solution of the multidimensional Stefan problem for quasilinear parabolic equations, Math. USSR-Sb., 40, 1981, pp. 157–178.

[328] Caffarelli, L.A., The regularity of free boundaries in higher dimensions, Acta Math., 139, 1977, pp. 155–184.

[329] Kamenomotskaja, S.L., On Stefan's problem, Mat. Sbornick, 53, 1961, pp. 489–514, (in Russian).

[330] Oleinik, O.A., A method of solution of the general Stefan problem, Soviet Math. Dokl., 1, 1960, pp. 1350–1354.

[331] Friedman, A., One dimensional Stefan problems with nonmonotone free boundary, Trans. Amer. Math. Soc., 133, 1968, pp. 89–114.

[332] Friedman, A., The Stefan problem in several space variables, Trans. Amer. Math. Soc., 133, 1968, pp. 51–87.

[333] Cannon, J.R., Henry, D.B. and Kotlow, D.B., Continuous differentiability of the free boundary for weak solutions of the Stefan problem, Bull. Amer. Math. Soc., 80, 1974, pp. 45–48.

[334] Cannon, J.R., Multiphase parabolic free boundary value problems, in: *Moving Boundary Problems*, (Eds: Wilson, D.G., Solomon, A.D. and Boggs, P.T.), Academic Press, New York, 1978, pp. 3–24.

[335] Crowley, A.B., On the weak solution of moving boundary problems, J. Inst. Maths Applics, 24, 1979, pp. 43–57.

[336] Cannon, J.R. and Yin, H.M., On the existence of the weak solution and the regularity of the free boundary to a one-dimensional two-phase degenerate Stefan problem, J. Differential Equations, 73, 1988, pp. 104–118.

[337] Matano, H., Nonincrease of the lap-number of a solution for a one-dimensional semilinear parabolic equation, Pub. Fac. Sci. Univ. Tokyo, Sec. 1A, 29, 1982, pp. 401–441.

[338] Xu, X., Regularity results for a one-dimensional two-phase degenerate Stefan problem, J. Differential Equations, 87, 1990, pp. 96–114.

[339] Meirmanov, A.M., An example of nonexistence of a classical solution of the Stefan problem. Periodic solutions, Soviet Math. Dokl, 23, 1981, pp. 564–566.

[340] Meirmanov, A.M., The structure of a generalized solution of the Stefan problem, Soviet Math. Dokl., 28, 1983, pp. 440—443.

[341] Niezgodka, M. and Pawlow, I., A generalized Stefan problem in several space variables, Appl. Math. Optimiz., 9, 1983, pp. 193—224.

[342] Di-Benedetto, E., Continuity of Weak solutions to certain singular parabolic equations, Annal. Math. Pura Appl., 80, 1982, pp. 131—177.

[343] Caffarelli, L.A. and Evans, L.C., Continuity of the temperature in the two-phase Stefan problem, Arch. Rat. Mech. Anal., 81, 1983, pp. 199—220.

[344] Rogers, J.C.W. and Berger, A.E., Some properties of the nonlinear semigroup for the problem $u_t - \nabla^2 f(u) = 0$, Nonlinear Anal. Theory Methods Appl., 8, 1984, pp. 909—939.

[345] Götz, I.G. and Zaltzman, B.B., Nonincrease of mushy region in a nonhomogeneous Stefan problem, Quart. Applied Mathematics, 49, 1991, pp. 741—746.

[346] Götz, I.G. and Zaltzman, B.B., On the behaviour of mushy region in a Stefan problem, International Series of Numerical Mathematics, 99, Birkhäuser-Verlag, Basel, 1991, pp. 155—163.

[347] Gustafson, B. and Mossino, J., Isoperimetric inequalities for the Stefan problem, SIAM J. Math. Anal., 20, 1989, pp. 1095—1108.

[348] Diaz, J.I., Fasano, A. and Meirmanov, A.M., On the disapperance of the mushy region in multidimensional Stefan problems, *Proc. Free Boundary Problems : Theory and Applications*, Montreal, 1990.

[349] Götz, I.G. and Baltzman, B., Some criteria for the disappearance of the mushy region in the Stefan problem, Quart. Applied Mathematics, 53, 1995, pp. 657—671.

[350] Borodin, M.A., On the solvability of the two-phase time-dependent Stefan problem, Soviet Math. Dokl., 25, 1982, pp. 469—471.

[351] Rothe, E., Two dimensional parabolic boundary value problems as the limit of one-dimensional boundary value problems, Math. Annal., 102, 1930, pp. 650—670, (in German).

[352] Athanasopoulos, I., Caffarelli, L. and Salsa, S., Regularity of the free boundary in parabolic phase-transition problems, Acta Math., 176, 1996, pp. 245—282.

[353] Kreyszig, E., *Introductory Functional Analysis with Applications*, John Wiley & Sons, New York, 1978.

[354] Conway, J.B., *Course in Functional Analysis*, Springer-Verlag, Berlin, 1990.

[355] Milne, R.D., *Applied Functional Analysis*, Pitman Advanced Publishing Program, Boston, 1980.

[356] Curtain, R.F. and Pritchard, A.J., *Functional Analysis in Modern Applied Mathematics*, Academic Press, London, 1977.

Supplementary References

[357] Tarzia, D.A., A bibliography on moving-free boundary problems for the heat-diffusion equation. The Stefan and related problems, No. 2, Rosario, 1999.

[358] Magenes, E., Verdi, C. and Visintin, A., Theoretical and numerical results on the two-phase Stefan problem, SIAM J. Numer. Anal., 26, 1989, pp. 1425–1438.

[359] Xu, X., A degerate Stefan-like problem with Joule's heating, SIAM J. Math. Anal., 23, 1992, pp. 1417–1438.

[360] Visintin, A., Two-scale Stefan problem, GAKUTO Internat. Series, Math. Sci. Appl., 7, 1995, pp. 405–424.

[361] Shi, P., Shillor, M. and Xu, X., Existence of a solution to the Stefan problem with Joule's heating, J. Differential Equations, 105, 1993, pp. 239–263.

[362] Xu. X. and Shillor, M., The Stefan problem with convection and Joule's heating, Advances Differential Equations, 2, 1997, pp. 667–691.

[363] Yu, W., A quasisteady Stefan problem with curvature correction and kinetic undercooling, J. Partial Differential Equations, 9, 1996, pp. 55–70.

[364] Zaltzman, B., Multidimensional two-phase quasistationary Stefan problem, Manuscripta–Math., 78, 1993, pp. 287–301.

[365] Friedman, A., The Stefan problem with Kinetic condition at the free boundary, Annal. Scoula Norm. Sup. Pisa, (4), 19, 1992, pp. 87–111.

[366] Zhao, J.N., A free boundary problem for quasilinear degenerate parabolic equation in several space dimensions, Notheast Math. J., 9, 1993, pp. 291–304.

[367] van-de-Fliert, B.W. and van-der-Hont, R., A generalized Stefan problem in a diffusion model with evaporation, SIAM J. Appl. Math., 60, 2000, pp. 1128–1136.

[368] Abergel, F., Hilhorst, D. and Issard-Roch, F., On a Stefan problem with surface tension in the neighborhood of a stationary solution, in : *Progress in Partial Differential Equations : Elliptic and Parabolic Problems*, Longman Sci. Tech., Harlow, 1992, pp. 149–155.

[369] Luckhaus, S., Solutions for the two-phase Stefan problem with Gibbs-Thomson law for the melting temperature, pp. 317–327; Evans, L.C. and Spruck, J., Motion of level sets by mean curvature, pp. 328–374; Sekerka, M., Problem with kinetic undercooling, pp. 413–471; in : *Fundamental Contributions to the Continuum Theory of Evolving Phase Interfaces in Solids*, (Eds: Ball, J.M., Kinderlehrer, D., Podio-Guidugli, P. and Slemrod, M.), Springer-Verlag, Berlin, 1999.

[370] Abergel, F., Hilhorst, D. and Issard-Roch, F., Local existence and uniqueness of a Stefan problem with surface tension, Applicable Analysis, 60, 1996, pp. 219–240.

[371] Suzuki, R., Boundedness of global solutions of one dimensional quasilinear degenerate parabolic equations, J. Math. Soc. Japan, 50, 1998, pp. 119–138.

[372] Fasano, A., Phase transition with supercooling, Boll. Univ. Mat. Italiana, 1 B, 8, 1998, pp. 49–69.

[373] Herrero, M.A., Medina E. and Velazquez, J.L, The birth of a cusp in the two-dimensional, undercooled Stefan problem, Quart. Applied Mathematics, 58, 2000, pp. 473–494.

[374] Yi, F. and Liu, J., Vanishing specific heat for the classical solutions of a multidimensional Stefan problem with kinetic condition, Quart. Applied Mathematics, 57, 1999, pp. 661–672.

[375] Pines, V., Chait, A., Zlatkowski, M. and Beckermann, C., Equiaxed dentritic solidification in supercooled melts, J. Crystal Growth, 197, 1999, pp. 355–363.

[376] Urbano, J.M., A singular-degenerate parabolic problem: regularity upto the Dirichlet boundary, *Free Boundary Problems : Theory and Applications* (chiba) I, GAKUTO Internat. Ser. Math. Sci. Appl., Gakkotosho, Tokyo, 2000, pp. 399–410.

[377] Garofalo, N. and Sartori, E., Symmetry in a free boundary problem for degenerate parabolic equations on unbounded domains, Proc. Amer. Math. Soc., 129, 2001, pp. 3603–3610.

[378] Danilov, V.G., Omelyanov, G.A., and Radkevich, E.V., Hugonist-type conditions and weak solutions to the phase-field system, Euro. J. Appl. Math., 10, 1999, pp. 55–77.

[379] Morosnu, C. and Motreanu, D., An extension of the nonlinear parabolic equation in Caginalp's model, Comm. Appl. Anal., 2, 1998, pp. 159–168.

[380] Colli, P. and Grasselli, M., Hyperbolic phase change problems in heat conduction with memory, Proc. Roy. Soc. Edinburgh, A, 123, 1993, pp. 571–592.

[381] Beregova, G.I. and Kirilich, V.M., The hyperbolic Stefan problem in a curvilinear sector, Ukranian Math. J., 12, 1997, pp. 1684–1689.

[382] Yang, T. and Yi, F., Global existence and uniqueness for a hyperbolic system with free boundary, Discrete Contin. Dynam. Systems, 7, 2001, pp. 763–780.

[383] Di-Benedetto, E. and Friedman, A., Conduction-convection problems with change of phase, J. Differential Equations, 62, 1984, pp. 129–185.

[384] Yi, F. and Shih, T.M., Stefan problem with convection, Appl. Math. Comput., 95, 1998, pp. 139–154.

[385] Yang, G.Z. and Zabaras, N., An adjoint method for the inverse design of solidification processes with natural convection, Int. J. Numer. Meth. Engng, 42, 1998, pp. 1121–1144.

[386] Neittaanmaki, P. and Tiba, D., An embedding of domains approach in free boundary problems and optimal design, SIAM J. Control Optim., 33, 1995, pp. 1587–1602.

[387] Dautov, R., Laitinen, E., Lapin, A. and Rekkilä, M., On the optimal cooling control for continuous casting problem, in: *Proc. of CHT* 01: *Advances in Computational Heat Transfer II, Palm Cove Queensland, Australia,* 20-25 May, 2001, pp. 1185–1192.

[388] Aiki, T. and Imai, H., Global existence of solutions to one-phase Stefan problems for semilinear parabolic equations, Annal. Mat. Pura Appl., (iv), 175, 1998, pp. 327–337.

[389] Caffarelli, L.A., A Harnack inequality approach to the regularity of free boundaries, Part II: Flat free boundaries are Lipschitz, Commun. Pure Appl. Math., 42, 1989, pp. 55–78.

[390] Borodin, M.A., Existence of the global classical solution for a two-phase Stefan problem, SIAM J. Math. Anal., 30, 1999, pp. 1264–1281.

[391] Guan, Z., A class of free boundary problem to quasi-linear parabolic equations, Acta Math. Appl. Sinica, 15, 1999, pp. 197–205.

[392] Brauner, C.M., Hulshof, J. and Lunardi, A., A general approach to stability in free boundary problems, J. Differential Equations, 164, 2000, pp. 16–48.

[393] Feireis, L.E. and Petzeltova, H., On the domain dependence of solutions to the two-phase Stefan problem, Applied Mathematics, 45, 2000, pp. 131–144.

[394] Fornari, L., Regularity of the free boundary for non degenerate phase transition problems of parabolic type, Rend. Sem. Mat. Univ. Padova, 104, 2000, pp. 27–42.

[395] Hoffman, K.H., Kenmochi, N. and Niezgodka, M., Large-time solutions of the two-phase Stefan problem with delay, J. Math. Anal. Appl., 170, 1992, pp. 433–448.

[396] Kamin, S. and Kersner, R., Disappearance of interfaces in finite time, Meccanica, 28, 1993, pp. 117–120.

Captions for Figures

Subject Index

www.ingramcontent.com/pod-product-compliance
Lightning Source LLC
LaVergne TN
LVHW011530200926
848484LV00001B/215
9780444510860